LOYAL TO THE LAND

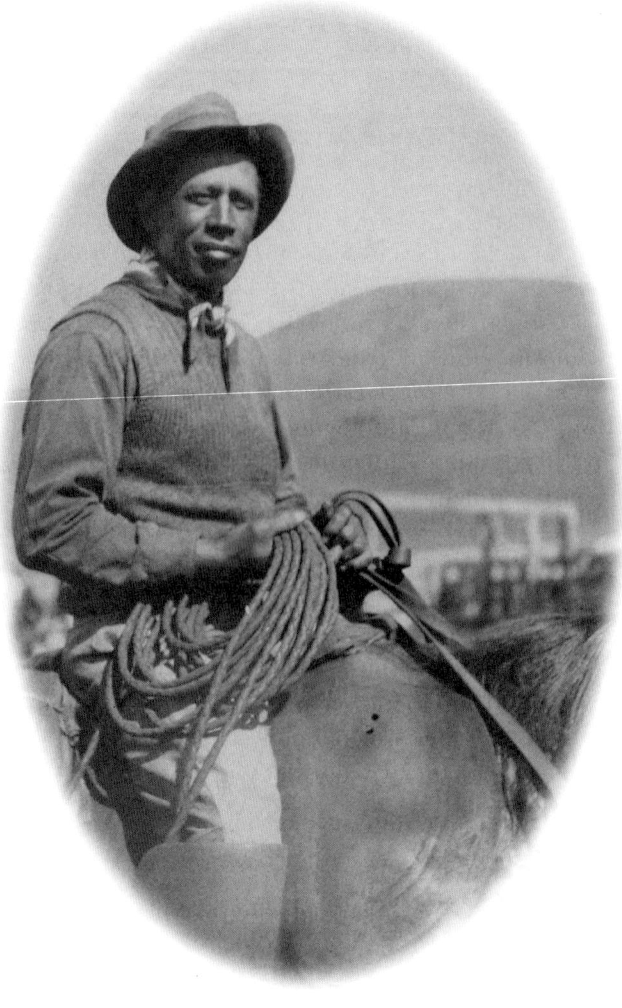

LOYAL TO THE LAND

The Legendary Parker Ranch, 750–1950

With kindest regards for Mrs. Yoshikami,

Aloha ʻĀina Paka

Aloha,

Dr. Billy Bergin

University of Hawaiʻi Press
Honolulu

Printed in the United States of America

08 07 06 05 04 03 6 5 4 3 2 1

Library of Congress Cataloging-in-Publication Data

Bergin, Billy.
Loyal to the land : the legendary Parker Ranch, 750–1950 / Billy Bergin.
p. cm.
Includes bibliographical references (p.) and index.
ISBN 0-8248-2692-2 (hardcover : alk. paper)
1. Parker Ranch (Hawaii). 2. Parker family.
3. Hawaii—History. I. Title.
DU629.P3 B47 2004
636'.01'09969—dc21
2003007993

Published with the support of the Maurice J. Sullivan & Family Fund in the University of Hawai'i Foundation.

Frontis photograph: Sam Kaehalio, a Parker *paniolo* from 1935–1945. Sam was not only a good cowhand but also a wrangler and trusted guide for Ranch special guests. Photo by Hartwell Carter, 1938.

University of Hawai'i Press books are printed on acid-free paper and meet the guidelines for permanence and durability of the Council on Library Resources.

Designed by Santos Barbasa Jr.

Printed by Thomson-Shore, Inc.

For the caregivers of Parker Ranch,
past, present, and future

CONTENTS

FOREWORD

It is an honor to introduce *Loyal to the Land* by Dr. Billy Bergin, which records his history of the Big Island's Parker Ranch. Dr. Billy Bergin is unusually qualified to write about the ranch's past and current history.

The evolution of the ranches in the West and in Hawai'i is a story of drastic and dynamic shifts in land control, ownership, and use. The movement started with the pioneer exploration of the West and the arrival of the missionaries in Hawai'i. In time, the immigrants took over the lands of the Native Americans in North America. In Hawai'i, the Great Mahele of 1848 changed the historic land control and ownership—forever.

In 1793, long before the Great Mahele, Captain Vancouver brought cattle to the Big Island of Hawai'i. As the herds multiplied, they presented an opportunity for John Parker to become engaged on the Big Island. Through his marriage to Kipikāne, he received land tenure that was the embryonic start of today's Parker Ranch.

As time marched on, the Great Mahele opened claims by the Parkers to a large acreage of land. Vancouver's cattle supplied the cattle inventory and the Great Mahele granted land control to launch the Parker Ranch.

In 1899 came A. W. Carter, the architect and creator of the modern Parker Ranch. History proved Carter to be a "man for the time." He was a visionary, planner, manager, and doer. A. W. Carter solidified and expanded the land base, increased the herd, improved the quality of the livestock and pastures, and trained an effective team to operate the vast ranch. As trustee, Carter was loyal to the Parker Ranch heirs, always working for their interest and not his own.

Dr. Billy Bergin is uniquely qualified to bring together the story of the people, land, agronomy, and livestock in this history of the beginning and development of the Parker Ranch, *Loyal to the Land.* He was born and grew up on the Big Island, learning to love its people, its land, and its agriculture. Upon completion of his veterinary training at Kansas State University, Dr. Bergin returned to dedicate his career to the health and care of the Big Island's cattle and horses. But he went much further with his love and understanding of its diverse population by contributing significantly to his clients' people management.

I can attest to Bergin's honesty and integrity. In our close association at Parker

Ranch, he voiced his opinions without political bias—whether you liked it or not.

I congratulate Dr. Billy Bergin for bringing forth the Bergin history of Parker Ranch in this wonderful book.

J. Gordon Lent
Lent, Burden Associates,
Modesto, California
Former General Manager of Parker Ranch,
Moloka'i Ranch, and Hana Ranch

PREFACE

My earliest memories of the days of horses and mules are of 1943–1944 in Laupāhoehoe, a coastal plantation village on the Big Island. I vividly recall the teams of mules drawing sleds used to transport cut sugarcane from the muddy fields to the slick, meandering flumes leading to the mill. But in 1949, my world was transformed when I was sent to the village of 'Umikoa, nestled in the heart of Kūka'iau Ranch, which spread across mountain reaches from Maulua to Hanaipoe. Horses and things Hawaiian then became a permanent part of my life.

Off and on for a year, I was immersed in a small Hawaiian family that included John Holi Ma'e, his wife Esther, and her mother Malia Pa'ahao Ako. Malia spoke not a word of English. Holi, a lead cowboy for Kūkai'au, had asked to assist in raising me after my father, a plantation physician, saved Esther's life a few years earlier. She was stricken with peritonitis after the tragic loss of a newborn son, their only child. I became a sort of *hānai* (foster child) for them from that point forward.

I have remained fascinated by the vast depth of history and culture carried forward by elders such as Holi, who walked comfortably that narrow path that joined the religion of old Hawai'i with that of the modern-day Mormon faith. He was a Hawaiian priest-practitioner at the same time as a Mormon elder. Other cowboys of that era walked this same path—notably, Robert Ke'akealani of Kona and 'Ioane Kanaka'ole of Ka'ū.

I was captivated by the Hawaiian language and how it became the universal language of the working cowboy, regardless of his ethnic makeup. Until the renaissance of Hawaiiana beginning in the 1980s, it was the *paniolo* who kept the language alive—all day, every day.

But the most engaging learning came from the conversations that went late into the night, as Holi brought forward the history of his forebears traveling by canoe to Waipi'o Valley, their new home. These early-day *paniolo* had emigrated to Waipi'o from the Kaupō area of Maui, bringing with them their experience, skills, and *ukana lio* (horse gear). I also learned how to gently interview Hawaiian elders without being offensive *(maha'oi)* or nosy *(nīele)* beyond courteous curiosity.

By age twelve, I was able—largely due to Holi's tutoring and coattails—to secure my first full-time summer job: I worked for W. H. Shipman, Ltd., at Pu'u'ō'ō Ranch under Tommy Lindsey. The learning process was intensified on this wild cattle operation. In later years, I went to work in Ka'ū at the former Ka'alu'alu, Kapāpala, and Keauhou Ranches under Johnny Pieper. Hawaiian cowboy folklore abounded, along with the intense use of *'ōlelo kanaka* (spoken Hawaiian). By high school, my summers were spent at Pu'uwa'awa'a, Hōlualoa, and Honomalino Ranches, working for Johnny Medeiros and among cowboy clansmen such as the Ke'akealani, Alapa'i, and Hao brothers. All through this full sweep of Big Island experiences, including Parker Ranch from 1949 to 1959, I absorbed the language, traditions, and skills of these wonderful mentors.

While I had written many short biographies of these great men, I was inspired to write a book about Parker Ranch because so many people in the Big Island ranching community have basic and close ties to 'Āina Paka (Parker lands). By this time I had wrapped up a twenty-five-year career as veterinarian for Parker Ranch; I had noted ranch happenings in detail and had carried on extensive independent research. The Parker Ranch story, in my view, needed to be told from the inside out.

Launching a comprehensive documentary history of this great ranch includes the preservation of facts and the acknowledgment of key people, while commending them for their contribution to Hawai'i's heritage. Binding these forces is the matrix of animal life, from the ubiquitous feral hog through the vast sheep flocks of Humu'ula—and the horse, which faithfully enables the loyal *paniolo* to ride a range dotted with varicolored cowherds.

Joseph Brennan, under the auspices of owner Richard Smart, wrote worthy and romantic vignettes about the Parker Ranch and its people in his book, *The Parker Ranch of Hawaii*. But the rich history of Hawai'i called for a comprehensive account of how Parker Ranch achieved its level of greatness in a global context. Much of this greatness can be attributed to a deeply embraced sense of stewardship—that human instinct that obligates us to look after ourselves, our animals, and the very environment that makes life the great privilege that it is. We are learning today that our fate as a civilization is directly tied to land and its health. Hence Richard Smart's official greeting was always *"aloha 'āina"* (love for the land), reflecting the values of his Polynesian forebears that stewardship of the land must remain foremost.

This book celebrates the caretakers of the land and the caregivers of the human and animal lives that collectively formed the basis of 'Āina Paka. The reader will note how Parker Ranch leadership tied the health of the environment to the welfare of animals and humans.

This book also covers several fields of interest, beginning with the history of Polynesian exploration of the Pacific Ocean. In search of documentary background as to the selection, preparation, and care of people and animals and provisions on these voyages, I traveled by ship to the far reaches of the Pacific, including Tasmania, New Zealand, Tahiti, Australia, and the Cook Islands. People with a specific interest in the early history of the Big Island will note the transformation of ancient lava flows into tillable fields over a span of hundreds of years. As animal life in the Islands grew to include livestock (pigs and chickens) as well as the edible companion animal, the dog, a vast increase in population occurred over the 1,100 years leading up to the arrival of cattle and horses.

Due attention is paid to the details of the profound Hispanic contribution beginning in 1832, as the Mexican *vaquero* mentored the budding *paniolo* from novitiate to world championship ropers by 1908. For indepth experience and research, I traveled again by ship to Mexico, seeking documentary support for the *vaquero* influence.

While the Parker Ranch name properly attributes the ranch birthright to John Palmer Parker I, readers of this book will come to understand that A. W. Carter deserves credit for the true building, maintaining—and, in fact, preservation—of this fine institution. Carter in turn attributed ranch greatness to the foundation names of Lindsey, Purdy, Bell, Stevens, and Spencer. Underpinning the growing success of the ranch, other great names surface in this book: Brighter, Vierra, Akau, Kimura, Kawai, Ah Fong, Kawamoto, Pacheco, Vredenburg, and Yamaguchi. Credit is given to the families in a multigenerational context.

For war history buffs, one can find the detailed impact of World War I on horse quality on the ranch, as the War Department officially designated Parker Ranch a Remount Station, leaving in its wake stallions of world-class excellence. World War II had an even greater impact on the ranch, as 40,000 Marines camped in the Waimea community in a bivouac, aptly christened Camp Tarawa. Epic journeys of the interisland, intercontinental, and international transport of livestock—and the historic evolution from schooners to steamships in this process—are given due attention.

In its closing memoirs, the narrative describes the end of an era, as A. W. Carter, passed the torch ever so tentatively to his son and successor. Hartwell Carter stood alone for a decade as major domo—until the prodigal son, Richard Smart, was home at last.

ACKNOWLEDGMENTS

Without a doubt, the first order of gratitude for completing this book project goes to my wife of forty years, Patricia Cockett Bergin, who transcribed the manuscript from legal tablets into computer format. Her diligence in attending to grammar, spelling, sentence flow, and manuscript continuity was humbling, her patience remarkable.

Paralleling Pat's efforts were those of my editor, Scott Stone. His firm, disciplined direction was regimental, while his warm, encouraging posture kept me focused on the light at the end of the tunnel. He has become a cherished friend in the process.

To Richard Smart I owe a debt of gratitude far beyond the kindness shown to my family and myself. It was Richard's mantra, *"Aloha 'āina,"* that inspired everyone's love for the ranch. This mantra also served as an inspiration for the title of these volumes.

To the ranch managers I knew personally and worked for—including Hartwell Carter, Dick Penhallow, Rally Greenwell, Jim Whitman, Gordon Lent, Walter Slater, Don Hanson, Charlie Kimura, Dave Ramos, and Robbie Hind—I extend my heartfelt respect and appreciation. All of you were great and special men, each making a unique contribution to the history and betterment of Parker Ranch. Many of you served as readers and critics of these manuscripts, and your candor was welcomed.

For the history of the fine horses on the ranch, I turned to Alex Penovaroff, Donnie DeSilva, Teddy Bell, and Randy Rieman for counsel, and there could be none better.

For the early dairy and beef cattle history, I looked to Yutaka Kimura, Rally Greenwell, and Charlie Kimura. Charlie was an invaluable source for the transition of history into the modern day.

For a grasp of the complex water systems of the ranch, I relied on the counsel of William Case, with Rally providing the early history. William also deserves credit for his photographic contributions.

Regarding the bountiful array of grasses, legumes, shrubs, weeds, and pests, Hisa Kimura was a steady and reliable background resource. As the agronomy component of the ranch matured, Hisa had teamed with Dr. Edward Hosaka and Josef Tamimi of the University of Hawai'i College of Tropical Agriculture. Dr. Tamimi also provided me with useful advice on soil and forage history. Earl Spence was vital to the later history of the agronomy component.

For saddlery and tack, no one could

match the counsel and contributions of Donnie DeSilva, a noted saddle, tack, bit, and spur maker who was mentored by the artists of old Hawai'i.

For Hawaiian language editing, Kepā Maly's assistance was invaluable. Special thanks go to Dr. Kalena Silva and Kaulana St. Clair for preparing the genealogy charts. To Andrea Adrian for her help with computer input during the final hours, I am grateful.

I am profoundly grateful to the many *kama'āina* families who willingly shared their family photographic collections and generously allowed me to replicate them for this text. For photographic identification, I sought and appreciated the help of Coco Vredenburg Hind, Mary Ha'ena Bell Lindsey, the three Kimuras—Yutaka, Hisa, and Tida—Rally and Pat Greenwell, and most notably, Jiro Yamaguchi. Robin Dutson and Matt Pierce deserve special thanks for their contribution in photographic reproductions.

Archival contributors were immensely important—notably, Becky Carter, Pat Clifford Penhallow, and Gordon Bryson of HPA, with his generations of history students who gleaned every fiber of information from the old Parker Ranch files. As a special reader of the manuscripts, Kiyome Yoshimatsu, Richard Smart's faithful secretary of many years, provided vital historical support.

Undoubtedly, the uniformly helpful support for genealogical research created the interwoven *'ohana* (family) matrix for this book on Parker Ranch. Again, Coco Hind, Tida Kimura, Tita Spielman, Aletha Lindsey Barkley, Bernie Lindsey Cacoulidis, Eva Lindsey Kealamakia, Stella Ka'au'a Akana, Godfrey Kainoa, Mona Malani, Alan "Burr" Ka'au'a, and Aloha Stevens Tanimoto were invaluable resources for this book.

And above all, *nui nui mahalo* to each of the many cowboys—numbering more than a hundred—whom I have interviewed over the last half century, as a result of my genuinely and incurably inquisitive Irish nature. To the man, each of you has a valid claim to coauthorship of this book, which is dedicated in your memory.

Kauka Billy Bergin

LOYAL TO THE LAND

INTRODUCTION

In the middle of nineteenth-century Hawai'i, a giant ranching enterprise was born of one man's drive and dedication. But the true history of Parker Ranch lies with the coming of men and animals to a group of islands that had been hidden for centuries.

Man and animals came together to these new islands, dependent on one other. No society has been known to exist without animal life—wild or domestic, or both—and in every society, certain individuals emerged as natural caregivers. They walk in the ancient mists of history, those shepherds of long ago, and their counterparts today continue the tradition of care on modern—often remote—ranches. Whether the caregivers used innate or acquired skills is unimportant; the salient point is that throughout history, animal welfare has been an accepted responsibility of mankind.

In the early years of the first millennium, the Hawaiian Islands were discovered by proud, hardy, courageous Polynesians in epic voyages that represented the final stages of what is arguably the greatest human dispersal in history, covering thousands of miles and years. Beginning in Central Asia, flowing through the turbulent continental wars and calamities, they arrived at last in Southeast Asia. They pushed off across a trackless ocean to make discovery after discovery, until their canoes had touched on almost all the beaches of the great Pacific, an ocean covering at least 60 percent of the world's surface. Hawai'i was the last major island group they found, but voyages continued between Hawai'i and what are now the islands of French Polynesia and New Zealand.

Preceding their arrival were birds, insects, plants, and seeds carried by birds, all borne on the strong Pacific winds and ocean currents. These prehuman colonizers, arriving by wind and wave, include species of 15 birds, 25 snails, 250 insects, and 275 plants. As an example, *'ōhi'a* seeds, readily windblown to newly emerged lava flows or established soils, became firmly rooted at opposite poles of the Pacific—Hawai'i and New Zealand. In Maori, the *'ōhi'a* is known as *rata* and is the official Christmas tree of New Zealand. The *naio* (false sandalwood) of Hawai'i is very much in evidence in New Zealand, where it is even known by the same name.

Bird feathers safely harbored seeds and insects with hooks, claws, burrs, and barbs with sticky or bristly surfaces. Birds also served as an internal transport medium among the islands by carrying viable seeds within their intestines. Depending on the species, digestive tract contents can be held up to 120 hours in flight, allowing ample time, with support from the wind, for transplanting to a remote island.

As an example, the *kōlea* (golden plover), which migrates north annually from the South Pacific and Australia, has a conformation (body characteristic) of profuse plumage that lends itself well to seed and insect transport. It is quite likely that the *kōlea* provided guidance to Polynesian voyagers heading northward to Hawai'i, assuring them that there was land ahead. Nainoa Thompson, master navigator, provides an eloquent perspective in David Eyre's *By Wind, By Wave:*

> We probably once had the innate ability to orient ourselves on the earth, but we've lost it. Now learning the navigation skills of the Polynesians means rediscovering what the animals know—whales, sea turtles, birds and insects—sometimes traveling thousands of miles to tiny islands to breed. If they're not precise they'll die. Of course the navigator uses techniques and bits of information accumulated over years of ocean experience, but he must also relearn the capacity to be fully attuned to his environment, fully concentrated on what's going on around him. He relies on the physical senses, and a sixth, spiritual, one as well.

Waves and ocean currents literally floated logs or mats of organic debris carrying seeds and insects to new territories via a scientific process known as rafting. These same wave and wind forces brought human explorers north to Hawai'i. The shores of Hawai'i, with its mild weather and absence of major predators, provided genuine sanctuary. This tropical sanctuary welcomed the arrival of the "Vikings of the Sunrise."

The voyagers brought with them companion and food animals—dogs, pigs, and chickens. They also brought the staples of their diet: fruits and vegetables and fiber plants—some sixteen species in all—along with seeds and seedlings to be nurtured in their new and distant homes. Each item included in those transoceanic journeys is worth noting, but the introduction of one creature was particularly significant: the pig.

To nurture a pig across thousands of kilometers of open ocean would be a difficult job even today, requiring careful monitoring. The feeding, breeding, birthing, and medicating of these porcine voyagers in a transpacific sailing venture presented no small responsibility. There were a number of voyages between Hawai'i and the islands of the Marquesas and Tahiti, and it is reasonable to assume that there was some acquired knowledge that was shared about the health and nutritional needs of pigs making the trips. Transporting a pig successfully over the seas to a new environment would require all the knowledge and skill that could be brought to bear, and it certainly ranks as an awesome achievement in animal husbandry.

It is true that dogs and chickens made the same trip, but getting pigs to Hawai'i in a healthy state had some inherent difficulties. The hog requires shade; it is unable to regulate its body temperature by sweating and reacts badly to periods of heat, especially from the merciless sun. In addition, the hog is unable to tolerate salt water; even brackish water, if it contains a high degree of salinity, can be toxic. In veterinary literature, deaths under these circumstances are described as "saline toxicity." Neither the dog nor the chicken needed as much care regarding water quality, shade, and ventilation.

For the Polynesians, the hog was worth the effort, and it took on cultural significance. The bonding of the pig to the Hawaiian culture, present even today, is mirrored in societies throughout the scattered islands of the South Pacific. The long ocean voyages served to strengthen the bond between people and pigs, and this was especially true in the new lands of Hawai'i.

Between the first arrival and the eighteenth century, both men and animals thrived. The dog was more of a social animal (though it was eaten), while the hog multiplied vigorously in the quiet forests, becoming both feral and plentiful. By 1778, when English explorer and sea captain James Cook discovered Hawai'i for the West, the large island of Hawai'i had an estimated human population of one hundred thousand and some three thousand swine.

For the men who were to make up the ranks of the Parker Ranch, the hog was an integral part of their lives, an important source of protein, an object of recreational pastime (hunting, feeding), and an icon of their religion as Kamapua'a, the pig demigod. This was true of the old Polynesians, who always regarded the pig with much more respect than it was accorded in the non-Polynesian world.

The first cattle came to the Islands in 1793, brought by sea captain George Vancouver, who had sailed with Cook and who was to spend three winters in Hawai'i. The cattle, a gift to Kamehameha I, caused a stir among royalty and commoners alike. The Hawaiians were bemused by these great beasts, calling them large hogs or, referring to their horns, "iron-eared *pipi*." The name devolved to *"pipi"*—a Hawaiianization of the English word "beef."

The original cattle did not fare well, not only because of the stress of sea transport but also because of a new diet of high-moisture and poor-quality grasses—and, perhaps, handlers who held the hog in higher esteem. But Kamehameha chose to establish this new species of animal by invoking a *kapu*—a law that carried the death penalty, in this case, for anyone killing the cattle. The stricture worked well indeed, for the cattle propagated far beyond expectations—and beyond practical levels of dealing with them. They plundered villages, gardens, and farms before another royal decree granted permission to shoot the offenders. Still their numbers continued to grow, and there was no end in sight to the pillaging of valued crops of taro, sweet potatoes, bananas, and the native forests and plant life.

In 1815, Kamehameha took steps to solve the problem. He hired an ex-sailor from Newton, Massachusetts, a capable cattleman, to bring the cattle population under control. John Palmer Parker was young, energetic, and determined. He had first seen the Islands in 1811, when he signed onboard a ship carrying sandalwood to China. The ship was blockaded in Canton for two years, and it wasn't until 1815 that young Parker returned to Hawai'i. That year he went into Kamehameha's employ as a supervisor of fishponds—not small, decorative ponds, but the large rock-walled coastal enclosures where Hawaiians practiced an effective aquaculture—and then Parker was assigned to handle the problem of the cattle.

In 1818 Parker married the chiefess, Rachel Keli'i Kipikāne Kaolohaka, great-granddaughter of Kamehameha I, and set up a homestead at Wai'āpuka, North Kohala. Parker began collecting cattle, working for another firm where he took his pay in cattle, and he established the nucleus of what became Parker Ranch in 1847. The ranch began with 2 acres that Parker bought from Kamehameha in 1818 for the token sum of ten dollars. Parker's marriage brought an additional 640 acres, a royal gift in recognition of his wife's status.

For many years, the wild cattle remained difficult to control. A census of livestock on the Big Island in 1851 counted 8,000 domestic cattle—but an astonishing 12,000 cattle in the wild. Parker and his descendants worked diligently to bring the cattle under supervision, in the process extending the ranch and setting up diversified but related operations. The magnitude of the ranch and the mingling customs and cultures that began with Parker's marriage to the chiefess led to one of the great cattle ranches under the U.S. flag.

The history of the growth and development of Parker Ranch for the approximately 150 years following its founding in 1847 centers on the animal caregivers, the fence men, drovers, herdsmen, milkers, and grooms—the cowboys who were indispensable in shaping this vast ranch. The story begins circa A.D. 750, as scientists agree that this marks the approximate advent in Hawai'i of animal life—the pig, dog, and chicken. The salient date of 1847 marks the birth of Parker Ranch. This volume ends in 1949 with the death of A. W. Carter, the great builder of the ranch, second only to John Palmer Parker himself.

With the respect due them, this book celebrates the leaders of the cowboy crews and their chiefs and foremen for their contributions to ranch history and success. Modern management—considered to be post-Carter (1950 to the present) and developed in a second volume—elevated the role and rank of some of the progeny of these men to superintendents and eventually livestock managers (with some less than notable exceptions). Earlier, they were elected by their peers for their horsemanship, knowledge of cattle and how to judge and handle them, and their understanding of the strengths and weaknesses of the *'āina*, the land. They were called *kanaka makua,* a Hawaiian term meaning "mature and respected person." Many of them were young men with natural leadership qualities who, by their conduct and maturity, earned the respect of their peers.

Of specific veterinary interest, especially in preventive medicine and animal health consultation, early Parker Ranch management sought advice from veterinarians beginning in the late 1800s. This subject is covered in detail. By 1895 a Hilo-based practitioner, H. B. Elliot, MRCVS, became the Big Island's first veterinarian to administer to sugar plantations and large ranches, including Parker Ranch. By 1905, ranch trustee A. W. Carter appealed to the territorial minister of agriculture to appoint a regulatory veterinarian for the agricultural community. Dr. A. W. Rowat subsequently was assigned temporarily to North and South Kohala in 1908. In 1920, Parker Ranch retained its first consulting veterinarian to prevent the introduction of tuberculosis and brucellosis across its borders. The veterinarian was W. L. Williams of Cornell University, who was also handed the job of obtaining disease-free breeding stock of the highest quality, transporting them across the United States and then the Pacific Ocean, and bringing them safely into a tropical environment. The task may not have been greater than the Polynesians' long haul with pigs, but it was formidable.

Professor Williams went on to develop a rudimentary but pragmatic animal health program. Since 1920, Parker Ranch has continued a vigorous program of consulting veterinarians on retainer to care for its animals. The owners, trustees, and administrators of the ranch have used veterinarians as an integral component of top-level management. This enabled the ranch to make significant early advances in livestock health and production, and it would take thirty years for the cattle industry in the continental United States to become aware of them.

Marketing Parker Ranch lamb and beef is worthy of a chapter of its own. The sea transport of cattle from the Big Island's Kawaihae Harbor to Honolulu continued over more than half a century, using more than a dozen vessels. The SS *Humu'ula* was one of the better known carriers, but it was preceded by schooners and steel-hulled steamships of a range of tonnage, capacity, and ownership.

Livestock shipping methods began to change. For many years, as late as the end of the 1940s, the conventional method of loading cattle was to *hō'au pipi*, meaning that a cowboy swam the cattle by lead rope to a slip where six or eight cattle were lashed to the sides of a boat. The boat then had to be jockeyed out to the ship, where overhead boom cranes lifted the steer onboard ship via the horns or a belly sling. History records the transition to a more efficient method. By the early 1950s, a new and improved shipping system came into being with use of barges towed by tugs. Moving cattle from the Big Island was limited to the ports of Hilo, Kailua, and Kawaihae, but all took on a new air of efficiency with the use of the tug-towed barges. This system introduced the use of cattle trucks that could get into remote locations to load the cattle.

The actual meat-purveying arm of Parker Ranch was the Hawai'i Meat Company, an entity nearly as rich in its history as the ranch itself, given the colorful individuals who ran the feedlot operations and the packinghouse. The influence of ranch trustee A. W. Carter in the birth of Hawai'i Meat Company typifies Carter. His vision in building an early lucrative market for the ranch's sheep and cattle, while meeting the marketing needs of other ranchers, was concomitant with his compassion for fellow ranchers who were suffering because of drought or wartime exigencies.

The history for which Parker Ranch is best known, however, concerns the sheep, cow, and horse populations—and certainly the cowboys. This book describes the evolution of the ranch horse, from feral descendants of the mounts delivered by Captain Richard J. Cleveland in 1803 in the *Lelia Byrd* to the improved Thoroughbred type during the ensuing century, ending with the advent of the quarter horse.

The story of the ranch cow takes on a comparable timeline of early and steady progress, from a rangy longhorn to an upstanding

Hereford dam. In the century between 1850, when the first purebred bull was purchased, and the mid-1950s, when the Carter influence was waning, the Parker Ranch cow was transformed into a well-framed Hereford with substantial ability to rustle for feed and water and to conceive and develop the lactation necessary to wean a 450-pound calf that would have the innate ability to convert grain into marbled beef at a reasonable rate.

Sheep and their impact on the ranch, in terms of wool and meat productivity, were overshadowed in importance by acquisitions of key sheep properties. The reasoning was revealed in the acquisition of the Pu'uloa Sheep and Stock Company (with 5,600 sheep) in 1903 and Humu'ula (with 23,000) in 1914—parcels that played major roles in ranch operations in the ensuing century when they were converted to cow-calf operations.

The cowboy, as the primary caregiver, shares the central theme here with the animals. Starting with the first cowboys, such as H. H. Jack Purdy, a native of Ireland (where the term "cowboy" originated), and certainly John Palmer Parker, this text traces the story of these foundation caregivers. Biographical sketches of these men are included, along with available photos. Included in the discussion of cowboy life are his gear—such as his saddle, tack, and spurs—and, in particular, the Hawaiian cowboy saddle and saddletree (frame) and how they evolved from the Mexican model into a prototype that has remained basically unchanged in 150 years. This first volume of *Loyal to the Land*—covering the period from A.D. 750 to 1950—concludes with the evolution of A. W. Carter's protégé and son, Hartwell, and his tentative leadership of the ranch.

Loyal to the Land serves as a celebration of the Parker Ranch animals and—especially—its cowboys. It is to these creatures great and small and to their cowboys and caregivers that this book is dedicated.

1

HAVAIKI to HAWAI'I

A.D. 200–750

From whence did people come to settle in Hawai'i? When did this occur? How did the travelers prepare for navigation and the storage of food for animals and themselves? Who would care for the animal and plant life? Where in world history does this voyage fit?

By the time of Christ, a people ultimately to be called Polynesians—"people of many islands"—discovered and settled the Marquesas Islands. In the early part of the first millennium, voyagers from the Marquesas would settle in Hawai'i, 4,100 kilometers to the northwest. In time, a second wave of voyagers from the Society Islands followed. From a global chronological perspective, the discovery voyages of the Polynesians began long before Columbus found America, Balboa the Pacific, and Magellan the Philippines.

Discovering Hawai'i from the Marquesas was a remarkable accomplishment. In their favor, these seafarers enjoyed the predictable trade winds, blowing northerly toward their destination at 20 degrees north latitude. Remarkable is an apt description of this marathon journey when one considers the expansive field of navigational exploration these voyagers conducted. The Stewart Islands at the southernmost tip of present-day New Zealand and Hawai'i in the north form the southern and northern extremes of Polynesia, respectively. Tonga on the west and Easter Island on the east are the western and eastern borders. Of the destinations noted, New Zealand presented perhaps the greatest sailing challenge, lying far south of the trade winds and within reach of the unpredictable effects of the Tasman Sea.

The four outposts of Polynesian settlement shared several common characteristics. Havaiki was their place of origin, taro was their food staple, and oral tradition was the medium through which all things important were passed.

Mythical Havaiki—often considered the common origin for all Polynesian people—has a different meaning to a Hawaiian, a Maori, or a Cook Islander. All Polynesians, however, point to the Bismarck Archipelago where ocean voyaging originated as people moved there from Southeast Asia. People of the Lapita culture fanned out through the Pacific Islands, arriving in Tonga and Samoa from Fiji around 1000 B.C. Migration continued, reaching the Marquesas and Society Islands between 200 B.C. and A.D. 300, or perhaps earlier. Forging on, the boldest of all voyagers reached Hawai'i, New Zealand,

and Easter Island shortly thereafter, thus completing their exploration of the broad Pacific Ocean. At the risk of generalizing for the sake of this text, the most accurate and nonmythical points of origin for the settlers of the Hawaiian Islands are the Society and the Marquesas Islands, with emphasis on the latter for the first arrivals.

These "Vikings of the Sunrise," as noted Maori anthropologist Te Rangi Hiroa (Sir Peter Buck) named them, carried a carefully selected gene pool with proven potency and optimistic prepotency. This means carrying forth the most desired human qualities and leaving the frailties in Havaiki. Both men and women shared in paddling these seafaring canoes. There is evidence of this in the presence of a particular groove on the lower surface of the collarbone of both sexes—the result of this kind of strenuous activity. These voyagers were physically fit, with a predisposition toward pioneering adventure.

The voyagers to all four outposts shared a common manifest: people and their food supplies, animals and their feed supplies, fresh water for both, and an inventory of seeds and seedlings, roots and tubers; added to this were religious and personal items. A significant member among the roster of pioneers was a spiritual leader—the *kahuna (tohunga* for the Maori*)*, or priest. These individuals served in a powerful advisory capacity throughout voyage preparations, during the voyage, and at disembarkation. They provided guidance in mind, body, and spirit for pioneer voyagers during the journey and, once landed, helped establish the cultural foundation upon which the new homeland would be based.

Regardless of the direction of the voyage, the *kahuna* would carry on three tenets of Polynesian culture: *mana, uku,* (*utu* in Maori), and *kapu* (*tapu* in Maori). These tenets provided three dimensions of mediation. *Mana,* forever the spiritual capital, gives authority via the intellect. *Mana* could be acquired or inherited, and in each case oral tradition provided the medium. In a Hawaiian context, *uku* refers to the gifting of materials or hospitality (positive), just as with the Maori. The latter people also carried a negative *uku,* where vengeance would be exacted. *Kapu,* like *mana,* was universal. It reflected social constraint—sacred laws to be followed, with often severe punishment for infractions.

It is likely that the crew followed a command hierarchy. After the priests in charge of navigation, the order of command for a vessel quite likely gravitated to food preparation and storage—a role that undoubtedly involved the women, although the preparation of taro was reserved to the men. They would also guard an impressive inventory of cultivated plants for food, fiber, and medicine. These plants, brought over an extensive time span, included:

- *'auhuhu*, whose leaves provided a narcotic/relaxant effect on fish
- *'awa*, noted for its narcotic effect in pain control, anxiety, and insomnia; this plant served a religious and cultural function as well
- *'awapuhi kuahiwi,* with leaves useful in healing lacerations and roots that served as an analgesic in oral (dental) and abdominal pain
- *hala,* with seeds and flowers that provided a cathartic (strong laxative) effect; aerial roots *('ule'ule),* when

steamed in ti leaves, were a medicinal source of vitamin B; and leaves were used for weaving and thatching

- *ipu,* whose species in general are a broad variety of gourds that provide functions as food (white pulp interior); receptacles (food, water, articles); musical instruments (rattles, drums); and medication (a variety with bitter white pulp interior)
- *ipu ʻawaʻawa,* whose leaves and shoots work as a sedative or cathartic depending on dosage given orally; topically, the leaves promote healing of skin lesions
- *kalo, talo,* or *taro,* which was primarily of food value; medicinal value for skin infections (leaves); catharsis *(hoʻo nahā—*roots*)*; and as a source for soothing stings (stems)
- *kī (ti),* with flowers that provided asthmatic relief; roots that when specially prepared provided a distinct alcoholic/intoxicating effect; leaves *(lāʻī)* that were useful in bandaging (with or without a poultice dressing), headache, and fever relief; also useful in food preparation
- *kō* (sugarcane), with roots useful in treating lacerations and a juice that was combined with medication as an herb sweetener
- *kukui,* having a leaf useful in wound treatment; a nut with profound laxative effect; oil when lit provided a source of light; bark was useful in treating urinary tract problems; and the green nut juice was used for treating thrush and other mouth sores
- *lau kāhi,* whose leaves were used externally to ripen and heal boils; used internally for diabetes; and fruit of food value
- *maiʻa* (banana), which went beyond the food value of the fruit; roots of medicinal value *(maiʻa malei);* leaves and trunks useful as *hāliʻi*—lining or covering the *imu* or ground ovens in cooking; sap that soothed open wounds; and the heart of the trunk passed orally to relieve choking
- *niu* (coconut), beyond its food and fiber value, provided a juice used for treating eye inflammations; shoots for treatment of lacerations; and oil as hair and skin cosmetic
- *noni,* a pungent fruit that worked profoundly as a poultice for wounds and boils, as did its leaves; a bark and leaves that could be boiled into a tonic for urinary tract and muscular-skeletal system problems
- *ʻohiʻa ʻai* (mountain apple), tree bark prepared as a tea for tonic effect on colds and cold sores
- *ʻōlena* (the turmeric plant), with bulbs that provided relief in ear, nose, and throat afflictions when used topically by dripping the juice into the ear, nose, or throat
- *pia* (arrowroot), tubers offered diarrhea relief; a starch used in cooking; and a powder with qualities of talc
- *ʻuala* (sweet potato), whose roots provided asthmatic relief or were used as an emetic to induce vomiting *(hoʻo luaʻi),* depending on how it was prepared; leaves prepared as a tea to provide a tonic effect or to induce lactation

• *uhi* (yam), which served primarily as a food source

• *'ulu* (breadfruit), which was predominantly used as a food source; gives off a sap with medicinal qualities for treating skin lacerations and infections (may have been introduced as recently as A.D. 1500); the sap was also used as a glue for catching birds and caulking canoes

Given the versatility of these products, it is readily evident that the plant life brought on board provided a veritable food pantry and well-stocked medicine cabinet, thus reinforcing the significant role of the seagoing caregiver of plants, animals, and humans.

Preservation of food stores provided an interesting challenge, due largely to the length of the voyages. Drying and/or salting are traditional in Polynesian culture. Access to fresh fish was a ready alternative, and large catches could be salted or sun dried for passage through periods of inclement weather or the doldrums. This system provided a long-lasting food resource.

It is likely that the women aboard were charged with the responsibility for animal husbandry—the health and welfare of the pig, dog, and chicken—the creatures that were successfully transported to Hawai'i's shores early on and during subsequent migrations. Although the dog and chicken are species not emphasized in this text, the rat apparently deserves a place in voyaging history—but only as a stowaway. In Maori beginnings, the rat was an edible creature considered a delicacy. These species have a varied history of survival in the four outposts of Polynesian destination. All four thrived in Hawai'i; only the dog and rat survived in New Zealand. Chickens—but neither pigs nor dogs—made it to Easter Island. Tonga also enjoyed the proliferation of all four species.

The Polynesian hog was a smaller model of the present day version, having an adult weight of about 50 pounds. It was typically cloven hoofed *(kapua'i māhele 'ia)* and shared the physiological strengths and weaknesses common to the species. This animal, through the ensuing centuries, has retained its focal prominence in Hawaiian culture. This reverence for the hog was not unique to Hawai'i. Pigs were the most highly valued animals in the entire Pacific, attested to in Oliver's phrase, "pigs are our hearts" *(Native Cultures).* This animal icon, in my own experience, was routinely used in the processes of *ho'opōmaika'i* (blessings), *ho'oponopono* (righting of a wrong), and *ho'oka'awale* (to cast off, exorcise). While the hog was often consumed in the first two processes, the ritual known as *ho'oka'awale* often consigned the *imu* as the permanent grave of the cooked pig—preferably a solid black, intact boar. The unopened *imu* at the blessing of the new arena at Paniolo Park on Parker Ranch stands as an example.

The pig possesses few physiological traits amenable to long voyaging over saltwater seas. An exception is that the hog's diet is easily provided, hogs being generally omnivorous. Some disadvantages inherent to the species are serious ones—especially their intolerance to heat and salt. The hog is highly sensitive to mild elevations in dietary salt, leaving oceanwater off limits as a source of drinking water. Their intolerance to heat relates to an inability to dissipate heat via their skin, as pores are absent.

Respiration is left to manage the cooling process, and shade and ventilation are vital in this function. Shade was likely a premium reserved for humans on the voyages, but sharing this luxury was necessary to safeguard the passage of pigs. Of course, consumption of pork en route—especially of those disinclined to complete the voyage for a variety of reasons—possibly provided some limited incentive to continue.

With an average gestation period of only three months, three weeks, and three days, depending on when she conceived, a sow could likely farrow (give birth) en route if conditions were amenable. Litter size was probably not large—perhaps four piglets per gestation. The actual farrowing, nursing, and rearing of these piglets presented no small task for the seafaring sow and caregiver. This conjures up memories of my youth, when Hawaiian elders told me stories of their own lactating mothers taking on the nursing of piglet runts to ensure their survival. One could argue that seeing a newborn piglet on his mother's opposing breast might be one's first exposure to "unfair competition"! But this exemplifies animal caregiving in a culture that held the pig as a very special and respected animal companion. The seafaring caregiver was ultimately responsible for the survival, growth, and propagation of the porcine species, and they were dedicated. Nursing of animals by Polynesian mothers also occurred in the canine species during times that the dog was considered a delicacy. Pups reared on human breast milk were especially savored. These specially raised canines were referred to as *'īlio poli,* and they were reserved for consumption by the *ali'i* (royalty).

Ka Lae (South Point), Hawai'i, with its towering cliffs reminiscent of Havaiki, paralleled by accessible landings at Honu'apo to the east, Wai'ahukini to the west, and Ka'alu'alu Beach nestled between, gave welcome to the forebears of the Hawaiian race. As a backdrop to Ka Lae, the expansive mass of Mauna Loa, an active volcano, might have lit up the night skies like a mountainous red lighthouse.

While the region adjacent to Ka Lae is traditionally referred to as the original landing site, it is just as likely that the voyagers were inclined to disembark in verdant valleys such as Waipi'o, Waimanu, and Pololū. The rain-endowed bay of Hilo, with its calm waters and multiple landing places, would have provided an appropriate reception to the Polynesians and their animals. The following quote is from a *Journal of the Polynesian Society* article entitled "Small Population Isolates: A Micro-Simulation Study":

> The numbers of people arriving on the initial voyage is unknown. Predictive studies indicate that 6 people (3 couples) would have had as low as a 33% chance of successfully producing a surviving population, while 14 people (7 couples of child-bearing age) would have had over an 80% chance of founding a colony that would last. Clearly, a double canoe could easily carry 14 people (Norma McArthur, quoted by Ross Cordy in *Exalted Sits the Chief*).

Disembarkation must have been a joyous occasion for man and animals, calling for rest and respite. Searching for food and water followed in short order. With sudden access to unlimited shoreline came the availability of shore fish, shellfish, and a variety of seaweed to bolster their newly prescribed diet.

While pigs, dogs, and chickens were transported food sources, another lesser-known source of protein came from wild birds.

As for animal life, the settlers found the hoary bat, insects, a variety of birds, including the duck *(koloa),* and possibly the *moa,* a large flightless bird. The proto-Polynesian word for duck, *"toloa,"* is still used in the Marquesas. As for the *moa,* recent findings of the skeletal remains of a flightless bird on the island of Hawai'i give credence to its existence. Whether it was indigenous, as in the case of New Zealand, or brought with the settlers from the south, remains to be determined. Helen Farmer, noted Smithsonian expert on the identification and study of Hawaiian birds, has stated that there were more than twenty species of flightless birds in the Islands before humans arrived—including ducks, geese, and ibises.

In New Zealand, the giant *moa* encountered by the Maori is said to have stood twice the height of a man. The smaller bush *moa*—likely the Hawaiian variety—is said to have been about 1.3 meters in height and to have weighed about 58 kilograms. *Moa* hunters of New Zealand could get about 37.5 kg of meat from a single bird—enough to feed about fifty people for a day. The *moa* was fair game, as it was able to outrun humans and swim short distances.

Looking for food, the settlers found similarities with their ancestral Havaiki, including edible ferns such as the *uluhe* (false staghorn) at higher elevations. Clearing wooded lands for planting crops was an immediate chore. Felled trees were used for firewood and building materials. Rudimentary tools were hewn from the wood. With these implements, fertile soils were prepared for immediate planting of transported roots, seeds, slips, and seedlings, with reasonable success.

In the interim between planting and harvesting, pressure was brought to bear on the food sources of their new home. Seafood, bird life, and wild plants were heavily in demand during this period. Harvesting their first crops reduced the need to rely solely on the natural environment. As domestic animal husbandry began to produce a consistent protein resource, less pressure was placed on bird and marine sources. Cultivated plant harvesting simultaneously relieved the pressure on the native flora.

On Hawai'i Island, the slopes of Mauna Kea were hunted for the *nēnē* (Hawaiian goose) and *'u'au* (dark rumped petrel) nestlings, while the plover *(kōlea)* and wild duck *(koloa)* were found in lower elevation forests, grasslands, and shores. The Saddle Road area of Pōhakuloa in present-day Hawai'i has produced bountiful archaeological evidence of large meat birds.

The native populations increased, possibly doubling every thirty years. In the ensuing centuries approaching the arrival of Captain Cook in 1778, the Hawaiian community grew from the initial population made up of the original voyagers to perhaps a hundred thousand people occupying the coasts and verdant plains and valleys of the Big Island. More than three thousand hogs lived in close harmony with their caregivers, while a comparable number dwelled within the expansive forests to form the ever-present feral droves of present-day Hawai'i. Just as did their human counterparts, they arrived relatively free of diseases. Brucellosis, pseudorabies, hog cholera, leptospirosis, and erysipelas were infectious agents that would not touch their immune systems until long after contact with Cook.

The Hawaiian community reigned free from the ravages of measles, rubella, chicken pox, smallpox, scarlet fever, mumps, tuberculosis, leprosy, gonorrhea, and syphilis until repeated foreign contact introduced them. With the coming of Cook and others, a series of rapid occurrences forever changed the complexion of Hawai'i's people, animals, plants, and environment.

2

EARLY FEATURES OF NORTH AND SOUTH KOHALA

750–1775

Deep soils and strong winds characterize the lands of Kohala. Forty miles of shoreline provide an attractive feature. Kohala Iloko is the ancient name for the windward lands east of 'Upolu Point. Steep cliffs, narrow reefs, and rough seas are characteristic of the Kohala Iloko shoreline. The deep valleys of Pololū and Honokāne contrast with the gentler central slopes, where permanent streams meander among shallow gulches and *'ōhi'a* forests abound in the upper elevations.

The leeward lands of Kohala—notably Kawaihae, along the shore, and the uplands of Waimea—are also characterized by strong winds. Waimea is said to be the point of origin of the fierce *mūmuku* winds that blow down from the Kohala Mountains and Mauna Kea, giving the area its windswept nature, characterized by the many leaning trees in evidence.

Leeward Kohala has four distinct zones of land terrain: Kohala *i waho, 'āina* Kawaihae, the coastal plains, *wao one,* and the Waimea plains, *kula.*

Kohala *i waho* of ancient Hawai'i ranged from 'Upolu Point to Kahuā, just north of Kawaihae, where dryland forests on

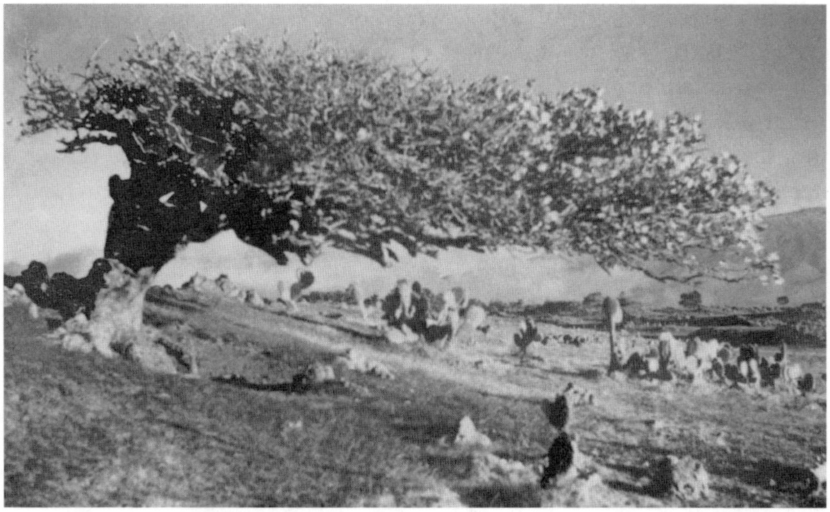

A gnarled and leaning native tree, showing the impact of the strong trade winds that come out of the northeast. Panini *cactus grows in the foreground; Mauna Kea is in the background. Photo taken in Lalamilo Paddock by General C. D. Herron, 1940.*

gradual slopes were commonplace. As much as 2–3 miles inland from shore were deep soils along the 30-inch rainfall line.

In the *'āina* of Kawaihae, the rainfall line and deep soils extended farther inland from a shoreline blessed with calm waters and multiple landing coves. For bountiful fishing, the reefs extended far beyond its long sandy shore, which would one day receive the island's first cattle, sheep, and horses.

South of Kawaihae are the lands that make up the coastal plains that by the *ahupua'a* system (basic unit of Hawaiian land tenure) extend from shore inland as far as 8–10 miles, bordering on the Waimea plain. The arid coastal plains of 'Ōuli, Lālāmilo, Kalāhuipua'ā, and 'Ānaeho'omalu, with less than 30-inch rainfall patterns, vary from grassland to barren lava fields that accommodate natural anchialine ponds along the shores. Clusters of housing units —*pā hale*—were located around the anchialine ponds and fishponds.

In the intermediate zone between the arid plains and verdant Waimea were rainfall zones that varied between 40 and 80 inches. It is at these elevations that the early inhabitants realized the limits of the taro versus the hardiness of the sweet potato. Homesites were commonplace at these middle elevations. Approaching Waimea, a shift back to the predominance of dryland taro occurred.

Some 10 miles inland from Kawaihae, the cool *'āina* Waimea located at the upper juncture of the valleys of Waimanu and Waipi'o, was likely the first leeward Hawai'i land settled. Within the Waimea *ahupua'a*, archaeological evidence suggests that this occurred around A.D. 800–1000. Waimea is further characterized as the only leeward Kohala land blessed at that time with the perennial streams of Keanu'i'omanō, Kohakōhau, and Waikoloa. Other attributes were deep soils, high rainfall of between 80 and 120 inches, and the coolness *(ma'ū)* of higher elevations (2,000–2,800 feet). 'Āina Waimea, with these natural features, became the cradle of early intensive upland farming—and later, the home of a historic livestock industry.

The population of Waimea became the most significant in density, scattered among fields adjacent to streams that provided year-round water for consumption, cleaning, cooking, and irrigation. The availability of dependable irrigation systems gave Waimea a unique advantage whereby both dryland and irrigated *kalo* (taro) could be grown.

Unwalled or low-walled plots of taro were built on flats, terraced slopes, or natural swales. The swaled lands were cultivated on the plains of Pu'u Pā near current-day Waimea airport and down to the 1,800-foot elevation. Dependable rain patterns prevailed, while Waimea's streams provided an additional water supply via canals that spread water over wide surface areas, allowing very gradual percolation into the soils beneath the *kalo* fields. The more eastern taro fields were served predominantly by rainfall moisture, while the more western fields were increasingly dependent on irrigation as a water source. These low-walled taro terraces can still be seen from the upper Kawaihae Road looking *mauka* (northeast) to the slopes of the Parker Ranch pastures of Lanikepu, Pu'u Kawaiwai, Pauahi, Kāmoa, and Laelae.

The early Waimea inhabitants resided

typically within a *pā hale* (fenced house lot) with a sleeping house and adjacent protected cooking facility. The *pā pōhaku* (stone wall) surrounded the *pā hale,* and likely included within was a *kīhāpai* (garden). The farming plot *(ʻāpana)* of the householder was located elsewhere within the agricultural zone of the respective *ahupuaʻa*. These prehistoric farmed areas have become known as the Waimea Field System. Evidence of Waimea's native Hawaiian activity is seen along the Kohala Mountain Road from several vantage points between the Hawaiʻi Preparatory Academy Upper School

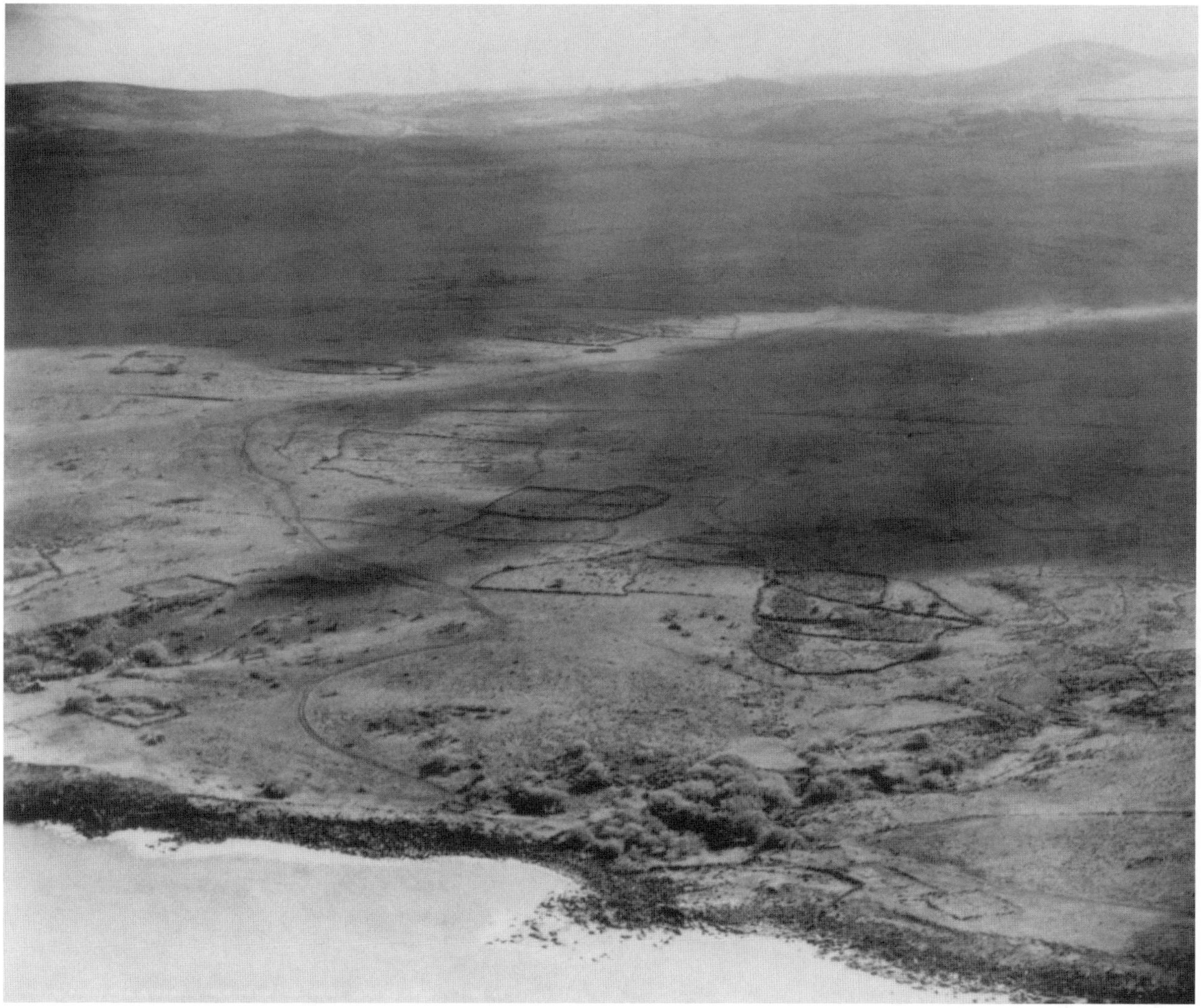

Aerial view of the South Kohala coast showing extensive stone wall configurations of dryland farming. LHMM 74.30.2.86, circa 1925.

campus and the cinder hill of Pu'u Kawaiwai. Remarkably, the *pā hale, pā pōhaku,* and *kuleana* are clearly outlined.

The chronology of Waimea's evolution as an early community is best demonstrated by the progressive expansion of the Waimea Field System. Archeological studies have revealed that rudimentary farming of the 'Ōuli flats between Lanikepu and Keanu'i'omanō Streams *makai* (toward the

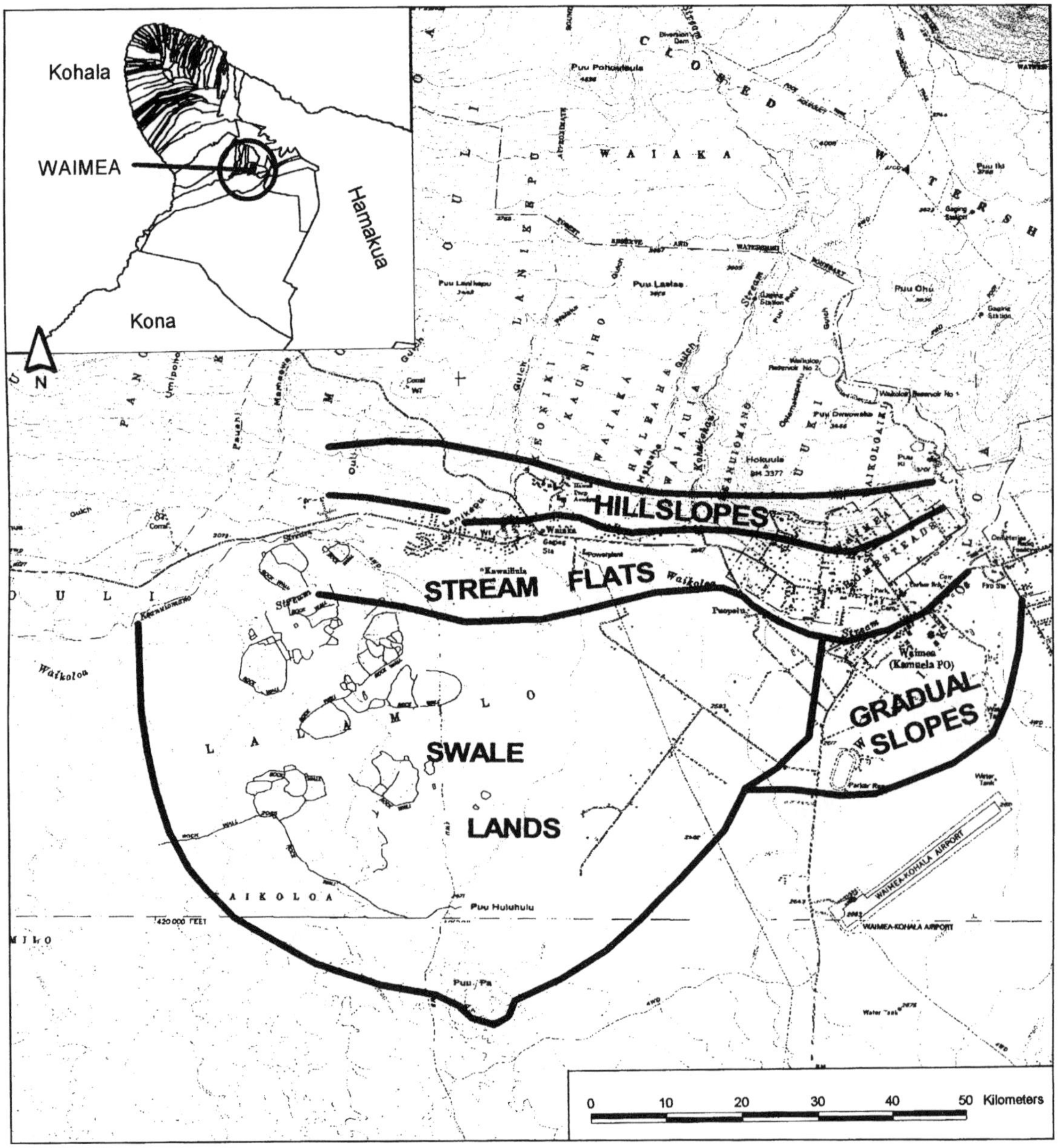

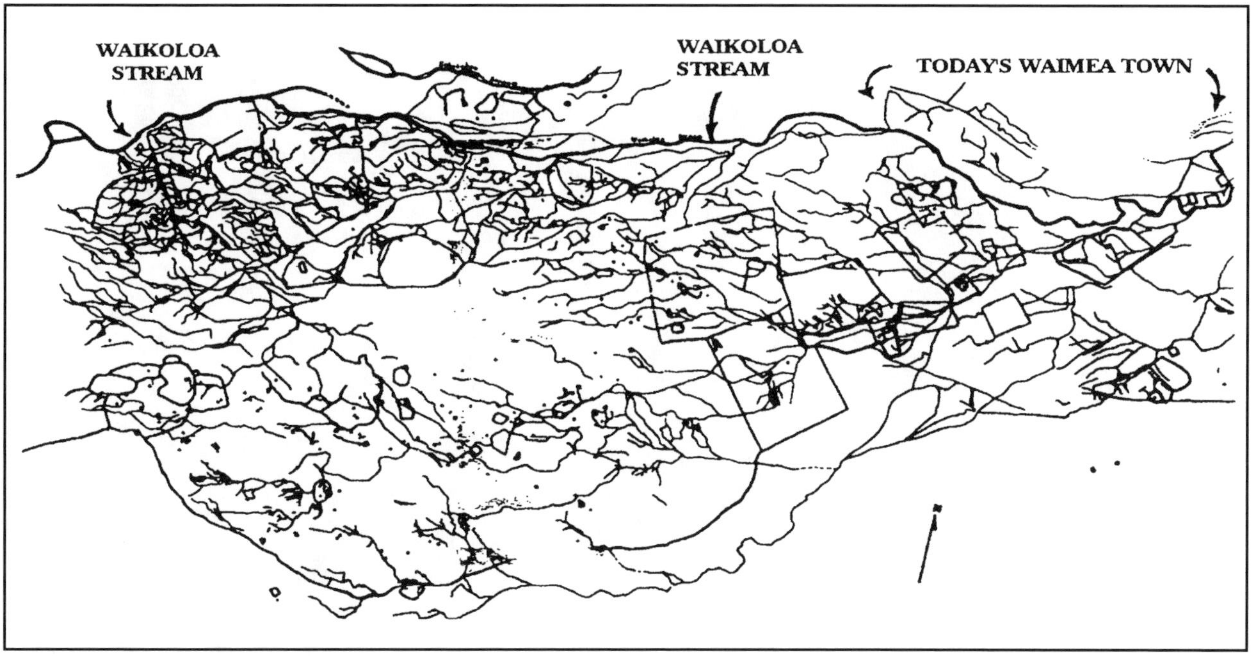

Opposite page: *Waimea Field System and Zones, showing considerable stone wall enclosures in the left* (makai) *portion of the swale lands of the 1600–1800 era. From Ross Cordy's* Exalted Sits the Chief. Above: *Map showing the irrigation canals* (pūnāwai), *house lots* (pā hale), *and animal pens* (pā pua'a). *The latter two are square and rectangular in shape. The irrigation canals reflect a vast networking of* pūnāwai *from the Waikoloa Stream. From Ross Cordy's* Exalted Sits the Chief.

ocean) of the airport began as early as A.D. 1100–1200. The southernmost swales reflect the presence of housing and agriculture about A.D. 1300–1400. Substantial evidence, however, points to the cultivation of walled, irrigated fields coupled with permanent habitation during the period from A.D. 1600 to 1800.

A marked increase in the extent of the agricultural fields in the 1600–1800 era is suggested in Ross Cordy's text, *Exalted Sits the Chief,* which coincides with two chapters in early Hawaiian *ali'i* history—the presence of Alapa'inui and Kalani'ōpu'u in the Waimea-Kawaihae area, coupled with the construction of Pu'ukoholā Heiau by Kamehameha. The pressure on both plant and animal food resources had peaked.

3

ARRIVAL OF ANIMALS AS GIFTS FIT FOR A KING

1775–1850

Since this text is about the Parker Ranch animals and their caregivers, those social aspects of the arrival of Captain James Cook in 1778 will be left for the reader to explore in the many available resources. At this juncture, however, it may be appropriate to note that the Big Island population is estimated to have been slightly over 105,000 native Hawaiians, with more than 23,000 of them living in North and South Kohala. Before we leave Cook, it is important to note that greater credence is being given to reports of Spanish galleons being shipwrecked in the Islands—for example, off the coast of Kaua'i as early as 1527—long before Cook's arrival. Spanish sailors are said to have married Polynesian women at that time. Further Spanish contact may have occurred around 1542 and in ensuing years. In *Old Time Hawaiians,* Mary S. Lawrence writes:

> Let us leave Kamehameha I and go back to the time when Umi's son was ruling. It was then that a strange ship was supposed to have been wrecked off the rocky coast in South Kona, Hawaii. A legend tells us that the captain and his sister were the only ones who were saved.
>
> The natives received them very kindly, and they spent the rest of their lives as Hawaiians. At that time—the sixteenth century—many ships were sailing the Pacific. We suspect that this ship was one of those, which were lost in severe storms and never heard from.
>
> In 1555, not many years later, another Spanish ship was supposed to have stopped at the islands on its way from Mexico to the Philippines. Juan Gaetano, the captain, located islands previously unknown upon his chart, but kept his discovery a secret.

In 1792, Captain George Vancouver arrived on the HMS *Discovery* and delivered a pair of goats to Ke'eaumoku, a Big Island chief. Their successful propagation was due to the chief's enthusiastic charge of this caprine pair, and it led to another Vancouver gift. The first sheep, arriving in 1793, amounted to four head—a pair of ewes, a ram and a lamb. That same year, Vancouver also delivered the first cattle—California longhorns—given to King Kamehameha I and unloaded at Kealakekua Bay, quite likely at Nāpo'opo'o Landing.

The cattle came from a Californian named Quandro. Because they were both blacks and reds, they may have been a subtype of longhorn cattle originating from three strains of Spanish cattle brought to the New World. It is unlikely that they were the Berrenda strain, which has a white body and black markings limited to the neck and ears. It is quite likely that Quandro sent both the Retinto (Jersey tan to cherry red) and the ancient black variety, the Ganado Prieto—the progeny of the Andalusian fighting bull.

An emergency landing at Kawaihae preceded the main landing at Kealakekua Bay. This was necessitated by the weakened condition of a bull and cow, prompting their dispatch by canoe to the shores of Kawaihae Bay. The Kealakekua delivery of five cows, a bull, two ewes, and a ram was less eventful owing to better condition of the stock. The low quality and quantity of feed provided to the livestock en route contributed to their poor condition on arrival. This may explain Vancouver's curious observation that while the cattle were transported to shore on the chief's canoe, they fed voraciously on fresh vegetables. In later arrivals, newborn calves of such malnourished dams were raised on fish—and they thrived. This was reported shortly after the delivery in 1794 of a bull, two cows, two bull calves, five rams, and five ewes to Kamehameha I at Kealakekua.

The tribulations of seafaring cattle, sheep, and goats mirrored the problems that faced the Polynesian transport of the hog more than a thousand years before Cook's discovery and Vancouver's gifts to the chief and king. The British, however, had some definite advantages in animal husbandry, especially in the area of storable concentrates (grains) and cured hay. Storing stock water in large oak barrels was a vast improvement over the Polynesian gourd. Furthermore, the route from California's coast was shorter than a Havaiki-to-Hawai'i journey. The British had sophisticated navigational instruments at their disposal and greater storage capacity, but their sailing vessels were slower than the double-hulled canoes.

The caregiver responsibility was still a serious one, particularly when the animal cargo was intended as gifts for a king. Sheep and goats are relatively docile, but longhorn cattle under the best of circumstances do not lend themselves to hands-on care. It is likely that rough travel conditions yielded mild, chronic dehydration that put the longhorns in a more socially compliant mood. Once the cattle were acclimatized to their new home, forage of the native vegetation with its copious variety restored them to a sound reproductive state.

The Infamous *Kapu*

Then Kamehameha I issued the infamous *kapu:* the law—carrying the death penalty—that would apply to anyone causing the demise of any cow, bull, calf, steer, or heifer yearling. While the term of ten years for a *kapu* is commonly cited, a period of twice that long fits the chronological time frame more aptly.[1] Such a moratorium was reliable insurance for solid-based propagation, but it resulted in some serious socioeconomic consequences.

One such side effect was the destruction of plant life, both cultivated and in the wild. For a thousand years preceding the arrival of cattle, hogs had been foraging (rooting and grazing) native plants, so early extinctive processes were already well under way. Now the bovine—a far more mobile quadruped with a nutritional requirement of about six adult hogs and equipped with the ability to browse on low-lying trees and plants inaccessible to the pig—accelerated the process.

No longer could the native Hawaiian safely farm taro, sweet potato, and so on without risk to his physical safety when the wild bullocks plundered his gardens *(kīhā-pai)*. This necessitated the construction of taller stone walls *(pā pōhaku)* to secure the *kīhāpai*. Many such enclosures can be viewed in parts of Ka'ū and Kona. They are especially visible in Waimea from the air or from higher elevations. The enclosures were needed to protect crops cultivated in the Waimea Field System during the time of the *kapu*. Kamehameha also ordered the building of a great wall, which stretched from north to south Kona, to keep the feral cattle out of the more densely populated areas.

The distinction of feral status can be seen as a double-edged sword for this species. The downside of its environmental and socioeconomic destruction is tempered by the species' role in the birth of the cattle industry and the folklore of the Hawaiian cowboy. Without the species' vast numbers, there would be no Hawaiian cattle industry and no image of that stoic, capable caregiver, the *paniolo*.

Arrival of the Horse

With the arrival of the horse in 1803 via Captain Cleveland's *Lelia Byrd*, a new mode of transportation was introduced. While oxen *(pipi kauō)* also served this purpose, the maturation and training of the draft ox came later.

The mare and foal landed at Kawaihae came from the Mission of Saint Joseph in San Borgia and were intended as gifts for Kamehameha I. John Young, a close advisor to the king, provided their immediate care. Young's counterpart, Isaac Davis, received another pair of horses that were similarly intended; however, these were disembarked on Maui. The native population reacted with awe, just as they did when they encountered cattle for the first time.

The growth in the number of horses was slowed by the failure of the beast to acclimatize to the Hawaiian diet of moist—but not necessarily nutritious—grasses. This failure to acclimate imposed a grace period of at least two years. Mineral nutrition probably lacked adequate phosphorous, which the Hawaiian Islands continue to experience into the twenty-first century. In the 1820s, a man on horseback was still a rare sight.

Horses were allowed to run and breed freely, and they thrived even without the protection of a *kapu*. This created a solid base to the siring side of the process. Inbreeding, as with cattle, sheep, goats, and hogs, was daily fare, and by the early 1830s there were feral horse herds on the slopes of Mauna Kea.

Sheep

Because of their warm woolen coats, sheep sought the higher elevations for propagation. Goats reproduced at any elevation, but they preferred arid country. By 1822, feral sheep were noted near the summit of Mauna Kea.

NOTE

1. Both *Thrum's Hawaiian Annual* of 1906 and John Turnbull's *Voyages in the Pacific* cite the ten-year *kapu*. Other references, including reports of growth in cattle numbers, cite a date that the *kapu* was lifted and John Palmer Parker received royal authorization to begin to trim the herds.

4

JOHN PALMER PARKER AND THE EARLY BEEF INDUSTRY

1815–1847

At the time of the landing of Hawai'i's first cattle by Vancouver, John Palmer Parker was the two-year-old son of Samuel and Ann Palmer Parker of Newton, Massachusetts. Samuel Parker and Ann Palmer had wed in 1770 and lived on Clark Street in Newton, where Sam was born and raised. The Parkers had been part of the Newton community since the mid-1600s, while the Palmers had arrived more recently from Ireland.

John Palmer Parker was born on May 1, 1790, which was the centennial of the town of Newton. He was the sixth and last child, the second surviving son. His brother Ebenezer remained in Newton, the hometown

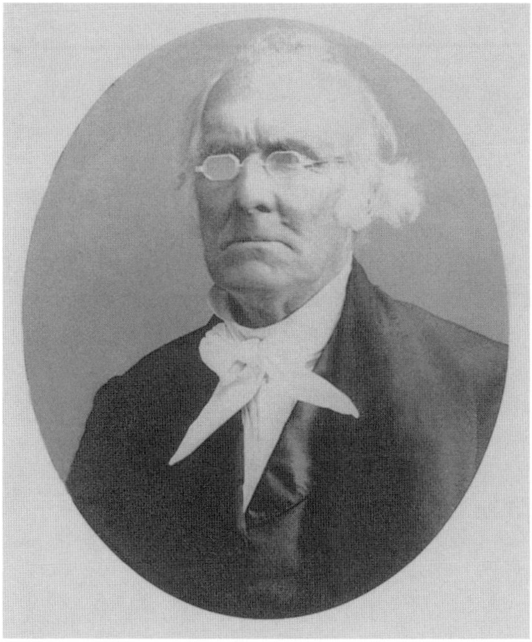

Samuel Parker IV, father of John Palmer Parker I. PPS Collection.

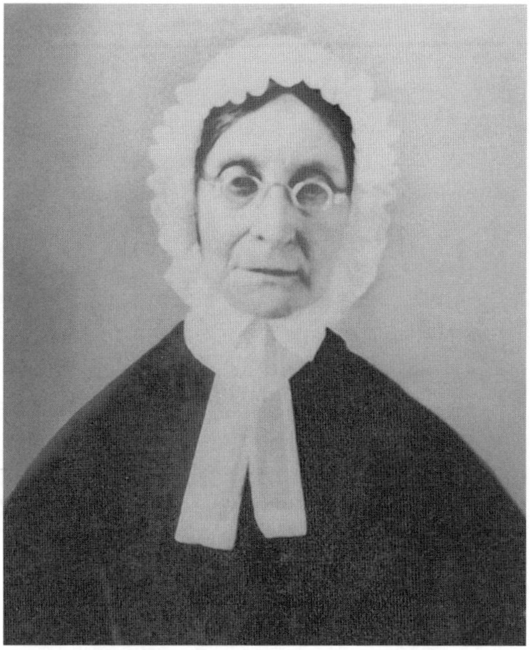

Ann Palmer Parker, mother of John Palmer Parker I. PPS Collection.

Parker genealogy chart, modified from Brennan's book, *The Parker Ranch,* traces the Kamehameha and Parker forebears through ensuing generations to the year 2002.

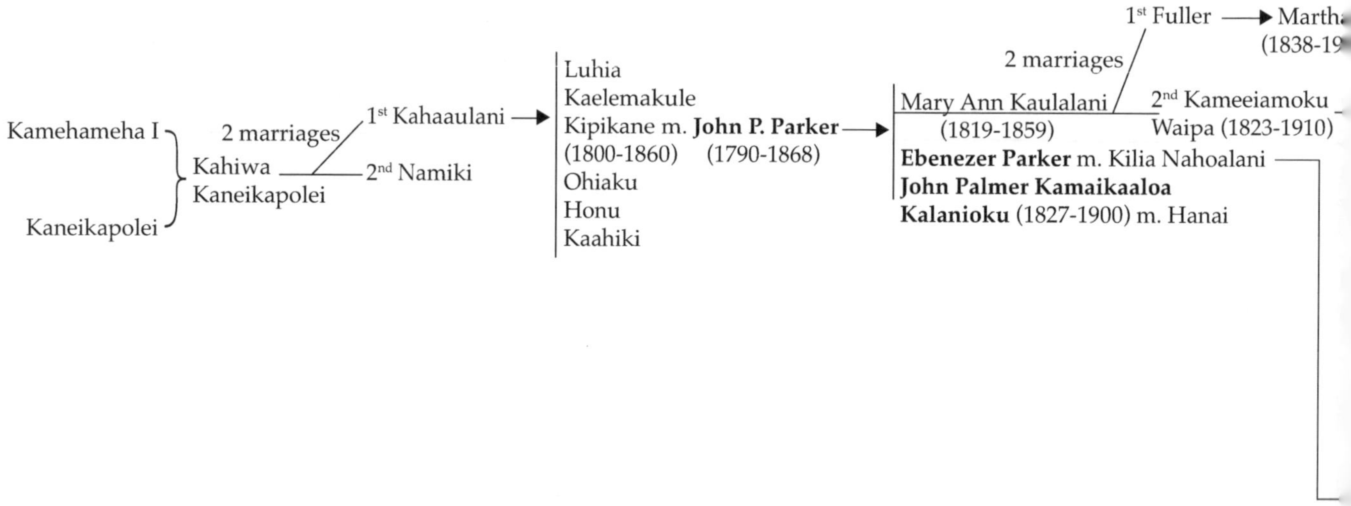

1st **J. S. Low** → (d. 1873)

John Stanley Low (1857-1912)
Eliza Davis Low (b. 1858)
Clementine Davis Low (b. 1861)
Ebenezer Parker Low m. Napoleon → (1864-1954)
Clara (b. 1866)
Mary Kipikane Low (b. 1868)
Hannah Pierce Low m. Robert Hind → (b. 1869)
George Kaina Low (b. 1872)

Sandford Dole Low (1888-1889)
Annabelle Dole Low (b. 1890)
Evelyn Woods Low (Eben) (b. 1892)
Caroline Kahalelaukoa Low (b. 1894)
Elizabeth Jessamin (Clorinda) Low (1895-1986)

Eben Fuller (1899-1925)
Laura Louise (1896-1922)
John Somes Low (b. 1900)
Sandford Dole Low (b. 1905)

R. Leighton I m. Marjorie Capps → R. L. Hind II (Bobby)
Margaret
Mona
Irma
Robson

R. L. Hind II (Bobby) / Florence Vredenburg (Coco) → **R. L. Hind III**

R. L. Hind III / Marty Benton → Hannah, R. L. Hind IV

: marriages

2nd **Kaaua Kekuawela** →
Stella Hailauaala Kaaua (b. 1874) m. James Hind → Eva, Maud, James
Archibald Cleghorn Kaaua (1896-1918) / Frances Eudora Spencer →

Archibald Spencer Kaaua m. Piilani Yates → Alan Cleghorn (1904-1950)
Edward Stanley (1908-1957)
John Low (Jack) (1912-1962)
Samuel Parker Kaaua (b. 1913)
Stella Servia Hailauaala Kaaua (b. 1819)

Alan Cleghorn / Marjorie Bergin → **Christopher**

Christopher / Iwalani Roque → Harry

William Ulumaheihei K. (1842-1905)
Kuakini (1844-1895)
Anna Parker (b. 1848)
Hannah (1849-1849)
Malie Kaonaeha (1852- 1895)
Grace Wahineikeili (1855-1895)
Robert Parker (1856-1937)
Kalilikane (1858-1955)
Keliiheleloa (1859-1881)

Ebenezer Christian (1850-1860)
Mary Ann Kaulalani II (1851-1909) — 2 marriages: 1st Woods → Bella, Palmer, Alice, Frank; 2nd Stillman → Charles
Samuel m. Panana Napela (1853-1920) (d. 1901)
Nancy Kameeualani (b. 1854)

Mabel
Samuel
Maud
Lucy m. Ackerman → Lucille, Pat, Maude; Alice, Mary J., James J.

Mary Kihilani m. Charles Maguire → Aileen m. Stillman → Thelma, Mary, Nancy (1871-1895)
Eva Kalanikauleleiaiwi m. James Frank Woods
Helen Umiokalani m. Carl Widemann → Parker (b. 1874)
John III m. Elizabeth Jane Dowsett → Thelma (1875-1893)
Harriet Kaonohilani (1876-1884)
Palmer Kuihelani (1878-1896)
Samuel Keaoulilani (1879-1934)
Ernest Napela (1884-1945)
James Kekoalii (1886-1962)

Parker — 2 marriages: 1st Hettie → Samuel; 2nd Marguerite → Parker, Panana

Thelma / Henry Gaillard Smart → **Richard** (1913-1992), Elizabeth Ella Smart (1914-1914)

Richard / Patricia Montague → **Antony Parker Smart Gaillard (Gil)** → Stephanie

of the family, maintaining its farming traditions. By 1892, Newton had a Parker Street.

John Palmer Parker attended Framingham Academy and excelled in mathematics. He went to sea on a whaling ship when he was eighteen. At nineteen he sailed to Hawai'i and became thoroughly enchanted. Parker easily assimilated the Hawaiian culture. His sincere and open posture made him comfortable among the *ali'i* as well as the *maka'āinana*. After a brief return to the sailing life, Parker made the Big Island his permanent residence in 1815—a timely coincidence, as that was the year the *kapu* on cattle was lifted.

Parker earned his appointment as the king's first authorized cattle hunter. He was able-bodied and capable and had behind him years of experience to prove his trustworthiness. In his favor was the king's knowledge that Parker possessed a musket with ammunition, along with the skill necessary for its effective use. His marriage to Chiefess Kipikāne undoubtedly placed Parker in an advantageous position.

While the industry expanded to include a host of other bullock hunters, Parker's role was businesslike and in keeping with the desires of the king. This continued after the passing of Kamehameha I in 1819. That year was eventful, as described by Janice Morgan in her 1991 monograph, *The Newton-Hawai'i Connection:*

> Another momentary footnote. Perhaps in the history of no other nation or society has one year been as seminally important as 1819 was in Hawaii. The year Ann Parker wrote the letter to her son imploring that "God give you grace to live a Godly life so if we never see one another in this world we may meet again in the Kingdom of Heaven . . ." and the year Parker began building his estate was the year the great monarch Kamehameha died, the year the missionaries arrived, the year the islands' old religion, culture and society crumbled, and the year Hawaii was opened wide to whaling and settlement. In short, Hawaii's world turned upside down and 29-year-old John Parker of Newton was there not only to witness but play a central role in the changes.

At the time of the king's appointment of Parker, Chief Ke'eaumoku had a retinue of hunters, and overlapping resources undoubtedly occurred. The use of bullock pits became commonplace. One report indicates there were as many as sixty operators in 1837. The year 1832 became a pivotal point in the Hawaiian cattle industry when, by invitation, the *ali'i* welcomed the arrival of the *vaquero,* whose mission involved not only bringing the vast wild cattle herds under control but training the native Hawaiian in the skills of the cowboy.

By the 1830s, the entire island of Hawai'i participated in the cattle industry. The districts of Kohala, Hāmākua, Kona, Ka'ū, and Puna provided fertile grounds for hunting wild cattle. The environs of Mauna Kea, Mauna Loa, the Kohala Mountains, and Hualālai afforded great forests for protection and propagation of wild cattle. Today only the forests of Hualālai lack wild cattle in the form of established herds.

Parker had the strategic advantage of access to the seaport of Kawaihae, concentrating his headquarters around his home in

Wai'āpuka, a *kuleana* (homestead) adjacent to Pololū Valley with hunting, farming, and fishing within close proximity. Waiapuka, which means "springwater," came with a grass shack *(hale mau'u)* and excellent farming potential. The soils, rich and tillable, produced superior crops of taro and garden vegetables in his well-designed and attended *kīhāpai*. As Parker concentrated his career more on ranching around Waimea, this became his permanent center of operations.

Waimea Foothills and Adjacent Plateaus: Post-Discovery

In *Exalted Sits the Chief*, Ross Cordy maintains that early recorded post-discovery history describes the plateau of Waimea as being forested with *'ōhi'a lehua (Metrosideros collina)*. 'Ala 'Ōhi'a (fragrance of *'ōhi'a*) was the euphonious name given to the district. Reverend Lorenzo Lyons frequently described his home as 'Ala 'Ōhi'a Nei (home of the fragrant *'ōhi'a lehua*). Between the overwhelming impact of cattle herds, coupled with drought and the Great Fire of 1918, the district of Waimea is known today more for its rolling hills and plains covered with a variety of grasses and shrubs and a minimum of trees. Not only were the *'ōhi'a* forests lost, but the *koa, koai'a, māmane, naio,* and a host of other trees as well.

The rain forests of the Kohala Mountains also had abundant *'ōhi'a lehua* trees. Despite alien species invasion, these stands remain today. Herbs, ferns, and mosses coexist in cooperative harmony, providing luxurious shade in the enduring undergrowth. These tropical jungles offer protection for the proliferation of humus, or *pulu,* that was significant for its water- and nutrient-retaining ability derived from decomposed vegetation. In exchange for shade, the *pulu* encircles the base of the tree, conserving additional moisture and nutrition. The early deforestation caused a decrease in rainfall and a reduced flow in streams, including the Waiaka, Kohakōhau, Waikoloa, Lanimaomao, and Keanu'i'omanō Streams.

Early settlers in the Waimea foothills and plains encountered vast areas covered in the bunchgrass *pili* in the *kula* (arid range) and the fuzzy-leafed *kākonakona* grass in shaded meadows. Shrubbery vegetation included *'ilima,* which was attractive to sheep and goats but not palatable for cattle. *'Īli'e'e, ko'oko'olau, 'u'ulei, 'auhuhu,* and *hi'aloa* were also described as shrubs with medicinal value, but they were minimally palatable to cattle even during dire drought conditions. Later, the prickly leafed *puakala* emerged, and it continues to be a noxious weed in certain areas to this day.

Pili grass was not hardy, given its shallow roots and need for periodic rest. Excessive grazing effectively uprooted the plant. Different varieties of panicum grass, especially *kākonakona,* were hardier. Other indigenous grasses found at more than 4,000-foot elevations drew the wild cattle to these mountain slopes. Notable among these are the *heu pueo* (owl fuzz), *pili uka,* and *kalamālō,* which were coarse but somewhat palatable. *Pūkiawe* (black-eyed Susan) was browsed only in times of drought, but crab grass *(kūkae pua'a),* introduced earlier, was found to be succulent in the mountain meadows of Waimea.

In 1835, Bermuda grass, *Cyanodon dactylon (mānienie)* was introduced to Hawai'i and was well suited to the lower elevation, drier plains and coastal ranges. This narrow-leafed perennial spreads both by seed and by sending out rootstocks that root at the joints along its stems. It is highly palatable, moderately nutritious, and drought resistant. Bermuda grass creates dense sod, which is highly effective in curbing the erosion of lighter coastal soils.

In 1840, Hilo grass *(Paspalum conjugatum)* came to the Islands. The town of Hilo has the dubious honor of being its first home. This hardy but relatively unpalatable grass covered the lower mid-elevation and wetter meadows of the ranch, especially on the windward side. Like *mānienie,* this grass spreads by means of seeds and runners sometimes extending for several yards, then rooting and sending out shoots from the nodes. While the rank Hilo grass possesses minimal nutritional value, it provided marginal feed for mature cattle, affording reproductive rates of moderate satisfaction. But breeding cattle could never fatten on Hilo grass, which was a well-known advantage of *mānienie. Mānienie* avoided the shaded base of the forest trees—even that of *kiawe* (algaroba or mesquite). Hilo grass in turn invaded the undergrowth of the mountain forests of native trees, ferns, and other delicate plants. The aggressive destruction of forest ground cover by Hilo grass coupled with the herds of wild cattle, fire, and drought actually contributed to the breakdown of the ecosystem and supplied only minimal nutritional value.

Other varieties of crabgrass were introduced onto the ranch during the era of John Palmer Parker II, along with varieties of bluegrass and fescues. *Honohono* grass (wandering Jew or *Paspalum dilatatum*), found in moist valleys, was highly palatable but its nutritional value was marginal. *Pualele* (sow thistle) was introduced and became widespread among the bunchgrass. This was palatable and provided moderately nutritious forage. The arrival of *pilipili,* or beggar tick plant, was relished forage that wreaked havoc on the sheep operation because its barbed awns attached tenaciously to the wool. *Pilipili* was also commonly found among the bunchgrasses on the frontal plains between the Kohala Mountains and the foothills of Mauna Kea. Even the pretty little wild dandelions, or *laulele,* were moderately significant cattle forage in the middle elevations of the ranch.

Legumes, a plant family that attracts (fixes) nitrogen from the environment, were also introduced fairly early in the nineteenth century. There are nitrogen-fixing bacteria in the nodules of legume roots. Soil nitrogen is then absorbed through the roots, traveling to the leaves and stems consumed by the cow. This in turn converts the plant protein to animal protein (beef).

The first and most notable legume was black medic. The nitrogen-fixing quality of legumes, which include clovers, provides for higher protein forage for livestock because nitrogen is one of the chemical building blocks of protein—that is, muscle and meat. Legumes also transfer nitrogen to neighboring vegetation. Black medic provided this valuable quality in a forage mixture made up of grasses of relatively marginal nutritional value. This legume prefers the leeward parts of the ranch at about 4,000-foot elevations, but it has been found as low as 2,000 feet and as high as 6,000 feet above sea level.

During these introductory periods leading up to the dawn of the twentieth century, wild geraniums and filaree *(hinahina)* became commonplace at the 4,000-foot elevation. These were probably accidental introductions. Filaree has significant nutritional value.

Of the native and introduced forages found in the pre-Carter era, specific dates and credits for introduction can only be assumed. It is not unreasonable to assume that introduced forages occurred accidentally via imported feed, seeds, and animals, in addition to birds. As a whole, such introductions were positive in that they replaced indigenous grasses that lacked the hardiness and drought resistance necessary for survival on the plains of Waimea.

In 1929, Professor L. A. Henke summarized progress in forage grasses and crops on the ranch in University of Hawai'i Bulletin No. 5, entitled "A Survey of Livestock in Hawaii":

> Notable work has been done in the past thirty years in improving the carrying and fattening capacity of the pastures by the introduction of desirable forage grasses from various parts of the world.
>
> Rescue brome grass *(Bromus unioloides)* is considered a very valuable and desirable grass, doing particularly well during the winter months at elevations of 2,500 feet and above. Perennial rye grass *(Lolium perenne)* is considered of second importance only to rescue brome grass. Among other important grasses and legumes are cocksfoot *(Dactylis glomerata)*, Kentucky blue grass *(Poa pratensis)*, Rhodes grass *(Chloris gayana)*, Natal red top *(Tricholaena rosea)*, Australian water grass *(Paspalum dilatatum)*, Bermuda grass *(Cynodon dactylon)*, white clover *(Melilotus officinalis)*.
>
> Bur clover *(Medicogo denticulate)* and purple vetch *(Vicia sativa)* do well on sections of the ranch. The Pigeon pea *(Cajanus indicus)* has not podded well where tried. Koa haole *(Leucaena glauca)* may do well on some of the lower, wind swept rocky land. Kikuyu grass *(Pennisetum clandestinum)* has been planted at 5,000 feet elevation and is doing well.
>
> Hilo grass *(Paspalum conjugatum)* does not fatten cattle well and is being gradually replaced by more valuable grasses.
>
> About 1,000 to 1,500 acres at 5,000 feet elevation are planted to corn each year, which yields from 30 to 35 bushels per acre. A special variety of corn has been developed by the Parker Ranch for this region.
>
> Oats harvested just before the dough stage makes an excellent hay.

Early Beef Industry: 1793–1847

Bullock hunting was primarily to provide meat, but hides and tallow from these animals grew into an export industry of its own. Much of the meat went to ships in the form of salted beef. Ships also carried the hides and tallow for marketing. Additionally, the local community kept back a limited amount of fresh beef and preserved it by drying or salting. A moderate amount of provisions went to the Royal Hawaiian Navy.

For the seafaring people of Hawai'i, salt was the only preservative available. Fortunately, it was plentiful, for the shores of the Islands allowed easy access to an abundant supply. Kegs of salt were carried to the mountains on the backs of both men and pack animals. With the return trip to shore, carrying the salted beef, it was a distance of 30 miles or more—at best, an arduous trip.

Given the challenge of producing wholesome and well-preserved beef, immediate salting of the freshly slaughtered beef was as close an approximation to meat hygiene as one could expect. But the skill and tidy habits of the butcher were vital to the end result of a safe product. One can imagine the delight of the galley chef aboard a ship hundreds of miles out of Kawaihae, upon opening a barrel of well-prepared salt beef to boil with cabbage to feed the hungry crew. Salt cures by an osmotic exchange of fluids within muscle cells. It also provides a profound disinfectant effect by lowering the pH level and elevating salinity to a level incompatible with the growth of contaminant organisms. "Corning" of beef is taking the process a step farther by including sodium nitrate (saltpeter) with sodium chloride. This further improves the preservation process, while giving the beef a pink tinge that has greater eye and taste appeal.

Salt beef and *pa'i 'ai* (hard-pounded, undiluted taro cakes) grew to become the mainstay of the *paniolo* diet, just as hardtack and beans were to the cowboy of the American West. Ironically, both salted beef and fresh corned beef have become pricey commodities in today's diet.

The salted beef, hide, and tallow export industry grew to become a major component of commerce. Forty to fifty-nine whaling ships called annually at Kawaihae in the mid-1850s, taking aboard 1,500 barrels of salt beef, 5,000 barrels of sweet potatoes, 1,200 bullock hides, and 35,000 pounds of tallow on an average. Between Waimea and Kawaihae, South Kohala became the center of the cattle industry.

Hispanic Influence on the Early Hawaiian Beef Industry: 1832 Forward

The Hispanic style of handling cattle combined the skill of the *vaquero* with a lariat and his mount's ability to apprehend the prey. Roped and tied for an overnight rest, the bullock was led to a central processing corral. Hunting with guns was gradually replaced by this system, which became known as the Spanish method, recalling those influences brought to the Islands by Mexican cowboys—*vaqueros*—who came to the Big Island at the invitation of the island's governor, John Adams (Kuakini). The *vaqueros* brought equestrian values and cattle-handling methods that originated in Spain and came to Hawai'i via North, Central, and South America.

Spanish settlers in "New Spain" (Mexico) arrived with experience and a culture that made it possible for them to convert arid lands to productive ones. Their forebears had turned dry, rugged pastoral lands in Spain to usable lands, and the knowledge was passed down. So were the gender-related work roles, equipment, apparel, food preparation techniques, and spiritual and medical knowledge and information. The ranch *(hacienda)* became a focal point, the social locus, and it was developed and spread throughout the American West. The culture became the genesis of America's greatest mythic hero—the cowboy.

Spanish and Anglo cultures merged as the latter adopted from the Spanish the knowledge, gear, apparel, and skills in managing large herds of cows from horseback. Concomitant with this was the Anglo penchant for developing laws of land ownership and business skills aimed at building a strong market economy.

Spain rightfully claims credit for the evolution of *El Rancho* as a social unit, much as John Palmer Parker I did for Waimea. The peninsula of Iberia was the cradle of the cattle ranching industry as early as the twelfth century in the *tierras* (lands) of Zamala and Salamanca in Leon and Avila and Segovia in old Castile. The swimming of cattle from shore to shipping vessels is said to have begun off the coast of the Spanish peninsula. The plains of Andalusia also played a significant role in developing cattle ranching when it was introduced to New Spain in the sixteenth century. The Andalusian rancher introduced the horse for managing cattle. He brought about the systems that include extensive grazing of extremely large unfenced pastoral tracts, periodic roundups in the form of rodeos, branding and marketing, and lengthy overland drives to new pastures.

By A.D. 1500, two types of ranching had developed in Spain. A seigneurial rancher used extensive grazing grounds in remote areas, much like Parker and the Waimea Grazing and Agriculture Company (WGAC) operations of the mid-1800s, an example of which were the lands of Ka'ohe and Humu'ula. The other type of ranching—municipal—was closely associated with communities. It is what Parker and WGAC operations of the mid-1850s merged into for processing and marketing in the village of Waimea. The Mesta or stockman's guild of Spain and Mexico also had its counterpart in Hawai'i, known as the Committee on Neat Cattle of the Royal Hawaiian Agricultural Society. Hawai'i had a territory-wide cattlemen's association by 1850.

When the Spaniards introduced cattle and horses to the Americas in the early 1500s, they also introduced the image, knowledge, equipment, and skills of the rancher to the Americas and the progenitor of the *vaquero*/cowboy. Some examples are the *gaucho* of the Argentine Pampas, southern Brazil, and Uruguay; the *vaquiero* (Portuguese) of northern Brazil; the *huaso* of Chile; the *llanero* of Venezuela; the *vaquero* of New Spain (Mexico), California, Arizona, and New Mexico; the buckaroo of North America and Canada, and the *paniolo* of Hawai'i—all brothers under the skin. These brothers share common characteristics of pride in their profession and cultural values and tend to be daring and independent. R. W. Slatta shares this perspective in his book, published in 1990, *Cowboys of the Americas:*

> Cowboys lived and worked primarily in frontier areas. Their actions played a major role in shaping our perception of frontiersmen. In some countries, notably the United States, Uruguay, and Argentina, cowboys became cultural symbols of national importance. In the northern and southern plains of Brazil (the sertao and pampa, respectively) and in the prairies of western Canada, the cowboy's impact remained regional. Hawaii is only now uncovering and recognizing its own cowboy culture.

Español (the Spanish language) was the origin of the term *"paniolo,"* the commonly used name for the Hawaiian cowboy.

Kossuth, Louzada, and Ramon are the names of the three men who first came to the Big Island. O'ahu, Maui, and Kaua'i were host to other Mexican cowboys. Kyle Shinseki provides some background in his master's thesis, "El Pueblo Mexicano de Hawai'i":

> The introduction of cattle and horses from Mexico would lead to the first wave of Mexican immigration to the islands. By the 1830's, the offspring of the cattle brought to Hawai'i by Vancouver had grown to such a number that, "they had become a menace to crops, cultivated taro fields and habitations, and even to the lives of the natives themselves." On April 4, 1831, the British Ship Harriet arrived in Honolulu and among its passengers was an expert at catching wild cattle, Mexican Joaquin Armas.
>
> Armas was born in San Diego, Alta California in 1809 and raised in Monterey, where his brother Felipe was born in 1810. Both boys became soldiers and vaqueros (cowboys) at an early age. Joaquin Armas was referred to by some as "the Spaniard," however he probably was mestizo since vaqueros came from the mestizo working class. In addition, Armas himself was described as being "five feet eight inches tall, with black hair and eyes, and a dark swarthy complexion." Armas worked for the Catholic Missions as a vaquero, but by the 1830's, there was a push for secularization of the Missions and breaking up large herds of cattle. It was at this time that Armas chose to leave on a whaleship bound for London.
>
> When Hawaiian King Kamehameha III learned that Armas was onboard the Harriet, he saw an opportunity to deal with the problem of the wild bullocks or cattle. Although at first Armas "had no desire to leave the ship," the King successfully convinced him to stay through promising him "his friendship and protection and a means of making a livelihood." After spending seven months on O'ahu, Armas was sent by the King to the village of Waimea, on the island of Hawai'i, where his efforts allowed for a significant increase in "beef catching" in January of 1832.
>
> During his stay at Waimea of nine years, Armas apparently spent most of his time alone, often working for periods of four or five months in the mountains suffering cold, hunger, and injuries due to the heavy demands of the King. Several other Mexicans worked as vaqueros at Waimea, with two arriving on April 27, 1833. Among these vaqueros were Frederico Ramon Baesa, who came to Hawai'i with his Yaqui wife Vincenta Romero and his sons Frederico (born ca. 1846 in Mexico) and Jose Ramon, who both married Native Hawaiian women. Jose Ramon is said to have married Mary Kainoakupuna Kalahohinu (born 1850 in Pololū, Island of Hawai'i) and together they had nine children, seven boys and two girls.
>
> The lives of these Mexican pioneers on the Island of Hawai'i is noted as being "necessarily simple to the point of starkness." One account describes them together at a thatched building that served as Waimea's store after a ten-day bullock-hunting expedition, lean-

ing against the counter and dressed in brightly colored ponchos, pantaloons, and long-spurred boots. Another account indicates that Joaquin Armas and other Mexicans traded at stores in Kawaihae and Waimea on the Island of Hawai'i.

The presence of these Mexican vaqueros was not limited to O'ahu and Hawai'i, but also extended to Maui and Kaua'i. Joaquin Armas was given land at Moanui (Lahaina), Maui in 1837, where he was made konohiki (chief) for the ahupua'a (political land division), conceivably the first political position held by a Mexican in Hawai'i. Apparently due to the prohibition of Catholicism from 1837 to 1839, Armas lost his cattle and horses in Waimea and moved to Lahaina, where on January 24, 1841, the first Catholic mass ever celebrated on Maui was observed in his home. In Lahaina, he was joined by his brother Felipe, and the two worked as store clerks for Joseph Mellish, a Catholic from London and also sold food and drink to sailors.

Joaquin Armas had a short-lived marriage with a Native Hawaiian woman while in Lahaina, from which he fathered a daughter, Luisa. Armas was "a well known and important figure in Lahaina in the 1840's," and had acquired 1,911 acres of land at Honokahua (north of Lahaina). Felipe married Mary Richardson, apparently the daughter of an Anglo-Saxon father and Native Hawaiian mother, and with her had at least two children, Maria and Lucia. The Armas brothers had left Hawai'i by 1848, apparently to deal with the consequences of the U.S. takeover of California. Although the exact date of their departure is uncertain, three citizens of Mexico left Hawai'i for California between May 1, 1845 and November 1, 1848.

However, the presence of the paniolos, as the Native Hawaiians referred to these Mexicans, was not to end with the departure of the Armas brothers. Another paniolo, Miguel Castro, originally from San Diego, Alta California was working as a vaquero at Kawaihae, on the Island of Hawai'i. On August 1, 1848, Castro signed an agreement with G.P Judd, the King's Minister of Finance where he would care for government cattle at Wailua, Kaua'i and in return the government would "build a house for Miguel to live in, and a house for storing milk, and pens for the cattle and calves."

Castro married a Native Hawaiian woman, Kikila, with whom he had several children and although he was Catholic, he and his family later converted to Protestantism. Castro was renowned for his rope-hurling as well as rope and saddle making, and taught his skills throughout Kaua'i. He later lived in Ko'olau, Kaua'i, where he was buried in an area known as Miguel's Valley, after he drowned in the Lumaha'i River in 1869. While Castro was ranching on Kaua'i, on November 28, 1850, a vaquero referred to as Huanu (Juan) Paniola, asked to be employed caring for cattle at Waimea, on the Island of Hawai'i. Other Mexicans, referred to as Kossuth and Boronda are said to have worked at Waimea at different times. Nonetheless,

> it seems that by 1859, the vaqueros were no longer riding the ranges at Waimea.

This may have been the third arrival of the Spaniards; as noted earlier, there are reports of Spanish ships visiting the Islands some two hundred years before Cook. Nonetheless, the *vaqueros* of 1832 shared a variety of skills, particularly in the area of horsemanship as it relates to cattle-handling practices. Their style of handling wild cattle—*buiso (pipi ʻāhiu)*—became known locally as the Spanish method. With the Spanish influence came their equipment—most significantly, a saddle that would gradually undergo a metamorphosis to become the Hawaiian tree saddle of today. The Hispanic influence on *paniolo* apparel, equipment, and tack left an indelible mark spanning more than 150 years.

The Spanish Method

Poʻowaiū / Alakaʻi

The use of a rawhide lasso drastically changed the movement of cattle closer to market and into confinement for eventual domestication. The Spanish method replaced the labor-intensive system of packing salt to the mountain and hauling kegs of salted beef *(iʻa kōpī* or *pipi miko)* back to shoreline communities. As part of the method, the bullock was roped in the remote range, snubbed up to a tree, and tethered snugly over a night or two for a cooling-off period. This process is called *poʻowaiū* by the *paniolo*. At the appropriate time, usually at daybreak, the bullock was shorn of his horn tips *(ʻoki hao),* then led *(alakaʻi)* by a single head rope or two to a central corral. These corrals were often located at a rudimentary camp or village *(kauhale).* At these strategic locations, the butchering *(kuaʻi lole)* and salting process *(kōpī)* culminated with the salted barrels of beef via ox *(pipi kauō)* or horse-drawn carts *(lio kauō)* being sent to the marketplaces, located in ports or communal gathering sites.

Kūpeʻe Style

Several variations of this method are usually associated with the particular terrain in which cattle were hunted. Trees to which the 800-pound bullock was tied were scarce in the arid plains of the north, west, and southern coastal areas of the Big Island. In these areas, the cowboys would hog-tie *(kūpeʻe)* the bullock overnight. This was done in a variety of ways that included binding three or even four feet. Such binding could be detrimental to the bullock's circulation, first overwrought with the shock of apprehension and then rendered static by the inability to move about. The practice also created undue pressure on the beast's underside. A variation common in the drier Hawaiian plains was to simply cross-tie the diagonally opposite fore and hind legs and let the bullock rise. This allowed him to move about in relatively open country, traveling a radius of perhaps 100 feet. The bullock was confused by the cross-tying effect but, once unhobbled, fresh enough for a snorty and speedy trip leading down to the corral *(pāʻeke)* for processing. The leading cowboy used his lasso *(kaula hoʻohei),* tied hard and fast to the saddle horn *(ʻōkumu)* on one end, and looped the other end over the bullock's horns to lead *(alakaʻi)* the bullock *(pipi ʻāhiu)* to the *pāʻeke,* just as was done with the *poʻowaiū* method.

Pini System

A less labor intensive and safer alternative was the *pini* system. With this system, after a night's cooling-off period, the bullock was pinned or joined with a closely tethered neck rope to a gentle ox for a slow six- to twelve-hour trip to the home corral.

Training oxen as leaders to which wild cattle were attached and led home is a time-honored art that originated in Spain and was brought to America. In old Mexico, the man who trained and managed these oxen *(cabestro)* was called *Caporal de Cabestro.* He taught the oxen their names and to know and keep their position in the band when traveling single file. The lead ox of the *cabestro* band was called *El Capitan.* Upon approaching a tethered wild bullock, the oxen, when called by name, moved alongside. Another command called for the *cabestro* to press against the wild bull to be yoked for a slow walk home in tandem.

A *pini* bullock or pin bull was usually a stag—a bull castrated after maturity. A *pini* could be an ox or a common steer, but in all cases was stout enough to ward off the frantic hooking, head-butting, and pounding that occurs in the first few minutes to hours after the bullock has been released from a tree or hobble. His handler often raised a *pini* from a calf to draft age.

Over the years, I gained insight into this process from two prominent *pini* men. The patriarch of the Loando family of South Kona was a Puerto Rican immigrant named Henry. He and his wife Antonia raised a large family as they lived and worked in various Hind family ranching operations on the west side of the Big Island, from Puakō to Honomalino. The raising, training, and maintenance of the *pini* bullock were family chores that began with orphaned calves. Keeping a *pini* candidate as an intact bull through maturity prior to castration gave him a heavy head and shoulders. Castration after maturation left the heavy framework but took away the temperament usually associated with bulls and allowed the ox to yield to verbal and physical commands without incident.

Another noted *pini* man was Charles Akiu of the original Kahuku Ranch. In the days leading up to the trip *mauka,* the cowboys would rope and tether the wild bulls to the trees in the *'ōhi'a* forests. Akiu would thereafter travel into the forests with a string of several *pini* led in mule train manner to locate the tethered bulls. Upon locating a bull, Akiu would dismount to pin the

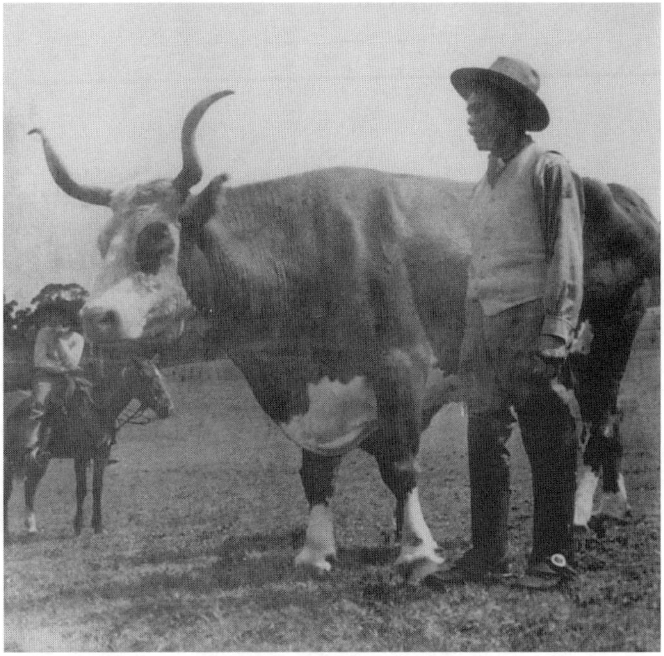

Ikua Purdy in retirement, with retired but favorite pini *ox, Babe. Taken at 'Ulupalakua Ranch, circa 1942. Martin Purdy Collection.*

pini to the wild bull. Uttering commands in Hawaiian, he would cause a select *pini* to sidle up to the bull and some immediate skirmishing would ensue, with the ox remaining quite passive. Once settled, Akiu would climb aboard the *pini* and deftly reach down to encircle the bull's neck with a secure bowline knot *(pōlena)* in the neck rope. Further commands brought the *pini* neck to neck with the bull. When this had taken place, the free end of the bull's neck rope was tied securely around the *pini's* neck. At this point, Akiu would slide off the free side of the *pini* and detach the *po'owaiū* rope that tethered the bull's neck to the tree the previous day. In a flurry of activity typical of range bulls fighting, the *pini* and bullock would engage in a head wrestle while Akiu mounted up and rode off in search of another wild bull for another *pini*. In a full

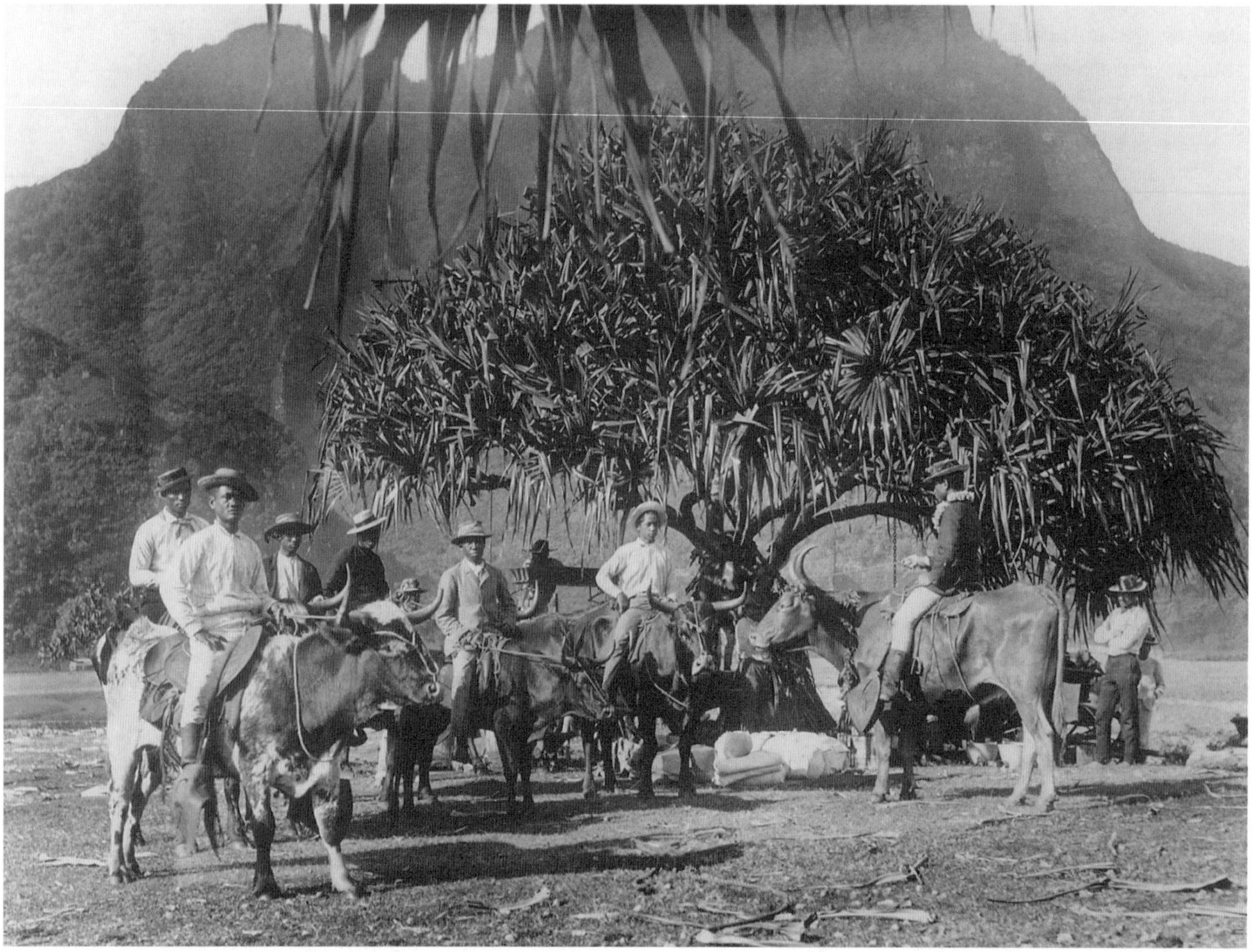

Paniolo *mounted on oxen in Haena, Kaua'i. 1892, Alfred Mitchell Collection, Bishop Museum. BM CA 36582.*

day's time, it was not unusual to pin eight or nine wild bulls. The remaining *pini* were reserved for another day's work. By sunrise, eight or nine pairs of *pini* and bulls were calmly waiting back in the home corral.

While the *vaquero* introduced the oxen as draft animals for carts or as mounts, it is likely that the *pini* system was an outgrowth of the Spanish *cabestro* method. The technique was obviously a great time saver and spared the cowboys and animals from wear and tear. It was also viewed as a relatively humane way of securing wild bulls. Leading wild cattle over long distances is not practical, as they overheat due their alarmed state. Even a distance of 2 miles on a cool day would be pushing it. In contrast, the labor-free movement via *pini* over a 10–12 mile stretch over a period of 24–30 hours made such distances achievable. Once broken to his neck rope, a wild bull is "sored" and gives to the lead of a *pini* readily. In the warmth of midday, the *pini* may choose to browse on grass or water before finding shady rest to chew his cud. His once fierce partner has no alternative but to follow suit. The *pini* system yielded a much higher survival rate than the *alaka'i* method where wild bulls are led briskly over relatively long distances.

Driving herds of feral cattle into traps was also labor intensive, but it was a highly productive process when all the variables worked. Remnants of these traps still remain on the northern and southern slopes of Mauna Kea at very high elevations. Such corrals served as destination sites for cowboys leading wild cattle or for *pini* bringing home their mates. The term *"pini"* still remains in the language of the *paniolo*. The unwise practice of leaving an animal alone *(mehameha)* in a corral for long periods is corrected by leaving a *pini.* Animals love company, especially of their own kind. Solitude, particularly out of view of a mate, brings on anxiety that can be severe in a feral or range animal. Frequently they will work themselves into an unnecessary frenzy. Leaving a *pini* is a thoughtful, caring gesture.

In recent times, the *pini* was reintroduced to Parker Ranch using domesticated nightingales (donkeys) from Keamuku. Nightingales were tethered to weanlings (colts) at the breaking pen to teach the latter to lead. This system was dropped when it was learned that the conventional methods were less labor intensive and more productive.

Role of the Ox (*Pipi Kauō*) as a Draft Animal

There is considerable evidence that the Mexicans had a primary influence on the use of the ox as a draft and riding animal in early Hawai'i. The design and construction of the vehicle for freight hauling clearly bears a striking resemblance to that of old Mexico, though constructed with native Hawaiian woods.

For wheels, the *vaquero* fashioned reasonably circular cross-sections of *koa* logs with centrally bored holes to receive axles. The deck was roughly planed planks of *koa* or *'ōhi'a,* while the walls were framed with *'ōhi'a* or *koai'a* uprights and bars. Cowhide was stretched over the framed sides and ends and allowed to dry intact, laced in place by stout strands *(ma'awe).*

The paired tongues (shafts) were commonly of *'ōhi'a* wood or *māmane,* with *wiliwili* wood being preferred when available. The yoke was hewn from *hau,* due to its light but strong character. *Milo* and *koa* were too heavy to be comfortably yoked behind the oxen horns, whereas the *hau* yokes were snugly lashed with rawhide *ma'awe*.

Axle grease was readily available in the form of *'aila pipi* (tallow), and the ox was controlled by reins *(laina)* converging on nose rings. Reins were of *aho* (fishing line) for their light weight and strength.

Native Hawaiian craftsmen, already skilled woodworkers, refined the *ka'a kauō* (draft cart) appropriately just as they perfected the Hawaiian tree saddle from the bulky Mexican *fuste*. Governor Kuakini provided the mandate for a roadway to be built between Waimea and Kawaihae; he commandeered forty native convicts found guilty of violating the Seventh Commandment. It was on this first rudimentary road that the *pipi kauō* would travel, taking beef, tallow, and hides to Kawaihae Bay.

The *Vaquero* Heritage

All of these influences of the Hispanic world blended into the life and career of John Palmer Parker as he and his *paniolo* shifted roles from hunters of wild cattle to gatherers and domesticators of cattle. Through these early efforts, the genesis of Parker Ranch was underway, and John Palmer Parker was its first caregiver.

The genetic influence of the Mexican cowboy is worth noting, and it should not be lost. It is not unusual to hear local families or individuals refer to a *vaquero* heritage. Given modern resources for genealogical research, it is possible to develop a flow chart that shows such relationships. People of Hawaiian ancestry—beginning with the establishment of the Hawaiian Homes commission in 1920 and continuing into the Hawaiian renaissance of recent years—are far more inclined to preserve a purely native genealogy, with little incentive to include other ethnicities. Yet a *vaquero*—or Spanish—heritage does exist. Given the very real possibility that Spanish ships visited or shipwrecked in Hawai'i long before Cook, the probable offspring of these sailors may well be the subject of oral tradition, history, and legends—but perhaps privately held in family memories.

The *vaquero* also had a profound influence because of the clothing he wore, the saddle he rode, and spurs and bridle he used.

5

EARLY FOUNDATION FAMILIES
1815–1868

As the Parker Ranch empire was building throughout the years, several families played key roles. John Palmer Parker I relied heavily on their skills, dependability, innate good character, and their abilities. Some of these relationships developed before the official birth of the ranch in 1847 and, for the purposes of this text, these early foundation families are identified as the progeny of those patriarchs who came to Hawai'i between Parker's own arrival in 1815 and his death in 1868. Falling within this time frame are the families of Bell, Fay, Lindsey, Purdy, and Stevens.

Marriage among these families over the ensuing century and a half created a fascinating *paniolo* history and established the continuity—and heritage—of the great ranch and its families. Parker's close cohorts included Jack Purdy and James Fay, who worked alongside one another harvesting wild cattle. The original Bell was a Scots engineer who sired offspring closely connected with the ranch, as did the Lindsey brothers, Thomas and George. The Lindseys were the sons of William Lindsey, a hatmaker in London. Fay's daughter, Mary Ka'ala Fay Lindsey, became the matriarch of an extensive and expanding Lindsey clan. Her first marriage to Thomas John Weston Lindsey produced but a single child, duly christened William Miller Seymour Lindsey. Her second marriage, to Thomas' brother George Kynaston Lindsey, resulted in ten more Lindseys. These men provided the foundation of the Lindsey clan.

James Fay

Born to an unremarkable English family, James Fay grew up on the rugged streets of London and in early adulthood became a thorny antagonist of the British government, a stand that landed him in prison. These were times of severe economic depression in the British Empire. Owing to overcrowded conditions in the prisons, political prisoners were often shipped to the penal colony at Botany Bay, Australia. Many of Hawai'i's early settlers found their way to the Islands after escaping from these penal colonies; among them was Fay.

Fay escaped with a pair of seaworthy mates, who commandeered a vessel and sailed northeast. Off the island of Hawai'i, the small ship was scuttled and the men made their way ashore. There is no record of what became of his mates, but Fay headed for the mountains of Kohala, Mauna Kea, and Mauna Loa. Hardworking by nature, Fay was hired by John Palmer Parker to help harvest wild cattle to supply beef, tallow, and hides for the maritime industry.

About 1830, Fay married Ka'ipukailai'ikapuokamehameha, daughter of the high chief Kailihiwalani, son of the governor of O'ahu. Leaving his bullock-hunting career behind, Fay settled on a homestead near the confluence of the Waiaka and Keanu'i'omanō Streams. His holdings expanded *mauka* to include the area of Wai'au'ia, where Anna Ranch exists today. Fay stayed active by livestock farming, keeping horses, cattle, sheep, hogs, and geese. In the forests of Hanaipoe, he operated sawmills processing *koa, 'ōhi'a, māmane, naio, nai'a,* and other trees.

The Fay marriage produced a daughter, Mary Ka'ala, who spent her childhood in the small hamlet that eventually became Waimea, the central town in the subdistrict of South Kohala. The Fays provided well for their only child, sending her to St. Andrews Priory on O'ahu, a private boarding school for girls.

The Lindsey Clan

While on O'ahu, Mary wed English seaman Thomas John Weston Lindsey on September 4, 1846, and they moved to her home in Waimea. In 1847, the same year that Parker Ranch began, William Miller Seymour Lindsey was born—their only child. Thomas Lindsey worked for John Palmer Parker for a brief period of time. Becoming aware that he had a possibly fatal affliction, he yearned to see his homeland once more. He departed alone for England, where he remained until his death.

While in England, Thomas, in a gesture not uncommon during those times, asked his younger brother, George Kynaston Lindsey, to venture to Hawai'i to wed and care for Mary. In 1850, George did indeed move to Hawai'i, and the mission was more than successful. Mary's second union, on October 7, 1853, produced ten children. The descendants of the eleven surviving children were the foundation from which the Lindsey clan evolved. James Fay and Ka'ipukai, through their daughter Mary Ka'ala, are credited for the strong matriarchal influence in this prominent family.

The foundation families noted here are those who played a significant role in the evolution of Parker Ranch. The intent is not to present a complete genealogical history of the progenitors, but rather a cowboy lineage that began with Parker's peers marrying Hawaiian women of varied distinction, down to the present-day men and women who continue to serve the ranch. The patriarchs came from the ranks of ship captains, engineers, sailors, bush rangers (Australian outlaws), escaped prisoners—and sometimes even rogues. The matriarchs often came from royal lineage of some significance.

The Thomas John Weston Lindsey Line

William Miller Seymour Lindsey (WMS), the only child of Thomas John Weston Lindsey with Mary Fay, fathered a large and historically significant family. He first went to work for John Palmer Parker I and, during his career, he witnessed the gradual shift of ranch leadership to John Palmer Parker II and the interim periods of both Paul Jarrett and Fred Carter. He died in 1905 after witnessing the dawn of A. W. Carter's half century of ranch leadership. On May 25, 1865, WMS married Kaluna Henrietta Ka'inapau, daughter of James Ukewio Ka'inapau *(ma-*

kuakāne) and Ha'a'lo'u A'a *(makuahine).* Ka'inapau Sr., a second-generation *paniolo*, died at age 103. Except for an eight-year period in Hāmākua, they spent their lives in Waimea where they produced a family of twelve children. Four children did not survive into adulthood, but seven of the surviving boys attained stature in ranch history. Thomas, John Kawananakoa, George Miller Seymour, James Ka'alu'ea, Thomas Kamaki, and Albert Uiha held varied posts in those early years.

The eldest son, Thomas (1871–1889) had a brief but colorful life. A cowboy by age fourteen, Thomas was a natural horseman and often was asked by his elders to "green break" their horses. As a horse trainer, he had unlimited potential. He wrote a sweet but melancholy song, "Hone Sakala" (Honeysuckle), that reflected his sadness when he visited his lover's home in Kawaihae Uka, only to find another man's jack boots at the doorstep. En route he had gathered a bouquet of honeysuckle roses for her as he rode the rugged slopes of Pu'u Lanikepu and Pu'u Kawaiwai. The long, lonesome ride home inspired him to write the lyrics and music for a song that became a Waimea favorite.

The entire Waimea village grieved young Tom's tragic death in 1889 during a cattle drive at Pu'u Anuanu on the northwest slope of Mauna Kea. He was barely eighteen, but his legacy lives on in the haunting strains of "Hone Sakala."

The second eldest, John Kawananakoa Lindsey, was born on June 30, 1873. Due to his tidy and slight stature, he became known as Keoni Li'ili'i or Keoni Poko, shortened to Poko Lindsey. In this text he will be referred to as Keoni Li'ili'i, although A. W. Carter fondly referred to him as "Johnnie."

Keoni Li'ili'i (1873–1947)

Beyond any doubt, Keoni Li'ili'i's name holds one of the highest distinctions in ranch history and the Lindsey *ohana* (family). While Anna Lindsey Perry-Fiske enjoyed the reputation of a hard-working socialite rancher, Keoni Li'ili'i quietly and modestly rose to prominence for his ability, prudence, and loyalty to the Parker and Carter families.

His small stature was in stark contrast

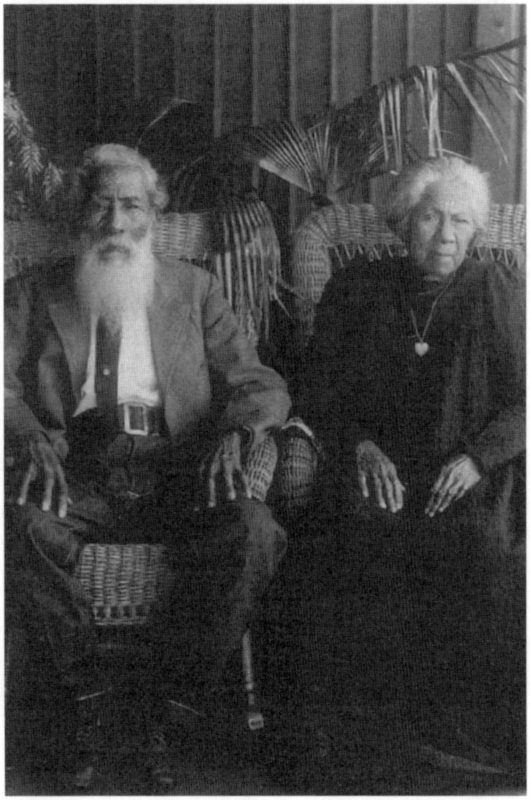

James Ukewio Ka'inapau and Ha'alo'u A'a, parents of Kaluna Henrietta Ka'inapau Lindsey, wife of W. M. S. Lindsey I. Elizabeth Lindsey Kimura Collection.

Genealogy chart of Thomas John Weston Lindsey and Mary Ka'ala Fay

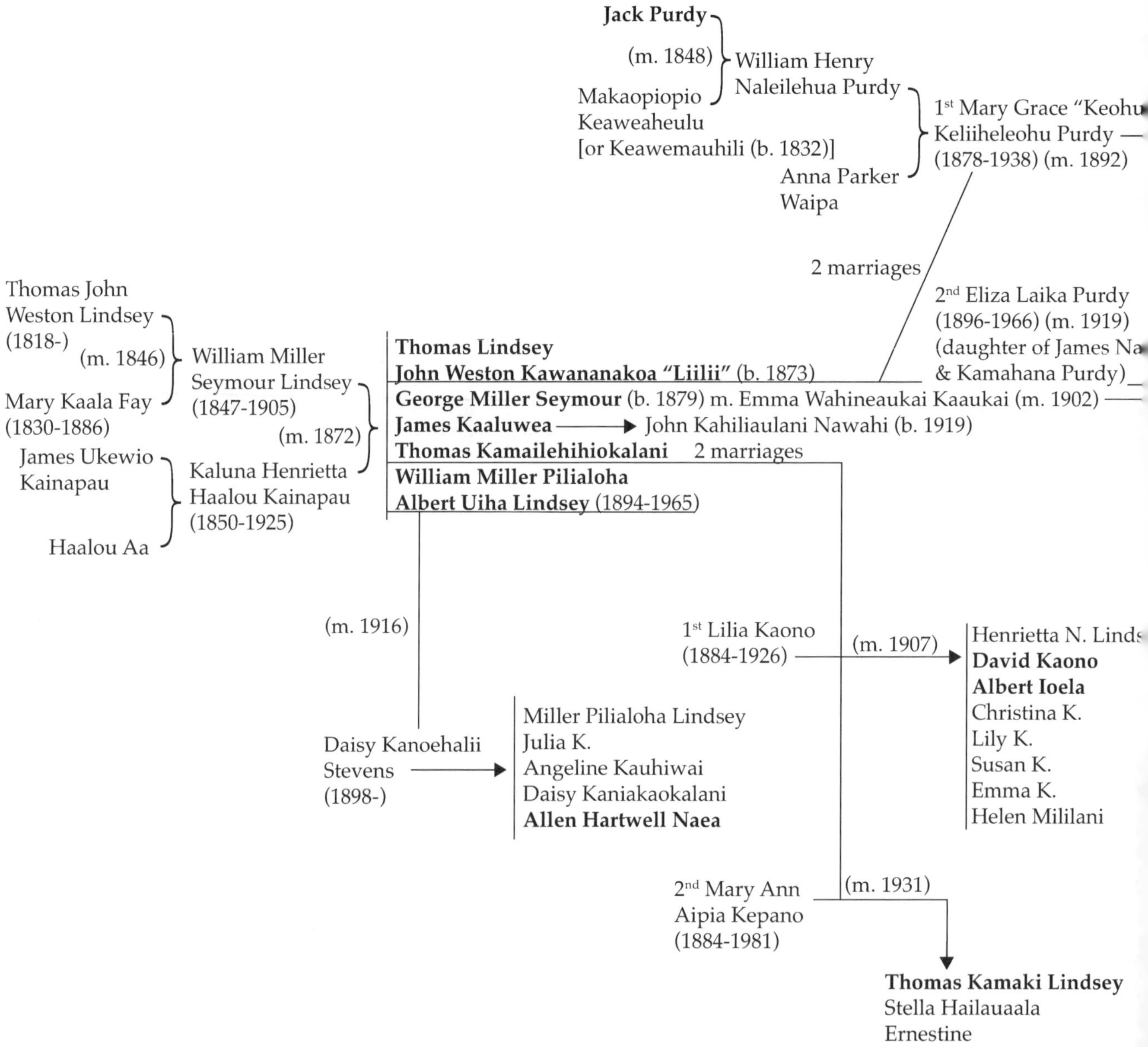

Mary Ann Lincoln
(1897-1945)

John Kuakini Lindsey
(1892-1983)
Annie Kameaalani
Mary Keaohuelani

Sarah Lindsey Taga
Laurence Bell Lincoln
Eva Kealamakia
Abigail Mary Kahalewai
Arthur Kuakini
Grace Keliiheleohupapa
Martin Campbell
Irene Loeyland Lindsey Fergerstrom
Ella Ilima Lindsey Kimura m. Samuel Kimura

Harry
Fergerstrom → Dale m. Gordon Roque

Loeyland Iwalani m. **Christopher Kaaua** → Harry Kaaua

Edwin Kealoha Lindsey
Elizabeth "Tida"
Esther "Dolly" Kanoho
Carrie
James Kimo Purdy
Hattie Kaluna
Rita Noenoe
Isabelle May
Louise Jean
Juliette Kahala

George Mahalopiha
James Samuel
John Kawananakoa
Eunice Keawe
Stella Haalou
Annabel Kaalani
Edwin Pelehelehonua

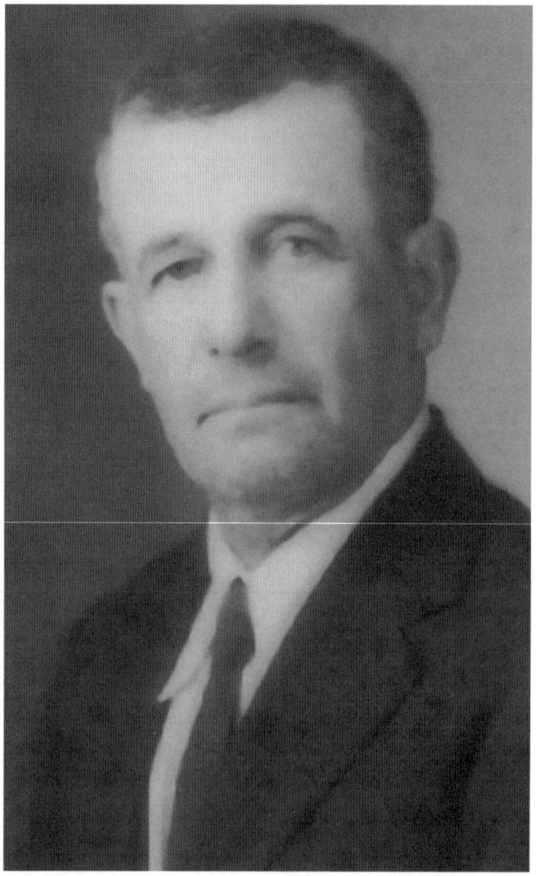

John Kawānanakoa Lindsey, also called Keoni Li'ili'i or Keoni Poko, who was often simply called "Poko." William Studio, Honolulu, 1938. PPS Collection.

to his reputation, which has endured over the 150 years of Parker Ranch history. He is celebrated in song—"Hanohano Wale no na Cowboy"—and in place names such as a stone corral in Waimea named Minuke Ole ("less than a minute"), celebrating his championship winning time in a 1907 matched steer roping contest against noted ropers that included Ikua Purdy.

Keoni's career began when he was about twelve years old. In 1904, W. A. Kinney, an attorney and former partner of A. W. Carter's in Honolulu, interviewed him and translated Keoni Li'ili'i's brief biographic recollections from Hawaiian to English:

> I am thirty-one years old. I was born at Waimea village, and have spent my life there.
>
> I never went away to school. I have been to Honolulu only once, and only for a few days, about a year ago.
>
> Also, I worked about two years for Horner above Hamakua [Kūka'iau Ranch/Umikoa] when I was a boy.
>
> I have known Parker Ranch from the time I was old enough to understand.
>
> When I was about twelve years old I commenced to turn out with the men and drive cattle on the Ranch and get paid for it. Before that, from the time I was seven or eight I used to go out with the men on the drives for the fun of it.
>
> When I was sixteen years old I began to work regularly for the Ranch and was paid by the month, not just by the day as when I went on the drives.
>
> When I first went on those drives the ranch was in charge of John Parker and an elderly native man named Kaho'o-hanohano, who was the head luna.
>
> In those early days there were no fences and no enclosures except the enclosures required to hold the cattle when they were driven in. There were no fences between the land of the Parker Ranch and the land of third parties, all cattle running in common.

Testimonies to his boyhood and cowboy abilities are offered by ranch head bookkeeper Frank Murray and by Henry Pieper, head teamster. Here are the recollections of Frank Murray:

> He could hardly read and write. He started roping cattle when he was about ten.
>
> Mr. Carter asked me to look back in the books for six years to see what Johnnie's average monthly number of days was. I went through those smelly old books. They were thick with mold. I got so interested I went back twenty years. Johnnie's average was TWENTY-SIX DAYS WORK A MONTH.
>
> Johnnie and I used to bet cigars on the average dressed weight of cattle on the droves he'd take to Kawaihae—120 head. We'd bet while they were on the hoof—the nearest ten pounds win.
>
> Johnnie got all the cigars; he'd guess from the whole drove, on hoof, the average dressed weight—and he'd always win.

Here are the reminiscences of Henry Pieper:

> Johnnie Lindsey could not read or write. As a little boy he would cry when they sent him to school. His father, old man William Lindsey, would take Johnnie behind him on the horse.
>
> In the wild country his father wouldn't have him on the horse while he was roping cattle, so he would put Johnnie up in a tree in the woods to stay until he got back.
>
> Soon Johnnie began to ride and rope on his own account.

Elsewhere, he mentions the heroic care that Keoni Li'ili'i gave a newly imported bull in 1902.

A. W. was clear and frequent in stating his admiration for Keoni, and the following quotes bear witness to the high regard in which he was held. The first quote is from a December 1903 letter from A. W. to Mrs. Knight (widow of John Parker Palmer; Aunt Tootsie was the mother of Thelma Parker), explaining his decision to promote Keoni Li'ili'i, who at the time was barely thirty years old, to head cowboy foreman:

> Three months ago I informed Tom Lindsey that he better resign as luna. When I was there he was always on hand and attended to his work, but when I left the place he was continually laying off and neglecting it.
>
> I put in his place Johnnie Lindsey, and he has made an excellent luna. In fact, I believe him to be a better man than Tom ever was, even in his best days.
>
> He is a hard working man, controls the boys perfectly, and takes an immense interest in his work. There is no man living that I would exchange for him in the position he now occupies.

The person he was replacing was a second cousin (old Tom was the eldest of the George Kynaston/Mary Fay line) who, in A. W.'s eyes, had outlived his usefulness as the lead foreman for Parker Ranch.

In that same year, A. W., in a letter to H. A. Baldwin of Maui, reflected on Keoni Li'ili'i's abilities as a horseman:

> My brother writes me that last week both gangs were out for wild horses, and that they struck a bunch by the Spring paddock. They could not get them down to the paeke. The horses turned back

> and went up the side of Mauna Kea towards Kemole.
>
> Johnnie was riding a little bay horse he uses in the pen on the annual drives. He soon left the rest of the boys behind, headed the horses down the mountain, and stayed with them across Waimea plain right into a corral in the village. This is the biggest ride on record in those parts. He must have stayed right with the wild horses to enable him to head them as he did. I have scaled off on the map the course that he took, and it is between 16 and 17 miles.

In earlier correspondence, A. W.'s brother, Fred Carter, reflects on Keoni Lindsey's ability as a polo horse trainer:

> I am sending by this SS Kinau a small pony that I got from Johnny Lindsey. It was his own. I traded him a horse that I valued at $95.00, and $5.00 in cash.
>
> This pony you will find quick and strong. John had him out a couple of Sundays with the mallet or polo stick. The pony is easy mouthed. I should be inclined to start easy with him however. He should be clipped immediately on arrival as his coat is very thick, and it is extremely cold up here.

In a 1905 memorandum to Tootsie, A. W. cautioned her to urge Thelma to be thoughtful and understanding of the deep loyalty the key men had toward the Parker Ranch and her family. In this memo, A. W. more than once refers to Johnnie and includes accolades for other Lindseys and Spencers:

> There is one statement in your letter, Tootsie, that bothers me somewhat, and that is that you are getting hardened towards the natives on the ranch and that you do not think that there is one that can be depended upon. There are some that can be depended upon, and I think are perfectly loyal, notwithstanding that strong influences have been brought to bear upon them to act differently. Johnnie Lindsey, Sam Spencer, George Lindsey, Willie Spencer, Willie Lindsey, Tom Lindsey—in fact all of the Lindseys and particularly Johnnie, are faithful and true to Thelma's interests. Then there is Frank Johnson, Bonnie Judd and Henry Pieper that are always loyal. I hope that you will teach Thelma to feel kindly towards these men, particularly the Hawaiians, and more particularly among the Hawaiians Johnnie Lindsey. He has certainly shown all characteristics that you would expect of one of more superior opportunities. She must never forget him, nor any of his people.

Many years later, A. W. admonished his son A. Hartwell Carter for not including Keoni Li'ili'i on a Honolulu trip for polo activities, and he made several other significant references:

> In mentioning the men coming down for polo you did not speak of Johnnie Lindsey.
>
> I wish you would include him always in any excursion of the part of any of the ranch employees. He is the only man living who was working for the ranch at the time I went there thirty-seven years ago.
>
> I know that he went to pieces a number of years ago but just the same, I do

not know whether you appreciate the work that he did in getting the Parker Ranch into shape after I took hold. No cowboy will ever, in the history of the ranch, be able to duplicate his work.

In addition to his regular work, he was the man that relieved me at the office nights at the time they tried to take the Parker Ranch by force. He slept under the counter with a six-shooter, with instructions from me to shoot anybody that attempted to break into the office. He would have done it and I felt safe when he took my place.

In his own handwriting, A. W. recorded both his sorrow and appreciation for Keoni Li'ili'i, who passed away on January 23, 1947:

> John Kawananakoa Lindsey—Born June 30, 1873—Died last night. Worked for Parker Ranch over 50 years—was head luna for a great many years, until age prevented, but he continued to work until the day before his death.
>
> He was a great cattleman. In the early days of wild cattle he was a wonderful roper. A great worker.
>
> His employer for these many years says, "No one could have been more faithful to his employer or could have taken more interest in the affairs of the Ranch to which he gave his life."

Children of Keoni Li'ili'i

Keoni Li'ili'i was survived by his second wife Eliza (Laika) Purdy and the children of two marriages. His early marriage to Mary Grace Keli'iheleohu "Keohu" Purdy in 1892 yielded a son, John Kuakini Lindsey, and daughters Annie and Mary. His second wife, Laika, bore him several children, notably Elizabeth (Tida), who married Hisao Kimura; he rose to become the agronomist for Parker Ranch. Other children of Laika who were on the ranch payroll were Edwin, the eldest, and James (Kimo), the youngest. Tida Kimura recalls their careers:

> My brother Edwin, first son of Johnny and Eliza at a very young age, was encouraged to follow in his father's footsteps as a cowboy. He thought he was quite enthusiastic when Mr. Carter gave him a saddle and bridle to acquaint himself as a Junior cowboy. Upon completion at Waimea School he began working as a cowboy. After couple years he realized he was not capable and efficient amongst the old timers. He resigned to work with the fence gang for couple of years. He enlisted into the National Guard before the Pearl Harbor attack. He served his term and then moved to Lanai and Honolulu where he worked for Hawaiian Pineapple Company. He passed away on Jan. 13, 1998.
>
> James Kimo, younger son, at age of 10 received a saddle and bridle from A. W. Carter to prepare him to follow into his father's footsteps. He did ride the range chasing cattle as an experience but somehow did not feel comfortable being a cowboy. Instead, he worked four years with the utility gang on P.R. He enlisted in the military, serving in the Korean War. On completion of his military service he worked for the government with the U.S. Geological Survey on the small islands of the Marshall Is., Eniwetok and the Johnson Island. He

returned in 1963 and became a full time employee of Kawaihae Terminals. He retired in 1989 and passed away on September 25, 1996. (Personal interview by author)

John Kuakini Lindsey

Commonly referred to by his Hawaiian name, Kuakini was also a slight man, but his piercing light blue eyes and his erect and conditioned physique gave him a certain air of competence and assurance. Raised in Waimea, he was sent off to Kamehameha School for Boys on O'ahu for secondary education. His first job after graduation was at the Hawai'i Meat Company on O'ahu.

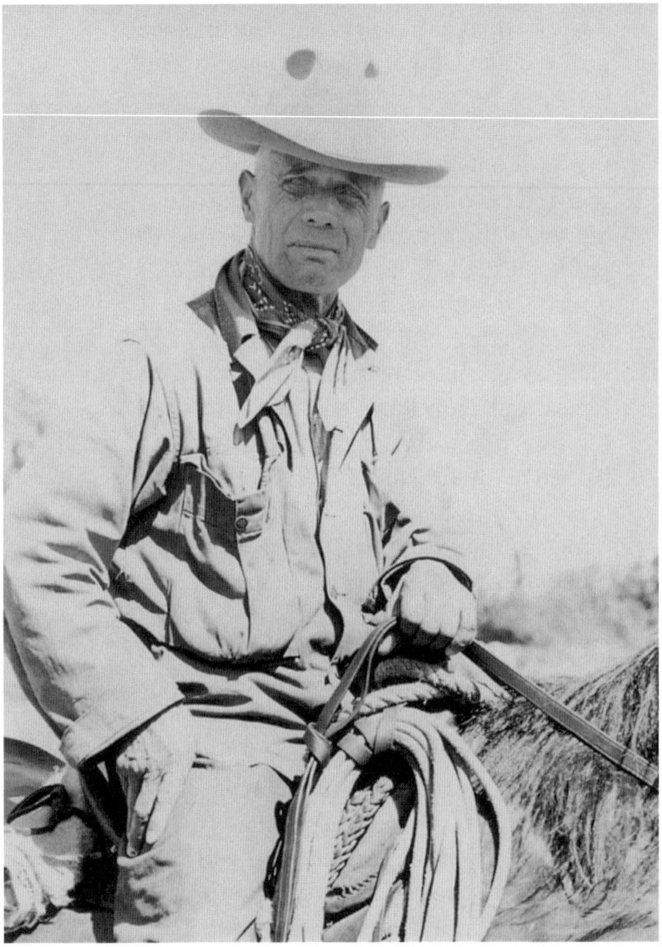

John Kuakini Lindsey on horseback, circa 1950. PPS Collection.

Yearning for his home and the cowboy life, he returned to Waimea and gradually worked his way up to the prestigious cowboy gang, where he spent many years. During his forty-nine years on the ranch, Kuakini worked in nearly every section of the outfit. Sixteen of those years were spent at Makahālau tending the large purebred Hereford herds. For an interim period, Kuakini worked at Kahuku Ranch with his uncle, James Ka'alu'ea. He returned to the main ranch and eventually became assistant cowboy foreman. He also served in leadership positions on other assignments, such as the fence, cactus/weed, and plaster (miscellaneous repair) gangs.

When Dick Penhallow was assistant manager and then manager of Parker Ranch, he took an interest in Kuakini, and they cultivated a close relationship. Kuakini's long ranch history, coupled with his inclination for dialogue on a variety of subjects, enhanced this relationship.

Even after his retirement on September 1, 1957, Kuakini continued to wear his traditional denim riding pants and neatly laced boots. For exercise he walked around Waimea town daily, often being stopped and queried by tourists impressed with his appearance. Kuakini died on October 13, 1983.

Through his eldest daughter Sarah, Kuakini acquired a grandson named Phillip Lindsey, who grew up on the ranch and spent his adult career there. Known as Chico by the community, he remained a faithful employee of the ranch in a wide variety of assignments, from fencing and carpentry to agronomy and utility gangs. On January 14, 2002, Chico decided to leave the ranch, along with a significant number of Parker Ranch employees, on what has become known as the infamous day of voluntary separation. His final day was March 31, 2002, and it marks the passing of an era and the end of nearly 150 years' involvement of the Thomas John Weston Lindsey clan with Parker Ranch.

Several years earlier, Samuel Kimura, head dispatcher and purchasing agent for the Ranch, retired from service. His wife, Ella, another daughter of Kuakini, brought a great deal of mystique and deep-rooted aloha to Parker Ranch, always fondly recalling the wonderful years preceding the death of Richard K. Smart.

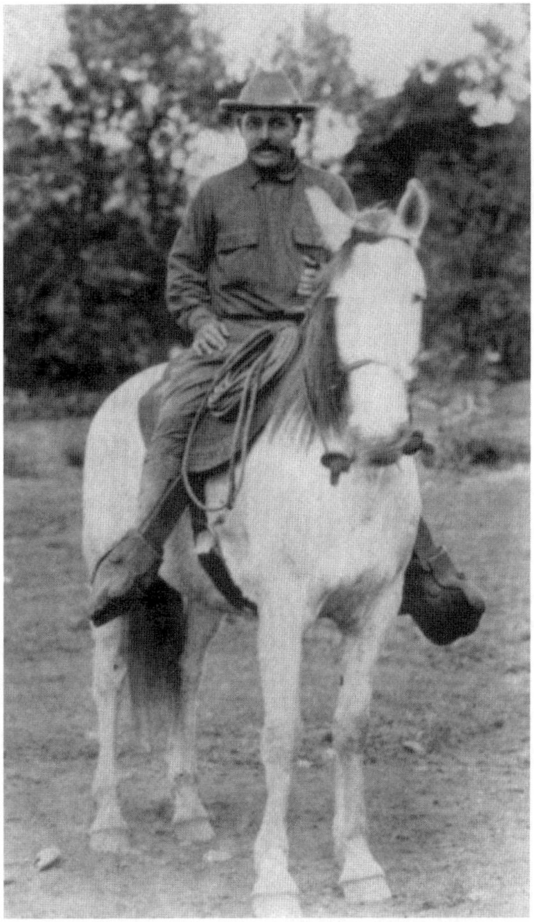

George Miller Seymour at work on horseback. PPS Collection.

George Miller Seymour Lindsey (1879–1917)

George Miller Seymour Lindsey was born the sixth child of William Miller Seymour Lindsey and his wife Kaluna. Although he died a young man of about thirty-eight, he had a large family of sons and daughters from his marriage to Emma Wahineaukai. Two of the children maintained close ties to Parker Ranch—his first son George, and later his daughter Annabelle, who married John Rickard of Waipunalei. Rickard became a chauffeur for the Carter family.

George Miller Seymour Lindsey's untimely death occurred on his way home from the stables for lunch when his gentle grey horse took a bad fall and rolled over on him. He died instantly. A. W. regretted the loss of a key staff member, especially one with a solid grasp of the horse-breeding program. Carter immediately arranged with the Kamehameha School for Boys to take both the second and third sons, James and John.

In his short career on the ranch, George Miller Seymour was a respected, loyal, and trustworthy employee. A. W. Carter held him in high regard. George passed these qualities on to son George Mahalopiha, who was first employed on the ranch at the draft stables overseen by German immigrant

Henry Pieper. With the early death of his father, George, at age fifteen, was given the additional challenge of becoming the "father" of his large family of younger siblings. Being an excellent caregiver of horses, George's responsibilities were gradually moved to the stallion station at Pukalani Stables. He married Thelma, the beautiful and talented daughter of skilled teamster Robert Victor Kailianu, originally from Kokoiki in North Kohala. Their union produced a family of five children who received a devout Mormon upbringing.

When I went to work at Pukalani Stables in 1958, George Mahalopiha was the foreman to whom I reported. Other men included Kanakanui Iokepa, James Merseberg and John Holi Ma'e, who spent his final years there before retirement.

Shortly afterward, George Mahalopiha also retired, dedicating the latter years of his life to his church, his beloved wife Thelma, and to his children. His death ended the George Miller Seymour Lindsey line of ranch employees.

James Ka'alu'ea Lindsey

James' true middle name was Kahalelaumāmane, but Jimmy Ka'alu'ea was the name by which he was known in the Waimea community. This name set him apart from other James Lindseys. He was also frequently referred to simply as Ka'alu'ea.

Ka'alu'ea was born October 5, 1882, the seventh child of WMS Lindsey and Henrietta Kaluna. By age six he was adept at roping the *keiki* (young) pigs in the family *kuleana* using a cord lariat. Entering employment on the ranch at age fifteen, his first assignment was at the Breaking Pen, and his pay was a dollar a day.

It was a good time to be a roughrider at the Breaking Pen because the fruit of A. W. Carter's enhanced horse-breeding program was just beginning to take shape. Ka'alu'ea was one of the four Lindsey boys who were assigned fresh colts to develop into polo ponies for A. W.'s developed market. This was a successful venture, and Ka'alu'ea played a vital part in this program. One of the most prestigious mounts from this venture was the polo horse Carry the News that went on to international fame. When the Parker Ranch acquired the Kahuku Ranch in Ka'ū, Ka'alu'ea was assigned there for a limited period.

Ka'alu'ea was often chosen by A. W. to be the stock tender for imported stallions. This meant time on the West Coast and O'ahu, as well as the final leg aboard the interisland steamers. It was Ka'alu'ea who shepherded the new stallions, Gladiator, a Percheron, and German Coach Endino, on the march from Kawaihae port to Pukalani.

During this period, Ka'alu'ea became enamored with Keoho, the wife of Keoni Li'ili'i, and he whisked her away. Much of Ka'alu'ea's work centered on the ranch headquarters and the Carter home, while his older brother Keoni Li'ili'i was leading the cowboy gang on long and remote missions. Ka'alu'ea and Keoho adopted a son, Johnny, while Keoni Li'ili'i went on to marry Laika Purdy and launch a whole new family. Despite some public discomfort in these trying interfamily relationships, the brothers maintained a reasonable relationship, basically living next door to one another. The children of both marriages grew up with mutual respect and understanding.

During his senior years on the ranch,

Ka'alu'ea's assignments included serving as a horseback guide to visiting dignitaries and government officials. He retired in 1931 after thirty-two years of service. His son Johnny did not work for Parker Ranch.

Thomas Kamaki Lindsey (1887–1959)

Like his older brothers, Kamaki was born to be a cowboy. Their upbringing was identical in terms of exposure to ranch life, but Kamaki had the luxury of older brothers who were the best teachers anyone could want.

Built slighter taller and more heavily muscled than Keoni Li'ili'i, Kamaki also had a more jovial—if not somewhat *kolohe* (mischievous)—personality. While his congeniality was occasionally overshadowed by his rascality, there is no question about the high position he holds in the cowboy ranks, most notably as a roper, both on the range and in the *pā'eke* (corral). Cowboy life started in his early teens. By early adulthood, years of experience, exposure, learning, and practice gave him a significant degree of expertise in handling horses and cattle.

In addition to routine cowboy work, Kamaki was given responsibility as a line rider *(nānā'āina* man*)* on the widespread sections of the ranch. Often the role of these men was to look beyond water and grass conditions to ascertain when cowherds would be calving, ready to be branded, or weaned. The timing of such chores was determined by A. W. and later Hartwell Carter, with firm dependence on the eyes of the master, the *nānā'āina* man. In her memoirs included in the Lindsey *'Ohana* Association monograph, "Na Mamo Aloha 'a Thomas John Weston Lindsey," prepared for the *'ohana* reunion in July 1999, Kamaki's daughter Helen Mililani Lindsey Roman provides detailed insight into her father's role on the ranch:

> Mr. Alfred Wellington Carter was the Manager of the Parker Ranch where my father and two brothers were employed. They were a conscientious and dedicated group. Papa was a cowboy and David and Albert were Cowboy Rough Riders . . . taming the wild horse for ranch and pleasure riding. This latter was done in a "Breaking Pen." Papa Kamaki would check on conditions of the cattle and land at Keeaumoku, Keanakolu, Humu'ula and other surrounding Ranch areas. Many times when he had to be away more than a day, the need for an extra horse was necessary to carry his blanket, food, clothes and other incidentals. There were many encounters with the undomesticated cattle in the chaparrals on the slopes of Mauna Kea and the ill-tempered pigs in the bushes. His only weapon was a Bowie type knife that had a peculiar point and sharpened on both edges. He also took along tools when he was "riding the fence" and certainly he would not forget his wad of chewing tobacco! He knew every tree and shrub in these areas he covered. We were not perplexed when his fellow cowboys dubbed him "Kamaki Kolea" . . . like the Plover that flies great distances annually.

Kamaki was first married to a native woman from Puakō named Lilialuluokekula Kaono, who brought seven children into the world before she died in a tragic automobile accident in 1926. The youngest child of this marriage was Helen Mililani.

The two eldest boys, David (Kaono) and Albert Ioela, became roughriders at the Breaking Pen. Two of the middle sisters, Lily and Emma, are noteworthy.

As a girl, Lily enjoyed helping her father when he wove *kaula 'ili* (skin rope), as well as horsehair neck ropes *(kaula 'ā'ī)* and girths *(kaula 'ōpū).* Lily married George Hind, and their daughter Odetta in turn married John Kaleimaiole (Keoni) Lindsey, whose early career was spent as a Parker Ranch cowboy before he moved on to construction work. Keoni was a distant cousin, having descended from the George Kynaston (great-grandson) side of the Lindsey *'ohana.* Lily later married Hartwell Carter's chauffeur, Tokohiro Yoshimatsu, and bore three children: Lloyd, June, and Judith.

Daughter Emma married Ralph Pi'ikoi King Jr., who had a one-day stint as a roughrider but quickly found his niche in the surveying department of the ranch under Stanley Wright and later James Murray. He remained in the survey department until his retirement. Ralph and Emma produced three daughters—Virginia, Priscilla, and Grace—who are noted for their beauty and singing talent, often in concert with their mother and aunts. Of the three daughters, only Priscilla married a cowboy, Henry Ke'ehikapa Ah Sam, son of senior Parker Ranch foreman Henry Ah Fong Ah Sam. Young Henry grew up to be a cowboy at Keamuku and eventually a foreman. He retired as superintendent of the Mauna Kea Division in 1989. Henry and Priscilla were the parents of three bright sons: Henry Jr. (Tippy), Ralph, and Rueben. Of these sons, Tippy became a cowboy working in various sections of the Mānā Division, where he excelled in technical aspects such as the artificial insemination of purebred herds. He left the ranch in 1987 to go to the Big Island Meat Company Feedlot in Pa'auilo, then to Waiki'i Ranch. After leaving Waiki'i Ranch, Tippy spent four years as a herdsman for the scenic Kualoa Ranch, then moved on to Kīpūkai Ranch in 1994 where he played a part in several movies. He returned to Waiki'i Ranch, where he remains today. When Tippy left Parker Ranch in 1987, it marked the end of the Kamaki Lindsey family presence on Parker payrolls.

Shortly after the loss of his first wife, Kamaki married a Hawaiian woman, Mary Ann (Meleana) Kepano of Kalaoa, North Kona, to whom eight more children were born. The eldest child was christened Thomas Kamaki Lindsey II. Thomas Jr. was called Tommy.

In the early 1930s, Kamaki left Parker Ranch and moved to Pu'uwa'awa'a Ranch, where he worked as a cowboy for Robert Leighton Hind I (grandfather of R. L. Robbie Hind III). Helen Mililani Lindsey Roman describes this transition in her memoirs:

> Our father, Thomas Kamaki Lindsey, left Parker Ranch to work for Mr. Robert Hind of Puuwaawaa Ranch in Kona, to do exactly what he was accustomed to doing on Parker Ranch. We, his five daughters and two sons did not agree with the change of school, home and everything in general. Mr. A. W. Carter to the rescue! Being the professional lawyer that he was, we were able to remain in Waimea and with Cousins Thelma and George Lindsey's approval we went to live with them. The Parker Ranch carpenters immediately added a

bath and three bedrooms to the mauka side of their home to accommodate us. (David and Albert continued to work on the Ranch but soon the city lights of Honolulu beckoned.) Mr. Hartwell Carter became our Guardian and his sister Barbara our disciplinarian and a true friend. When cousin Thelma reported our indignant behavior and other problems to her, she was always receptive and we got a long lecture in the "open air" green Packard with the two folding back seats. When it would rain, the car would have to stop so we could put up the curtains. The long lectures continued on the highways of Puukapu, Kona side, and Kawaihae road until she felt that everything she said had finally registered appropriately. They did.

Kamaki served as foreman at Pu'uwa'awa'a, at which time his eldest son Tommy took up the profession before entering the service. Kamaki moved on to Hu'ehu'e Ranch, which was being managed by a former longtime Parker Ranch overseer, Theodore Vredenburg. Kamaki's roping expertise was used at Kailua pier when cattle from Hu'ehu'e Ranch (then under the Trusteeship of A. W. and A. Hartwell Carter), Pu'uwa'awa'a, and even Parker Ranch were shipped in the traditional *ho'āu pipi* fashion. During this period, Tommy went to work at the Parker Ranch Breaking Pen under the foremanship of George Purdy. Shortly thereafter, Tommy married Kealalaumoani (Keala) Haleamau of Kalaoa near Hu'ehu'e Ranch. Tommy left Parker Ranch for that ranch, where he spent his entire career as a cowboy until his retirement. Tommy was the only one of the Kamaki/Kepano line to work for Parker Ranch.

As my family returned to Hawai'i in 1968 and worked at the Hualālai Ranch Veterinary Clinic, we were privileged to become closely acquainted with the Tommy Lindsey and Haleamau families. Keala's brothers Karin and David were gifted with beautiful voices, as was their elder sister. Karin's sons Gary and Kevin are major players in the modern Hawaiian music scene, especially slack key guitar *(kī hō 'alu).* At this writing, Tommy is enjoying retirement with frequent reminiscences of his days at Parker Ranch and the kind leadership shown by foreman George Purdy.

Kamaki died in Kona in 1959, nearly thirty years after leaving Parker Ranch.

Albert Uiha Lindsey (1894–1965)

Uiha was the twelfth and last child of William Miller Seymour Lindsey and Kaluna Lindsey. With the typical Waimea upbringing of those times, Uiha quickly immersed himself in cowboy activities, even during his boyhood. Showing promise, he was sent to Honolulu for secondary education.

After returning to Waimea and being employed by the ranch, Uiha married Daisy Kanoehali'i Stevens in 1916, the third child of Robert Na'ea Stevens Sr. and Harriet Ka'ualoku'okamaile Lincoln. The marriage produced five children. Of the two sons, Miller, the eldest, had a short career on Parker Ranch, then moved on to plantation life. His brother Alan Hartwell (Uku) spent his entire career as a cowboy on 'Āina Paka. For a brief period, Daisy's grandson Irwin Ching worked as a cowboy in Keamuku.

During his career with the ranch, Uiha

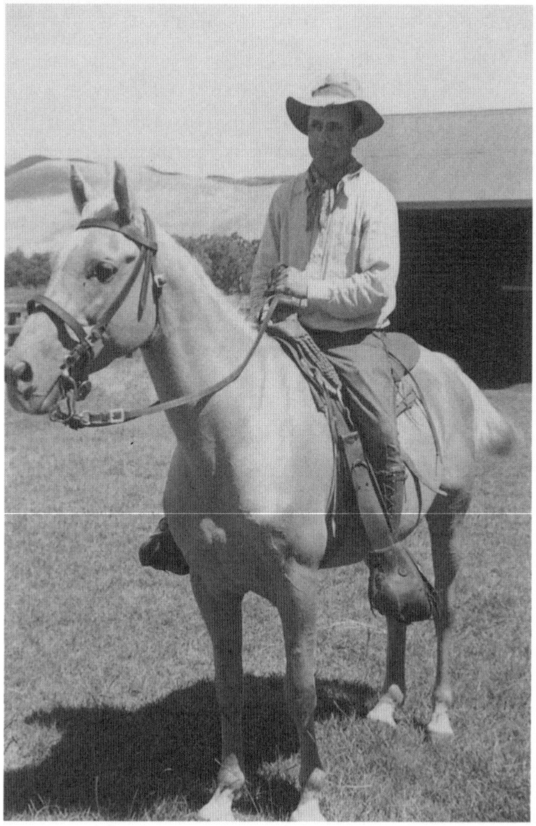

Albert Uiha Lindsey on horseback at Pukalani Stables. PPS Collection.

developed a good—but arms-length—relationship with A. W. Carter. His relationship with management improved in 1937 as Hartwell gradually assumed management responsibility for the ranch. In the meantime, Uiha earned a reputation as a skilled, knowledgeable cowboy with a firm grasp on the doings of the ranch as a whole, sometimes in areas beyond livestock. A. W. valued Uiha as a knowledgeable ranch hand and cowboy working through supervisory channels, while Hartwell, with a less direct management style, was inclined to use him as a source of information. Uiha eventually rose to become cowboy foreman, albeit for a brief period.

During the transitional years between A. W. and Hartwell, Uiha's foremanship ended abruptly. This happened shortly after a crisis involving a herd of shipping cattle that were being walked to Kawaihae from Pu'uiki Corral near the base of Pu'u Kawaiwai and the Old Kawaihae Road. Pu'uiki Corral was an overnight holding pen, built in the round to avoid crowding and bruising of the fat cattle immediately before shipment. The incident involved the loss of only one steer (a broken leg), but the method in which it was reported troubled A. W., and Uiha was relieved of his foremanship. While dismissing Uiha was perhaps an extreme reaction, it was typical of A. W., who was very sensitive to the issue of proper cattle handling practices.

Uiha left for O'ahu shortly thereafter, selling off most of his cherished horse gear, saving only what his cowboy son Uku could use. Uiha's O'ahu career involved working at a dairy, which was a great letdown after being a lead *paniolo* on the exalted Parker Ranch. When Hartwell Carter assumed more day-to-day control over the ranch, Uiha was given a job as *nānā'āina* man, for which he was eminently qualified. His role as a source of information was enhanced; ranch activities in general were subject to his surveillance. This role was rather evident through the late 1950s when I worked at Pukalani Stables, riding green colts into the surrounding paddocks, occasionally as far out as Holoholokū Hill. One could see Uiha off in the distance, perched on a knoll, as still as a rock for hours at a time.

Even in his twilight years in the saddle, Uiha struck a handsome pose on horseback in his riding breeches and high-laced boots, buttoned-up long-sleeve denim shirt under a Levi jacket, and creased felt Stetson cocked jauntily aside his head.

His son Uku grew into a cowboy known for his hard work and horsemanship, eventually becoming assistant foreman at Waiki'i under Daniel Kaniho. A severe stroke on the job forced a medical retirement, during which he made a significant recovery by sheer determination. Uku remains active in community athletic organizations and the Lindsey *'Ohana* Association. A marriage to Mary Pacheco, daughter of Joe Pacheco and Betsy Ma'ina'aupo, produced four daughters. Uku's third daughter, Aletha (Peewee Barkley), a noted horsewoman, maintains the family tradition.

Allen Hartwell Lindsey (Uku) is the last of the Albert Uiha Lindsey line to serve in the ranks of the Parker Ranch.

The George Kynaston Line of Lindseys

The marriage of Mary Ka'ala Fay to her brother-in-law George Kynaston Lindsey following the death of her first husband, Thomas, took place in 1853. In two years, the first child of this union arrived. The baby was named Thomas Weston Lindsey, after his deceased uncle. After this child, nine more followed: Emma Angeline, Elizabeth Amelia, Mary Pomelia, Angeline Constance, George Lindsey, James Fay, Edwin Robert Kingston, Charles Robert, and William Kingston. Of the ten children, Thomas Weston, Emma Angeline, and James Fay won significant places in ranch history, with Emma's being matriarchal.

Thomas Weston Lindsey (1855–1912)

Thomas Weston was born during the formative years of the ranch. By his teens, he demonstrated a natural talent as a rising cowhand. He aptly qualified as *kanaka makua* in the eyes of his peers, as well as John Palmer Parker. These leadership qualities eventually allowed him to follow Ulumaheihei Waipa and Kaho'ohanohano as cowboy foreman. As the years rolled by, Thomas was somewhat upstaged by Keoni Li'ili'i when A. W. Carter made a leadership change and old Tom became a *nānā'āina* man. Old Tom's enduring contribution, however, was his fostering of the cowboy tradition through several generations of his progeny.

In 1875, young Tom married a Kaua'i *kanaka wahine* named Becky "Beke" Kaipukai Fredenberg. They had eight children. Their first son, William Miller Seymour Lindsey II, was born in the same year as their marriage. Although he grew up to work on the ranch, he eventually developed his own holdings. William Miller Seymour's most significant contribution was the family he sired with Mary Leialoha Rose—notably, Anna Leialoha Lindsey, the Anna Lindsey Perry-Fiske of Anna Ranch fame. Anna also had a brother, Charles Kalaniali'i, and a sister.

Six of the seven children that followed—Charles Notley I, Thomas Weston Kinsley Lindsey, Elizabeth Rosy, Samuel Frederick Nia, Christine Pomelia, and Mary Helen—all developed significant relationships with Parker Ranch.

Charles Notley I

Charles Notley I (Kale Kuli I) married Fanny Chesebro during the period that he actively

Genealogy chart of George Kynaston Lindsey and Mary Ka'ala Fay

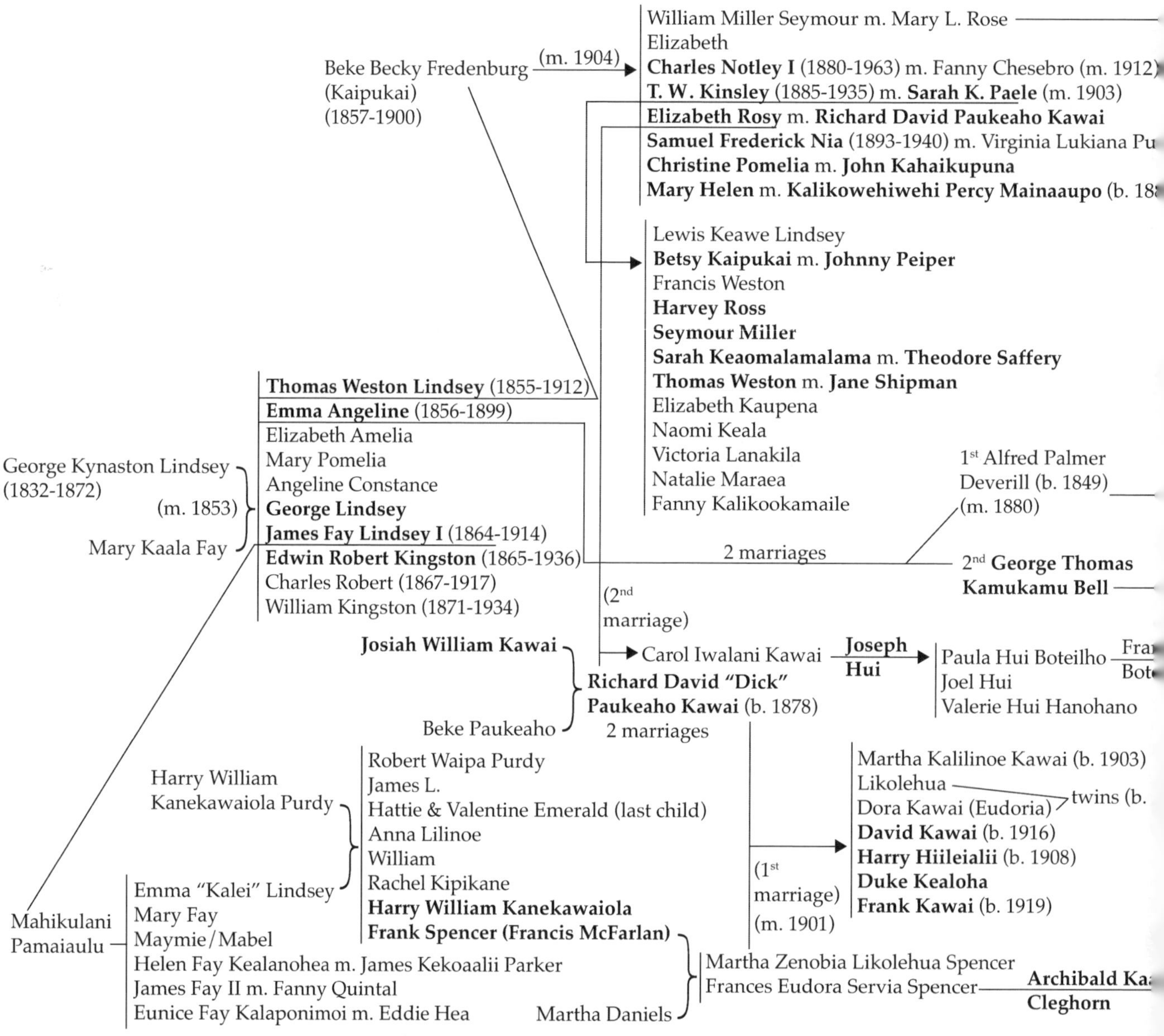

William M. S. Lindsey (Sheriff)
Anna Leialoha Lindsey Perry-Fiske
Charles Kalanialii Lindsey —son→ Richard

Charles Notley II (b. 1912) m. Cozie Perry →
- **Leroy** (b. 1934) m. Patricia T. Ozawa
- **Paul** (b. 1936) 2 marriages — 1st Ruby Feliciano → Paula, **Paul Kaleimaiole**, **Eric Kahiliaulani**, Betsy Kaiu; 2nd Roberta Fergerstrom

John Kaleimaiole (b. 1914)
Samuel Frederick m. Jennie Piilani Bell →
- Norman Molo
- **Calvin Pika** (b. 1938)
- **Gordon Aloha Harbor** (b. 1942) m. Hazel Hanato
- **Clifford** (b. 1943) m. Gail Sandra Solle
- Bernadette Piilani
- Jonathan

Thomas Milimili (b. 1918) m. Juliet Nathaniel
Joseph Chesebro
Robert Kamaile
Fanny Leialoha
James Fay IV (b. 1924)
Elizabeth Kamaile —son→ **Dennis Dean**
Fred Maluhia
Edwin Keao (b. 1932)
Agren →
- **Herbert Miller Semoa "Dead"**
- **David "Dango"** (b. 1916)

→
- **William K. L. W. Mainaaupo "Billy Boy Lindsey"** (1906-1968) m. Elodie Helbush
- **Betsy Leinani Mainaaupo** m. **Joseph Pacheco**
- Thomas Weston Mainaaupo
- Rose, Marion, Ellen
- Mary Kealoha
- Percy Kaliko
- John Henry

Arthur Deverill
Lydia Sarah m. **Joseph "Snow" Herman Maertens** —15 children→
- Emma Angeline
- Joseph Booth
- Herman (1908-1908) > twins
- **Fredrick** (1908-1984) m. Grace Caires
- Katherine Lydia
- Josephine Kahalelaukoa
- Mabel Laola
- Marjorie Hermina
- **Joseph L.** (1916-1940)
- **Alfred** (1918-1919)
- Marietta Cassey
- Elizabeth Maile
- Florence Keala
- Lydia Pauline
- **Arthur "Buddy"** (b. 1927) m. Hazel J. Angelo

Helen Lindsey
Angeline m. Walter Foss Sanborn → Percy Sanborn

[L]acy Kawahine-[K]anaumoku Boteilho →
- Maria "Maraea"
- **Alexander Bell** (1896-1965) (m. 1918) Antonia Cordeiro →
  - **Violet** m. **Antone Silva Ramos** → Anthony Jonathan Ramos; **David Bruce Ramos** (b. 1947) m. Jade Lyn Andrade
  - **Alexander** (1921-1996) m. Annie K. Purdy
  - **Theodore** (1923-2002) m. Piilani I. Matsu
  - Gilbert
  - Stanley
  - Dorothy
  - Theodora

Archie Spencer Kaaua m. Genevive Piilani Yates →
- Alan Cleghorn Kaaua (with Marjorie Bergin) → **Christopher Kaaua**
- Archie Kaaua
- Lani Kaaua Brus

Edward Stanley Kaaua
Jack Low Kaaua
Samuel Parker Kaaua
Stella Hailauaala Kaaua

rode the ranges of Waimea; he sired twelve children. His eldest son, Charles Notley II, carried on the cowboy tradition and eventually shifted to fence work, in which he was clearly recognized as a steady and productive foreman. Part of his cowboy years were spent at the Second Gate Section, living in a line cabin that eventually became the Department of Hawaiian Home Lands homestead where Sarah Kaupiko Ruiz now resides, about 6 miles east of Waimea on the old road. He married a pretty Hispanic girl named Cozie Perry and they had two sons, Leroy and Paul (Cookie). Leroy worked as a cowboy at the Breaking Pen and Pukalani Stables, while Cookie became a policeman after a few years of working in the fence gang under his father. Cookie and his wife, Ruby Feliciano of Kohala, had two sons, Paul Jr. (Kaleimaiole) and Eric Kahiliaulani, both of whom became skilled cowboys on the modern Parker Ranch. Kalei eventually became the ranch saddle maker. Both men were furloughed in the 1990s after injuries sustained on the ranch.

Charles Notley II's younger brother, John Kaleimaiole, began a cowboy career but left for construction work as noted earlier.

Like Keoni, third brother Samuel

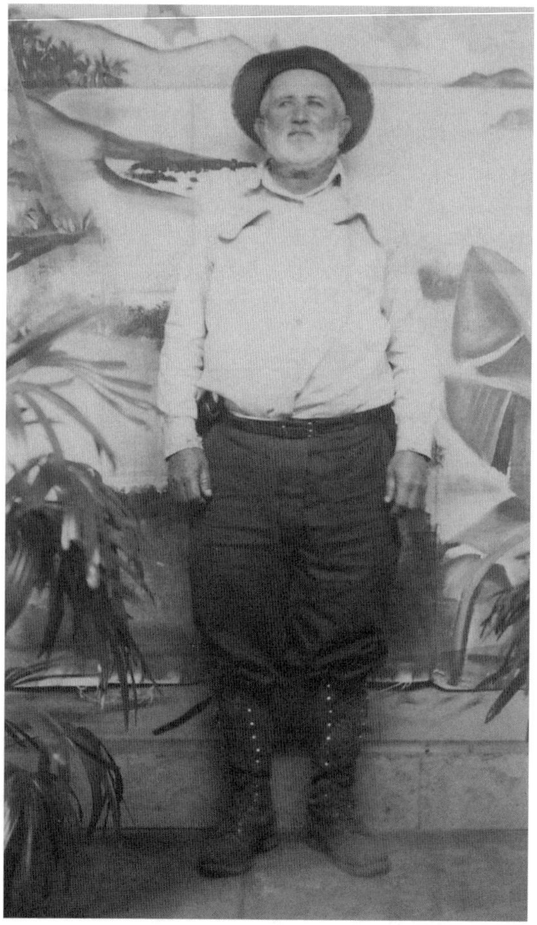

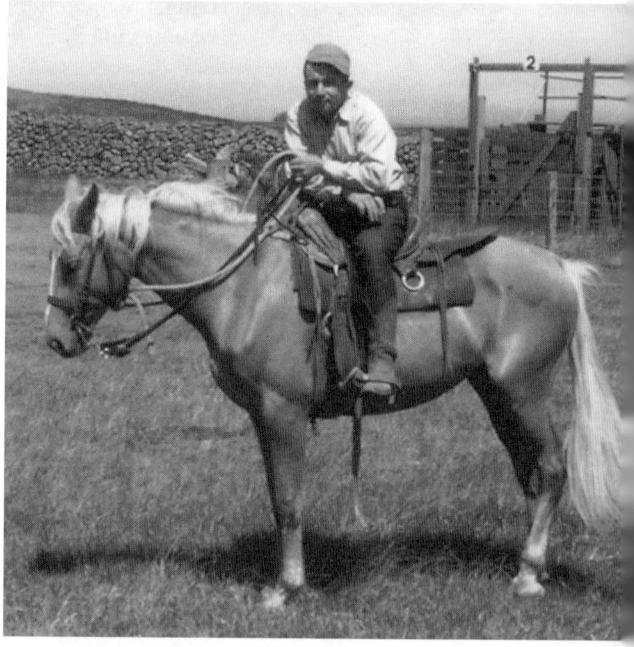

Left: *Charles Notley Lindsey I in a studio setting. Eric Lindsey Collection.*
Above: *Charles Notley Lindsey II on a palomino horse at Pu'uhihale Corral. In the background are World War II bucking chutes. A. W. Carter Collection.*

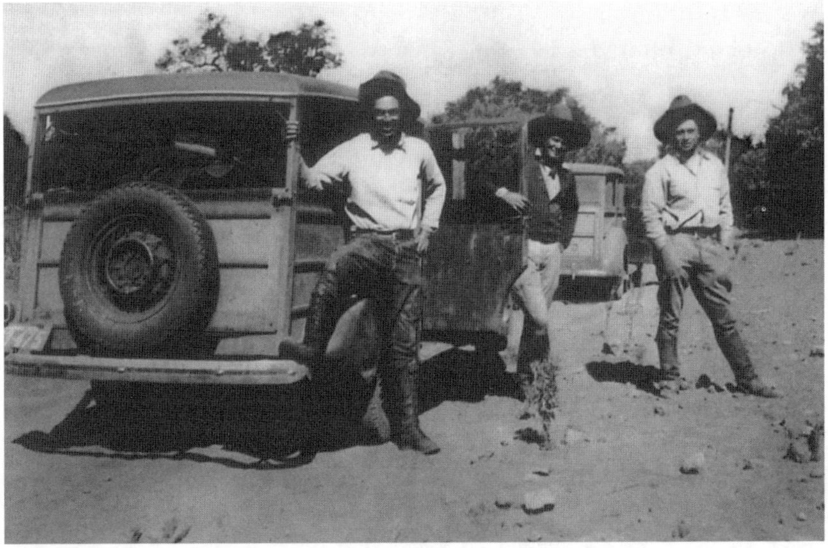

John Kaleimaiole Lindsey (right) with his cousin David Dango Lindsey (center) and John Samoa Lekelesa on their way to Humu'ula through Omaokoili, standing by Ford "banana wagons." Circa 1939. PPS Collections.

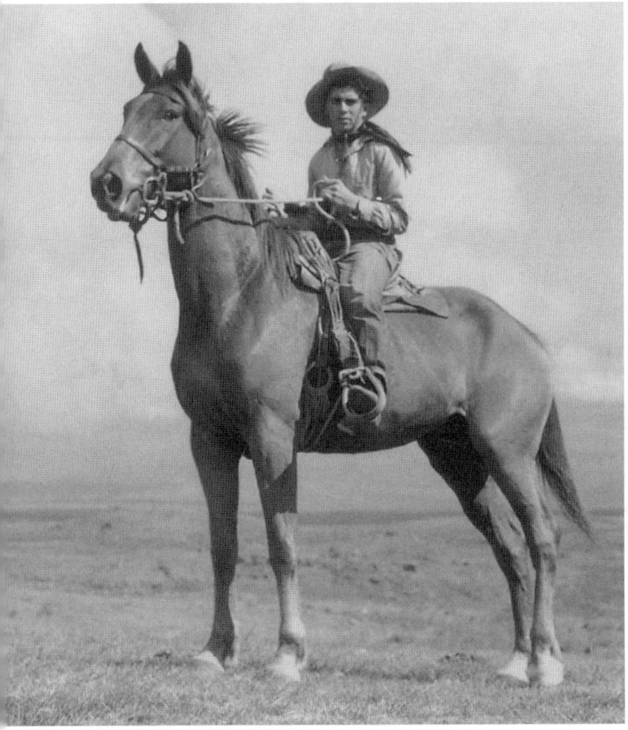

Thomas Milimili "Bear" Lindsey as a rough-rider on a Thoroughbred colt out on the range. PPS Collection.

Frederick left for the construction boom of the 1950s and 1960s after a stint on the ranch. His marriage to Jennie Piilani Bell, great-great-granddaughter of patriarch Scotsman William Bell, produced five children, two of them rising to foremanships after returning to the ranch following construction careers. Calvin (Pika), the second oldest, became foreman of the building department, and Gordon, third oldest, became foreman of the fence gang, succeeding his uncle Charles. Unfortunately, both men died young—Pika in 1991 and Gordon in 1996.

Another Charles Notley I son, Thomas Milimili (Bear) Lindsey, started his career at the Breaking Pen but moved to Hilo when he married Juliet Nathaniel. This marriage is responsible for the famed Lindsey influence in the Haili Church Choir and the Haili Volleyball Tournament that has brought athletic acclaim to Hilo.

A sister, Elizabeth Kamaile Leinani, bore a son, Dennis Dean, who remained on

Parker Ranch in the utility gang until he accepted the early separation package offered by Parker Ranch in January 2002. Dean's departure ended the Charles Notley II line at the ranch.

The youngest of the Charles Notley I line, Edwin Keao, worked the fence gang under his brother Charlie, but he was seriously injured in a grenade blast while working Range I. He left for a career in construction shortly thereafter.

T. W. Kinsley Lindsey

T. W. Kinsley Lindsey was the fourth child of Thomas Weston and Beke Lindsey. He grew up with an interest in mechanical work, becoming involved early in the far-reaching water systems of the ranch. He was a *kanaka makua* teenager when A. W. took the ranch reins in hand. Because in these times the ranch had several Tom Lindseys on board, he was nicknamed Tom Pipe, a name he kept until his death at

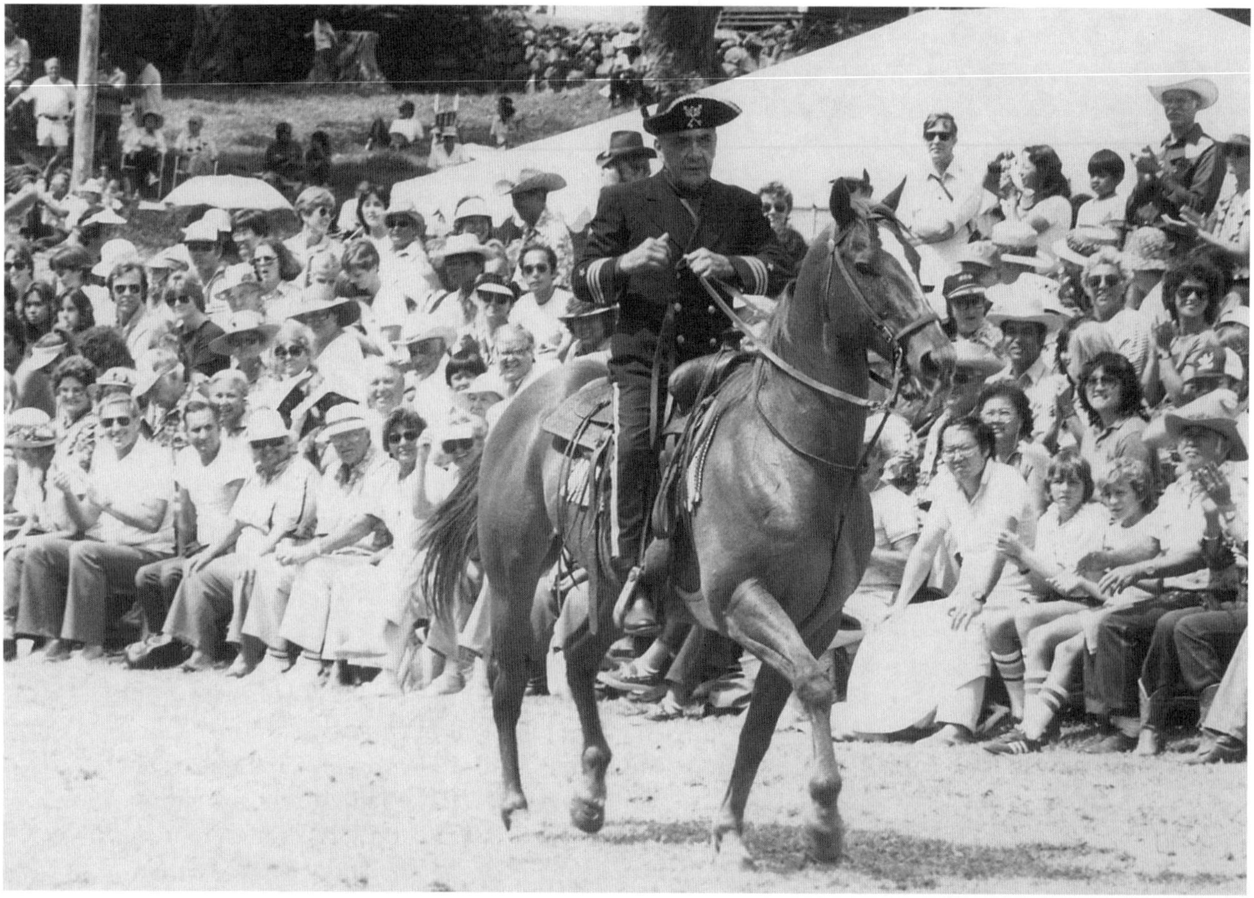

Thomas Weston Lindsey, Pu'u'ō'ō Ranch boss, riding in the 1978 Old Hawai'i on Horseback pageant on his fine parade horse, Mountain Bear, depicting the original Captain Thomas John Weston Lindsey. W. C. Bergin Collection.

Pōhakuloa in 1935. His marriage to Sarah K. Paele of Kokoiki, North Kohala, resulted in a dozen children, the second oldest of whom, Betsy Ka'ipukai, wed Johnny Pieper, son of the German teamster Henry Pieper. Her brothers Harvey Ross and Seymour Miller started as young horse breakers under James Palaika but left for careers in construction on O'ahu prior to World War II. Middle sister Sarah (sixth child) grew up to marry a well-liked Parker Ranch cowboy, Theodore Saffrey. Her next younger brother, Thomas Weston, grew up to be a wild and wooly cowboy, leaving the ranks of Parker Ranch at the decisive prodding of A.W. Carter. Tommy went on to work for W. H. Shipman Ltd. where he enjoyed a solid career, retiring in 1978. He died at his homestead in Kea'au in 1987.

Elizabeth Rosy

Elizabeth, fifth child of TW and Beke, developed into an attractive woman who chose not to rush into marriage. At about age forty-two, she married Richard (Dick) David Paukeaho Kawai, a skilled *kama-'āina* woodworker employed on the ranch as a cabinetmaker and carpenter. This marriage produced a daughter, Carol Iwalani Kawai, whose family has maintained enduring ties to the ranch rank and file to this day. Carol married a strikingly handsome man, Joe Hui, a native of Kawaihae, a noted horse breaker and cowboy. Their marriage produced two equally attractive daughters, Paula Hui Boteilho and Valerie Pualani Hui Hanohano. Paula's daughter Kacy, a skilled rodeo horsewoman, continues at the ranch in the accounting department, carrying the respectable genetic coding maternally of Lindsey/Kawai/Spencer and paternally the Hui and Awaa lines. This lineage, with the Boteilho name noteworthy in livestock circles, demonstrates the convergence of some of the most prominent *'ohana paniolo*.

Samuel Fredrick Ni'a

Samuel Ni'a was the sixth child of TW and Beke and, in his marriage to Virginia Lukiana Purdy, daughter of William Naleilehua Purdy, produced twelve children. Two sons entered cowboy life on the ranch: Herbert "Dead" Miller Semoa Lindsey and David "Dango" Lindsey both worked at the Breaking Pen and as cowboys. Both men moved on to other careers.

Christine Pomelia

Christine Pomelia was the seventh child of Thomas and Beke. In early adulthood, she married John Kauwewahine Kahaikupuna, son of Kahaikupuna and Mai'i. John Kahaikupuna was a noted horseman. Their daughter, Christine Pomelia Kahaikupuna, married Ronald (Yamada) Lindsey.

Mary Helen

The eighth and last child of the TW/Beke marriage was Mary Helen. She in turn married a charismatic *paniolo* named Kalikowehiwehi Percy Ma'ina'aupo, son of Joseph Po Ma'ina'aupo of Mānā, Hawai'i. While the marriage produced ten children, it was Kaliko's image as a *paniolo,* underscored by his cowboy talent, that made him a frequently photographed ranch hand. He wore chaps when his peers wore leggings. Even in his later years, he sat erect on his mount; that of itself was pleasing to the eye of a knowledgeable horseman. In the waning years of his retirement, he would camp out with the Shipman cowboys, led by Thomas Weston Lindsey.

The eldest child of the marriage, William Billy Boy Kalikowehiwehi Ma'ina-'aupo, had an intermittent career on Parker Ranch, usually with the horse-breeding program. For whatever reason, Billy Boy changed his last name to Lindsey. While none of his siblings had strong cowboy ties, his younger sister Betsy Leinani married Joe Pacheco. A daughter of this union, Mary Pacheco, married Uiha Lindsey's son, Allan Hartwell, and remained on Parker Ranch until her husband's retirement.

Thus ends the Parker Ranch progeny of the marriage of Thomas to Beke Fredenberg. His next sister in line, Emma Angeline, married twice and launched a long succession of descendants, many of whom worked for Parker Ranch in both the Deverill and Bell branches. David Ramos, former executive vice president of the ranch, is a descendant of the Bell/Lindsey line.

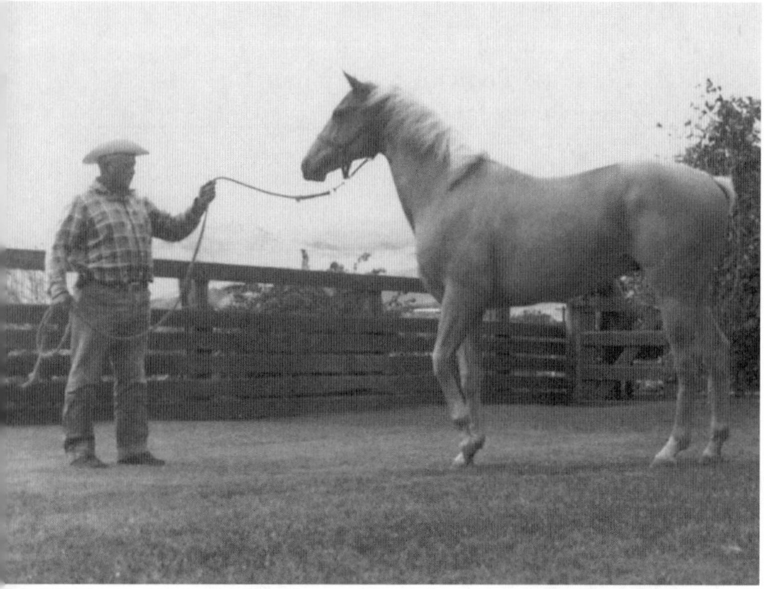

Billy Boy Ma'ina'aupō at Pukalani Stables, showing a palomino gelding that was exported to Japan. Hartwell Carter Collection.

Emma Angeline Lindsey (1856–1899)

Emma Angeline was the second child born to George Kynaston and Beke Fredenberg Lindsey. At age twenty-four she married Alfred Palmer Deverill.

Alfred and his brother, William Edward, arrived in the Islands after 1862 as bearers of gifts from Queen Victoria of England for the son of Queen Emma and Kamehameha IV, aptly named Albert Edward after the husband of the English queen. The Deverill brothers settled in the Islands despite the outcome of their mission—young Prince Albert Edward Kamehameha had died.

Alfred Deverill worked at the Humu'ula Sheep Station Company for the Haneberg brothers, in addition to having his own horse business. Attracted to Emma Angeline, Alfred married her in 1880 and they produced five children. Their fourth child, a daughter named Lydia, married Joseph "Joe Snow" Herman Maertens, Honolulu-born son of German immigrant William Ludwig Maertens. Lydia and Joe Maertens had fifteen children, two of whom developed ties with Parker Ranch.

The first of the Maertens boys to cowboy for Parker Ranch was Frederick, the surviving twin of sons born in 1908 (Herman died as a baby). Frederick worked at various tasks, including training mules at John Purdy Stables. He went on to become a member of the cowboy gang, a career he maintained until his retirement. He acquired a nickname early in his childhood. He was called Napier (pronounced "Nepia") by everyone from the fact that, during his boyhood, he would tag along after Alex Sandy Napier Sr., the herdsman at Makahālau where the Maertens family lived. He was also called Nesso by the cowboys after an accident that happened when he was aggressively chasing a fat hog during a cattle drive in Ka'ala, near Ahualoa. His horse tripped *(hinapū)* and fell, and Fred was propelled into the brush where a twig impaled itself deep into one of his nostrils. Hunting during a cattle drive was frowned upon, especially if John Samoa Lekelesa or Willie Kaniho were the foremen, but Fred was unable to conceal this episode due to the twig's stubborn lodging up his nose. Fortunately a cowboy cohort was able to extricate the foreign body, but the story of the nasal impalement became quite a tale. "Nasal" was localized in pronunciation to "nesso," a name that John Samoa routinely used in reference to Fredrick Maertens. Fred died in 1984.

In 1916, Joe and Lydia had another boy who was destined for cowboy life, albeit a short one. Joseph Ludwig grew into a strapping, tall, square-shouldered horseman who spent a lengthy period at the Breaking Pen under Albert Uiha Lindsey. Joe was a productive horse trainer in the time leading up to World War II, when A. W. was aggressively marketing army remounts as well as polo and cowboy horses. Joe had ridden out of the Breaking Pen corrals into the surrounding paddock on what appeared to be a well-started (early trained) colt when the horse suddenly slipped and fell with Joe, whose head and neck were badly injured. He died at twenty-four on April 4, 1940, after a short but colorful cowboy career. A. W. Carter expressed his grief in a memo to the family:

Joe Maertens Death

> Everyone who knew young Joe considered him a friend. He was a boy that was cheerful and friendly. He was a hard worker and everything that one should look for in a young man. To me his death is a distinct loss of a fine young man and I shall miss him. To you it is a terrible loss and I wish to express to you both my sincerest sympathy. Such a thing is beyond understanding.

The youngest child of the fifteen children was Arthur Buddy Maertens, a colorful fellow, avid horseman, and community volunteer. His daughter Betty was an accountant at Parker Ranch for a short time. She married Earl Spence, who became ranch agronomist before leaving for Kahuku Ranch.

This brings to a close the Emma Angeline Lindsey–Alfred Deverill line. In 1899, Emma Angeline married George Thomas William Kamukamu Bell and set the course for a long line of successive Parker Ranch–employed descendants. George Thomas William Kamukamu Bell was the son of a Scottish engineer who landed in Kohala. He had been previously married, producing a daughter, Hattie, the mother of Thomas Bell. Tom Bell's father was Irish teamster Tom Gallagher, but young Tom kept his mother's surname.

Maria Maraea was the firstborn (1893) of this marriage. Alexander Bell, who was born in 1896 in South Kohala, followed her.

James Fay Lindsey I

James Fay Lindsey was the seventh child born to George Kynaston and Beke Fredenberg. The year was 1864. Specifics are scarce as to his exact role on the ranch. At twenty, he wed a pure Hawaiian woman from Waimea, Mahikulani Pamaiaulu, in the home of Rev. Lorenzo Lyons. They produced six children, three of whom had significant ties to Parker Ranch. The eldest, Emma Kalei, married Harry William Kanekawaiola Purdy Sr., who was a son of Anna Parker Waipa (granddaughter of John Palmer Parker I) and William Harry Purdy Jr.

Emma Kalei was educated at Kohala Seminary, where she was noted for beautiful penmanship. Emma spoke eloquent English, as well as being fluent in Hawaiian and Japanese. Her middle name had fifty-one letters and *okina* (glottal stops), including four independent chantings of Kalei:

Kaleiho'okie—heightened

Kaleikaukeha—on high

Kalei palaua—Kapalawa a'o Keopuolani—daughter of

Kalei 'i'o Puolani—wife of Puolani

The Purdys were blessed with eleven children, none of whom developed strong Parker Ranch ties.

The fourth daughter of the marriage was Helen Fay Kealanohea Lindsey, who later married James Kekoalī'i Parker, son of Sam Parker Sr. and Napela Panana. Helen was a composer of several melodies, including "Akaka Falls."

Fifth child and only son—James Fay Lindsey II—had an early cowboy career on the ranch but moved to O'ahu where he married Fanny Quintal. This marriage yielded eleven children, many of whom became the Moloka'i branch of the family. One was James Fay, a career wildlife ranger with the territorial and state governments of Hawai'i. Many of this side of the family were employed in construction on O'ahu—notably, Steven and George. Sons of both of these men returned to their ancestral home in Waimea in the 1970s.

The Bell Foundation Family

William Peter Bell was born in Scotland in 1817, two years after John Palmer Parker I made Hawai'i his permanent home. Of the five foundation patriarchs, Bell came to Hawai'i the best educated, a product of the Glasgow educational system. He served as a sugar mill engineer and later as a business accountant in the district of North Kohala.

Bell arrived in Kohala around 1835 to begin a career in the early sugar industry and married a native Hawaiian woman named Kipikāne in 1841. She was native to a village in Kohala near the present-day Kohala High and Elementary School in Kapa'au. There is no evidence to confirm that she was related to her likely namesake, Keli'ikipikāne (shortened to Kipikāne), who married John Palmer Parker I nearly a generation earlier. Research is warranted to determine whether she did have a genetic relationship to Keli'ikipikāne.

The Bells were blessed with six children: William Peter Jr., in 1842, Rebecca Lepeka in 1846, George Thomas William Kamukamu in 1847, John in 1848, Mary Ann in 1850, and in 1851, Henry. All this occurred within nine years. Two of these children, William Peter Jr. and George T. W. K., had ties to Parker Ranch.

Both Rebecca and Mary Ann Bell were married to George Kawaeho'olaua Lincoln consecutively, from whom a series of Lincoln offspring became ranch people. William Peter Jr.'s marriage and the family it produced represent a fascinating cultural and ethnic journey. George T. W. K. Bell married Ani (Emma Angeline Lindsey) and began the Alex Bell lineage that has had a profound and enduring influence in Parker Ranch rank and file.

William Peter Bell Jr.: The Kainoa Connection

This connection marks the convergence of native Hawaiian bloodlines with those of Scotland, Chile, and Mexico that form the common matrix of two prominent ranching families—Bell and Kainoa.

In his marriage to Waine'enui Kaluaiele, William Peter Bell had a family of four, split evenly between girls and boys. The eldest, Mary Mele Bell, eventually bore ten children—the first seven of whom were fathered by Charles Anderson, but none of whom developed direct relationships with Parker Ranch, unlike their half siblings who came later. The later family came from the second marriage to a Chilean caballero named Francisco (Frank) Palakiko Caitano. The middle child, Matilda, died as an infant. The youngest, Delphine, once a foreman for Wight Estate's Puakea, later became a cowboy at Parker Ranch–Pu'uhue, old Kahuā Ranch, and Pu'uwa'awa'a Ranch. Delphine was still a Pu'uwa'awa'a ranch hand in the early 1930s when my father, William N. Bergin, M.D., administered medical care for the community of North Kona.

The eldest child, Rose, married into a long line of *paniolo* when she wed Peter Kalāho'olewa Kainoa. He descended from the rich heritage that blended native blood of old Hawai'i with that of old Mexico. From the native Hawaiian side came a woman of Pololū Valley named Malia Ka'inoakupuna Kalahohinu, born in 1850, who married a Waimea-born son of one of the first *vaquero* to settle here. Frederico Ramón Baesa and his Yaqui wife, Vincenta Romero, had moved to Waimea on April 27, 1833. One son, José Ramón, grew into an accomplished caballero in the Kohala area. He joined with Malia in a marriage that bore nine children. Of their seven sons, one stood apart in horsemanship skills. His name was Francisco (Palakiko) Kamaka Kainoa, taking an abbreviated version of his mother's maiden name. He was commonly called Frank. Coincidental with these times, two Hawaiian women from Kalapana came courting. Margaret (Makalika) Makaē Makua married Frank Kainoa, and her younger sister Mary Ann (Meleana) wed a German-Hawaiian *paniolo* named Antone Waltjen who worked at the Humu'ula Sheep Station Company.

The marriage of Frank and Makalika resulted in four sons and a daughter, Theresa Kelekia. All four sons became *paniolo* in the truest sense of the word. The eldest, Sylvester Kiliwilio, became a cowboy at Old Kahuā for a brief period, while the youngest brother, Lui, worked for R. L. Leighton Hind at Maliu (Puakea Ranch) before moving on to Pu'uw'awa'a Ranch. Lui spent most of his career there, having four sons that kept up the *paniolo* tradition. Lui Jr. worked at Kahuku Ranch, eventually becoming a tractor operator under the management of Freddy Rice, while David and William worked for the Hinds at Pu'uwa-'awa'a. Youngest brother Adam was a cowboy at Pu'u'ō'ō Ranch during the time that I worked for its owner, W. H. Shipman, Ltd.

The third son, Peter Kalāho'olewa, worked under James Kimo Greenwell at the

Genealogy chart of Bell/Kainoa

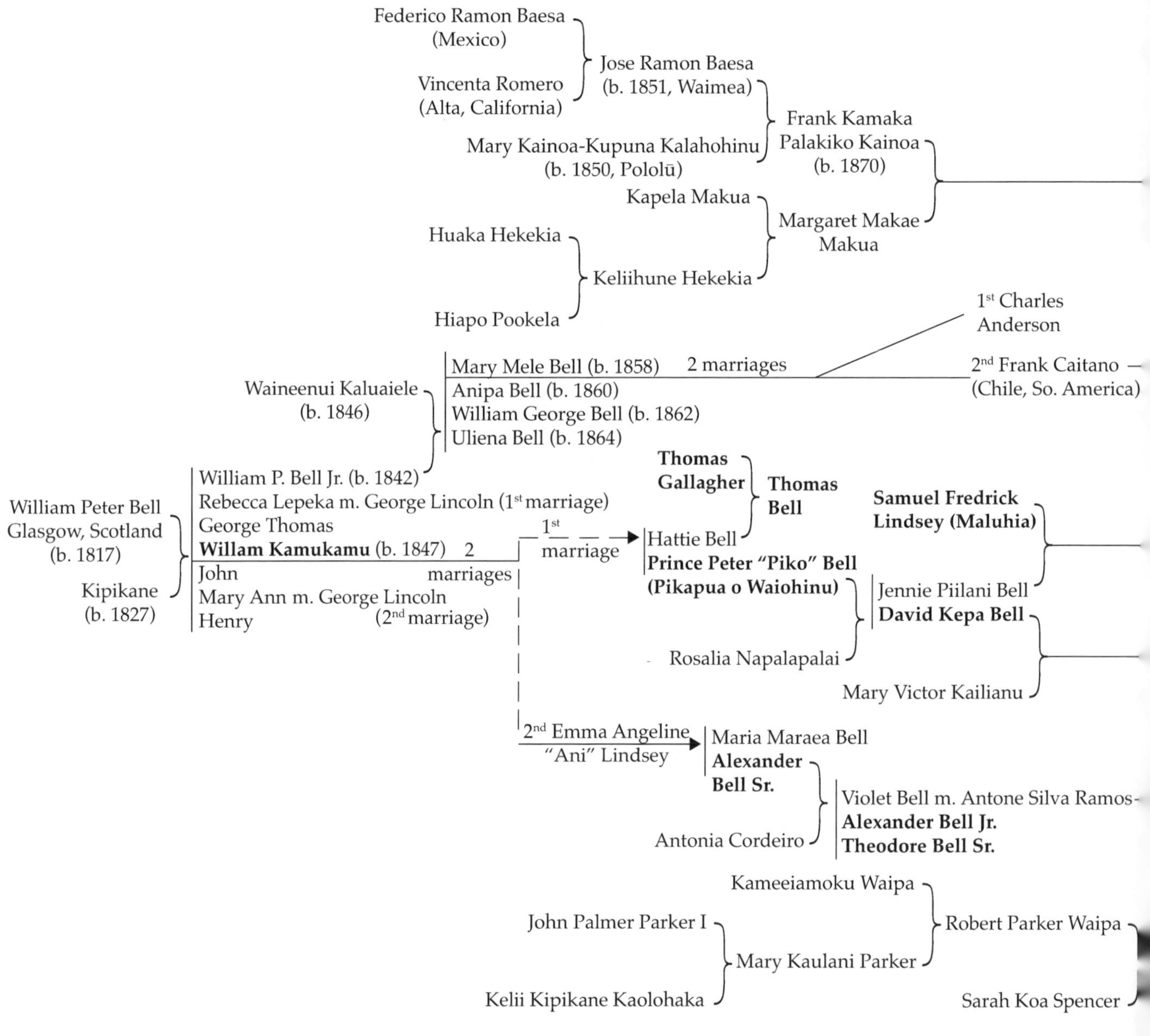

Theresa Kelekia Kainoa
Sylvester Kiliwilio Kainoa
Lui Kainoa → Lui Jr., David, William, Adam Kainoa
Gabriel Kainoa Malani
Peter Kalahoolewa Kainoa (b. 1890)

15 Children → Frank Kainoa, Peter Jr., John Kelekolio Kainoa m. Jennie Dinah Peahi → **John Kainoa**, **Godfrey Kainoa**; Sam Kainoa, Francis Palakiko, **Charles Kale Kainoa**

Rose (b. 1892)
Matilda (b. 1894)
Delphine Caitano (b. 1896)

Norman
Calvin
Gordon
Bernie
Clifford

Thelma m. **Willie Kaniho Jr.** → Klem Keolani Kaniho Sr.
Samuel
Marian m. **Fred "Bolo" Lindsey**
Shirley m. **Keao Lindsey**

Klem Keolani Kaniho Sr. + Janice Napua Parker → Klem Keolani Kaniho Jr.

Klem Keolani Kaniho Jr. + Jamie Loke Giltner → William Makaniniau Kaniho (son), Kelsey Kauikeolani Kaniho (son)

Anthony Jonathan Ramos
David Bruce Ramos

Mary Rodrigues + Stephen Keolani Parker → Samson Parker

Mary Choy + Samson Parker → Janice Napua Parker

Pu'uhue Section of Parker Ranch, left to work at Kahuā Ranch for a few years, and returned to Parker to work at Makahālau. After another five years, he again left for Kahuā, where he remained until his retirement. Peter was a noted *hō'au* (swim) roper for both ranches as they shipped cattle out of Kawaihae. His descendants developed productive careers on both ranches. The youngest brother, Gabriel John, who as an infant was adopted by a childless Kawaihae Uka couple named Malani, grew up to be a career *paniolo* for Parker Ranch in the Keanakolu Section of Humu'ula.

Peter K. Kainoa

Peter K. Kainoa spent his life in North Kohala, homesteading in Ka'auhuhu but working for both Kahuā and Parker to the south. Peter was a noted *paniolo*, a great roper both on the range and in the arena. He fathered fifteen children with Rose Caitano to make up the Mexican, Chilean, Scottish, and Hawaiian Kainoa line. Five children did not survive infancy. Two of the surviving children developed significant relationships with Parker Ranch.

The second to last child was Charles, who spent most of his childhood around Kahuā Ranch, where he was first employed in his late teens. Parker Ranch later hired him, and it was there that he earned the nickname Black Hawk, after a heroic comic book character. What feat he performed to earn the nickname, no one—including Black Hawk himself—is able to recall. His career on the ranch was with the water department, where he served loyally and diligently under William Case, son of former ranch veterinarian Dr. Leonard Case. Black Hawk retired from the ranch and is frequently seen walking about town with his daughter and his dog, Whitey.

Of Peter's sons, John Kelekolio (Johnny) Kainoa as a boy worked alongside his dad at the Pu'uhue Section of Parker Ranch when it was under the supervision of James Kimo Greenwell Sr. in 1934. As a young adult, Johnny married Jennie Dinah Peahi, daughter of a W. H. Greenwell Ranch family. Johnny then became a lifetime employee of Kahuā Ranch. By this time his father Peter, Uncle Sylvester Kiliwilio, and brother Frank were all Kahuā employees. Johnny was known for his horsemanship, cowboy skills, and rodeo victories. He was a consummate horseman noted for training racing and rodeo horses. The Kainoas had ten children—four boys and six girls. During the late 1950s and early 60s, his daughters Jennie and Olinda were popular barrel rac-

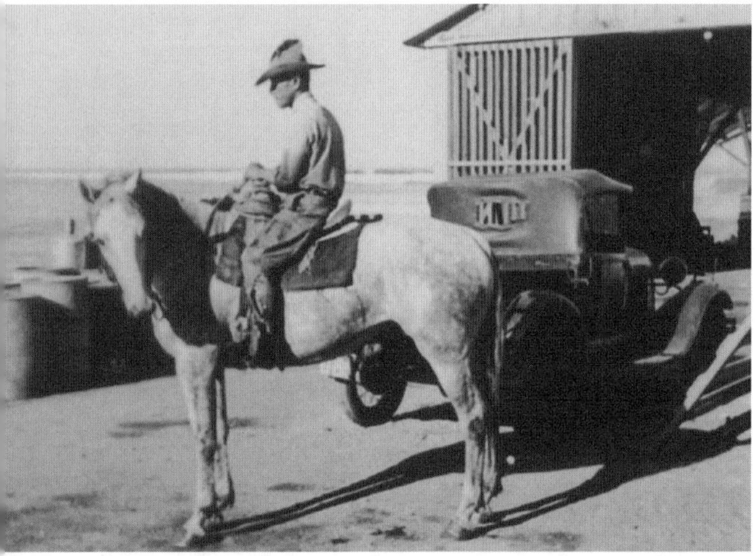

Peter Kalāho'olewa Kainoa on his horse, Blue, a gift from Buster Brown, whose Model A pickup is parked near the Kawaihae Warf. Peter is mounted on a shipping or hō'au *saddletree. W. C. Bergin Collection.*

ing and roping competitors in Big Island rodeo events. Their greatest fan was their mother, Jennie, who could be heard from the bleachers cheering on her daughters and husband Johnny, who routinely won the calf roping event at the Hawaii Saddle Club Rodeos on Memorial Day weekends. A son, John, who wasn't quite old enough to compete, was a robust toddler.

After serving in the armed services, eldest son Johnny Jr. worked as a cowboy for the McCandless Ranch in South Kona under Alfred Medeiros Jr. In 1970 he moved to Parker Ranch, where he served for fifteen years, first as a cowboy at Waiemi Station under foreman Jiro Yamaguchi and superintendent Henry Ah Sam. Johnny Jr. was already a skilled commercial truck driver, and when the opportunity arose he elected to change his assignment to operate one of Parker Ranch's semi trucks. He later left the ranch to work for Deluz Trucking, where he remains today. He continues to haul Parker Ranch cattle, as this function became outsourced.

Godfrey, the youngest son, became the scion of the Kainoa clan. He was popularly known as Jingles after his affinity for wearing the *kēpā pele* (jingle-bobbed spurs) as part of his childhood apparel. By the time he was nine years old, he was often seen at the Ho'olulu Park racetrack in Hilo, as the jockey for Kahuā Ranch's racing team, mounted on Johnny's fast bay Thoroughbred mare, Kawaihae. Godfrey participated in the sport of rodeo while working for Kahuā summers and weekends.

Fulfilling a lifetime dream of Johnny Sr. to acquire a Hawaiian Homestead ranch lease, all ten children worked alongside their parents in fencing the 300-acre lot they were awarded in Nienie above Ahualoa. Every spare moment and dollar was put into the Kainoa Ranch, which carried the joined JK brand acquired from elderly horseman John Kahaikupuna. During his formative years, the family ranch was quietly deeded to Jingles, who was already an established junior cowboy at Kahuā. Godfrey entered the Navy, but his father's death called him home. Through the kind intervention of Kahuā Ranch manager Monte Richards, Godfrey was given permanent release from service and began the long journey home for his father's funeral. Within weeks of his father's death in 1974, Godfrey chose Parker Ranch as his new home. This talented, charismatic leader was one of several employees who left Parker Ranch on February 28, 2002, as part of the early severance program, on the twenty-eighth anniversary of his Parker Ranch career. His departure marked the end of a remarkable ranch career and a tremendous loss for the ranch. He was only forty-seven years old at the time. As history moves in circles, Godfrey joyfully returned on September 9, 2002, to his *kulaīwi hānau* (birthplace)—Kahuā Ranch—as livestock manager.

Gabriel Malani

Gabriel John Kainoa Malani had an early cowboy career in Kawaihae Uka, but he found his bride in a native of Laupāhoehoe Point, Maraea Kāolowī, who had attended Kohala Seminary. Maraea whisked Gabriel home to her Laupāhoehoe *kuleana* near the point. Her nephew John Rickard, who chauffered for the Carter family, suggested Gabriel as a cowboy prospect to A. W. Carter. A. W. acted upon the suggestion, as

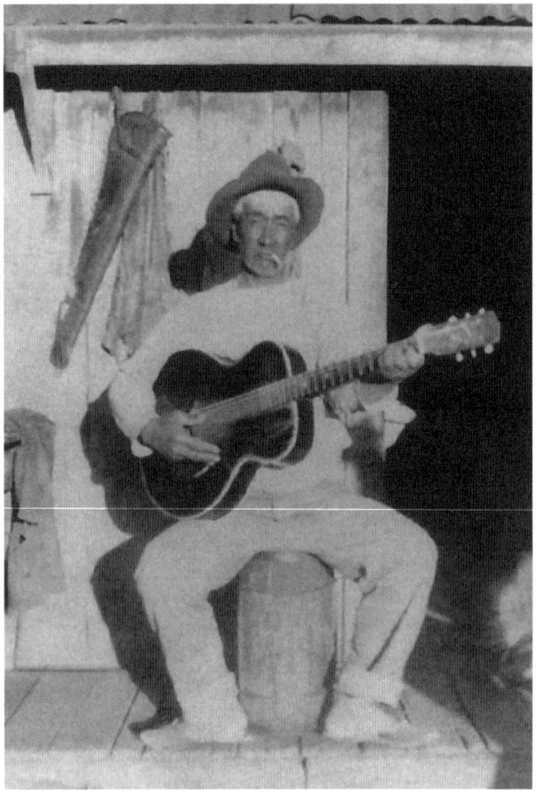

Gabriel Malani playing slack key guitar at sunset at Keanakolu House. Mona Malani Collection.

the fee simple ranch lands of Keanakolu extended *makai* to the canefields above Laupāhoehoe village. Gabriel's boss was Willie Kaniho.

Gabriel lived at the Keanakolu camp during the week, riding home to Laupāhoehoe Point on Fridays with a pack animal carrying the bounty of the mountain—pork and mutton. Sunday afternoon, his large family would send him on his way back into the mountain, bearing salted and dried fish, along with the produce of the family *kīhāpai,* in which taro, sweet potato, onions, and beans flourished.

My father, then a physician at Laupāhoehoe Hospital, described Gabriel as a quiet gentleman, not of great stature. Maraea, however, he described as a woman of large frame and strong mind who was musically talented. After delivering several of the Malani children at their home, my father, after much cajoling, talked Maraea into coming to Laupāhoehoe Hospital to have her next child. Nearing the birth of Baby Joe, with her water bag broken, she called on a crank telephone to alert my father that she was coming in. Expecting some family member to accompany her, my father recalled her pulling up alone on her noisy model T. Baby Joe was born within the hour in the sterile confines of the delivery room—a healthy and robust baby. It was within the next hour that my father, looking out the window of the maternity ward where a bed was being prepared for her, saw Maraea pulling out in her Model T with Baby Joe wrapped in swaddling blankets (*kīhei pili),* headed back down the winding road to Laupāhoehoe Point.

My father spoke of the deep-seated fear that native Hawaiians had for hospitals at that time. Given the rampant tuberculosis and leprosy (Hansen's Disease) that had spread throughout the Islands, a hospital stay usually was a death knell for them—or, in the case of leprosy, banishment to Kalaupapa on Moloka'i. His enduring effort to relieve these people of such fear was the basis for counseling Maraea to have her baby in the hospital instead of at home. He also showed compassion in understanding that Maraea's rush for home simply reflected the lingering fear among so many people of that time that any trip to the hospital was a final voyage. In her ensuing deliveries,

Maraea came to trust both my father and the system, and a lasting bond of friendship was shared between them.

None of the Malani children sought a career in livestock, but they are noted for their athletic prowess and careers in law enforcement. Laupāhoehoe remains the *kulaīwi hānau* (birthplace) of the *'ohana* Malani.

George Thomas William Kamukamu Bell Sr. (1847)

The middle brother of five siblings, George Bell Sr. had several children in a first marriage, including a daughter, Hattie. An Irish teamster named Tom Gallagher successfully courted Hattie, and they had two sons, Thomas and Walter, the latter of whom moved to Honolulu as a child. Thomas remained in Waimea and kept his mother's surname. Young Thomas' upbringing included working on Parker Ranch as a cowboy, where he became well known as a horseman. In fact, one of his horses was sold to an Argentine polo player in 1925 for the princely sum of $30,000. Not only was this a tribute to his ability as a trainer, but it was also a credit to A. W.'s horse-breeding master plan to supply the military and foreign markets. Unfortunately, Tom and A. W. had a parting of the ways and he went to work for W. H. Shipman, Ltd., as the foreman for Pu'u'ō'ō Ranch, adjacent to the Humu'ula Sheep Station. Tom Bell was often called Gallagher after his father, distinguishing him from several other Tom Bells that came from North and South Kohala. After Shipman, Tom moved on to Kāpāpala Ranch where he retired in 1979, taking his calico horse with him. He died a few years later at his Glenwood homestead, survived by his charming wife, Elizabeth Rickard Bell.

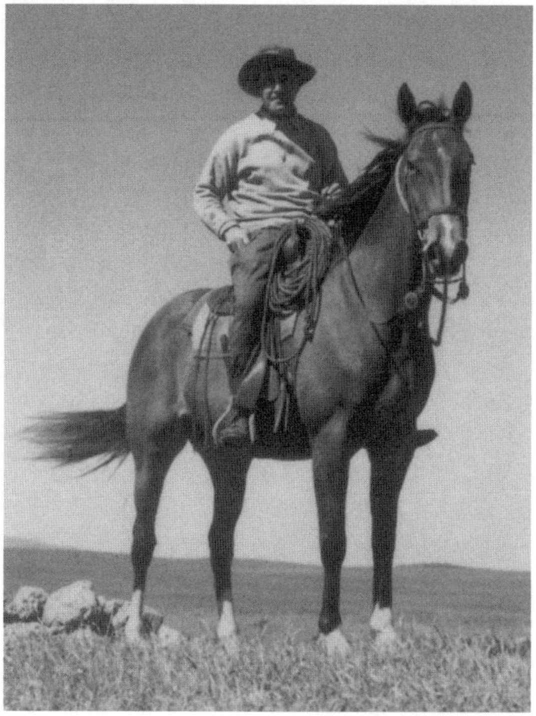

Alex Bell Sr. on horseback in Holoholokū, 1940. PPS Collection.

Besides Hattie, George T. W. K. Bell Sr. had a son, Prince Peter Pika Pua o Waiohinu ("flower vase of Waiohinu") Piko Bell, who married Waipi'o Valley native Rosalia Nāpalapalai. The marriage shifted this line of Bells to Waipi'o, where the couple's offspring were born and raised. One of the children, David Kepā Bell, became a longtime Parker Ranch employee. He married Mary Kailianu, daughter of teamster Robert Victor Kailianu, and they had six children. One of their daughters, Thelma, married Willie Kaniho Jr. (Wili Keiki). One of their sons, Klem Kaniho, had a son named Klem Jr., who continues the tradition of employment at Parker Ranch even as he continues the lineage of William Peter Bell of Scotland. Through his mother Janice, Klem Jr. carries Parker blood descending from Robert Parker, son of Kame'eiamoku Waipa and Mary Kaulalani Parker, only daughter of John Palmer Parker I.

George T. W. K. Bell later married Emma Angeline Lindsey, who bore a son, Alex Bell I, in 1896 in Waimea. His early years were spent at Mānā, partly because it was his father's work station and also because it was the home of his half siblings, including Hattie, mother of Tom Bell. When the Homestead Act came about, the Bells moved to their Pu'ukapu property behind Fukushima Store (currently a Chinese restaurant, Don's *Pake* Restaurant).

Alex was mechanically inclined from an early age, which the ranch recognized as an asset. After his education he became a truck driver for the ranch. As he matured, his responsibilities included chauffeuring for Hartwell Carter and Mrs. Knight (Aunt Tootsie), grandmother of Richard Smart. Alex spent a short period in the army during World War I. Returning to the ranch, his mechanical aptitude was put to good use in the Waiki'i farming department under the direction of Wilmot Vredenburg Sr. Taking care of the machinery that harvested corn and hay demanded the kind of mechanical competence that Alex possessed. For a brief time he returned to Waimea village doing freight drayage, so that his children, Alex Jr., Theodore, and Violet, could attend school. He returned to Waiki'i permanently in June 1934, where he eventually became superintendent, succeeding Donald MacAlister. Their family now included Stanley, Gilbert, and Dorothy. Alex was a quiet, private person but one who had a rather raucous laugh, a trait passed on to his son, Alex II. In addition to sharing his father's boisterous laugh, Alex II shared his love for equipment and trucking. He eventually became the trucking foreman for Parker Ranch. Alex II and his siblings, Theodore and Violet, were far more gregarious and outgoing than their father.

Alex II and Teddy both had colorful careers on Parker Ranch, although the younger brother's tenure was intermittent.

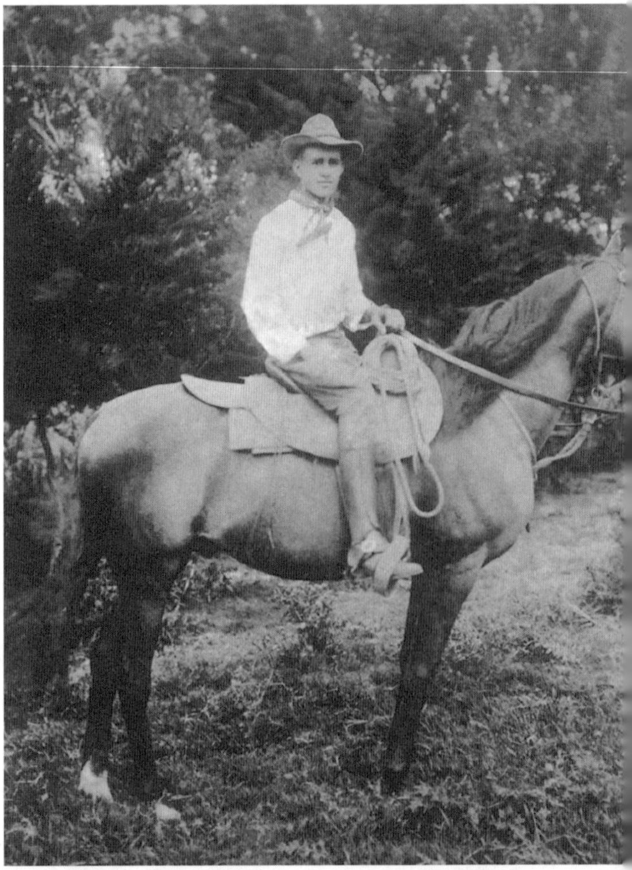

Tom Gallagher Bell on a polo mount he trained. It was sold to an Argentine polo player for $30,000 in 1929. W. C. Bergin Collection.

Teddy's first employment was at Parker Ranch in Waiki'i cutting corn suckers and later checking cattle and fences in Keamuku, but he was hired away by Pu'u'ō'ō Ranch, run by his cousin, Tom Bell. Wild cattle hunts were daily fare. After serving in the army from 1944 to 1947, Teddy returned to the ranch, where he joined the cowboy gang. In 1960, when Dick Penhallow became manager, Teddy was made horse foreman. Shortly after Penhallow's departure in 1962, Teddy left to work in the construction industry that was just beginning a long boom period. In 1972 Teddy returned to the cowboy life he loved, working for Dan Kaniho at Waiki'i Station until his retirement. Teddy died peacefully in July 2002.

One of Alex's sisters, Violet, carried the Bell line forward through her marriage to Tony Ramos and the birth of her sons, Tony Jr. and David. Both graduated from Kamehameha Schools, where Tony Jr. returned for a long career as one of the school administrators. David graduated from Seattle University in Washington and became a CPA in the large accounting firm of Haskins and Sells. Ralph Dobbins, who was eventually appointed Richard Smart's trustee, headed the firm. David frequently audited the Parker Ranch books, giving him an entrée to Parker Ranch, the ancestral home of his Bell relatives. David rose to become business manager of the ranch and eventually executive vice president of the Richard Smart Trust, a position he held through the January 14, 2002 separation activities. His decision to leave Parker Ranch as part of the voluntary separation package leaves Klem Kaniho Jr. as the single representative to carry on the Bell tradition of Parker Ranch employment. Klem Jr. also is the last of the Parker line on the ranch.

Charles Stevens Family

The Stevens family may be the most colorful of the Parker Ranch foundation families. This family line has a deeper, more spiritual connection through its maternal history's direct blood relationship to John Palmer Parker's wife, Chiefess Keli'ikipikāne.

Little is known about the early history of Charles Stevens, other than that he came to Hawai'i as an American whaler. During the winter months when, as now, whales migrated from the colder Alaskan environment to the warmer Pacific temperatures, the shores of the Big Island served whaling men as a disembarkation point and a place for whaling sprees. Kawaihae served as a convenient port for restoring the ships' pantries.

On one of these visits, Stevens rekindled the torch he carried for a native Hawaiian chiefess named Keali'iwahamana. There was a complication, however, as she was already married to Ka'ā'īpu'upu'u and had a daughter, Kiniwahamana. But Keali'iwahamana was so taken by Stevens and his romantic overtures that in 1848 she left to live with him.

Keali'iwahamana, which means "the royalty who speaks with spiritual force," was a daughter of Keali'i o Moana, chief of the Maui and Puna lines. Her mother was Na'ehuahiku, daughter of Lonomakua, a chief of the Big Island's ruling house of Keawe. Born in Kohala around 1830, she was raised in an era when Western ways were being imposed on Hawai'i. Christianity had a marked influence on her, despite her allegiance to the teaching of the

Genealogy chart of Charles Stevens

Jack Purdy

John Davis

Fannie Davis

Fannie Purdy (Kaleleoili) "Pane"

Kuahine Haa

Captain Charles Stevens

Kealiiwahamana

James "Kimokiwini" Umi-a-Liloa Stevens (Umaumakea)
Emma Kaoo m. William Jarrett (brother of Paul Lapaula Jarrett)

Harriet Lincoln (Kalokuokamaile)

Robert Naeahuau (b. 1875)
Katie Lahilahi (b. 1876)
James Naeahuau (b. 1877)
George Haa (b. 1881)
Fannie Kaleleoili (b. 1882)
John Keonaona (b. 1885)
Lily Kamakanoe (b. 1886)
Mattie Kalani (b. 1888)
Emily Kahaleuinihapa (b. 1889)
Julia Kaulawenaokalani (b. 1893)
Ned Stevens (b. 1895)
Sarah Kahailani (b. 1898)

Robert Naeahuau Jr.
Daisy Kanoehalii Stevens

Albert Uiha Lindsey

Miller Pilialoha Lindsey
Angeline Kauhiwai
Julia K.
Daisy Kamakokalani
Allen Naea Hartwell Lindsey

Joseph Pacheco

Betsy L. Mainaauao

Mary L. Pacheco

Aletha Leinani Lindsey

John "Makua" Stevens
David
William

Robert
Gilbert

Andrew Stevens m. Jennie DeSilva

Charles Stevens

Annie Lincoln

Lydia
Kay
Charles "Kale"
Walter
Aloha
Edna

2nd marriage

John "Makua" Stevens

Nellie
Hugh
Strickland
Winnie
Wanda

ancients. While she embraced the verses of the *Paipala Hemolele o Jesu Kristo* (Holy Bible of Jesus Christ), the *pono* of old Hawai'i still stirred in her heart and mind. While she enjoyed singing the Christian *himeni* (hymns), she quietly invoked chants of her *ali'i* lineage. Above all else, she was a good mother to her brood, including daughter Kiniwahamana, as well as another son and daughter she bore by Charles Stevens, the *selamoku* (sailor).

Their firstborn was a boy named James Umi-a-Liloa Stevens. His Hawaiian name designated his descent from famous kings of Hawai'i Island, Umi and his father Liloa. His English name came from James Russell, who managed the Līhu'e Sheep and Stock Company near Pu'u'ōpelu. James was born in 1853, and a sister, Emma Kao'o, followed within three years.

Despite their mother's objections, James and Emma were whisked away to O'ahu in their early years. There he was placed in a Roman Catholic seminary in 'Āhuimanu near the Ko'olau mountains, and a Mrs. O'Neal kept Emma. Keali'iwahamana kidnapped her son and brought him home to Mānā, where he was placed under the direct care of John Palmer Parker II and his wife, Hanai. Hanai took great care of James.

James grew up in Mānā until schooling called him to Kohala, where he attended Elias Bond Special School for Boys. Although Parker saw to it that he was well educated, it was ranch chores such as curing and stacking hides that developed a sound work ethic in the young man. He grew into a strapping young man more than six feet tall. Returning to the ranch, he spent most of his adult life as a cowhand. For a brief

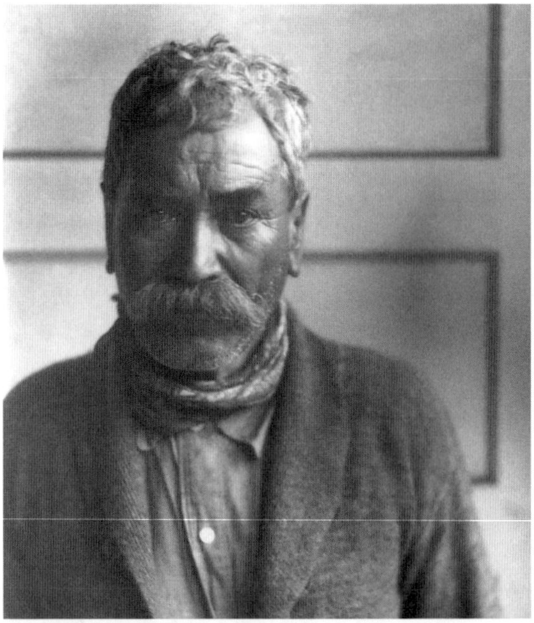

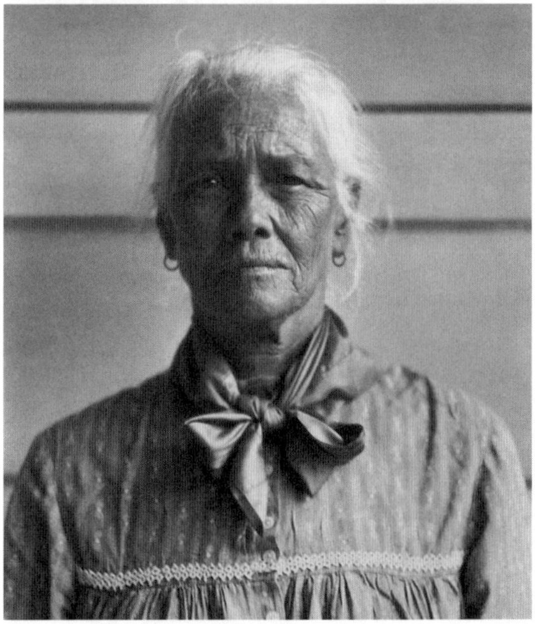

Top: *James Kimo Kiwini Stevens in 1920. Sullivan Collection. B. M. CN 4534.*
Above: *Fannie Pane Purdy Stevens, wife of James, a noted horsewoman, in 1920. Sullivan Collection, B. M. CN 4536.*

period, James worked at the Roy and Johnson Ranches in Kona.

In his early years on the ranch, James was mentored by Ulumaheihei Waipa, who groomed him to become an excellent cowboy and horseman. Waipa and Stevens had deep respect for one another, and before long James was promoted to head cowboy under his mentor. As a young adult he became enamored with a young horsewoman, Fannie Kaleleoili Purdy, daughter of Jack Purdy and Fannie Davis.

Fannie, also known as Pane, and James, who was commonly known as Kimo Kiwini, were married in 1874. Their love of horses reinforced by this union launched Pane on a victorious path as a jockey on the fine racing mounts that James trained. James was a skilled steer roper, placing third in a twenty-seven-man matched roping at a Hilo rodeo and placing well ahead of Ikua Purdy in 1908. This was quite a feat for anyone at the time.

Emma went on to marry William H. Jarrett, brother of Paul Lapa'ula Jarrett, who served as Parker Ranch manager in the late 1890s.

The Stevens couple raised a family of thirteen cowboys and cowgirls. Typical of the times, nearly all of the children were passed on to relatives who needed or wanted children to *lawe hānai* (raise as their own) or to elders to look after—*kahu hānai.* This practice of sharing one's children with others was not only acceptable but quite commonly done in the Hawaiian culture. Many of the children returned to their parents when they were old enough to be of help to them.

For the Stevens family, learning to ride was an early priority. This meant that they had to be able to learn to ride a horse, feed the pigs and chickens, and perhaps chop some firewood. Solely, her parents raised only one child—Lily Kamakanoe. Lily's children and grandchildren rose to prominence as leaders of Parker Ranch. This will be detailed in a later section. Their mother, Fannie, retained her striking beauty well into her senior years.

The eldest child, Robert Na'ea, was an accomplished violinist. He married Harriet Haleaka Kalokuokamaile Lincoln. Harriet brought substantial *ali'i* lineage to this union. Her mother was Luki Keawemauhili of royal linkage to Keawe. Her father was George Washington Lincoln, who was actually part Hawaiian despite his patriotic name. Harriet was a statuesque woman, measuring over six feet in height. Robert Na'ea was known for his quiet demeanor, though his Hawaiian name ironically means "violent temper." It was given to him by Queen Emma after her father, Chief Na'ea. A name thus given by an *ali'i* is considered *kapu* (sacred), and it was handed down through subsequent generations.

Robert and Harriet had only two children—Robert Na'ea II and Daisy Kanoehali'i. Robert Na'ea married Juliet Bell. This marriage produced six children, one of whom was Napua Stevens, who became a well-known musical personality. Napua was a shortened version of her christened name, Kawaiola Kao'ionapuapi'ilani, the former of which means "the perfumed water" and the latter "the finest flower in the bouquet of Pi'ilani."

Daisy K. Stevens, Robert Na'ea's sister, wed Albert Uiha Lindsey. One of their two sons, Allen Hartwell, was given the family name Na'ea. Like his forebears of that name, he is level headed with a balanced temperament.

The first daughter of James and Fannie Stevens was Katie Lahilahi (slender), who grew up to become a songwriter. One of her compositions was "Pua Kamaka Hala," reflecting her Waimea childhood memories. Of the Stevens girls, Katie was the most accomplished horsewoman. She started at the age of eight helping her father cowboy. This came with her riding *kaupu* style, placed in the saddle between her father and the *'ōkumu* (pommel). Seated in this position, she had an excellent vantage point to appreciate the serious skill of roping *'āhiu pipi*. When the bull was being tethered by her dismounted father, it was Katie that kept the horse reined back, keeping the bullock securely snubbed to the *'ōhi'a* tree.

For Katie, life in Waimea was not meant to be. On O'ahu she met and married a native Hawaiian who was also of royal lineage, James Ka'ihi'ihikapuokalani, which means "the sacred majestic *kapu* of the heavenly born." He was referred to as James Kaihi 'Ī'ī, the son of John Papa 'Ī'ī, a nobleman who advised both Kamehameha III and IV. James 'Ī'ī's grandfather was High Chief Papa, councilor to Kamehameha the Great, and who also served as a guardian of Liholiho (Kamehameha II). James 'Ī'ī was a skilled linotype operator and worked for the Advertising Publishing Company for forty years until his death in 1937. James 'Ī'ī also brought musical aptitude to this marriage with Katie Lahilahi. This trait proved very durable with the musical accomplishments of their children and grandchildren, such as Victoria Vicky 'Ī'ī Rodrigues and Nina Keali'iwahamana. Although Katie Lahilahi's musical compositions were never published and remained within family circles, the *paniolo* theme was woven through her lyrics.

Lily Kamakanoe, the third daughter of James and Fannie, was the prime maternal progenitor of the Stevens men, who were held in high esteem as cowboy leaders of Parker Ranch. Lily had only one son, who was named Charles after his great-grandfather, the whaler. A description of young Charles' father has never been provided, but he was likely of good height and appearance as his son was classically tall, fair, and handsome. Young Charles, however, was not inclined toward life as a cowboy, preferring mechanized agriculture instead. He was one of the ranch's first bulldozer operators, operating a D-2 vintage Caterpillar.

Charles married Annie Lincoln, and Waiki'i became their home, where he was foreman of the corn farms. Charles and Annie had six children: Lydia Kamakanoe, Katherine Kalei, Edna, and their first son, Charles Jr., were born while at this quaint village on Mauna Kea's western slope. Charles III was born in 1929 and Walter in 1931. Tragedy struck when their father contracted pulmonary tuberculosis and the family moved to Waimea, where Walter was born. It was here that their last daughter, who was aptly named Aloha, also was born. Charles Sr. required hospitalization in Hilo at the Pu'umaile Sanatorium, located beyond Keaukaha at Kings Landing. He died at Pu'umaile, leaving Annie with their six children.

It was at this point in the Stevens family history that John Makua Keonaona Stevens Jr. entered the scene. A first cousin to Charles Sr., Makua was the son of John Keonaona Stevens, immediate older brother of Lily Kamakanoe. Makua was already an established cowboy and a highly respected horseman when he wed the widow Annie Lincoln Stevens. He and Annie went on to have five more children: Wanda, Winifred, Nellie, Hugh, and Strickland, none of whom developed strong career ties to Parker Ranch, although it was considered their *'āina hānau* (birthplace).

Makua was quite rugged compared to his cousin, but he raised all eleven children as one family. For young Charles (Kale) and brother Walter (Wala), life at Waiki'i launched their careers on the ranch in cattle and horse operations, respectively. Kale took an early interest in cattle production while mentored by legendary Willie Kaniho. Walter was born to love horses, and from Makua he gained vital information and training, already possessing immense patience. Sister Aloha rivaled her brothers' strikingly handsome appearance. Her adult career was largely in the visitor industry section of Parker Ranch. Kale and Walter were career Parker Ranch cowboys, save for their years of military service.

Kale eventually became the foreman of the Waiki'i Station, and Walter went on to work with the purebred herd during the decade of the 1950s. Walter Stevens is considered by many old-timers to be the finest horseman the ranch produced in its 150-year history, and they often credit his horsemanship to his stepfather Makua.

Both Stevens brothers became superintendents under Gordon Lent's leadership team and continued to rise to any challenge. Sadly, the brothers died of cancer in 1994, four months to the day of one another—and both had recently retired. *Aloha no . . . aloha menemene.*

Makua had a brother, William, who also had descendants that were involved with Parker Ranch. William Stevens had sons Gilbert, also called Gunga, and Robert, who both spent time as cowboys in the Pu'uhue Section of Parker Ranch. Eventually they both opted for careers in the construction and hotel industries. Still *paniolo* at heart, they often camped with the cowboys during branding times at Humu'ula when their cousin Walter supervised this section in the 1980s. Gunga's son Andrew became a Parker Ranch cowboy in 1989, stationed at the Kohala Division under foreman Wallace Ching. Andrew, with his athletic nature and endowed with a love for horses, cattle, and music of old Hawai'i, is a natural cowboy.

Andrew married coworker Jennie DeSilva, daughter of Mabel Waltjen and Frank DeSilva of Hilo. Mabel's brother Fredrick Jr., a quiet, pleasant individual, worked at the Breaking Pen when I was at neighboring Pukalani Stables in 1958. Fredrick Sr. had these children in his marriage to a native Hawaiian woman of Kalapana, Meleana Makua, sister of Maka'ē, who was an early maternal benchmark of the Kainoa family of North and South Kohala. While Fred Sr.'s work was in the Humu'ula area with the Civilian Conservation Corps program of the 1930s, his father, Antone, a German immigrant, worked for the Haneberg brothers at Humu'ula Sheep Station in the 1880s as a ranch hand. Andrew and Jennie left in 1994 for other pursuits in the ranching industry.

Maternal Influences

The foundation families had peripheral relationships that were largely maternal in nature, such as the Lincoln and Spencer *wahine.* This is not to say that neither family produced significant men of Parker Ranch. Certainly Solomon and Bob Lincoln and Samuel and the Francis Spencers played respectable roles on the ranch.

But the Lincoln and Spencer *wahine* had a more enduring influence in the full span of ranch history. Annie Lincoln's marriages to Charles and later John Makua Stevens served to put that family name into nearly daily usage over the last seventy-five years.

Two Spencer girls, Martha Likolehua Zenobia and Frances Servia Eudora, were married to Dick Kawai and Archibald Cleghorn Ka'au'a, respectively.

There are other maternal influences that enhanced ranch production, history, and even inheritance, and some of these surfaced in the Parker Family itself. While John Palmer Parker II and Hanai lost their only son Samuel, their adoption of the four children of Parker's deceased brother, Ebenezer, preserved the Parker lineage, and—because of these maternal influences—it still exists in the new millennium. The sister of both Parker men, Mary Ann Kaulalani, in her brief marriage to George Moses Fuller and her subsequent wedding to Kame'e'iamoku Waipa, is especially significant.

Martha Kekapa'oka'ahumanu Fuller was the only product of Mary's first marriage. Martha's matriarchal influence expanded via two marriages—first to John Somes Low and later to a native Hawaiian, Ka'au'a Kekuawela. As was the case for many Hawaiian surnames during these times, they reversed names. Therefore, this line became known as Ka'au'a.

Martha's marriage to Low produced seven children, two of whom were noted figures in Hawaii's early cattle industry. Eben and Hannah were destined for historic significance—Eben for his colorful and multifaceted life story and Hannah for spawning a prominent line of ranchers in her marriage to Robert Hind. John Somes Low later died in the Nevada silver mining era.

The first child of the Low-Hind union was Robert Leighton Hind I, who brought Pu'uwa'awa'a Ranch forward as a major stake in the cattle industry of the territory from the early 1900s. The only child that came from his marriage to Marjorie Capps was R. L. Hind II, known as Bobby. As a young adult, Bobby courted the stunningly attractive daughter of the manager of neighboring Hu'ehu'e Ranch, Theodore Vredenburg, formerly of Parker Ranch. Her name was Florence, but she grew to adulthood with the nickname Coco. She and Bobby moved on to McCandless Ranch in South Kona, where he brought that outfit into modern management. Their eldest son, R. L. Robbie Hind III, grew up on McCandless, where he worked summers while attending Hawai'i Preparatory Academy in Waimea during the school year. After college Robbie returned to McCandless and worked his way up to the manager position. Richard Smart tapped Robbie for the Parker Ranch livestock manager position, which he accepted in 1984. Richard especially valued Robbie's work experience, coupled with the combined Parker and Vredenburg roots that firmly founded his character. He left the ranch with the Voluntary Separation Program on January 14, 2002, after eighteen years of service with the ranch. He stood tall as an eighth-generation Parker descendant.

Martha's second marriage with Ka'au'a produced a boy and a girl. The daughter, Stella, married James Hind, but none of their three children—Eva, Maud, or James—developed strong ties to Parker Ranch. James Hind was the brother of Robert Hind, who wed Hannah Low.

The son of the Ka'au'a union was Archibald Cleghorn, born on February 18, 1876. Archie grew up in Waimea and was attracted to a cowboy career, a line of work for which he possessed a natural talent, particularly in horsemanship and cattle work. A. W. made periodic references to Archie's ability and dependable nature. Archie had his share of horse accidents in the line of duty, but this never dampened his enthusiasm for the cowboy life. His third-place win in the 1908 Frontier Days Championship Rodeo is testimony to that.

Archie's marriage to Frances Eudora Servia Spencer produced five children: Archie Spencer, Edward Stanley (Blue), Jack Low, Samuel Parker, and Stella Ha'ilaua'ala. While none of these children became cowboys by trade, Archie and Stella enjoyed a significant relationship with Parker Ranch. Archie became a civil engineer and, during a long career with the Castle and Cook's Land Department, found time to become a pivotal player between the territorial government and several private industries, including ranching. He was also heavily involved as a land consultant for Parker Ranch, notably in the area of lease transactions. Archie's marriage to Pi'ilani Yates produced two sons who had a significant interest in the livestock industry. While neither brother worked directly for Parker Ranch, their interest in the outfit has endured over many years, especially in their desire to preserve ranch history. Archie, the eldest, has preserved his family's Parker heritage with bountiful albums of photos and documents, while Alan has a limitless mental capacity for ranch facts and figures.

Alan married my sister Marjorie (named after Marjorie Capps Hind), and their brief union produced a daughter and son. Kathleen Pi'ilani makes her home in Tacoma, Washington, and her brother, Chris, worked in the Mānā Division of Parker Ranch for several years before leaving the cowboy life for a better one. Chris is an eighth-generation Parker descendant.

Archie and Alan's aunt Stella, a tidy and attractive woman, married Albert Akana while living on O'ahu. They moved to Waimea in 1960, where Albert worked as project manager for the Department of Hawaiian Home Lands. Stella had a brief but meaningful career working directly for Richard Smart in the Parker Ranch office. She was instrumental in producing the *Paka Paniolo* circular, which began as a company newspaper and grew into a slick historic document, especially when read forty years later. Stella's family has a homestead ranch in Waimea in addition to the family-operated Akana Petroleum Company.

Mary Ann Parker's Second Marriage

It was with serious parental regret that Mary Ann Parker left Fuller, a successful Waimea merchant. But Kame'e'iamoku Waipa, tall, handsome, and of distinguished *ali'i* lineage, was the man Mary wanted as a lifetime companion. Their marriage resulted in eight children. The eldest, Ulumaheihei, rose to become the first head cowboy on the ranch and is remembered in A. W. Carter's memoirs. He left the ranch just

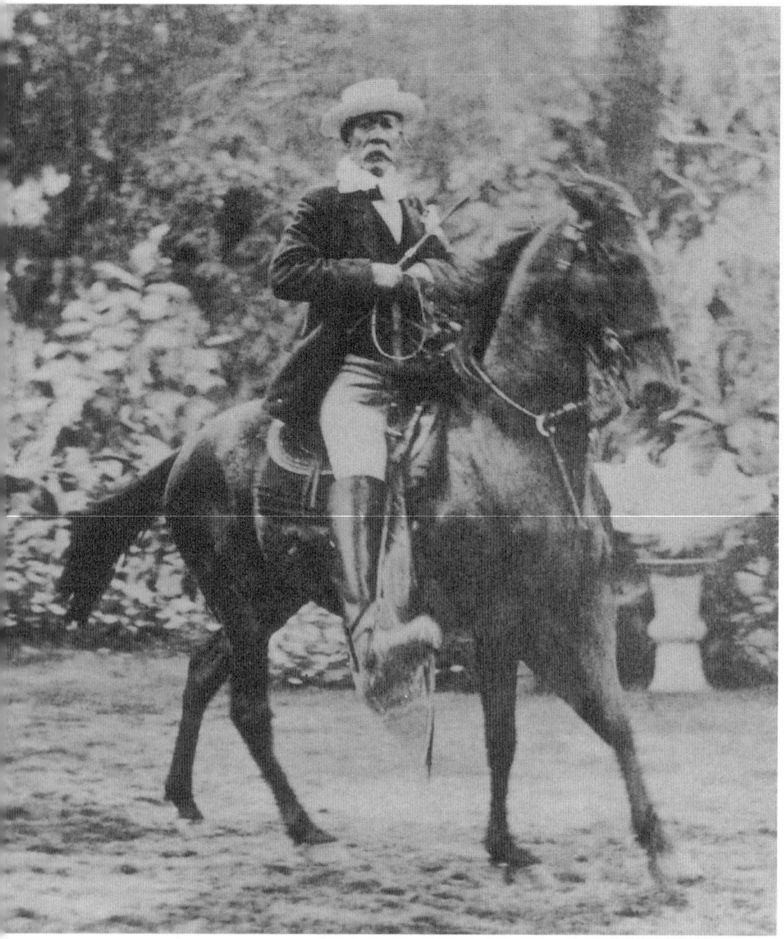

William Ulumaheihei Waipa, one of Parker's early and prominent cowboy foremen, a paniolo nui. *He left the ranch to become a sheriff in Honolulu and was killed in a horse fall not long after this photo was taken. Alan Ka'a'ua Collection.*

prior to the overthrow of the kingdom in 1893. On O'ahu, he became a sheriff and was actively serving in that capacity at the time of the overthrow.

Ulumaheihei's sister, Anna K. P. Waipa, married a son of Jack Purdy's, William Harry Naleilehua, who sired a large clan of prominent Purdys.

The W. W. Jack (Harry) Purdy Family

There are many opinions as to what was Jack Purdy's real name. The initials have been represented both as William Warren and William Wallace. Lloyd Ching, a Purdy descendant and avid genealogist of North and South Kohala families, opines that the initials represent Jack's true first and last names—William Wallace. Ching further offers that the surname "Purdy" was borrowed from a shipmate as they journeyed from Botany Bay, Australia, to Hawai'i. The real Purdy reportedly went on to the Northwest after a short stay in the Islands. There he married an Indian princess and is said to have started a genuine clan of Purdys that stemmed partly from Native American stock.

For the purposes of this book, I accept Jack Purdy as the name he preferred and the role he played: an early pioneer in Hawai'i's rich ranching history. Purdy was not alone in coming to Hawai'i with an imperfect past. Many of his peers arrived in Hawai'i with similar imperfections to bear. Like many who left their motherlands, it was the desire for freedom and opportunity that brought them to these shores.

Jack's childhood was spent in Londonderry, Northern Ireland. As early as age ten, he boarded a sailing ship as a cabin boy. It is not clear whether this was due to preadolescent problems with the law or, as with many of his peers at that time, to escape British tyranny and the political unrest that often resulted in the jailing or deportation of a dissident Irishman.

Some writings glorify Purdy as an adventurous whaler who jumped ship in Kawaihae. But the existence and use of so many aliases support the more broadly accepted theory that he was an exiled fugitive of a penal colony—likely Botany Bay.

Purdy's early career in the Waimea area paralleled that of John Palmer Parker. Both became major forces in the depopulation of wild cattle, frequently working in tandem. On several occasions they both worked for the Waimea Grazing and Agricultural Company, headed by W. H. Green. While Parker developed a businesslike approach to the opportunities in the fledgling livestock industry, Purdy took pride in his role as a skilled, aggressive, and intense bullock hunter. Purdy was a rowdy living the rugged life typical of his peers in the early American West. Jack also served as the patriarch of a large, varied clan of offspring via a series of marriages.

Jack Purdy's First Marriage

During his colorful career, considerable history was made by the progeny of Purdy. While not all four unions were formal marital relationships, each produced offspring.

In 1832 Purdy took his first wife, who was of royal lineage. Keawe-ma'u-hili was the daughter of Keawe-a-heulu and Ka'akau. A year later, a daughter—Hannah—was born. It is of coincidental significance that Hannah chose to marry Alfred Wellington Carter in 1861, a man eight years her junior. Carter was the son of Captain Joseph Oliver Carter, grandfather of Parker Ranch's A. W. Carter. Hannah's husband was born at sea, off the shores of Tahiti in 1841 aboard the ship *Caliope,* but he grew up on O'ahu. Hannah and Carter had eight children, none of whom developed ties with the ranch.

In 1838 Jack and Keawe-ma'u-hili had a son, William Harry Naleilehua Purdy, who was born at the Po'o Kanaka homestead near Mānā. He grew to become a respected *paniolo* as well as the father of notable cowboys from his two marriages.

W. H. N. Purdy First Weds Rose Kaleimoku Kamaile

Purdy and Rose Kamaile were married in 1858. The marriage produced a single son, William Naleilehua, who grew up to be a cowboy working for Sam Parker at Humu-'ula Sheep Station Company before entering a political career. He became a county supervisor for the Hāmākua District of the Big Island. William Naleilehua sired a large family in his marriage to Hattie Kaleopa'a Lakananui Kaho'owaiwai. One of their daughters, Virginia Luciana (Lukiana) Purdy, married into the Lindsey family when she wed Samuel Frederick Ni'a.

In addition to Virginia, a second child developed a noteworthy role on the ranch—John Kaukokalani Purdy. Nicknamed JP, he was born in Hāmākua in 1891 and soon developed an interest in horses. By age fifteen he was cowboying in the First Gate Section of Parker Ranch for a salary of $15 per month. He was then called to the Pā'auhau Section of the ranch to work under the tutelage of Charles N. K. "Kale Kuli I" Lindsey, where his natural skill with horses and mules was further refined. This led him to an assignment back at the ranch headquarters as a trainer of draft horses and mules at the Pieper Stables, aptly named after German teamster Henry Pieper. JP's gentle touch and enduring patience coupled with a progressive attitude seemed to seal his fate and career, but he was not ready to settle quite yet.

JP took a three-year government assign-

Genealogy chart of Jack Purdy

Jack Purdy
aka: Harry Hale

- 1st — Keawemauhili (m. 1832) →
  - Hannah Purdy (b. 1833) } (m. 1861) Alfred W. Carter*
  - **William Harry Naleileihua Purdy** — 2 marriages
    - 1st Rose Kamaile →
      - **William Naleilehua Purdy** } Hattie Kahoowaiwai
    - 2nd Anna Kamakaila Parker Waipa →
      - **Samuel Paukeaho Parker Pur**
      - Mary K. Purdy
      - **Ikua Purdy**
      - Mary Grace "Keoho" 2 marri
      - William Robert
      - Kaniu
      - **Harry William Kane Kawaiol**
      - **Robert Keala**
      - Harriet-Hattie
  - **George Purdy**
- 2nd — Keoho Makaopiopio** (m. 1848) →
  - James Naea m. Kalilikane Kamahana → Eliza "Laika" Purdy m. **John K. "Keoni Liilii" Lindsey**
  - **Samuel Purdy**
- 3rd — Fannie K. Davis (m. 1853) →
  - Mary Purdy m. Alex Pahao
  - **John** (b. 1856) m. Hoohuli Pa →
    - Kini Purdy
    - **John Purdy** m. Annie Ikaia (1st marriage) (m. 1909)
    - Kalama
    - Lucy
  - Fannie "Pane" Kaleleoili Purdy m. **James Stevens** "Kimo Kiwini", "Umi-a-Liloa"
- 4th — Hoka (m. 1873 – no children)

Jack Purdy Last Marriage

- Annie Purdy
- Esther
- Virginia Lukinia
- **John Kaukokalani (JP)**
- Hattie
- William Naleilehua Jr.
- Benjamin
- Jesse
- James
- Maggie

Annie Ikaia (her 2nd marriage) →
- Hannah Purdy m. **John Lekelesa** (Samoa)
- Mary m. John Waihee Sr.
- Annie m. **Alex Bell Sr.**

William J. Kawai →
- Miriam
- **Josiah Kawai** m. Hiwa Lindsey
- **William Kawai "Willie"**

Margaret Napuupahee Peneamina →
- William Lumaheihei Purdy (b. 1901)
- George "Kauhi" (b. 1904)
- Hattie Leilehua (b. 1905)
- Nellie (b. 1906)
- **Harry "Kahuku"** (b. 1907)
- Wallace Halulu (b. 1909)
- Margaret (b. 1910) m. Louis Hapakuka - Ulupalakua
- George Mauae (b. 1911)
- Daniel Purdy (b. 1916) m. Nancy DeLima - Ulupalakua
- Cecilia Kalili (b. 1920) m. James Hapakuka - Ulupalakua
- **Martin Ikua** (b. 1924) m. Frances DeLima then Doris Nolan
- Annie Hiiaka (b. 1926) m. Walter Ritte

1st John K. "Poko" Lindsey

2nd **James Kaaluea**

→ **George Purdy** (b. 1918) + Annie Kawai → **George Purdy Jr.**

*Cousin of Parker Ranch guardian A. W. Carter

** Keoho appears on the genealogy of Martin Ikua Purdy as his great-grandmother

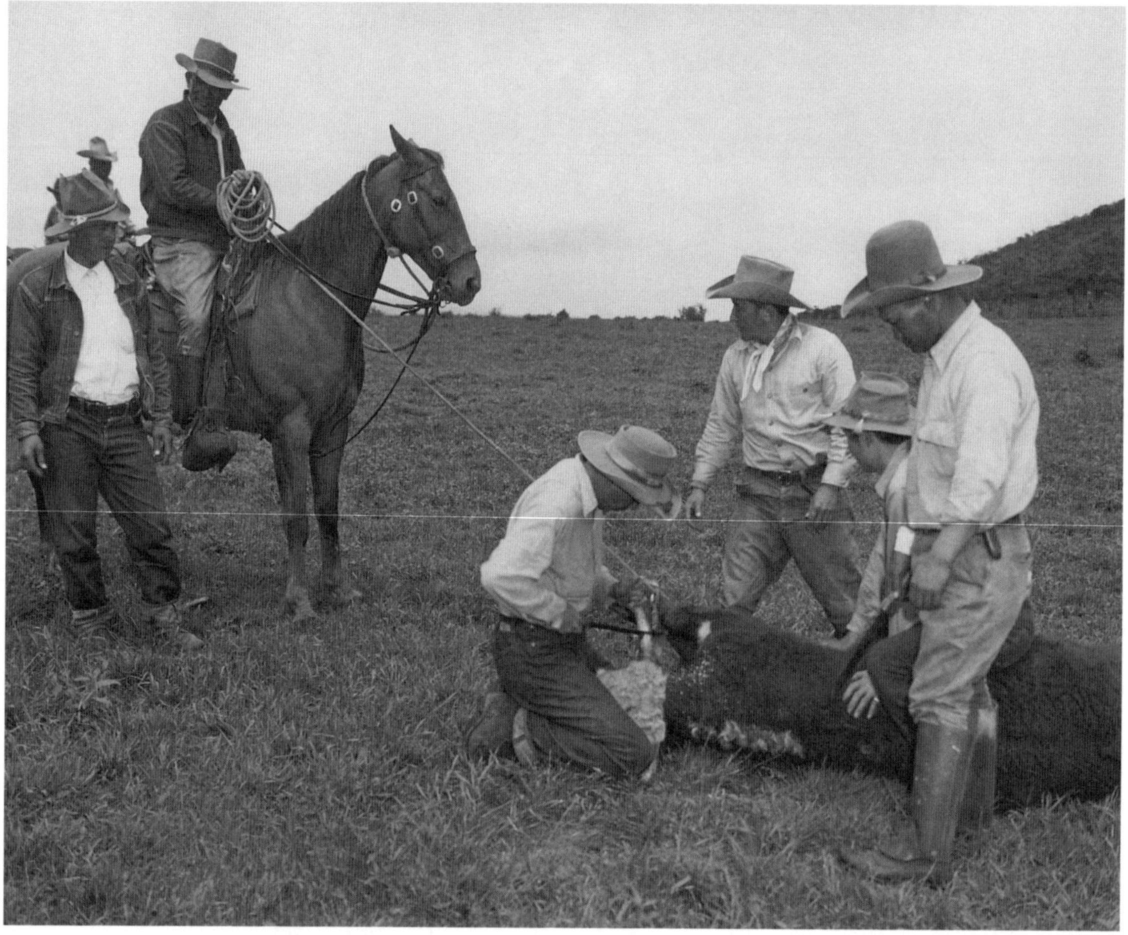

John J. P. Purdy, mounted at Pu'umahoelua Corral in Waiki'i as the cowboys dehorn a hao komo *cow. Mounted behind JP is Kaliko Ma'ina'aupō. Standing next to the horse is James Merseberg; Shiro Fujitani wields the horn saw; Bob Sakado is looking at the horse; Jiro Yamaguchi is holding down the cow; and Willie Palaika (Brighter) is observing. Hartwell Carter Collection.*

ment with the U.S. Geological Survey team exploring Mauna Kea and Mauna Loa. While he relished the high salary ($126 a month), JP's heart was with the ranch, to which he returned. Henry Pieper was in his twilight years, and the training stables were under the direction of George Miller Seymour Lindsey. George gave JP considerable latitude in spending spare time helping the cowboy gang, while he groomed him to eventually become his successor.

Again JP needed to move on, joining the U.S. Army in 1917 shortly after marrying Susanna Kukana Lujan, an attractive Hispanic-Hawaiian woman of royal lineage. Before leaving the service to return to the stables, he had been responsible for the quartermaster stables at Schofield Barracks.

Returning to the stables in 1919, JP quickly moved up to assistant foreman. The unfortunate and early accidental death of George M. S. Lindsey brought JP forward as foreman—a task for which he was well groomed. JP's enduring presence and reputation as a teamster/trainer soon led to his workplace being christened Purdy Stable. As mechanization decreased reliance on draft animals, Purdy Stable was converted to the ranch saddlery, with Sam "Puna" Kauwe and his nephew John Kauwe as the craftsmen. JP moved on to a variety of assignments, including the cowboy gang and *nānā'āina* work.

JP was a jovial person who enjoyed music and ballroom dancing. Up to his final days, he never lost the *kolohe* (mischievous) twinkle in his eyes. This twinkle was passed on to his vivacious daughter Kukana. JP retired in 1953 after accumulating forty-three years of service with the ranch.

W. H. N. Purdy's Second Marriage: Anna Kamakaila Parker Waipa

A widower in 1866, W. H. N. Purdy married into the Parker dynasty within a few years. His bride was Anna Kamakaila Parker Waipa, granddaughter of John Palmer Parker I. Together they had nine children, many of whom developed strong ties with Parker Ranch and Hawai'i's ranching industry. Samuel Paukeaho Parker Purdy, their first child, was born in 1871.

Sam Purdy became a prominent *paniolo,* particularly as a roper in both the arena and on the range. Sam's name appeared on the rosters of matched ropings at Ho'olulu Park in Hilo in 1907 and at Parker Ranch between 1912 and 1915. In his later years, Sam was known throughout the territory for his fine skin ropes *(kaula 'ili).* A. W. Carter thought highly of Sam for his honest and diligent service to the ranch. Leading up to his death in 1936, A. W. was frequently at his bedside.

Sam married Annie Keali'iha'aheo Ikaia (her second marriage) and in the span of their union produced four daughters: Hannah Heanu (1916), Mary P. (1918), Grace (died in infancy), and Annie Kamaka'aimoku (1924). Hannah married John (Samoa) Lekelesa, a tall ranch cowboy who eventually rose to become long-standing foreman of the cowboy gang. Mary P. married John Waihe'e, whose son John later became governor of the State of Hawai'i. Annie married Alex Bell Jr., and he also became a ranch foreman, supervising the trucking department.

In 1872, Sam's younger sister Mary K. Purdy married into the small but respected Kawai family of Waimea. William J. Kawai and Mary K. Purdy had two girls and a boy. Their son William earned a long-standing place in ranch cowboy history. His lifetime nickname was Willie.

Born in 1909, Willie was fortunate to grow up around people who saw promise in the wiry youngster. Especially observant of Willie's cowboy skills was Keoni Li'ili'i Lindsey. The two even looked alike. Willie Kaniho, another great cowboy, also watched young Kawai mature into a junior cowboy. By age fifteen, Willie was already on the payroll, covering the countryside gathering and branding cattle. Both A. W. and Hartwell saw promise in the young man. While their expectations of eventually moving him into a leadership role were never realized, Willie did excel as a cowboy, based on his natural instinct for handling cattle with

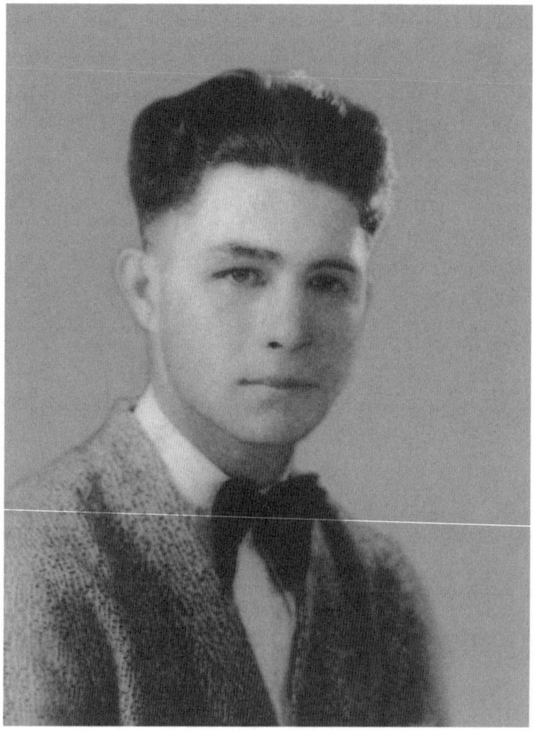

Willie Kawai. Williams Studio Portrait, 1938.

the least amount of excitement or unnecessary noise or activity. This quiet manner of working cattle was the Carter way of handling livestock.

Willie was quite a teaser, and his sense of humor was pervasive. He was of slight stature like Keoni Li'ili'i but more gregarious in nature. He had a few unconventional traits, one of which was the casual maintenance of his tack. His saddle always appeared to be starving for oil, even to the point of the tailpiece curling upward. Another practice frustrated his coworkers at branding corrals. Routinely, the roper faces his horse to the calf while branding is done to give the rider full vision of the process while his rope tightly secures the hind legs. Willie, on the other hand, routinely faced the *'ōkole* (rear end) of the horse to the calf, putting the ground crew at risk of a horse kick. After half a century of doing it his way, Uncle Willie retired from the Waiki'i Ranch Section of the ranch, where he spent his twilight years under the wing of Dan Kaniho. Willie's departure ended the Mary K. Purdy line of this *'ohana.*

Ikua Purdy

Ikua, third child of W. H. N. and Anna K. P. Waipa, brought fame to Hawai'i from around the world. Born on Christmas Eve in 1873 at Mānā, Ikua was destined to become the most celebrated cowboy Hawai'i has ever produced. While he is most remembered for his territorial and world titles as a steer roper, Ikua's performance as a ranch cowboy and foreman are discrete but equally important contributions he made.

It is understandable that most of Hawai'i associates Ikua with 'Ulupalakua Ranch on Maui, where he spent his later adult career. It is important to note, however, that Ikua played a strong role in Parker Ranch leadership leading up to his moving on. His role on the ranch had several aspects, the most profound of which was the mentoring of young and potential leaders of the sprawling operation. He also demonstrated deep loyalty to 'Āina Paka.

Willie Kaniho Sr., cowboy, foreman, and charismatic leader of great men, worked as a young cowboy alongside Ikua. In Willie's memoirs, Purdy is noted as having the most profound impact on molding his career. These were the early years of the twentieth century, when Humu'ula was still considered the outback and patches of wild cattle still plagued the countryside.

Whether it was dealing with sheep or cattle, domestic or wild, Willie paid close attention to Ikua's natural ability to gather and secure livestock with the least amount of stress on man and animal. As a roper, Purdy was never witnessed to miss a loop—a compliment rarely expressed between knowing cattlemen.

Ikua's deep loyalty to Parker Ranch was appreciated by A. W. in the midst of a hostile coup attempted by Eben Low and others. Carter had ridden early one morning to meet the cowboys driving the herds of the Pā'auhau Section. Addressing the men as to the risks involved in the attempted takeover of the Ranch, A. W. said that he would understand if some of the men chose to go with Low, who was related to many of the cowboys, including Ikua. Ikua was the first to step forward to state his enduring allegiance and loyalty to Parker Ranch and Carter's leadership. To the man, the rest of the cowboy gang followed suit. Carter never forgot this moment of truth and loyalty demonstrated by the men—and most especially, Ikua's forthright posture.

Ikua began his cowboy career for the Parker family at Mānā. He and his family moved to Keanakolu under the ownership of Sam Parker Sr. prior to its acquisition as part of the Humu'ula purchase in 1914. Ikua married Margaret Keala Manu Napu'u Pahe'e in 1900. The name Peneamina, from her father's first name, Benjamin, was often included as part of her name. She was commonly called Keala Amina. They started their family in 1901 with the birth of William Ulumaheihei, followed by George Kauhi'aimoku'okama (1904), Hattie Leialoha (1905), Nellie (1906), Harry Kahuku (1907), Wallace Halulu (1909), Margaret (1910), George Mauae (1911), Daniel (1916), Martin Ikua (1918), Cecilia Kalili (1920), and Nancy Hi'iaka (1926). The average interval between children was less than two years.

His fifth and sixth children, Wallace Halulu and Nellie, were born at Keanakolu. By the time of the Cheyenne Frontier Days Rodeo in 1908, Ikua and his family had moved to Humu'ula and they remained there through the 1914 acquisition by A. W. Carter. Ikua was now employed by Parker Ranch as head cowboy at Humu'ula under Ernest Haineki Campbell. One of Ikua's protégé's, the young Willie Kaniho, joined Ikua and a third ranch cowboy, Ka'inapau Ka'ilikini, in a move to Princeville Ranch in Hanalei on the island of Kaua'i. Two of Ikua's children, Martin Ikua and Cecilia Kalili, were born there. Their wages increased from $12 per month to $75 per month, which was quite an enticement. After a few years of Hanalei's rainy weather, Ikua was offered a foreman's position at Frank Woods' ranch at Kahuā, but a fine offer from F. F. Baldwin for a better position at 'Ulupalakua drew him to the Valley Isle of Maui.

Ikua's children all developed strong ties to the ranching industry. Aside from the youngest son, Martin Ikua, who spent most of his career with Parker Ranch, all of the children were associated with 'Ulupalakua Ranch, either through marriage or employment. All of the boys began as *paniolo* on the Maui outfit, with Daniel remaining through retirement. Of the daughters, Margaret married Louis Hapakuka and Cecilia Kalili wed his brother James—both 'Ulupalakua *paniolo.* Daniel and Martin married sisters Nancy and Frances DeLima,

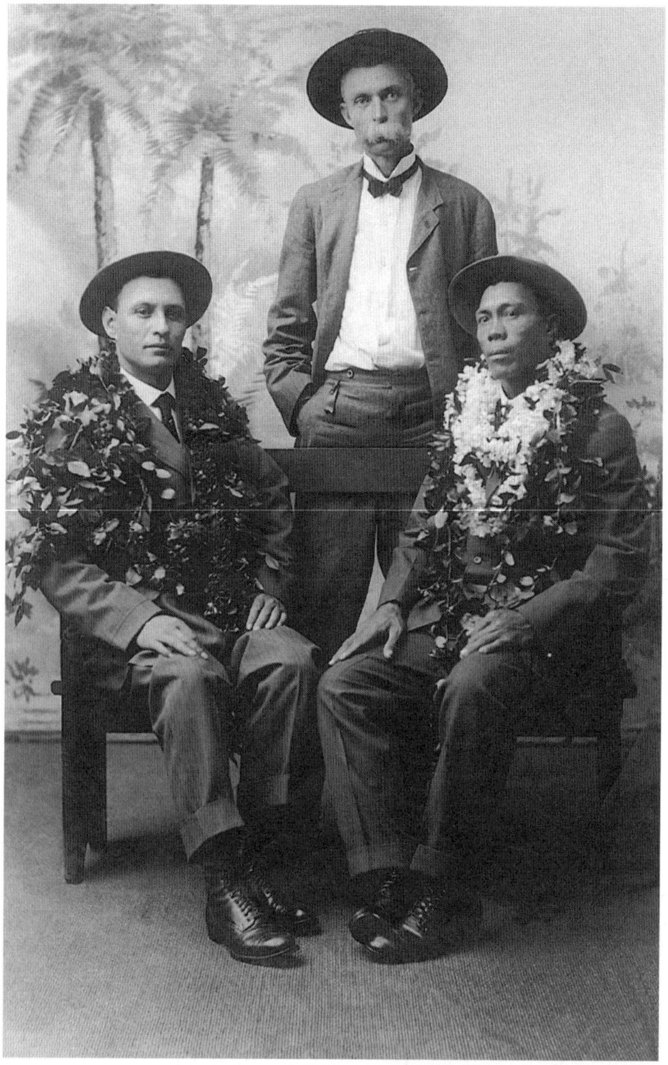

Ikua Purdy (right) seated with Archie Ka'au'a during victory celebrations on returning from Cheyenne, Wyoming, in 1908. Eben Low standing. Alan Ka'a'ua Collection.

respectively. They are aunts of well-known Hawai'i comedian Frank DeLima.

Martin Ikua of Parker Ranch was born in Hanalei, Kauai, in 1918. By the time he was ready to enter school his family had moved to Maui, where he attended Pu'unēnē School and graduated from the eighth grade. By age seventeen he was already working on Parker Ranch as a junior cowboy.

Martin married Frances Kukula'i DeLima, the daughter of John and Frances Ulualoha Wilcox DeLima. They had four children, none of whom developed ties to the ranch. As a widower, Martin married Doris Nolan in 1994.

In Waimea, Martin is commonly called by his Hawaiian middle name, Ikua. Many older folks are unaware of Martin being his first name. When I worked on the ranch as a trainee in 1958, Ikua was with the cowboy gang. When I returned as the ranch veterinarian in 1970, Ikua was still working as a cowboy and had aged very little. Not tall, but of moderately stout physique, even through his twilight years Ikua often outpaced younger cowboys in flanking (knocking down) the calves at branding.

Although never the roper that his father was, Ikua matched his father's athleticism in typical sports like baseball. This prepotency in athletic ability was passed on to Martin's children and grandchildren in sports like football and track. Young Ikua coached the Waimea Sons of Hawai'i basketball team and the Waimea Rangers baseball team. He retired from the ranch in 1980 and has enjoyed senior citizenry with Doris, performing at community activities and traveling frequently. Martin's retirement ended the Purdy lineage on Parker Ranch.

Today, as testimony to his contribution to Hawai'i's *paniolo* heritage and culture, a heroic bronze of Ikua Purdy, commemorating his world championship, has been erected at the Parker Ranch Center in Waimea.

Remaining Children of W. H. N. Purdy and Annie K. P. Parker

Ikua's younger sister immediately following him was Mary Grace Keli'iheleohu "Keohu," who married Keoni Li'ili'i and later James Ka'aluaea Lindsey. Her offspring are detailed under the section on the Lindsey family.

Keohu's younger sister Kaniu married a Parker employee, Moses Kōki of the hotel family. Their daughter Laulani married the well-known and respected Robert Kihoi, a teacher at Waimea School. A Kihoi grandson, Kevin Kaiwi Kihoi, served as a noted ranch cowboy at Humu'ula in the mid-1980s.

Ikua's brother Harry William Kanekawaiola Purdy married into the Lindsey family when he wed Anna (Emma) Kalei, daughter of James F. Lindsey I in 1880. Anna and Harry had a family of eleven children, many of whom died in infancy. One son, however, Harry W. K. Purdy Jr., became a Parker Ranch roughrider before moving on to a cowboy's life on Moloka'i.

The remaining children of W. H. N. Purdy and Annie K. P.—namely Robert Keala and Harriet—did not develop significant relationships with the ranch.

Jack Purdy's third child with Keawema'uhili was named George, born only a year before Purdy entered into his second marriage. Baby George was born in 1847, the year Parker Ranch was launched. Young George grew up more interested in construction than in livestock, but he served the ranch well, earning the nickname Keokihana because of his diligence at work. He often worked alone, even in remote reaches of the ranch. On corral-building assignments, Keokihana camped on-site until project completion. Upon his return to headquarters, A. W. Carter would be summoned for final inspection. In 1876 he married Grace Wahineikeili, but none of their children developed any relationship with the ranch. Keokihana was the last child from the union of Jack Purdy and Keawema'uhili.

Jack Purdy's Second Marriage: Keoho Makaopioio

Jack and Keoho were married in 1848 when she was twenty and he was a robust forty-eight-year-old cowboy. Their first child, James Na'ea, was born a year later. James Na'ea Purdy entered ranch employment briefly before marrying Kalilikane Kamahana and making Pu'uanahulu their home. Their only child, Eliza Laika, was born there in 1896. At age twenty-three, Eliza became the second wife of Keoni Li'ili'i Lindsey. The large progeny that resulted from this marriage is detailed in the section on the Thomas Weston Lindsey family in this chapter.

James Na'ea had only a brother, Samuel, and little is known of his relationship with the ranch. He died in 1906.

Jack Purdy's Third Marriage: Fannie K. Davis

Five years after his marriage to Keoho, Jack took up with Fannie Davis, daughter of John Davis and High Chiefess Kuahine Ha'a. Fannie was a striking twenty-seven-year-old part Hawaiian of royal lineage and Jack, at age 53, was still raring to go. Their firstborn, Mary, in 1855, did not develop any ties to the ranch, unlike her younger sister Fannie Pane Kaleleoili, who married James Stevens. Their detailed lineage is found under the discussion of the Steven's family. Only one son, John, came from this union with Fannie Davis. As an adult, John married Ho'ohuli Pa and they produced four children: Kini, John Jr., Lucy, and Kalama. One of the sons, John Purdy Jr., reinforced ties with the ranch beyond employment; he married Anna Keli'iha'aheo Ikaia (her first marriage) and produced three sons. One of these sons, George, was destined for prominence in Parker Ranch *paniolo* history. He was born in 1913. In his early years of parenthood, John Purdy died and Annie Ikaia married his cousin, Sam Paukeaho Parker Purdy. Details on their offspring are provided under the union of William Harry Naleilehua and Anna K. P. Waipa.

George was a credit to his father, John, while raised at Mānā and schooled at Kukuihaele. Going home for the weekend meant a 10-mile walk to Mānā, a challenge for the good student that he was. His family moved from Mānā to Pu'ukikoni in 1920 when the new dairy was built.

At age fifteen, George entered the ranch rolls picking corn at Waiki'i. He was assigned to the Plaster Gang, which got its name from the abrasive nature of hand-picking and shucking corn that often required taping workers' hands to protect them. Despite its unpleasantness, picking corn was full-time employment that provided meat, milk, poi, and housing—help sorely needed by his family.

The Plaster Gang was later more aptly named the Utility Gang, as the work often included fencing, pipeline installation, grass planting, and weed control. After three years of exemplary duty, George was handpicked by A. W. to go to the Breaking Pen as a roughrider under the tutelage of James Palaika, foreman. George developed great admiration for his boss Palaika, often describing him as a great horseman and cowboy.

George's skills as a horseman were naturally acquired, but Palaika's mentoring fine-tuned these abilities, especially reinforcing patience as the key to successful horse training. George was called to serve in the elite ranks of the cowboy gang at the very first available opening.

In the cowboy gang, George's abilities quickly placed him in key roles under the foremanship of Willie Kaniho. In moving large herds, he held the key role of *alaka'i* (lead) rider, whose job is to draw the cattle to follow his lead. Early on, George's cattle-handling skills earned him a place as one of the *hō'au* ropers for swimming cattle through the surf to offshore boats.

George never shied from the more

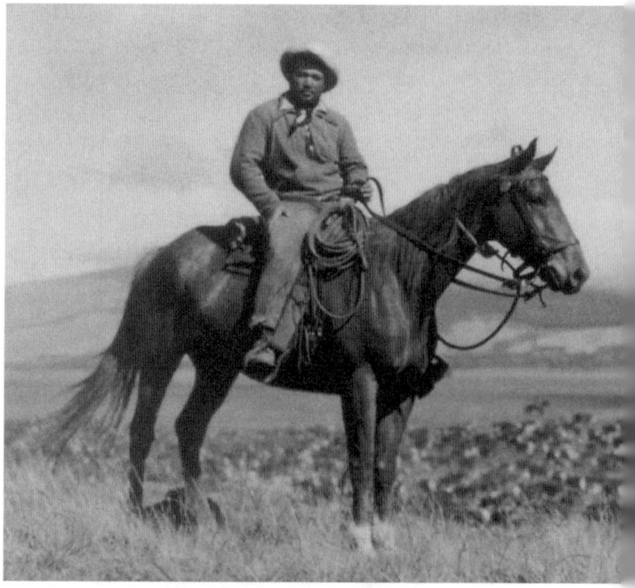

George Purdy on horseback in Holoholokū, 1939. A. W. Carter Collection.

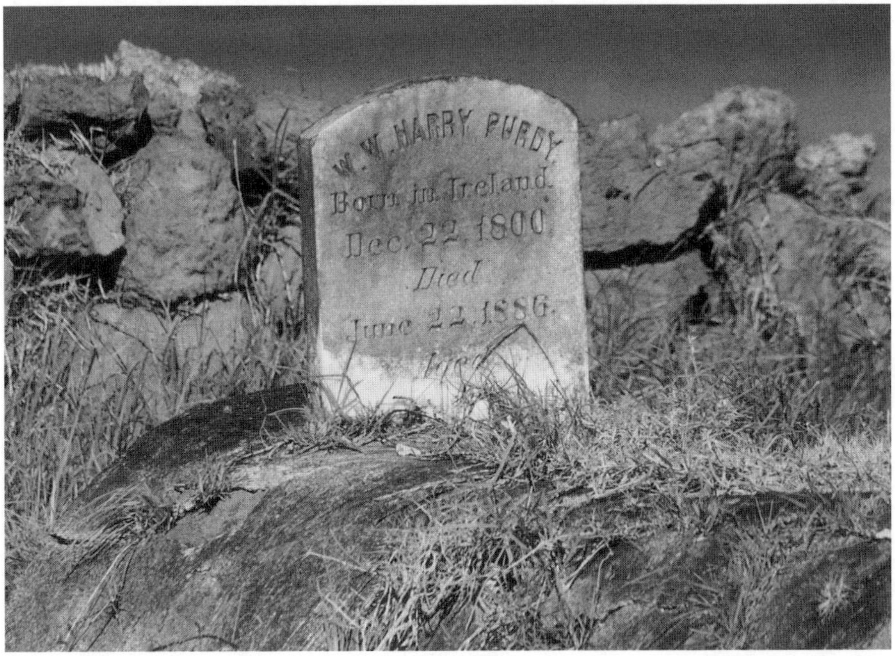

Jack Purdy's headstone, Parker Cemetery, Mānā. Iwalani Ka'a'ua Collection.

mundane tasks, willingly camping at Humu'ula during shearing season. In 1945 he returned to the Breaking Pen as its foreman, remaining there for twenty years. Many a young cowboy took his boot camp at the Breaking Pen under George's wing. His former charges consistently praise his ability as a coach and horseman. In 1965 he was transferred to the Racetrack Stables, where he halter-broke colts in his twilight years.

It was at this time that I came to the ranch. The new ranch business manager, Bob Eastman, furloughed many old-timers, including George. Many of George's peers were saddened to be asked to leave, just months away from age sixty-five or rounding off employment to an even fifty years of service—a goal many broken-down, loyal *paniolo* sought to achieve. George's departure ended the tenure of the John Purdy–Ho'ohuli Pa line. He was a great-grandson of Jack Purdy.

Jack Purdy's final marital relationship involved a native woman named Hoka in 1873. No children came from this marriage, but the companionship was welcome, as Jack spent the remaining thirteen years of his life in her care and comfort.

So ends the Purdy dynasty as a foundation family that helped build the famed Parker Ranch.

6

PANIOLO APPAREL AND HORSE GEAR

1831 Forward

Some have romanticized the clothing of the early *vaquero.* But the seventeenth-century Mexican cowboy was, in fact, considered merely a laborer on horseback and held in low esteem by the Spanish *haciendado* (ranch owner). The only clothing he owned was what he wore. His hat was made of straw and his footgear was minimal. The *charro* shoe, which was low heeled and thin soled, was pointed at the toe and not even ankle high.

In the following century, about the time that cattle came to Hawai'i, the role of the *vaquero* was enhanced in the social order of Mexico and California. This rise in stature grew out of appreciation for his expertise in horsemanship *(charreria),* roping, and cattle handling practices. His new stature was reflected in a marked improvement in his appearance. When the *vaquero* came to Hawai'i, Mexican cowboy apparel had evolved in a characteristic and functional nature.

By this time, the original *vaquero* sombrero, with its three-foot diameter and curled-up brim, had evolved into one with a cone-shaped crown and stiff, flat brim of moderate width. Made of leather *(vaqueteada* or *pāpale 'ili),* hard felt, or woven palm fiber, the hat was held in place in wind or speedy pursuit by a chin strap *(barbiquejo* or *kaula 'auwae).* Tipping the hat at a rakish angle was commonplace. The emigrating *vaquero* quickly adapted to the *pāpale lauhala* (woven pandanus hat), with its wide brim *(ka'e)* and high crown *(ipu)* giving it a distinctive *paniolo* flavor. The *pāpale lauhala* was made locally and provided a headpiece that was light and well ventilated. This was especially appreciated during cattle roundups on the coastal plains *(kahakai)* and in desert country *(wao one).*

Beneath the *pāpale,* the *vaquero* wore a bandana tied firmly over a head of hair that was brushed back, parted in the middle, and plaited into a single braid *(huelo hili).* Extensive sideburns blended into a bushy mustache or full beard *('umi 'umi).* The bandana eventually became a neckerchief *(hainakā 'ā'ī)* to the native cowboy. In modern *paniolo* life it occasionally serves as head cover during especially heavy and tedious cattle processing such as branding, weaning, and shipping.

The *vaquero* arrived with a shirt made of heavy cotton. Long sleeves were a necessity to protect against the heavy brush environment in which they worked. Generally, it was a pullover for simplicity, the button-down version coming with later affordability. Typically white when new, the shirt quickly soiled into tan or varying shades of brown. A navy blue cotton cloth called *'ainakini* also came into use. The *vaquero* and his *paniolo* counterpart eventually adopted a hardy denim *(kelamoku* or *selamoku)* fabric that was the standard of the maritime industry. *"Selamoku"* is the corruption of "sailor ship"—hence the nickname for denim worn by seamen. John Palmer Parker provided the blue denim for his cowhands to use as work clothes and brown denim for use on the Sabbath.

While Levi Strauss of San Francisco fabricated the heavy canvas denim trousers for the gold rush miners, the early sugar and beef industries of Hawai'i began to use a trademark heavy-duty cotton twill fabric for the worker that was a checkered, block *(palaka)* print shirt. Whalers from New England brought the fabric, which was commonly referred to as *palaka*, to Hawai'i. The blue and white block print was interwoven into the fabric and not printed. In New England the fabric was known as Cranston or Sturbridge plaid, and it is still used there as an upholstery fabric. It became quite popular in the early Japanese community, as it closely resembled their native *Kasuri* fabric. These long-sleeved shirts were hardy and inexpensive. Today the *palaka* is more reminiscent of the *paniolo* than associated with its plantation origin, especially on the Big Island. *Palaka* fabric has become fashionable and expensive. It was even used in political circles during the 1970s, when Palaka Power became a campaign theme.

The breeches *(sota* or *lole wāwae pōkole)* of the *vaquero* were of soft leather laced up the sides, fitting snugly enough to hold themselves up or secured by a belly sash *(faja* or *kā' ai)* that was often red in color. The leather breeches were soon abandoned owing to the rainy, humid conditions of the Islands. They were replaced by the *kelamoku* trousers, styled as knee breeches. It is not uncommon to find the *kā'ai*, a vestige of the Spanish past, used by professional Hawaiian entertainers or as preferred attire for formal occasions in the form of a red sash with long-sleeved white shirt and white trousers.

For early Parker Ranch cowboys employed in the A. W. and Hartwell Carter era (1900–1960), appearance was important. Thus the ranch store warehoused the blue denim and khaki fabric—with tailoring done locally as a cottage industry—to give uniformity to cowboy apparel. A Chinese tailor named Goo Han made the khaki riding breeches for Hartwell Carter, Willie Kaniho, and Yutaka Kimura. He also made them in denim, which was preferred by Harry Kawai, Sam Liana, and most of the cowboy gang.

Beneath the breeches, the *vaquero* wore long and baggy pantaloons *(pantalones* or *lole wāwae pū ha'uha'u)* that covered the knees and calf in a bell-bottomed fashion to permit easy foot and leg movement. Such limb freedom was basic to *vaquero* horsemanship, providing appropriate cues with spurs and leg pressure. A triangular piece of fabric sewn into the base of the bell-bottomed trousers was called a *pi'i*.

Below the knee, the pantaloons were covered by leather or rawhide leggings *(likini wāwae* or *polainas)* that covered the tops of heavy buckskin jackboots *(kāma'a puki* or *bota alta)*. Leggings were readily adopted by the *paniolo* and remained traditional cowboy apparel.

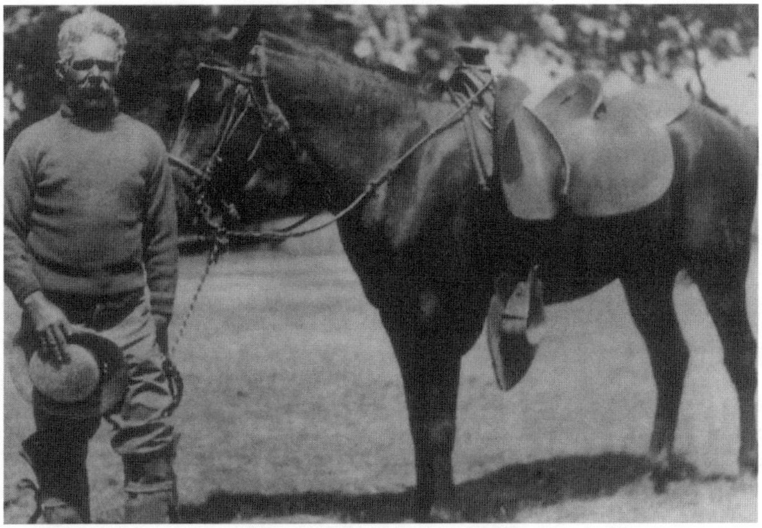

Left: *James Stevens wearing tied-on* bota-*type leggings. His mount is outfitted with a horsehair neck rope, a Spanish* vaquero *bit and bridle, a curled-back* lala, *tailpiece, and pancake* tapaderos. *Barnes King Collection.*

Below: *Cowboys branding. The rider with tall leggings has a saddle with a square* lala. *A. W. Carter Collection.*

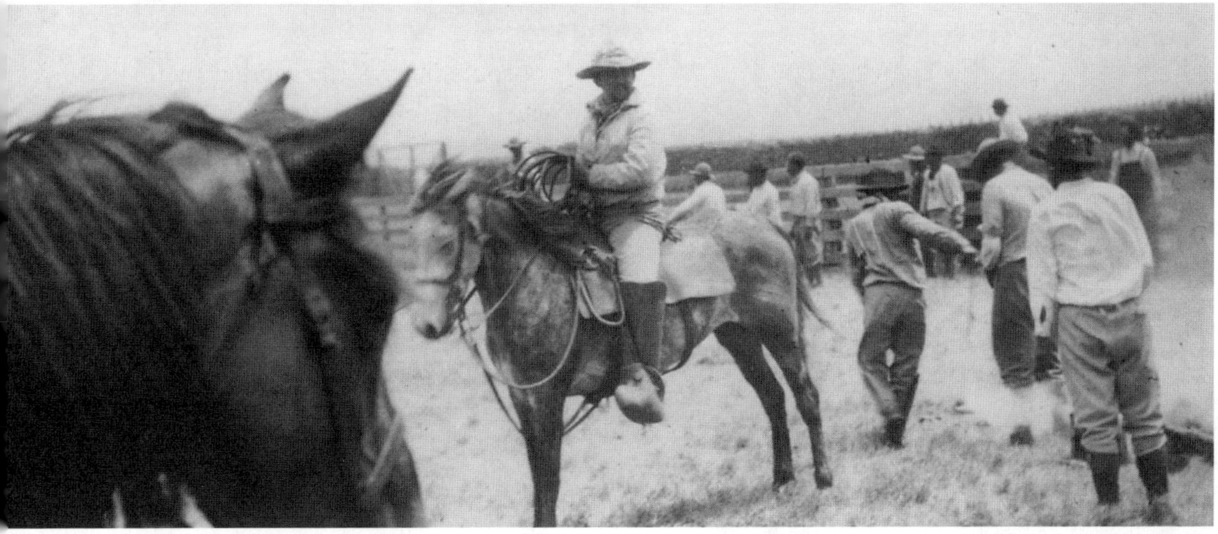

The *vaquero* brought his jacket *(chaqueta* or *iakeke),* made of leather or heavy cotton fabric. Humidity and rainfall caused the demise of the leather garment, and while cotton jackets continued in use, heavy denim became the fabric of choice for the *paniolo.*

Outer wear of the Hispanic cowboy was his beloved shawl (serape or *kīhei*), which was always tossed rakishly over the shoulder, steadfastly kept in place no matter how rigorous the roundup. The poncho effect of this garment gave way to a more practical full-length raincoat *(kuka 'aila* or *gabardina),* tailor-made of heavy cotton that was liberally coated with melted tallow (*'aila pipi* or *sebo)*—or in modern times, linseed oil. The *kuka 'aila* served all the purposes of the serape save for the rakish look reminiscent of old Mexico. A long-slickered horseman, however, still gives off a characteristic cowboy look.

Early Chaps

The early Mexican cowboy rigged his saddle with chap guards, or *armas*. Large, flared, and stiff, these heavy leather skirts were actually hung from the saddle forks and were laced to the front border of the stirrup leather. Later, smaller versions *(armitas* or *polainas)* were attached to the rider's legs, becoming the precursors of the *vaquero's chaparreras*—the chaps of the North American cowboy.

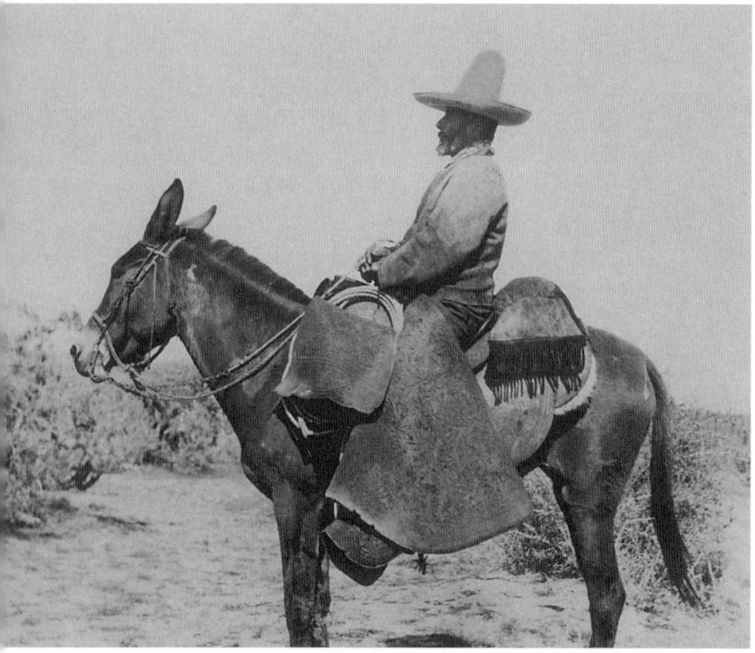

This photo of a vaquero *illustrates a headstall with* feador *and* mecate, armas *(chap guards),* cantina, mochilla, *and* anquitera *(with serape tied over it). Rio Grande Historical Collections, New Mexico State University Library Ms 259 "Mexican Vaquero."*

There is little evidence that the *armas, armitas,* and *chaparreras* took an early place in *paniolo* gear, unlike the leggings *(likini wāwae),* which were commonly used in Hawai'i—particularly with the well-fitting riding pants (jodhpurs)—as late as 1950. *Chaparreras* (chaps) gradually replaced the leggings during the same period that riding breeches were gradually replaced by typical blue jeans for the *Paka paniolo* (Parker cowboy). Chaps are still in use today, especially during very wet times or storms and at high elevations such as Humu'ula country.

Spurs

A pair of spurs bridges the gap between *vaquero* apparel and that of his horse *(caballo* or *lio).* Frances Allyn Olmstead, in *Incidents of a Whaling Voyage,* describes observations of a June 29, 1840, visit to the Waimea store of Mr. French:

> We came up with a collection of thatched houses towards the principal one of which we directed our steps which was a store belonging to Mr. French. . . . Here a novel scene presented itself to us. In front of the door a bright fire was blazing in a cavity in the earthen floor displaying in strong light the dark features of the natives congregated around it in their grotesque attitudes. Immediately back of these a group of fine looking men in a peculiar costume were leaning against the counter of the store. Some of them were Spaniards from California and they were all attired in the poncho, an oblong blanket of various brilliant colors hav-

ing a hole in the middle through which the head is thrust. The pantaloons are open from the knee downwards along the outside seam. A pair of boots armed with prodigiously long spurs completed their costume. They were bullock hunters that roam the mountains and had just returned from an expedition of eight or ten days in which they had been very successful.

On August 26 of the same year that Olmstead visited the Islands, W. H. Nichols, a third-generation Kentuckian, was born. A century later, on November 15, 1940, a great-grandson would be born in Laupāhoehoe, Hawai'i—William Carthage Bergin—whose destiny would include a deep relationship with Parker Ranch, where he "earned his spurs."

The spur of a horseman is the linchpin that connects his mind and limbs to the mount beneath him, providing the cue to pivot, turn, side-pass (lateral movement), and especially to accelerate.

Spurs and bits have held for me a lifetime fascination. I have collected these items since boyhood. As a young adult working on Big Island ranches as a cowboy and later as veterinarian, such artifacts were often given to me by *kama'āina* (native born or longtime residents) who were aware of my keen interest. The photos included here depict part of a historic collection, and they illustrate well the Hispanic connection. The spur also has ties to old Spain, where it appeared around 700 B.C. and eventually became a status symbol for the wealthy gentry and powerful knights.

The *vaquero* going ashore with his horse gear *(montura* or *ukana lio)* wore

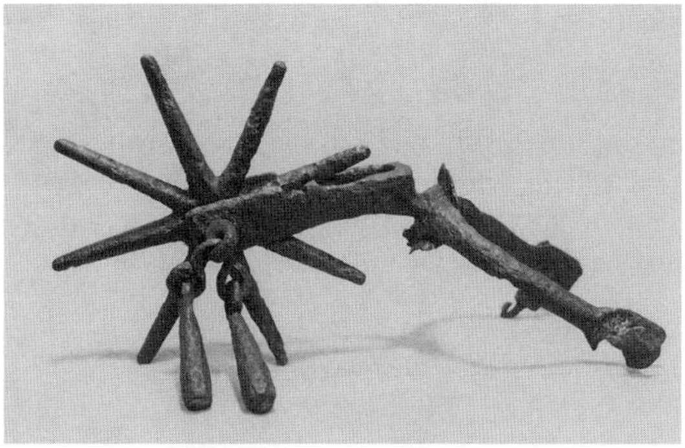

Spanish colonial spur with jingle bobs, 1700–1760. Robin Dutson photo; W. C. Bergin Collection.

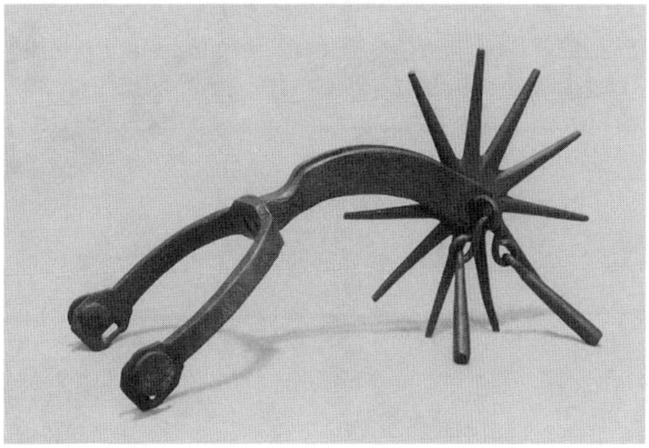

California vaquero *spur with jingle bobs, 1780–1800. Robin Dutson photo; W. C. Bergin Collection.*

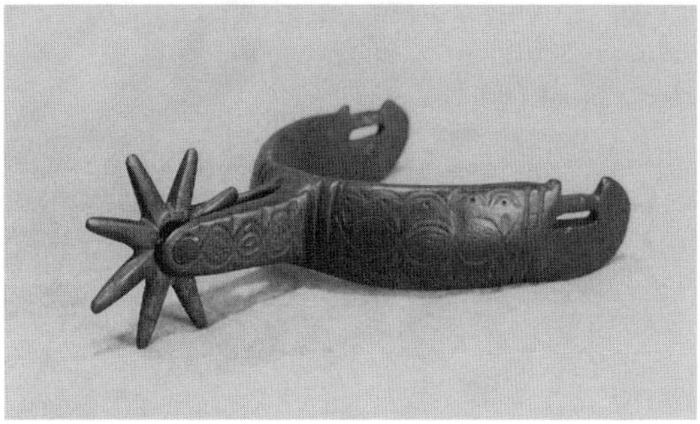

Wide heel-banded "Chihuahua" type spur, blending Texas-Mexico influences, 1800–1850. Robin Dutson photo; W. C. Bergin Collection.

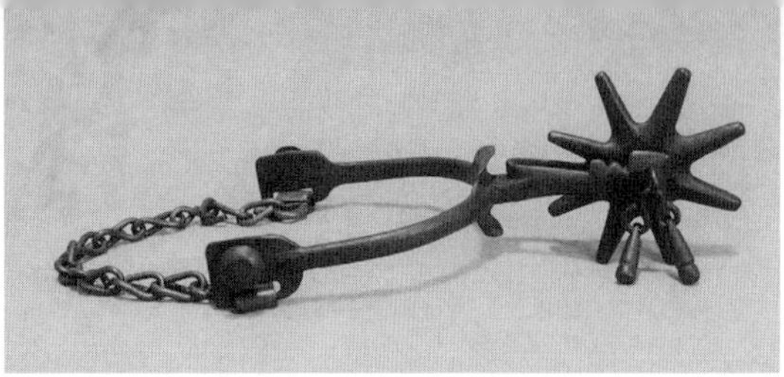

Narrow-banded southwest spur with jingle bobs and rodete, *which is the flattened plate at rear of spur for security and boot heel protection. Robin Dutson photo; W. C. Bergin Collection.*

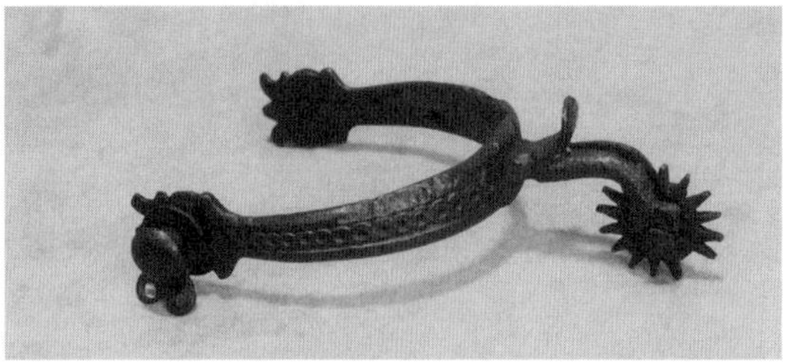

Popular Parker Ranch cowboy spur with chap guard and dropped shank, made by August Buermann. Robin Dutson photo; W. C. Bergin Collection.

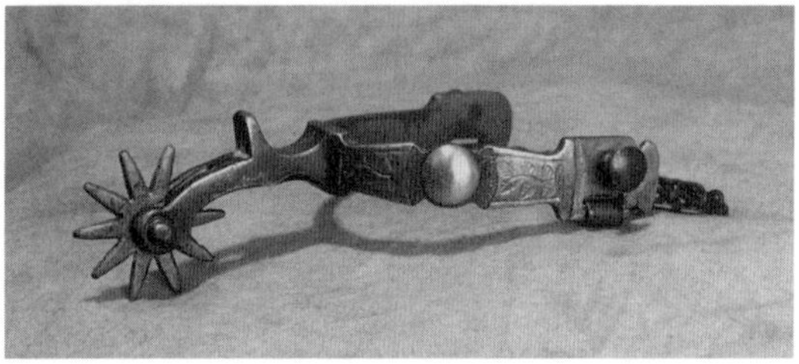

Another popular ranch spur made of brass with Monel buttons and custom-made ten-point rowel, also made by August Buermann. Robin Dutson photo; W. C. Bergin Collection.

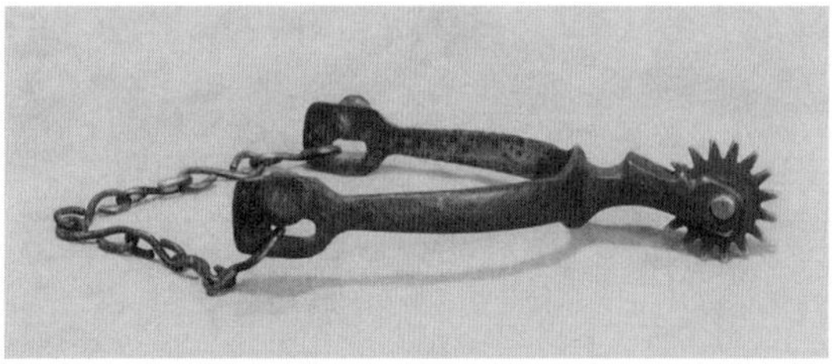

An ʻāmala *(blacksmith) spur made by Minato, a Waimea craftsman of the 1920–1940 era. Robin Dutson photo; W. C. Bergin Collection.*

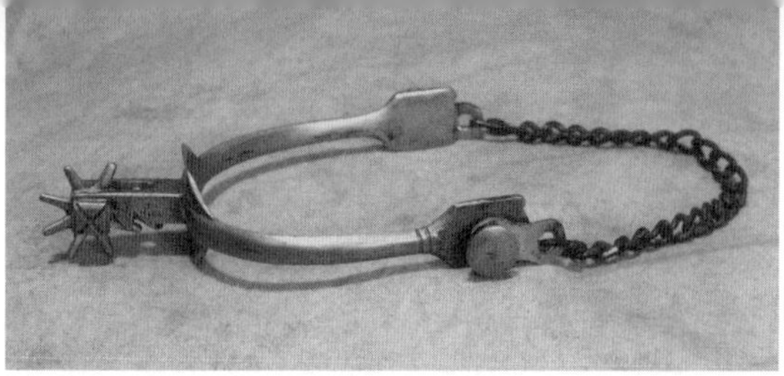

A highly cherished ʻāmala *spur made by Maui loriner Manuel Sebastian Santos for Parker Ranch cowboys and roughriders as gifts from Hartwell Carter in 1935. Robin Dutson photo; W. C. Bergin Collection.*

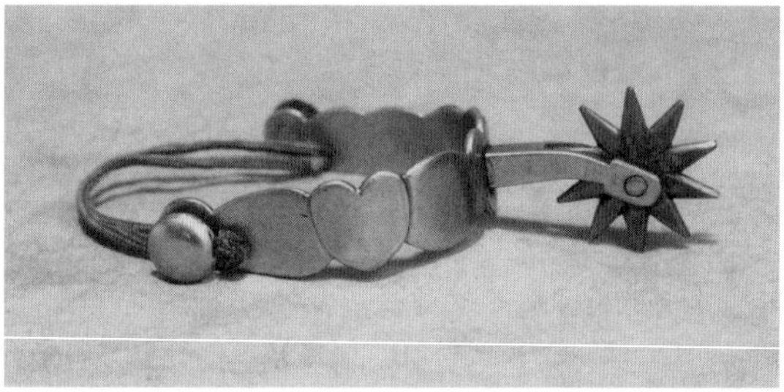

A Monel spur made by Donald G. DeSilva, Parker Ranch horse foreman and noted loriner and saddle maker. Robin Dutson photo; W. C. Bergin Collection.

spurs *(espuela* or *kēpā)* typical of his motherland, bearing large rowels (spiked wheels). The spur styles, however, probably represented two geographic areas of Mexico—Amozoc and Sinaloa/Sonora. The latter region was represented by a spur design described as Spanish Colonial, which had a relatively narrow heel band.

The wide heel-banded spur was common to Amozoc in the state of Puebla south of Mexico City. This was the heart of spur and bit making in old Mexico. This is where the conquistadors trained the Aztec people to become blacksmiths, to forge the bits, spurs, and armor *(ʻāmala).* To the Hawaiian cowboy, *ʻāmala* refers to a type of bit, spur, or tool made by a blacksmith *(kuʻi hao)* with his forge *(kapuahi),* anvil *(kua hao),* hammer *(hāmale),* and so on. An *ʻāmala* bit is so made.

The Amozocan style of crafting bits and spurs blended Moorish, Spanish, Mexican, and Aztec features. By 1800 there were more than three hundred bit and spur makers in Amozoc. In later years, craftsmen of Northern Mexico and California blended these styles with those of Colima and Guadalajara. The jingle bobs (two plummet-shaped bobs of metal hung adjacent to the rowel creating a pleasant chime-like sound when the horse and rider move) were common attachments. These are referred to as *pajados* or *campana* in Spanish and *kēpā pele* (spur bells) in Hawaiian. To create the chime-like sound, the *kēpā pele* were made of metal that was different from the spur rowels. While the *kēpā pele* emitted soothing, jingling sounds for the horse and rider, the chime effect carried over to the slack key *(kī hō'alu)* guitar of the *paniolo,* as in the tune "Maui Chimes."

Prior to the advent of the *vaquero,* horsemanship in Hawai'i used the standard English equestrian gear. Bits were commonly snaffles or pelhams, and the spurs were a refined jockey type with or without rowels. As the *vaquero* rode off into the sunset of his Hawaiian career, local blacksmiths gradually refined his large-roweled Mexican spur.

The loriners (makers of bits and spurs) of eighteenth-century Mexico were limited to using steel of varying grades. Although they used silver purely as decoration, copper was a metal that was used not only for its attractive appearance at low cost but also for adding a quality of "sweetness" to the bits. During the same era, the English were well advanced in the technology of metalworking and mass production. They supplied bits to the American colonies and horse communities for more than two centuries.

By the nineteenth century, wrought iron came into use as a popular metal, often referred to as "black," "Swedish," or "Norway" iron. This commercial form of iron was more malleable (softer and easier to work with anvil and hammer) than steel but comparable in strength.

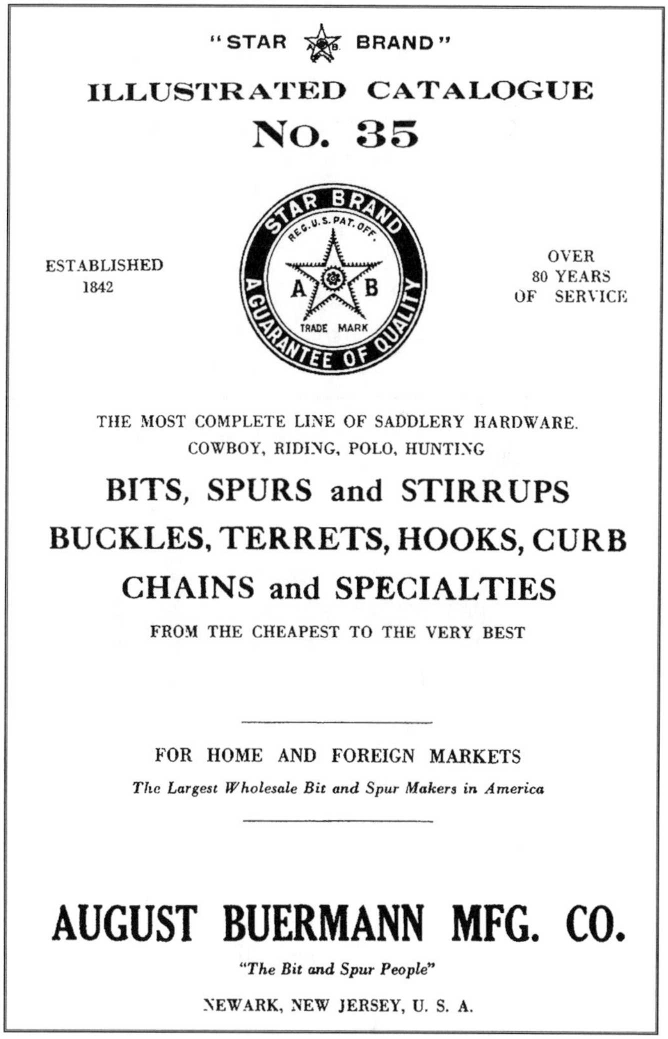
"STAR BRAND"

ILLUSTRATED CATALOGUE

NO. 35

ESTABLISHED 1842

STAR BRAND REG. U.S. PAT. OFF. A B TRADE MARK A GUARANTEE OF QUALITY

OVER 80 YEARS OF SERVICE

THE MOST COMPLETE LINE OF SADDLERY HARDWARE.

COWBOY, RIDING, POLO, HUNTING

BITS, SPURS and STIRRUPS

BUCKLES, TERRETS, HOOKS, CURB CHAINS and SPECIALTIES

FROM THE CHEAPEST TO THE VERY BEST

FOR HOME AND FOREIGN MARKETS

The Largest Wholesale Bit and Spur Makers in America

AUGUST BUERMANN MFG. CO.

"The Bit and Spur People"

NEWARK, NEW JERSEY, U. S. A.

August Buermann catalog cover. NMI.

Monel metal, the ultimate choice of metals for the loriner, was created in 1917 by Ambrose Monell, an American businessman. By 1950 it was popular in the American Southwest due to its light, bright, rust-resistant nature. Monel is an alloy of 67 percent nickel, 28 percent copper, 3 percent iron, and 2 percent manganese. The metal, sought after by the Hawaiian loriner for its rust resistance, was quite costly and hard to obtain until World War II, when aircraft component parts were discarded as surplus. It has since become scarce and costly once again, and hence it is rarely used.

Stainless steel also came into common use during the early 1900s. Because it is made of 12–14 percent chromium, it has a rust- and corrosion-resistant quality. Today, grades of stainless steel vary greatly, some of which are barely rust resistant. Brass *(keleawe)*, due to its relative softness and heavy weight, was used more for decorative and accessory purposes, such as for mouthpieces of bits and rowels of spurs.

Malleable cast iron became another popular metal for the loriner and lent itself to mass production, which was important in military equestrian tack and equipment needs. But cast iron bits and spurs were brittle and easily broken. To the early local blacksmith of the Islands, common "black iron" was popular because of its availability and affordability.

By the early 1900s the production of bits and spurs by Hawaiian blacksmiths was common. Parker Ranch cowboys of this era were supplied with spurs made by two South Kona Chinese blacksmiths—Chung Wai of Keālia and On Tai of ʻŌlelomoana. Later, in the 1920s and 30s, a Japanese blacksmith named Minato made spurs in the Waimea community. Around World War I, Parker Ranch used mainland spurs from August Buermann of Newark, New Jersey. These were called Star brand spurs.

Another South Kona bit and spur maker of note was Frank Gouveia of Hōlualoa, McCandless, and Hu'ehu'e Ranches. In 1939, Hartwell Carter commissioned Maui spur maker Manuel Sebastian Santos to make special Monel spurs for the cowboy and roughrider groups of the ranch to present as Christmas gifts.

While many of the Parker Ranch horsemen had access to store-bought models such as Star (North and Judd), Anchor (Buermann), and Crockett brands, the *ʻāmala* spurs of the makers described above were most precious to them.

After World War II, when stainless steel became the popular metal used in spur making, spurs took on a broad variety of styles that were practical in design. The more ornate models were used on the ranch only for special occasions such as parades, horse races, and so forth. Today, a Parker Ranch cowboy can be found wearing anything from an inexpensive China-made spur to a quality Ricardo or the cherished Monel spurs of his forebears. Spurs are considered a personal item of a cowboy's apparel—his bond to his horse.

The Knife Case

The traditional knife case *(ʻili wahī* or *cubierta)* worn by the cowboys likely had its birth on Parker Ranch in the days of wild cattle.

Belt-worn knife sheaths are occasionally visible in early 1900 photographs of Parker Ranch cowboys and probably held a folded knife. By the early 1930s, Parker

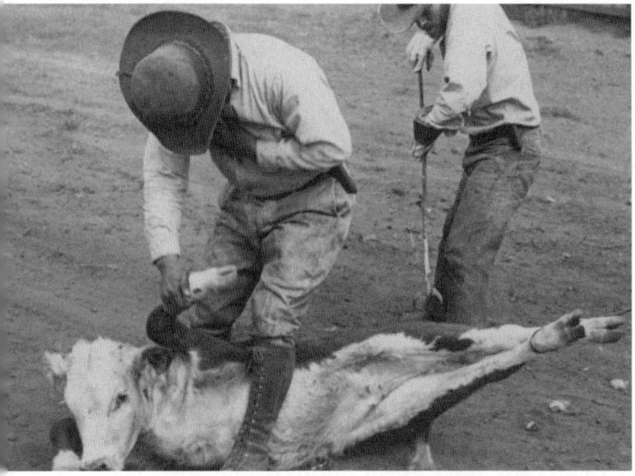

Both cowboys have belt-mounted knife cases. They were worn on the left hip by the right-handed cowboy. A. W. Carter Collection.

Ranch cowboys uniformly wore their cutlery not in a pocket, but in an external knife case. Some *paniolo* wore a fixed-blade knife and case on the shank of their boot, legging, or chaps. Typically, a cowboy used his knife to make ear and dewlap marks (the dewlap is the loose fold of skin hanging from the brisket between the forelegs), for calf, bull, and hog castration, and for butchering beef and game animals—usually feral goats, pigs, and sheep. Wild dogs also were quickly dispatched when a crafty roper snared the lamb eating cur.

Undoubtedly the most serious call for a quick blade came with cutting one's own lariat when entangled in a wreck involving a wild bullock. While cowboys of the American West routinely carried knives made with small folding blades that laid comfortably in their pockets, the cowboys of old California, Mexico, and South America wore a larger fixed-blade, daggerlike knife on their waist via a sash. In some areas, a machete was carried in a saddle-mounted scabbard much like the *paniolo* saw knife. *Paniolo* occasionally carried a saddle-mounted knife case that included a skinning knife and steel *(kila)* in addition to their knife cases.

The practicality of wearing a knife outside the trousers can be appreciated in dire circumstances such as when a cowboy is trapped beneath a fallen horse and pinned against a tree or the ground. In such a predicament, a pocketknife may be inaccessible, whereas a knife case mounted on the belt or boot is within ready reach.

The knife sheath was made of leather

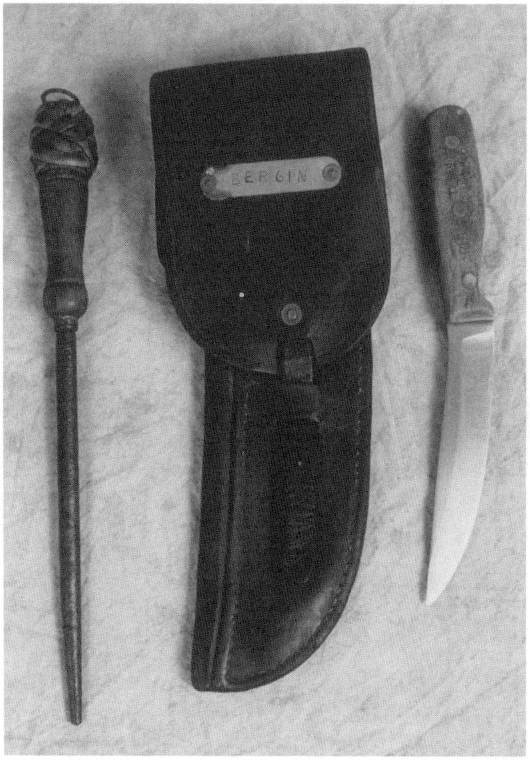

Skinning knife and steel (kila) *with leather case. These were mounted on saddles only on appropriate occasions. Robin Dutson photo.*

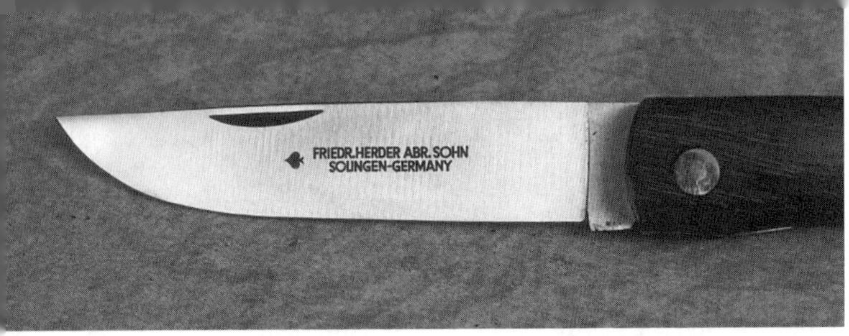

A Solingen knife with inscription of the maker on blade. Robin Dutson photo.

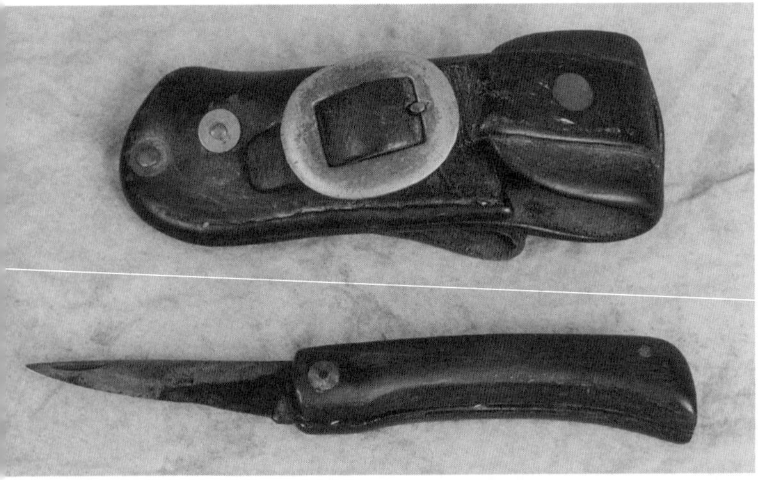

A well-used knife and case belonging to Sonny Ke'akealani. These endured over two decades of hard use. Robin Dutson photo.

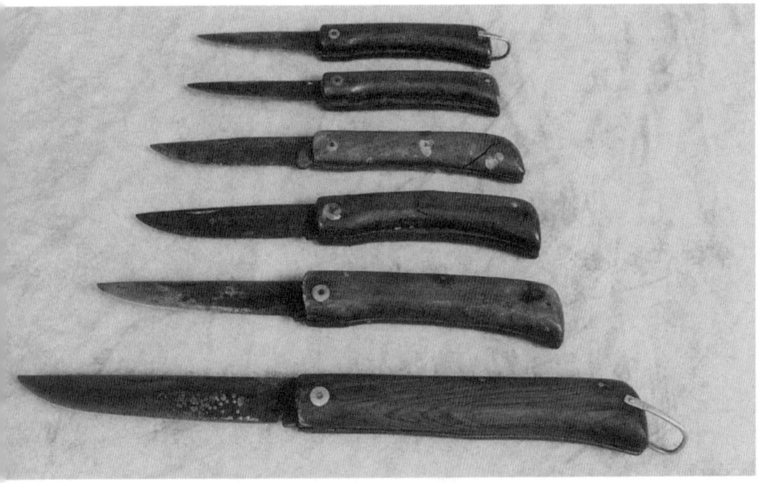

A half century of cowboy cutlery: Jiro Yamaguchi's knives, which saw fifty-three years of Parker Ranch service. Robin Dutson photo.

or rawhide and often had a buckle closure. Variations included a slip-through loop closure or none at all. The knife case was secured to the cowboy by his belt. If he was right-handed *(lima 'ākau),* it was usually worn on the left hip to prevent it from snagging the rope when roping, leading a wild bullock, or dragging a calf to the branding fire. The left-handed *(lima hema)* roper wore his knife on the right hip.

Early-day *paniolo* cutlery makers are not known, but by the early 1930s the Solingen knife, made in Germany of excellent metal, became the preferred knife. The smooth, round wooden grips (sides) lent themselves to carving one's initials for identification purposes. The grips were safe in all weather conditions, unlike metal sides. The Solingen knife remained the choice of the Hawaiian cowboy for another half century.

After World War II the practice of wearing an external knife case spread from Parker Ranch to neighboring ranches and islands. In 1952, Shipman Ranch cowboys routinely kept these knives in external cases. Like the Kona cowboys, these men dealt with wild cattle on a daily basis, and a handy knife for safety made sense.

By this time in *paniolo* history, the sharpening steel *(kila)* was added to the knife and its case as a practical accessory. This required adding a parallel slot to the rear of the knife case to accommodate the *kila.* This made in-the-field knife sharpening easy, especially when skinning a fat *laho'ole* (castrated male hog).

The *kila* had to be a short 5- or 6-inch version of a butcher's steel, which was usually 12 to 16 inches in length. The shorter steel was necessary to enable it to lie neatly alongside the knife case without excessive protrusion. A single long butcher's steel was frequently cut into two or three pieces for the same number of cowboys to fashion their own distinctive models with a variety of handles. Cow horn and deer antler handles were attractive and easy to grip. Wooden handles were also common, making identification via initials relatively easy. Aside from its sharpening function, the *kila* served as an awl for splicing rope or rawhide and punching holes in leather, as a hoof pick *('ōhiki lio)*, or as a tail-cleaning spike.

By 1980 the practice of strapping an attractive knife case to one's side spread to the mainland United States in cowboy circles and journeyman trades. These knife cases were often basket stamped or flower carved to give them a real buckaroo look and feel.

Local as well as mainland cowboys of today often use the multipurpose cutlery made by Leatherman or Gerber. These knives are made of excellent metal, offering superior blade quality and hardiness. While these knives are heavier, their versatility outweighs their density. A lifetime warranty makes them practical and affordable.

The Swiss Army knife never was used extensively by the *paniolo* because of its small blades and light, refined construction. They do, however, frequently boast other accessories that may include three screwdrivers, a file, awls of two different sizes, a can opener, and needle-nosed pliers with wire-cutting jaws.

It is remarkable that this early 1900s tradition of native Parker Ranch cowboy gear gradually spread across the United States by the 1970s, and wearing a knife case is now truly considered "cowboy."

The Lariat *(Kaula Ho'ohei)*

This tool of the cowboy trade traces its origin to old Mexico, where it was called *la reata,* which was eventually anglicized into the term "lariat," the familiar name in cowboy language. To the *paniolo* it is the *kaula ho'ohei,* with the rawhide version called the *kaula 'ili,* or skin rope. It is the cowboy's catch rope, used to lasso *(lazo* or *ho'ohei)* an animal. It is referred to as a lass rope by some.

The lariat was an improvement over its predecessor, the Mexican "catch line," whose loop was suspended at the end of a long pole; the rider would approach the cow and snag the head with the loop. An even less desirable means of apprehending a cow was the use of a hocking knife, or *desjarretadera*—a method brought from Spain to Mexico and outlawed by the late 1500s because of its cruelty. The rider would approach the bullock from behind and use the crescent-shaped hocking blade to sever the hamstrings of the animal. Neither of these techniques surfaced in Hawai'i.

With his *montura* (horse gear duffel, *ukana lio*), the arriving *vaquero* brought his *reata* and style of roping and rope making—tasks that were easily grasped by the native Hawaiian. Being hunters of wild cattle, their lariats were long—often 40 to 60 feet (6–10 fathoms or *anana; brazada* in Spanish)—and made of braided rawhide. The *reata* could be made of four, six, or eight plaits *(ma'awe)* of rawhide strips. The *honda (kīpuka)* through which the tail or *piko* of the rope is passed to form a loop was made of rawhide with an attached "burner" to preserve the integrity of the individual

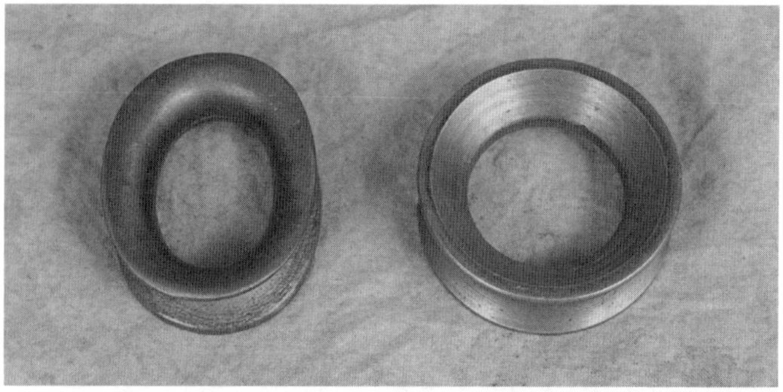

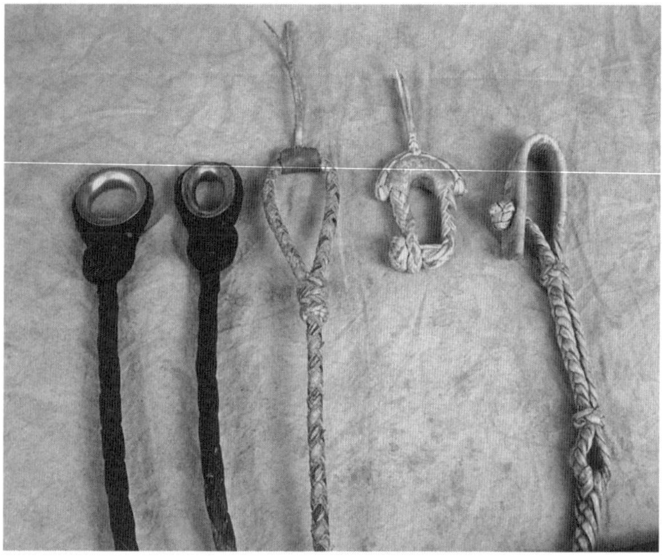

Top: *Brass* hondas, *round and oval. Robin Dutson photo.* Above: Kaula'ili *(rawhide ropes) with a variety of* hondas, *the brass oval being the most popular. Robin Dutson photo. Dr. Billy Bergin Collection.*

plaits under heavy roping use. The term "*honda*" is derived from the Spanish word *hondón,* which means "eye of the needle." By the turn of the twentieth century, the *paniolo* came to prefer a brass oval or round grooved ring for the *honda* due to its compatibility with rawhide, its durability because it outlasted several ropes, and its added weight—often necessary when roping in windy country, shaking out *(mōhala)* a new loop, or slackening *(hōʻo ʻalu)* the rope from a cow's legs. The post–World War II period brought Monel and stainless steel into use, while the brass oval rings went out of production by the 1950s. Today a premium is paid for the original oval brass *honda*—if they can even be found—as they proved to be of the ideal weight, size, and durability.

The *paniolo* on Parker Ranch had a preference for the four-plaited *(hili hā)* skin rope, although some were of six plaits *(hili ʻeono).* Certain men were particularly noted as craftsmen, and they supplied lariats to other cowboys. The real test of a man's weaving technique was to hold a strand with both hands about 3 inches apart and push the rope toward itself. If spaces appeared between the plaits, the weave was too loose and the rope would lack a supple body. These could be rebraided, giving body and tautness to the rope's character. Sam Purdy was the master weaver of fine-quality skin ropes. A. W. frequently attested to this fact.

A. W. was highly respectful of Sam Purdy's skills. In his early *paniolo* years, he was known for his roping skills on and off the ranch. Sam was noted as one of the best weavers of the skin rope, *kaula ʻili.* In 1944, Lieutenant A. J. McMurray, AAA Gun Battery, contacted Carter requesting that Sam make him one of his fine lariats. The officer was aware of Sam's excellent craftsmanship from previous Camp Tarawa assignments. Carter regretfully responded to the request with the sad news of Sam's recent passing from an aggressive form of oral cancer. A. W. had visited Sam at bedside in his final days. It was during this visit that Sam gave A. W. the cherished old Bible of the Parker family that had been handed down to him.

For the rawhide lariat, the hide from an aged cow or steer was preferred, owing to its uniformity in thickness and maturity. Brindle cattle were especially sought after for their hides— appropriately dressed, salted, and aged for drying and curing—in making *kaula 'ili.*

Arnold Rojas, noted *vaquero* authority, provides a description of the Mexican *reata*-making process in *California Vaquero:*

> When the old-timer needed a reata he first selected a steer of the proper age and condition, and when the moon was in the right sign for starting a rope, slaughtered it and removed the hide with great care. The strength and pliability of the reata depended as much on the animal's condition and on the phase of the moon, as in the braiding which was done in winter when the strands se prestaban (were workable), since they were damp and could be pulled tighter when braiding. The strongest ropes were made from hides taken from steers that had died from hunger. These reatas turned black when braided and in use and needed constant tallowing. There is little fat on a beast that starves to death in winter. The best hides came from cattle bred on the coast, there being much less sun and more shade than in desert country where the hot dry climate burned out the hide.

At the appropriate time, the Hawaiian craftsman would gently shave the hair from the outer surface and fat from the inner surface of the cured hide. The next step was the peripheral starting point of an inward spiral carving of a single three-eighths to one-half-inch strip of rawhide until the center of the hide was reached. The corners of the neck and limbs were given to other uses. Great skill is attached to the stripping process, as consistency of width of the strip is of utmost importance.

The craftsman would then stretch this long strip for days to further develop body and reduce elasticity. Then the long strip was cut into four equal lengths to form the individual *ma'awe* (strands or plaits). Each *ma'awe* was carefully beveled with a sharp blade to produce a smooth and binding quality as they were braided.

In the braiding process, the key was in alternating a firm, consistent pull between braid wraps to provide a tight weave, smooth handle or feel, and supple body to the rope. The end or *piko* of the rope appeared dovetailed and smooth to avoid wrecks when the cow pulled the loop through the roper's grasp. The *honda* end looped upon itself or encircled a metal *honda.* In some cases, a Turk's head knot joined the loop back to its origin. The rope was again stretched extensively before oiling.

Care of the *kaula 'ili* was exceptional, using beef tallow liberally followed by exposing and stretching it in the sun for deep penetration into and between the *ma'awe.* This also kept the rope somewhat moisture resistant, but only to a limited extent. The use of a *kaula 'ili* in very wet conditions betrayed its shortcomings: it would become too slick, elastic, and very heavy. The *kaula ho'ohei* was attached to the right saddle fork by a rawhide strap *('ili kaula)* that looped over the saddle horn after a couple of wraps. While in his saddle, a cowboy kept his slicker extended over the skin rope during rainy weather. He often took his rope home evenings and on weekends if his saddle locker was not secure from two- or four-legged *'iole* (rats).

While sisal, manila, and cotton ropes occasionally replaced the *kaula 'ili* for some cowboys, by the mid-1960s the nylon lariat became very popular for its durability, handling, and low cost. The *kaula 'ili* has since been relegated to a historic and decorative status—a role it surely earned.

Roping Bullocks and Calves

To the uninformed, roping of animals can conjure thoughts of inhumane restraint. The more dramatic form of such roping is steer jerking, where the roper casts his slack—after roping the steer by its horns—wraps it around the animal's opposite flank, and rides off in a diagonal direction. This gathers the steer's hind legs beneath him as his head is drawn up to his side, and down he goes. The rider quickly dismounts and hogties the momentarily downed steer.

The *vaquero* passed this skill to the *paniolo,* who put it to good use in hunting *pipi 'āhiu* (wild cattle) in the *wao one* (desert) and *kahakai* (coastal plains), where trees were scarce for tethering *(po'o waiū).* Once downed, the wild bullock was secured *kūpe'e* (hogtied) style. In *California Vaquero,* Arnold Rojas aptly describes the skill and art of such roping:

> Cattle were seldom hurt when roped and thrown by an expert lazador who knew how to choriar (play) the reata, as an expert angler plays the reel and line when a game fish is hooked. The vaquero judged the jerks by letting the turns round the horn slip or hold so the beast would fall softly, as tumblers in a circus roll and fall without getting hurt. Like all artistic souls, the old-timer took great pride in his work and could lay a full-grown bull on the ground as gently as a mother would lay her babe in a cradle. By taking one turn to stop it he could release the turn when the animal was thrown, giving it a chance to break its fall.
>
> The man who hurt an animal when roping it was considered very clumsy and not classed as a roper. For the reata is an intricate tool that is handled best when there is co-ordination of eye and arm. Old-timers say, "The hondo commands the loop, the loop the coils, and the throw is decided by the way one whirls and casts."

The *Kīhele* Loop

While typical range and arena roping is done with an overhead clockwise swing of the loop *('ahele),* Parker *paniolo* historically roped branding calves *(pipi keiki)* in a unique and humane manner. Whether this was passed down from the *vaquero* is not known. The roper's counterclockwise (reverse) swing parallels his horse's side rather than overhead. He sidles up behind a calf and casts a large loop over the *pipi keiki* head, allowing it to step through to the depth of its flanks *(kīhele).* At this time, the roper simultaneously gathers his slack as the loop passes down over the calf's heels, while he wraps his rope *(hāwili, dar la vuelta)* around the saddle horn *('ōkumu)* and turns his horse toward the branding fire, to which the calf is dragged on its side by

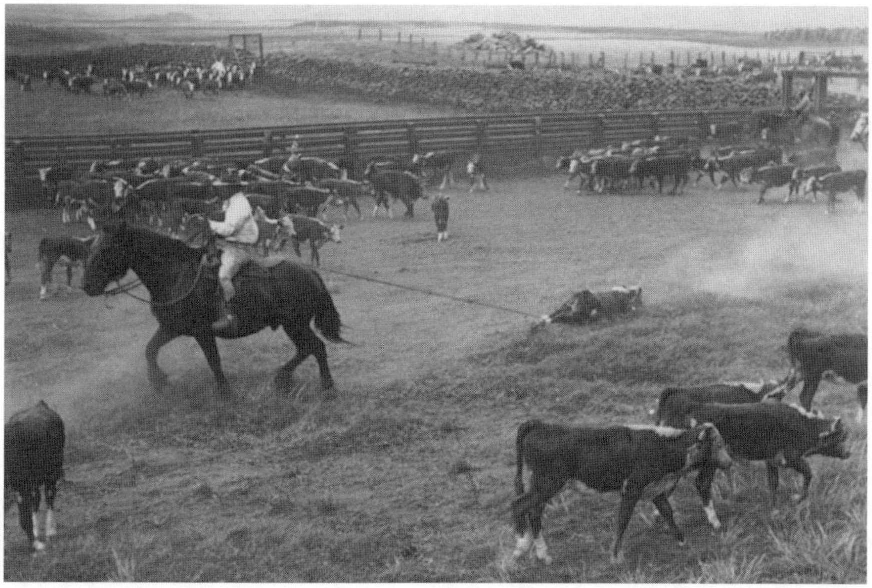

Calf being pulled to the branding fire by both *heels, a standard practice at Parker Ranch. The horse is a black half-Percheron gelding working in the Pu'uhihale Corral. A. W. Carter Collection.*

both heels. This is the most humane form of calf branding I have ever witnessed, even when compared to chute branding, because of the rapid, gentle process, mandating both heels to be in the "handle" for dragging calves to the branding location. Once the calf is branded, the lariat is immediately slackened, allowing the calf to dash through the gate to its mother's side. In 1974, some 1,501 calves were branded at the Keanahu'ihu'i Corral in half a day using fifteen men. Keanahu'ihu'i, which means "cold cave," is an area in the foothills of Waiki'i. Some old-timers call it Keoni Huihui, as John Lindsey (Keoni Li'ili'i) was known to stage the cowboy gang there before they launched the *'āhiu pipi* drive. After lunch, the "mothered-up" herd was returned to its native pasture in Pu'uanuanu. Since my first formal employ with Parker Ranch in 1958, I have never witnessed a calf being dragged to the branding site by a single leg.

Apparently the *kīhele* method was not a practice on the old Parker Ranch. Johnny Pieper made this observation in the years that Parker Ranch leased the Hawaiian Ranch, where he had been a longtime foreman. He related that when he left Parker Ranch in the mid 1930s, calves were still being roped by the head or hind legs *('ahele)* and wrestled to the ground for branding. Johnny was impressed with the fluid and gentle system described as *kīhele.*

Paniolo Horse Gear

Just as the spur was the link between man and horse, *paniolo* horse gear was equally significant and evolved over time. The neck rope or *kaula 'ā'ī* used by the *paniolo* in catching his mount is a distinctively Hawaiian piece of equipment.

The Neck Rope: *Kaula 'Ā'ī*

Somewhat unique to the *paniolo* is the use of a special neck rope, the *kaula 'ā'ī.* About 22 feet long and five-eighths of an inch in diameter, the neck rope is used to catch a mount in the corral and in leading. The neck rope is also used to hitch the horse to a post or tree for saddling, resting, and so on. This traditional use is in contrast to the custom of the cowboy of the American West, who used a lariat to catch the horse and, when bridled, the horse was hitched to a post or rail with the reins or simply hobbled.

The *kaula ʻāʻī* is possibly a vestige of the *vaquero* tack. The Mexican cowboy brought a *mecate,* a twisted horsehair rope of comparable length and diameter that doubled as a lead rope and reins when tied to the heel knot of the *bosal.* The *bosal* was a stiffly braided rawhide noseband used in early horse training that was suspended on the horse's head via a headstall and *fiador.* The latter fixture was a double-corded throatlatch that, when combined with the *mecate,* allowed firm lead-rope control over the mount.

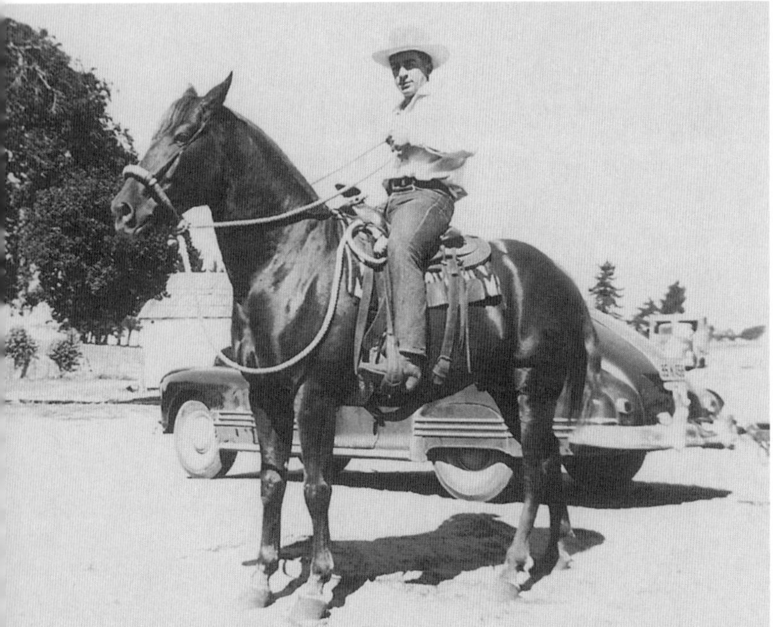

Typical bosal *and* mecate. *James Rodriguez Sr. on Lucky Tom, noted Parker Ranch cow horse sire. Rodriguez was a world champion team roper, as was his son James Jr. Hartwell Carter Collection.*

The *paniolo* of old developed a simple and practical alternative, combining the *fiador* and *mecate* into a single neck rope of similar length and diameter. They retained horsehair—*hulu ʻāʻī* and *hulu huelo*—as the material, as well as the specially knotted ends. These included a crown knot on the horse end *(poʻo, botón)* and a spliced tassel *(piko, colilla* or *cascabel)* on the rider's end.

The horse end of the *kaula ʻāʻī* is looped around the mount's throatlatch and tied in a nonchoke bowline knot *(pōlena).* The *piko* is passed fully through the noseband of the bridled horse and pulled into several coils, forming the lead rope. In mounting, the *paniolo* ties the coiled lead rope to the saddle strings attached to the left fork of the saddle above the *ʻaweʻawe.* These coils may be tied in reverse, with the *piko* in a forward direction, which keeps it from creeping between the cowboy's thigh and saddle during vigorous cattle work. The chafing effect when the *piko* is tied in a backward direction is painful and distracting to the cowboy.

When dismounting, the prudent *paniolo* routinely untied these coils and held them firmly in his left hand when stepping to the ground, as a spooked colt often chooses this occasion to bolt *(āhāʻi).*

In the early 1900s, horsehair was still in use, along with rawhide *(kaula ʻili),* for making neck ropes. Cotton was used to some extent but never became popular because it is moisture absorbing, weak, and

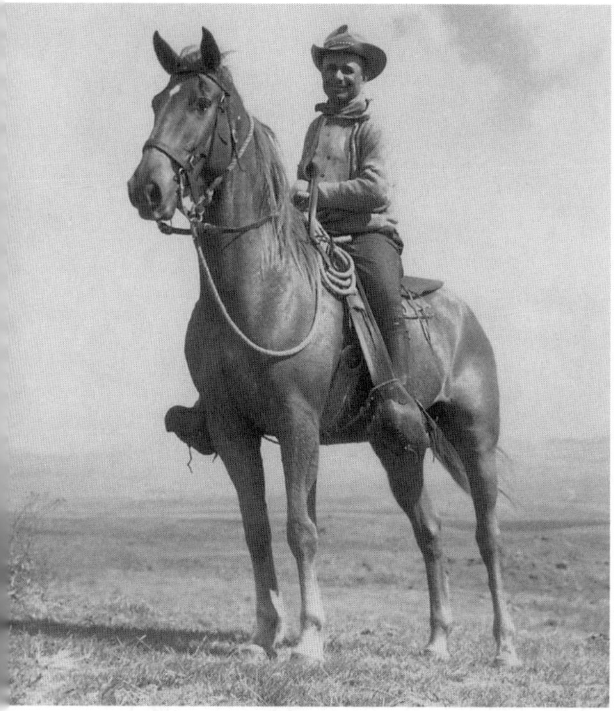

Typical kaula ʻāʻī *(neck rope) on a sorrel gelding, with cowboy foreman Uiha Lindsey aboard. Hartwell Carter Collection.*

easily frayed. Soft nylon combinations came into vogue in the 1960s after Manila rope had enjoyed greatest popularity in the first half of the twentieth century. Soft nylon continues to be the preferred neck rope material today.

The *kaula ʻāʻī* provides other functions, most commonly as an emergency *poʻo waiū* rope to tether a wild bullock to a tree or hogtie him to the ground. Typically, however, a wild cattle hunter carries as many as a half dozen heavier-grade tie ropes for this purpose. The *poʻo waiū* rope has a length and diameter slightly greater than the *kaula ʻāʻī* and is looped with a *kipuka (hondo)* on one end in place of the crown knot. The looped end was used in forming a nonchoke halterlike effect when tying the bullock's head to a tree. The traditional *kaula ʻāʻī* remains in use today by Parker Ranch *paniolo.*

Much like the spur, the headgear of the horse has a long history. People in every culture that used horses for transportation needed the bit and spur. The Moors of North Africa invaded Spain in the eighth century, bringing with them their horses, saddlery, and tack. As they did with the spur, the Spanish further improved the functionality of these instruments of horsemanship, while retaining their artful appearance. Artistic elements were enhanced by the inlaying of gold and silver in a process referred to as damascening, which originated with the Arabs of Damascus and came to Spain via the North African Moors. This inlaying was representative of the life of the affluent or knighted horseman, not the common equestrian.

Once established in New Spain (Mexico) as a craft of its own, inlaying became popular in the mid-1500s as major discoveries of silver lodes were found north of present-day Mexico City. Damascening does not appear to have reached Hawaiʻi to influence bit and spur designing locally, probably because the *vaquero* entering Hawaiʻi in the 1830s were of working-class mestizo origin (half Spanish and half Native Indian).

The Bridle Bit: *Kaula Waha Hao*

The importance of the bit to the cowboy exceeds that of the spur. The bit governs a horse's head and neck setting, which in turn controls balance, direction, slowing, and stopping. Generally speaking, bits of all persuasions have three components: a pair of cheek pieces that parallel the mouth on each side and a mouthpiece that connects the two. Bits commonly used in early and modern ranch history are of two basic types: the snaffle and the curb.

The snaffle had its origin in ancient Celtic culture, with an eventual evolution into the Irish bridoon, which was probably the type of bit that was introduced with the horse to the Hawaiian Islands by Captain Cleveland in 1803. The cheek pieces of a snaffle are usually rings of some sort, joined by a mouthpiece with or without a linked center.

The Hawaiian cowboy commonly refers to the snaffle as a broken bit because of its jointed mouthpiece. More traditional *paniolo* refer to this bit as a *kaula waha hao kēkake* (often shortened to *kaula kēkake*), or donkey bit, as its use is most common in that species.

Under typical ranch conditions, the snaffle was the most popular bit for training young horses, sensitizing the mouth while limbering up the neck for eventual neck reining. The reins used with a snaffle were often made of manila, cotton, leather strapping, or rawhide rope. The snaffle exerts pressure predominantly on the bars of the horse's mouth. The bars are gum-lined bone spaces between the incisor teeth and premolars of the right and left internal (oral) mandibles (jawbones), referred to as the "bit hole." The hinged nature of the snaffle allows a rider to apply pressure (with a single rein) to one bar at a time for turning or to both bars (with both reins) for stopping or slowing the animal. Both the snaffle and the curb bit provide secondary pressure to the corners of the mouth and body of the tongue.

The reins used with a snaffle may be straight (from bit rings to rider's hands) or block *(palaka)* lines. Block lines get their name from the "block and tackle" effect of their being threaded through the snaffle rings and tied back to the rigging rings *(lina 'awe'awe)* of the saddle. This gives the rider great leverage over the horse's mouth and was routinely used at the Parker Ranch breaking pens, where horses were started as five- or six-year-old broncs. *Palaka* lines disappeared from the ranch scene in the late 1960s, when colts were started as two- and three-year-olds and improved training methods were introduced.

The curb bit mainly provides dual leverage action by exerting pressure on the bars of the mouth and lower jaw via a curb strap. The curb strap may be of rawhide, leather, or chain and extends from one cheek piece (bridle ring) of the bit, under the chin, and on to the opposite bridle ring, lying in a depression called the chin groove. Pressure on the reins provides the lever action, with the bars of the mouth being the fulcrum.

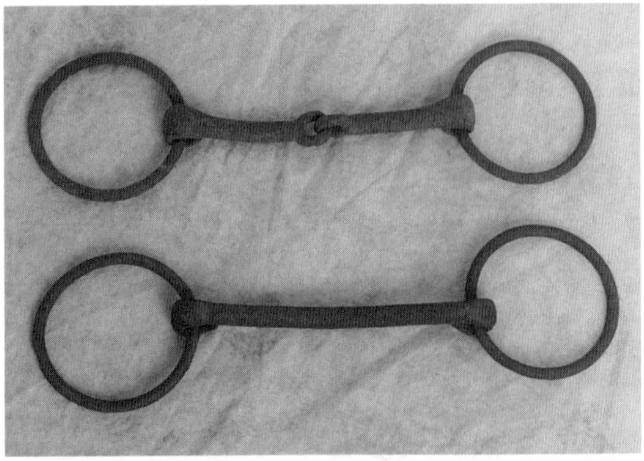

Typical snaffle bits: jointed (broken) and straight mouthpieces, the latter used more for wagon horses. Robin Dutson photo; W. C. Bergin Collection.

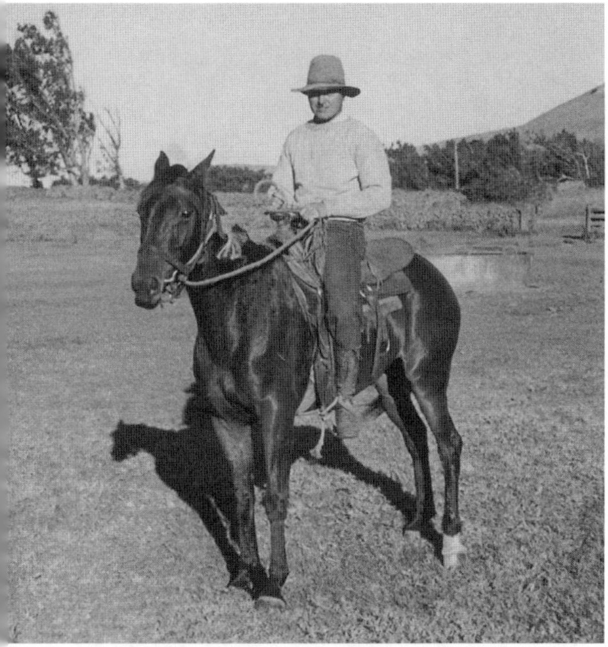

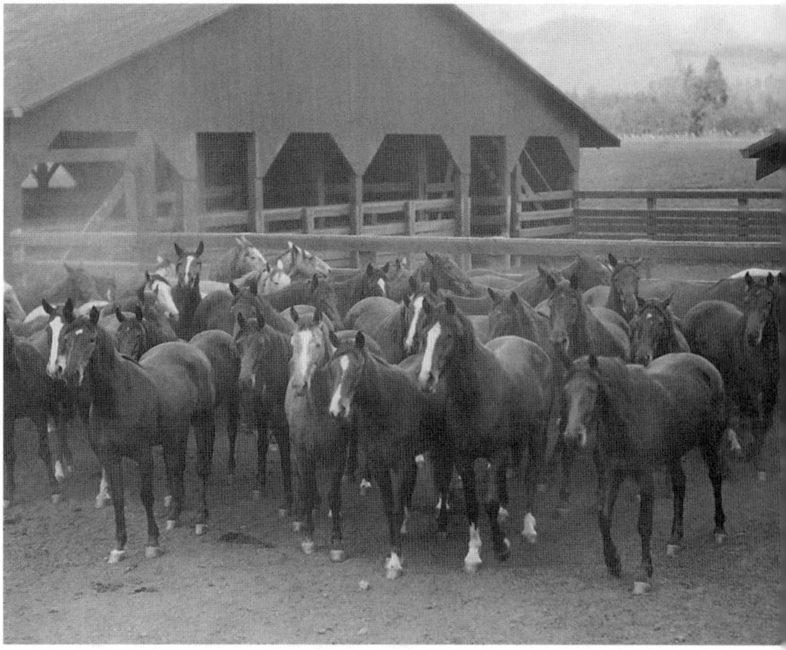

Five- to six-year-old green-broken horses at the Breaking Pen. Hartwell Carter Collection.

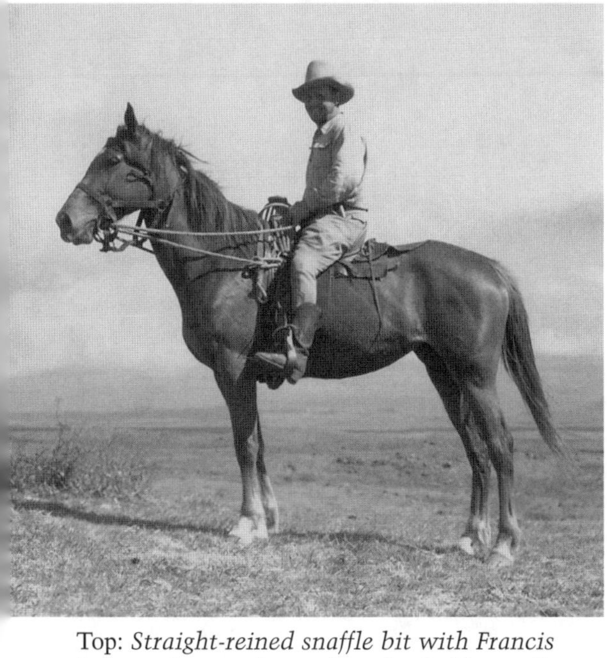

Top: *Straight-reined snaffle bit with Francis Pakana Spencer Jr., on horseback. Hartwell Carter Collection.* Above: Palaka *line usage; Joe Maertens aboard. Hartwell Carter Collection.*

To the Hawaiian cowboy, a curb bit is called a straight, standard, stand, or big bit. Traditional *paniolo* refer to the bridle as a *kaula waha* and the bit alone as a *kaula waha hao.* The Spanish word for bit is *bocado.* A bitted bridle is referred to as *bocado del freno.*

The cheek pieces of a curb bit are usually elongated shanks rather than the rings of a snaffle. The cheek pieces are usually made of flat metal or a rod bar with a bridle cheek ring at the top and rein ring at the bottom. The shank of the cheek piece is that section between the mouthpiece and the rein ring and provides most of the leverage by its pressure on the chin groove via the chinstrap. The purchase of the cheek piece is that section between the mouthpiece and the bridle or cheek ring. Under rein pressure, the purchase applies pressure on the poll *(piko)* of the horse's head.

The mouthpiece of a curb is placed within the mouth and is suspended on each side via the headstall (bridle). There is a wide variety of mouthpieces, from simple straight to curved bars with or without a port at the center. The port is the arched centerpiece of the mouthpiece. It can be low or high in its arch, with severity regulated by its height. It can be a plain inverted *U* with or without a cricket (roller), an addition that—when stroked by the horse's tongue—makes a soft whirring sound that provides a calming effect. Salivation is also enhanced, keeping the mouth supple. The port can be highly ornate, as in the spade bits.

The mouthpiece can be permanently attached to the lateral cheek pieces. These are called solid jaw bits. Bits in which the mouthpiece is hinged to the cheek pieces are known as loose jaw, swivel cheek, or spring bits. Loose jaw bits originated in Spain and Mexico and were likely the first curb-type bits brought by the *vaquero* to the Islands. Some of these were undoubtedly spade-type bits. The term "spade" refers to the spoonlike upward extension of the port of the mouthpiece.

Blacksmiths of the Islands began making curb-type bits by the mid-1850s. The more garish Spanish appearance of the spade-type bit and the large rowel spur was softened by the *ku'i hao* (blacksmiths). Hawaiian craftsmen gradually Hawaiianized the horse gear via metamorphosis from Hispanic embellishment. The mouthpieces of the bits were more humane, the rowels of the spurs were reduced to a practical size, and the enlarged saddle horn *(cabeza)* of old

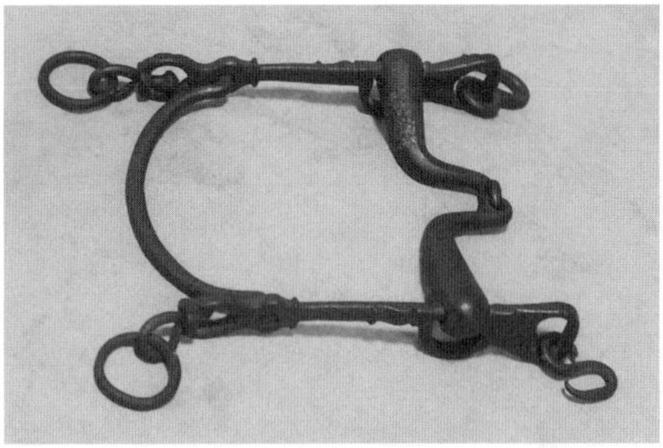

Early vaquero *loose check bit with a snaffle port. Robin Dutson photo; W. C. Bergin Collection.*

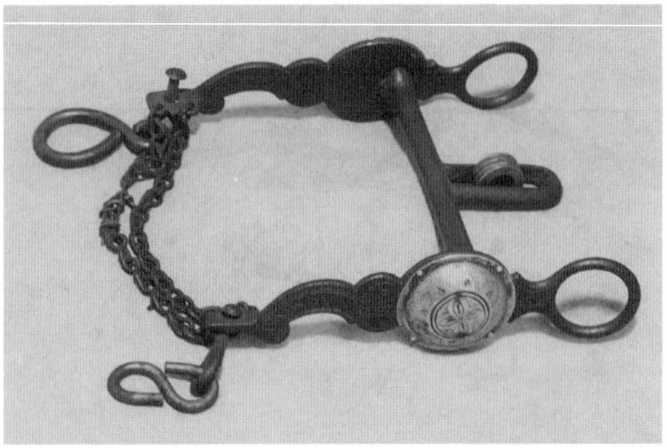

California loose cheek bit with a cricket port. Robin Dutson photo; W. C. Bergin Collection.

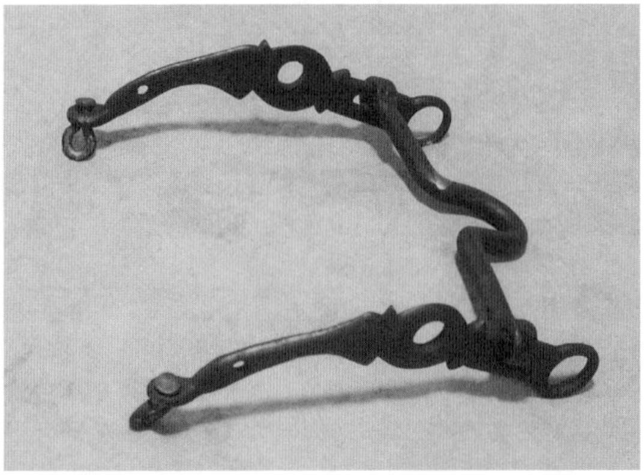

Loose cheek 'āmala *bit made of iron by Minato. Robin Dutson photo; W. C. Bergin Collection.*

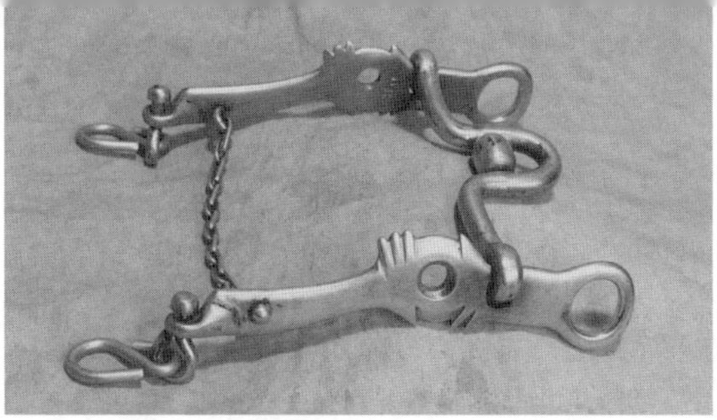

Loose cheek 'āmala bit made of Monel by Frank Gouveia. Robin Dutson photo; W. C. Bergin Collection.

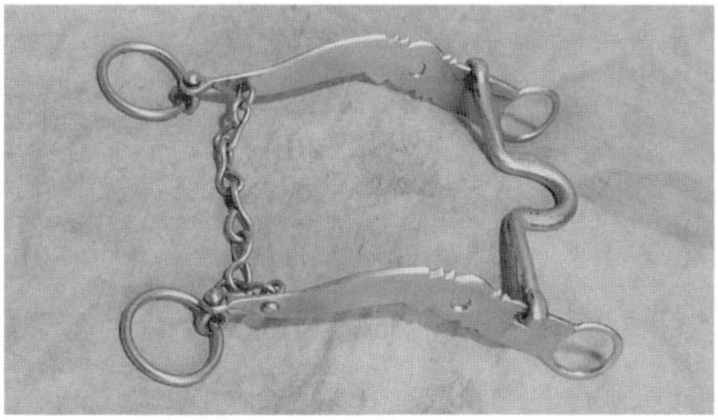

This is referred to on Maui as the "Parker Ranch Bit," from a 1920 bit taken to Maui when Edwin Robert Kingston Lindsey (son of William M. S. and Henrietta Kaluna Lindsey) moved to Pu'unēnē Ranch from Parker Ranch. Noted Maui master blacksmith Kaloi Cockett developed the Maui pattern, which was commonly used by loriners such as Manuel Sebastian Santos. Robin Dutson photo; W. C. Bergin Collection.

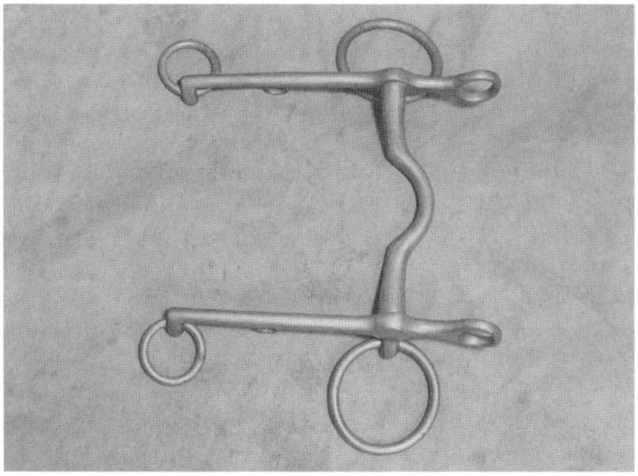

Stainless steel English military bit with loose side rings for a second pair of reins. These are called Pelham bits. Robin Dutson photo; W. C. Bergin Collection.

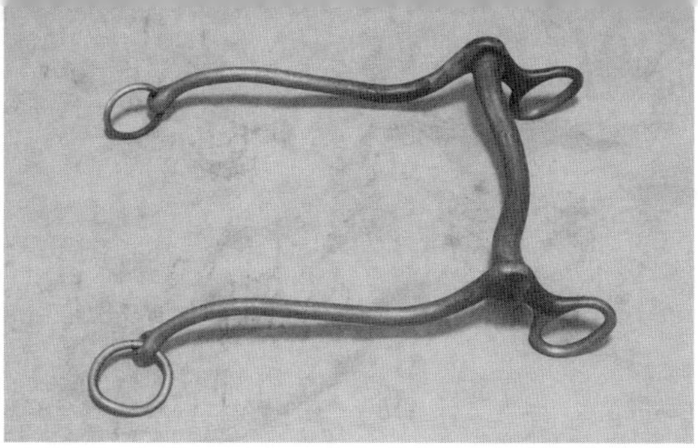

Stainless steel walking horse bit with a mildly arched mouthpiece and long shanks. Robin Dutson photo; W. C. Bergin Collection.

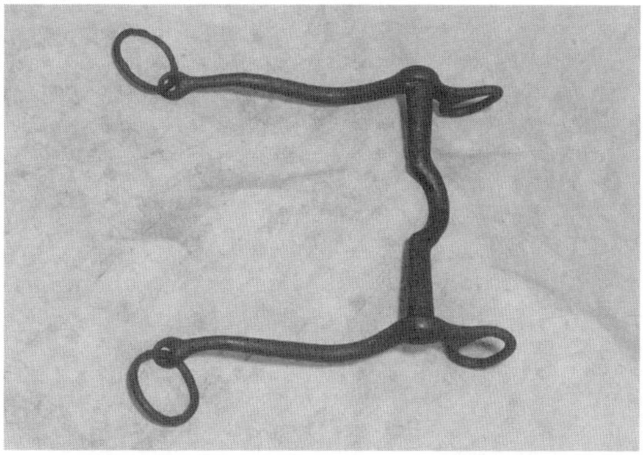

The Parker Ranch bit or pony bit, very popular on the ranch. It was made in eight or nine different variations for the U.S. military between 1860 and 1880 by Royal E. Whitman. Robin Dutson photo; W. C. Bergin Collection.

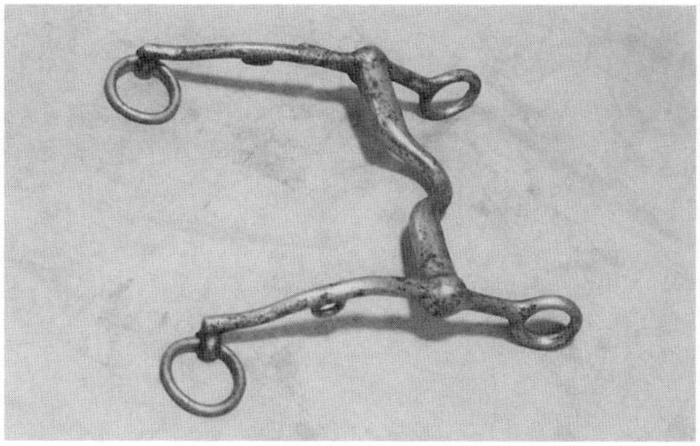

Another popular bit on the ranch, also of military origin, stamped RIA for Rock Island Arsenal. Shanks were curved slightly by Parker Ranch blacksmiths to give more of a grazing bit effect. These bits were custom made for Parker Ranch during Dick Penhallow's management. Robin Dutson photo; W. C. Bergin Collection.

Mexico was replaced by a more streamlined dally horn *('ōkumu)*. Traditional *paniolo* craftsmen have maintained their characteristic standards for well over a hundred years.

The spur and bit makers (loriners) of Hawai'i were typical blacksmiths, making everything from horse gear to gate hinges. Certain of these men became particularly noted for their equestrian-related products. At the turn of the nineteenth century, as World War I was approaching, the influence of the cavalry standard for horse equipment came into being. This high standard of horse gear required by the military extended into World War II, placing emphasis on quality and function. It influenced the bit and spur maker of Hawai'i, especially with the advent of Monel and, eventually, stainless steel.

The most notable product that the early war years brought to Hawaiian horsemen—and particularly Parker Ranch—was a simple metal cavalry bit that was a refined grazing-type model with a low port. There were several generations of this bit made by Royal E. Whitman between 1860 and 1880. It became the favorite mouthpiece of many local horsemen, but to Parker Ranch it became standard issue, and eventually it was given the name Parker Ranch bit. Some refer to it as the pony bit because of its refined makeup. Today in the new millennium, some complain that when corroded, the mouthpiece became abrasive to the tongue and jaws of horses. The old Parker Ranch bits, however, are still cherished by the discriminating horseman, their forebears of over a century ago having enjoyed the identical product.

Despite consistent use of the Parker Ranch bit, there were several prominent ranch horsemen who used locally made bits. Some notable bit makers supplying Parker Ranch men included Frank Gouveia of Kona, Manuel Miranda of Kūka'iau Ranch, and Louis Garcia and Manuel Sebastian Santos of Maui. Another loriner of Maui, Hichiro Karimoto, was actually born on Parker Ranch, a nephew of ranch blacksmith Duke Karimoto, who made stylish bits and spurs until his death in 1999.

Maui's influence on bits and spurs began in earnest in 1920 when Edwin (Robert Kingston) Lindsey of Waimea moved to the HC&S Ranch on Maui. He took with him a loose cheek bit with a pattern that Maui horsemen found appealing. Edmund Kaloi Cockett, the plantation master blacksmith, began reproducing the pattern, which by its popularity spread among Maui blacksmiths, including Manuel Sebastian Santos and others. Even today, the loose cheek blacksmith bits are referred to as the Waimea bits or Parker bits.

Donnie DeSilva, retired Parker Ranch horse foreman and noted craftsman, is also a bit and spur maker. The brothers Tadao and Yasuo (Yanko) Nakata were Parker Ranch cowboys as well as bit and spur makers.

Changing the Bit

The transition of a horse in training from a snaffle to a curb bit in the language of the Parker Ranch cowboy is to "change bit." The more notable horsemen of the ranch gained their stature largely by consistently producing well-reined horses in this process. There were three basic methods to changing the bit, and Mexican influence prevailed, especially with the use of the hackamore *(haquima)*.

The direct method simply replaced the

snaffle with the curb bit at any appropriate point of progress in the colt's training. This was done with no additional reins, and the success of this method was limited to the "soft" hands of highly experienced reinsmen *(rienadore),* who used such a horse for light work under calm conditions.

The *pūnuku* (noseband or *bosal)* method had two subtypes; both used an extra set of reins *(dos riendas)* from the noseband or *bosal.* In both cases, typical reins attach to the bit. Commonly, the noseband of a Parker Ranch bridle was made of stout 1-inch leather placed above the bit loops of the headstall and fastened with a brass oval buckle. The noseband could be tightened to a snug fit, to which usually rope, rawhide, or leather reins were attached. These paralleled the curb-bit reins, and all four were used to make gentle turns and stops under calm training conditions.

Another method uses the *bosal.* The *bosal* is a stiff, braided rawhide noseband that could be used as a full hackamore beneath the headstall or threaded through the noseband loops of the cheek pieces of the headstall. Rope reins are typically tied to the rounded knot of the *bosal* that lies behind the chin. The rider again has four reins to work the horse with bit and *bosal* in place, and selective control of rein pressure is allowed. In both the noseband and *bosal* methods, the selective use of the two rein sets gives the horse a gentler transition. Also, in a wreck or the sudden need to turn or stop, heavier pressure on the noseband/*bosal* reins allows adequate control, while the bit reins can be left to mild use. A light touch on the bit under the more tense work situations goes a long way toward preserving the integrity of the mouth of a young horse.

The third form of transition comes via the use of a shanked snaffle as an interim bit between the snaffle and curb types. Rather than simple rings for cheek pieces, the snaffle has swivel (hinged) shanks that are usually short (3–6 inches). Some of these are referred to as Tom Thumb bits due to the short shanks. Others have English or Argentine names. In all cases, they use a chinstrap that—coupled with short-shanked cheek pieces and a jointed mouthpiece—forms a very meaningful compromise between the snaffle and the curb bit.

Changing the bit is a critical juncture in the career of a young horse. In the hands of a less-experienced or impatient horseman, a calloused mouth results. Such horses are difficult to salvage from being hard mouthed *(uaua).* On the other hand, the finesse of a good reinsmen—in addition to launching a productive career for the mount—underscores his reputation.

The Bridle: *Kaula Waha*

The bridle *(kaula waha* or *freno)* of a horse has the primary function of suspending the bit in the mouth of the horse and linking it to the rider via the reins. With its noseband, it can provide a halter effect when the neck rope *(kaula 'ā'ī)* is passed through it after being tied neatly in a bowline *(pōlena)* knot around the horse's throatlatch. The halter effect is vital when tying or leading a horse or making him face you. The reins *(laina* or *rienda)* transmit the control function to the rider.

The bridle is made of seven major parts: the brow band *(kaula lae),* the cheek straps *(kaula papālina),* poll straps *(kaula po'o),* throat strap *(kaula pu'u),* chin strap *(kaula 'auwae),* noseband *(pūnuku),* and reins *(laina).* Except for the chinstrap, the bridle is made of 1-inch harness leather of top-grade quality. The chinstrap is usually made of rawhide.

It is likely that the tack that accompanied Captain Cleveland's landed horses was of an English design and included bridles with the above-described components. They were leather bridles and reins, whose useful life was shortened in Hawai'i's highly humid and salty environment. With horse numbers growing rapidly, especially on O'ahu, the availability of English tack and saddlery was to be expected given the presence of craftsmen.

As the *vaquero* arrived with tack made mostly of rawhide, the presence of leather tack was limited to the stables of royalty and the affluent. Local craftsmen became increasingly attracted to rawhide, especially with Mexican mentoring of braiding and weaving and the availability of salted cowhides. When properly treated and prepared, rawhide made a superior alternative given its weather-resistant nature.

The Spanish influence on leather craftsmanship persevered through the ensuing century, a fact confirmed at the 1876 Philadelphia Centennial Exposition. A Hispanic saddle maker from the Fred Lyman Saddlery of Hilo won first place for a bridle and saddle entry.

As a boy, I observed the high-quality leatherwork in tack made by Manuel Miranda of Kūka'iau, John Kauwe of Parker Ranch, and the Rodriguez brothers of Pa'auilo Mauka and Hilo. The cowboy with the best-conditioned tack was a frequent user of neat's-foot oil, a product made by Nourse and Company. As the leather aged from a rich russet to a brown-black, it became increasingly attractive to the eye and pliable to the horse and rider. A well-cared-for headstall-martingale combination may last a decade or two, depending on the cowboy.

Upon his retirement first from Kūka-'iau and then Parker Ranch, I saw John Holi Ma'e, a great but frugal cowboy, revert to many of the old ways. Over the thirty years of his postretirement, he made his own tack with rawhide—some braided, some flat. He routinely seasoned rawhide straps by using them heavily as cinch straps *('ili huki* or *látigo).* When properly seasoned, he would cut 1-inch straps and make headstalls and martingales, which were then put to heavy use. It was not long before his tack looked, felt, and performed like it was cured leather. I still have his tack, and with periodic application of *'aila pipi* (tallow), the supple strength lives on. I would guess that until Green and Company built the tanneries in Waimea, the importation of cured leather hides had limited availability; so, like Holi, the *paniolo* of old did quite well with *'ili pipi* (cowhide).

Paniolo also made tack from remnants of their old, worn *kaula 'ili.* Making reins from these old ropes was commonly done, as the well-worn *kaula 'ili* was soft, smooth, and pliable. Occasionally, full headstalls and martingales were made of old *kaula 'ili.*

As Parker Ranch grew, so did auxiliary support services such as saddle and tack production. A style that was characteristic of the ranch evolved. These characteristic styles exist today and are referred to as the "Parker Ranch Standard." Hence we have the typical ranch headstall, made of seven components, each individually adjustable (except for the brow band). Each cheek strap has an upper and lower buckle adjustment as well as a passage loop above the latter to accommodate the noseband, which has its own buckle adjustment. The brow band is roomy in length and has long riveted loops on each end to accommodate both the poll *(kaula po'o)* and throat strap *(kaula pu'u)*. The poll strap has no buckles, but is of copious length, with numerous adjustment holes on each end. The poll strap enters the buckle of the upper portion of the cheek strap on each side of the horse head in the area of the temples *(maha)*. Threaded behind the poll strap and through the side loops of the brow band is the throat strap, with its buckle coming beneath the throat latch to meet its counterpart on the left cheek of the horse's head.

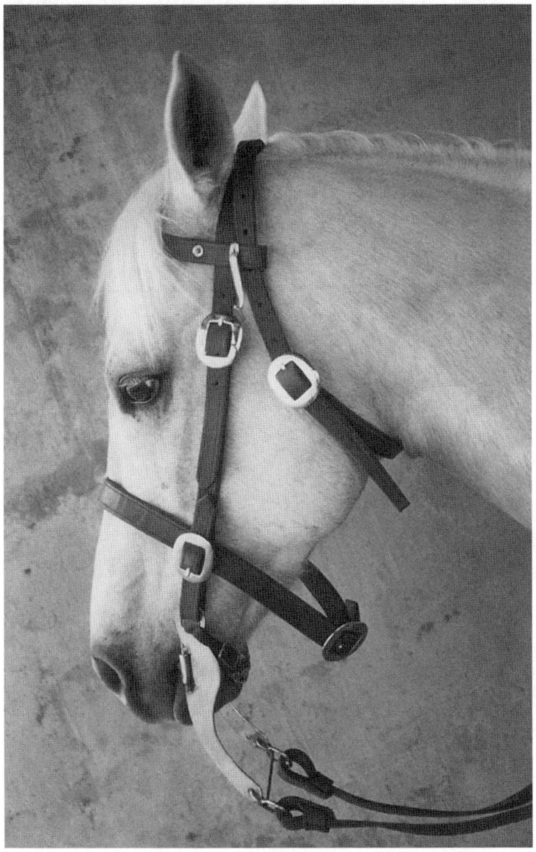

The Parker Ranch Standard headstall, with heavy-duty 1-inch harness leather strapping and flat, oval buckles. Photo by W. C. Bergin.

The lower portion of the cheek strap also has its buckle adjustment for looping through the bridle ring of the bit. Often, an extra strip of leather lines the inside of the buckle adjustment of the lower cheek piece, serving as a shield against the wear and tear of the metallic ring of the bit.

The *kaula 'auwae* or chin strap *(varvada)* has been described as a key factor in leveraging control of the horse via a curb bit. Sometimes called a curb strap, this component of the bridle was made of rawhide due to its strength, pliability, and availability. The Mexican ring bit *(chileno)* had the ring itself curving along the chin groove, serving in a rather severe manner, especially in the hands of an abusive horseman. There is little evidence that the *chileno* bit was accepted as *paniolo* tack.

The *paniolo* fashioned the *kaula 'auwae* by folding a 14–16-inch strap of rawhide (one-half to three-quarter inches wide) lengthwise and threading it though each cheek ring where it rests aside the chin groove. The balance of the strap is threaded back through a slit on the near left *(hema)* side with a simple clove hitch, binding the *kaula 'auwae* at its desired length and tautness. To be adjusted correctly, two fingers should easily slide beneath the chin groove and the *kaula 'auwae* itself when the horse is relaxed. Often a spare *kaula 'auwae* is tied loosely between the rein rings or simply hung from either ring. A fatigued *kaula 'auwae* is likely to break under severe cattle-working conditions or when a horse happens to *'oala* (buck) or *'ōkupe* (stumble). In all cases, a broken chinstrap leaves the rider with no "brakes," and this is a terrifying state. A spare chinstrap comes in handy at such times, but in a pinch, a cowboy can cut off part of a saddle string *('ili ma'awe)* or stirrup lace. Of course, one must first get control of the horse.

The *paniolo akahele* (prudent cowboy) periodically checks the *kaula 'auwae,* changing it at the slightest sign of wear, as he does with the *'ili huki (látigo), kaula 'ōpū* (girth or front saddle cinch), and *kaula keko* (back cinch), with the care of the reins *(laina)* notwithstanding.

On Parker Ranch, just like the *paniolo kahiko* (old-time *paniolo*), cowboys of the present day rarely use chinstraps of a severe nature. Yet while the humane flat nickel chain chinstraps of English origin were not uncommon, today they are hard to find, as production of this durable chinstrap ceased after World War II. Inexpensive nylon and chain chin straps, as well as leather and chain combinations, can be found on the ranch. The chains, however, are susceptible to rust and corrode quickly. After 150 years, rawhide is still the choice of the prudent *paniolo* for *kaula 'auwae.*

Blinders

Blinders *(pani maka* or *tapaojo)* are a bridle accessory, and they were used on unbroken, green-broken, or unruly horses in years gone by. When the cowboy of the old West and the *paniolo kahiko* had to deal with aged mustang-type horses, sliding the blinders down over the eyes would cause any bronc to freeze up. It allowed saddling, mounting, doctoring, grooming, shoeing, or packing with gear to be accomplished with reasonable safety for the horseman.

The *pani maka* is made of pliable but heavy leather. It measures about 16 inches in length and 4–5 inches wide. Often the middle is narrower and sides wider to create a broader eye-patch effect. Each end of the *pani maka* has a pair of 1-inch slots through which a three-quarter-inch throatlatch strap *(kaula pu'u)* with an adjustable buckle is passed. The *pani maka* is placed over the brow band *(kaula lae),* with the throatlatch placed over the poll strap *(kaula po'o).* The throatlatch strap of the *pani maka* surrounds and buckles over that of the bridle and fits snugly. Before mounting or dismounting, the *paniolo* reaches over the horse's poll and pulls or pushes down each side of the *pani maka,* completely controlling any visible access for the horse. Once mounted, the cowboy reaches forward and pulls the blinders above the horse's eyes where they rest on the forehead. In dismounting, the process is reversed. As the horse increases its gentleness, only the left side is blinded with the *pani maka.* Once full and mutual trust is achieved, the *pani maka* is no longer needed.

With a *pani maka,* dismounting a green-broken colt in the middle of Holoholokū, once a single paddock of 16,000 acres, could easily be accomplished. When pulled down, the *pani maka* composes a colt and safeguards against bolting as a runaway. But more commonly it allows the cowboy to mount and dismount a bronc while avoiding being cow-kicked. A horse cow kicks by reaching forward with the hind leg. A well-placed cow-kick usually strikes a cowboy near the right pocket, hip, or thigh and can be both painful and traumatic. A *pani maka* eliminates front, lateral, and rear visual access—but not the horse's suspicion; therefore, some degree of snorting *(kani ka ihu)* still occurs. Even when using a *pani maka,* a wise horseman never lets his guard down.

A shying horse, especially when approached by a needle-toting veterinarian, can be calmed by a *paniolo* gently placing a hand over the eye of the horse on the side being approached by the *kauka lio* (horse doctor). This creates a temporary calming effect, appreciated by the veterinarian.

The *pani maka* is rarely used today. Around 1960, foal crops began to be handled periodically between weaning, yearling, and two-year-old stages. Thus green-breaking begins with gentler two to three year olds as opposed to arrogant five to seven year olds. The trust between the horse and master was enhanced by earlier starting and gentler horsemanship.

The Martingale

Joining the bridle via the noseband to the saddle girth is another aspect of horse gear called a martingale or *makīnikela* by the Hawaiian cowboy. For the *vaquero,* the martingale is *el pretal,* which also serves as a breast collar. In addition to the noseband-to-girth strap (tie-down), the martingale includes a breast collar that encircles the shoulders of the horse to keep the tie-down in place against the chest. The breast collar component usually joins the tie-down at a mid-chest ring—usually of brass and about 2.5 inches in diameter—attached to a breastplate of leather. The upper portion of the tie-down has an adjustable loop (via a flat, oval buckle) for the noseband to pass through and a second in-line buckle for adjustment. Often this tie-down portion ends at the middle ring, and the second portion then extends down between the horse's legs to form a loop through which the girth is passed when saddling up. The girth portion of the tie-down also has a buckle for length adjustment. The tie-down component may also be a simple adjustable strap from the noseband to girth with a "keeper" effect, through which it slides at the breastplate ring. The breast collar also has an adjustment buckle placed midway up the left shoulder portion. A properly placed and adjusted martingale should lie comfortably in the juncture of the neck and shoulder of the horse to keep the tie-down close to the horse's chest.

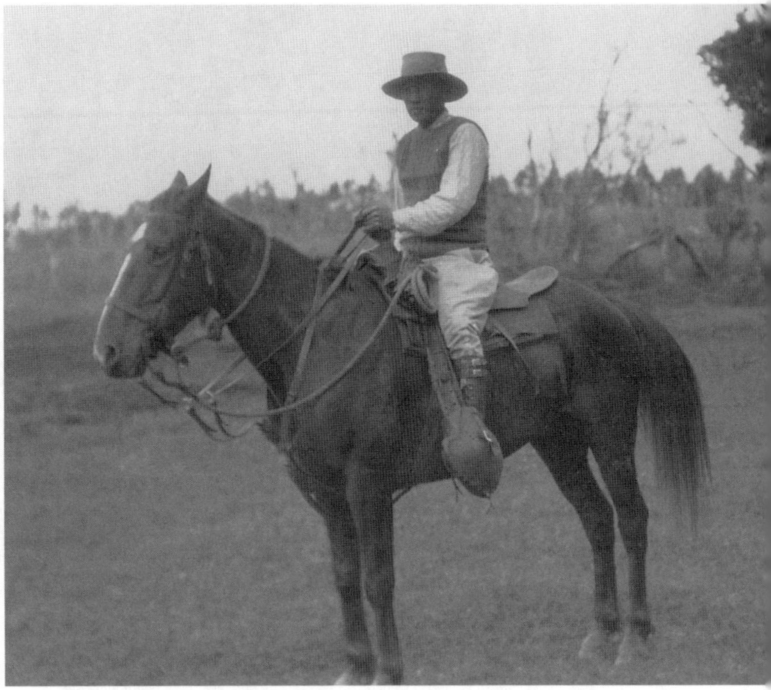

Yutaka Kimura using a running martingale, short-eared tapaderos, *and riding pants. Use of woolen vests was not uncommon. Hartwell Carter Collection.*

A variation on the typical martingale is the running martingale, whereby the single tie-down portion to the noseband is replaced by two shorter straps that end in smaller rings, through which the reins are passed. Some horsemen use these in the transition period between snaffle and curb bits.

Intended usually as a means of controlling high-headedness, or rearing of the horse's head, the tie-down portion is adjusted to give some freedom to the animal. Extension of the head and neck are vital to running and jumping, and overly taut tie-downs can be detrimental to necessary function. Some cowboys are inclined to use the tie-down alone, without a breast collar. The tie-down adjustment for a cowboy horse is always more liberal than that used in a rodeo arena, where flat, clear footing prevails. A looser tie-down allows the cowboy horse to lunge and leap over logs and other obstacles with relative ease.

The typical Parker Ranch martingale, like the bridle, was made of heavy harness leather usually about an inch in diameter. Like the headstall, multiple buckle adjustment points reflected a broad range of horse breeds, sizes, and types used on the ranch. In olden-day Parker Ranch, a cowboy's gear might have been on a Thoroughbred one day, a Morgan the next, and—between branding *(kuni pipi)* and shipping *(hōʻau pipi)*—a part Percheron mount for the last days of the week.

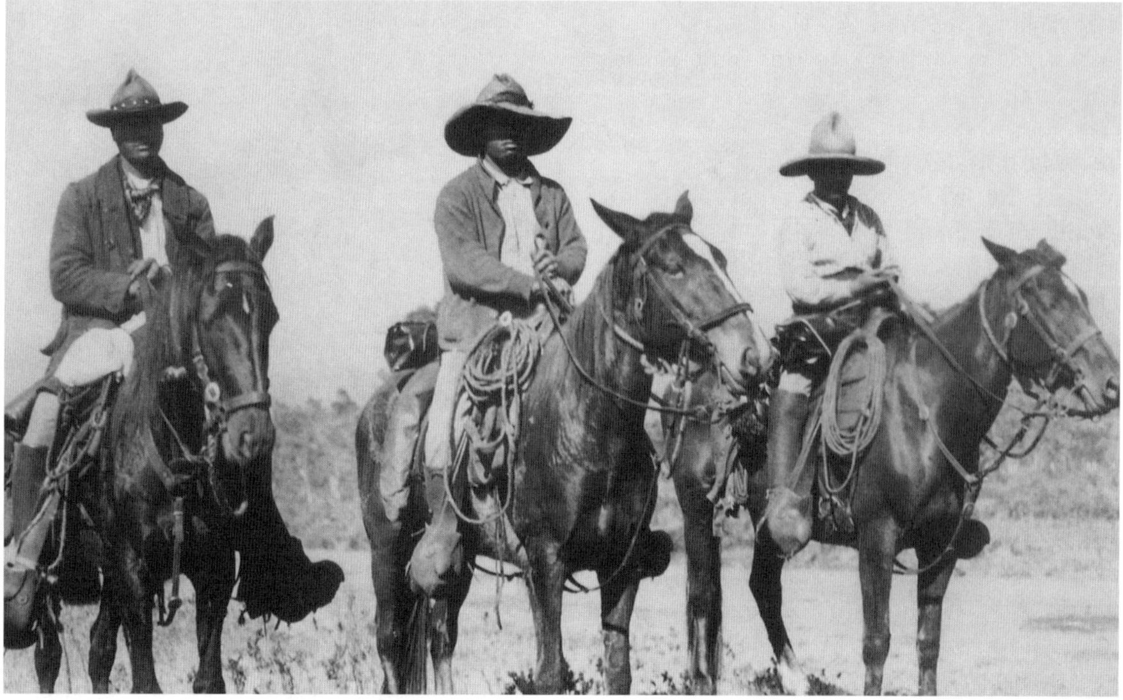

Three paniolo *using martingales, one with tie-down only. The black rubberized slickers, made by B. F. Goodrich, were used before the current yellow version was developed. A. W. Carter Collection.*

The Hawaiian Cowboy Saddle

The Hawaiian cowboy saddle *(noho lio),* nicknamed the Hawaiian tree, has roots deep in Mexico, a practical metamorphosis of the *vaquero* saddle brought to the Islands in the 1830s. Technically, a saddletree *(fuste* or *noho lāʻau)* is the wooden body to which rawhide or leather is applied in producing a complete unit.

The basic components of a saddletree begin with the fork at the front, which is an inverted Y that is referred to as a *campaña* by the *vaquero* and *hāmana* by the *paniolo.* The withers of the horse refer to the front portion of the back that forms a significant ridge as it articulates with the neck. The fork arches over the withers of the horse and is appropriately elevated to accommodate the spine bones. The open area between the fork that forms the arch is called the gullet. The upper portion of the fork is referred to as the pommel, upon which the saddle horn *(cabeza* or *ʻōkumu)* is located. The lower portion of the fork is attached to a pair of parallel, flat side boards called bars *(tablas* or *iama)* that lie on the horse's back from the withers area to the loin. Whether the bars are set high or low on a horse's back is determined by the height and width of the gullet. Early horses of Hawaiʻi had high withers.

The cantle *(teja* or *poe)* is the back wall of the saddle seat, contoured to a rider's bottom and attached laterally to the bars below. As with the fork, the cantle is attached by dowel pinning *(kui lāʻau)* and glue *(kolū* or *mea hoʻopipili).* The saddletree is covered with rawhide sewn *(humuhumu)* tightly with goatskin *(ʻili kao)* suturing.

As in the bit and spur, the Mexican saddle traces back to the horse gear of the conquistador in his conquest of Mexico in 1519. Over the ensuing three centuries, the saddle maker of New Spain made significant and practical changes to the saddle of the Spanish warlord. The high, ornate (but not functional) pommel was replaced with a large ball-shaped saddle horn. The steep, high, and perpendicular cantle was lowered, angled back to 45 degrees, and contoured to accommodate the *vaquero's* bottom conformation. These changes created a functional saddle horn to which one's rope could be fastened (dallied or tied hard and fast) and a seat and cantle that became user friendly. The tightly-bound conquistador, between a steep pommel and high wrap-around cantle, invited entrapment in the saddle should his mount buck, stumble, or fall. To the working *vaquero,* already challenged by wild cattle hunts and unruly horses, quick saddle exit was a must, both for tying down his prey and avoiding being rolled upon by his falling mount.

By the time the *vaquero* ventured to Hawaiʻi, the saddle horn was flattened with a wide cap and had developed somewhat of a neck between the pommel and the horn itself.

The Mexican saddle and its tropical cousin, the Hawaiian tree, are similar. First, both have a rawhide-covered tree with no attached leather covering the horn fork, bars, or cantle. Leather use was limited in both saddles to make the skirt and tailpiece (Hawaiian tree only), stirrup leathers, and *tapaderos.* Both are light, hardy, and comfortable saddles despite their rugged appearance. Both had prominent dally horns before the typical cowboy of the West realized their practical value.

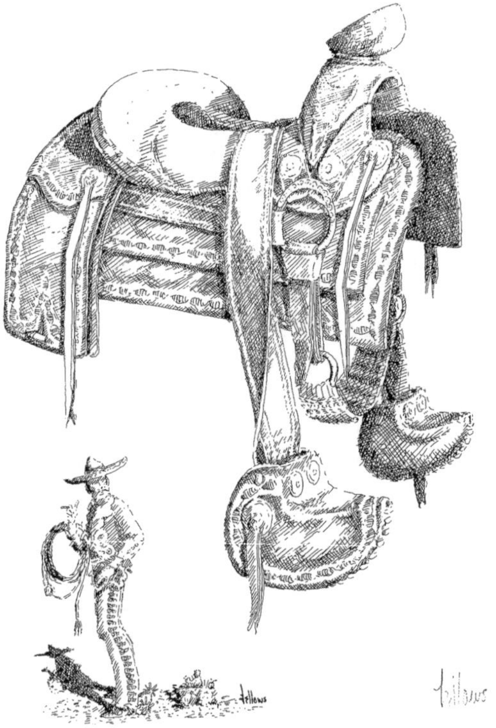

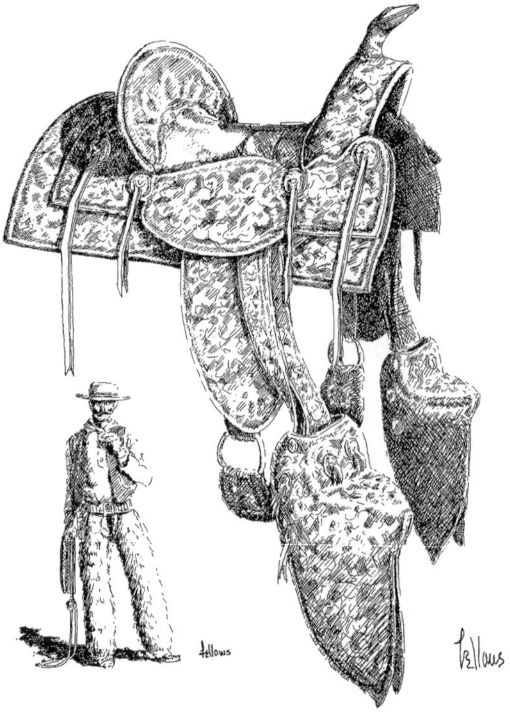

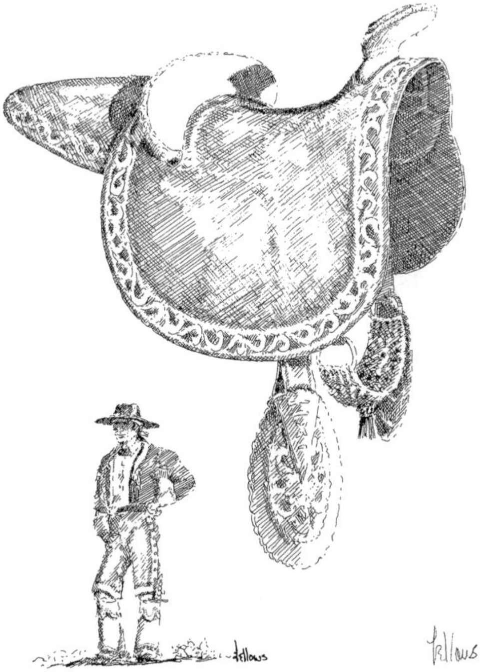

Top Left: *Saddle of old Mexico.* Cabeza de bala *is the term used to describe the ball-shaped saddle horn. Fred Fellows print.*

Top right: *California stock saddle, 1865–1935. Fred Fellows print.*

Left: *California* vaquero *saddle.* Cabeza de zaldivar *is the term used to describe the platelike saddle horn. Fred Fellows print.*

Opposite page: *Manufacturing Harness Co. advertisement, Honolulu.*

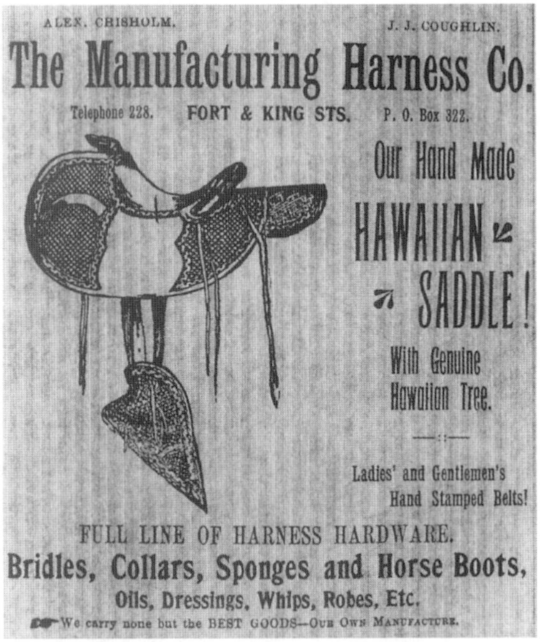

There are differences also—mainly in the saddle horn and rigging. The Hawaiian saddle maker lengthened the neck and refined the head of the saddle horn, a design that remained unchanged for over a century and a half. He also added a saddle skirt, whereas the *vaquero* saddletree was placed directly on horse's backs over thick padding.

Rigging—referring to those fixtures that attach the saddle to the horse—is common to all cowboy saddles. In the Mexican saddle, the front rigging was formed by a stout leather strap loop connecting the fork to the rigging ring *(argolla).* Early American saddle makers called this method the Sam Stagg rigging. The rigging ring could be adjusted front or back by the *contrareata,* or the rear rigging strap, which joins the two rings by wrapping around the back of the cantle with a loop for adjusting. The Hawaiian tree retained this adjustable rigging strap in the form of a three-strand *(hili kolu)* braided strap called the *'ili hope* or rear rigging strap, also with an adjustment loop.

The adjustable rigging controlled by the lengthening of the *contrareata* or *'ili hope* is a feature unique to the Mexican and Hawaiian tree saddles. Cowboy saddles typical of the American West were permanently rigged with single or double rigging rings. The single rigging could be a center-fire model with the rings centered midway between the fork and the cantle. This system was rare in Hawai'i. More commonly, the early Hawaiian cowboy used a three-quarter single rigging whereby the rings were placed forward a quarter of the distance between the fork and cantle. The 3.5- to 4-inch-wide *'awe'awe* rings had the stirrup leather *(arzones* or *'ili ke'ehi)* threaded through them, keeping the stirrups in a forward seat position and giving the cowboy an excellent riding posture. Another feature of this rigging/stirrup leather relationship is the very clear access the *paniolo* had to the cinch straps *('ili huki* or *látigo)* when saddling up or unsaddling. When the left stirrup is thrown over the horse's back for cinching, access to the *'ili huki* is unencumbered. When not threaded through the rigging ring, the rider must thread the cinch strap in between the stirrup leathers as they lie aside the horse. Very tight cinching is difficult this way, and when the mission calls for roping a wild bull, a loose cinch is a high risk.

The advantage of adjustable rigging was the option to move the rigging ring forward by lengthening the *contrareata* or *'ili hope* to a fully forward position directly

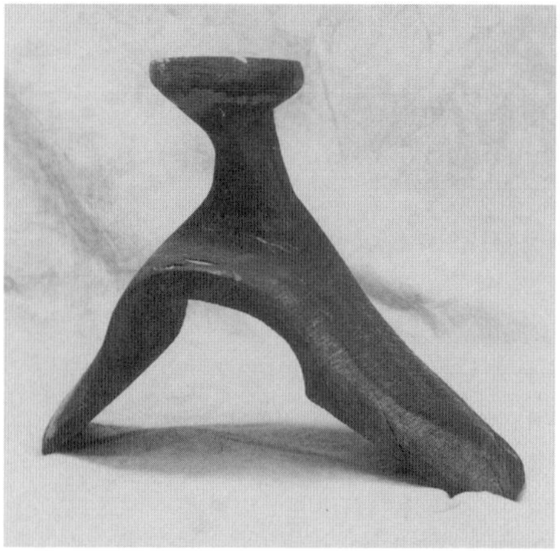

Barren saddle-fork Hawaiian tree made from sumac (neneleau). *Robin Dutson photo; W. C. Bergin Collection.*

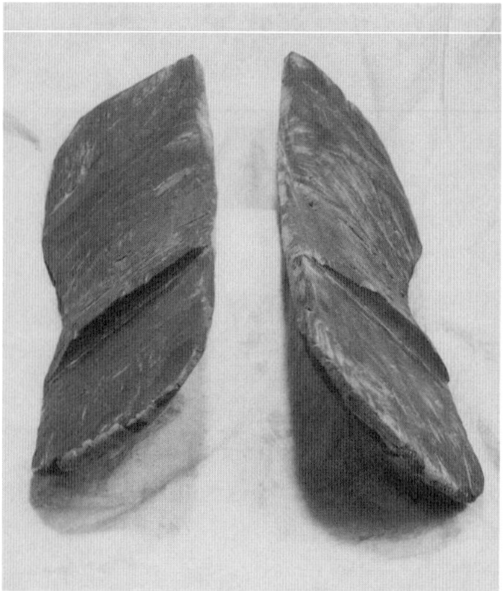

Barren rough-hewn saddle bars made from neneleau *or sumac. Robin Dutson photo; W. C. Bergin Collection.*

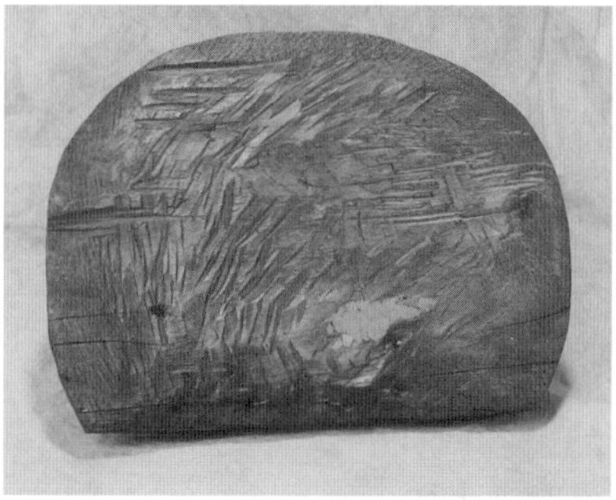

Barren rough-hewn saddle cantle made from sumac. Robin Dutson photo; W. C. Bergin Collection.

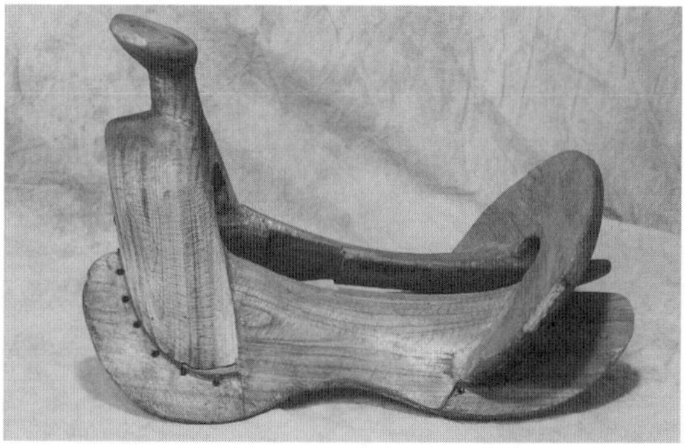

Barren assembled saddletree complete. Robin Dutson photo; W. C. Bergin Collection.

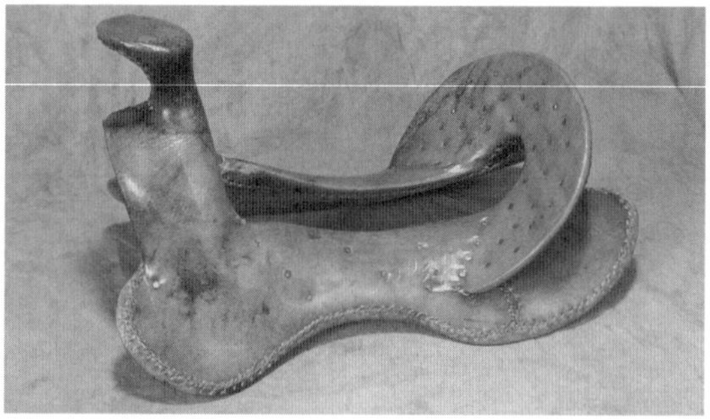

Rawhide-covered saddletree sutured with goat-skin lacing. Robin Dutson photo; W. C. Bergin Collection.

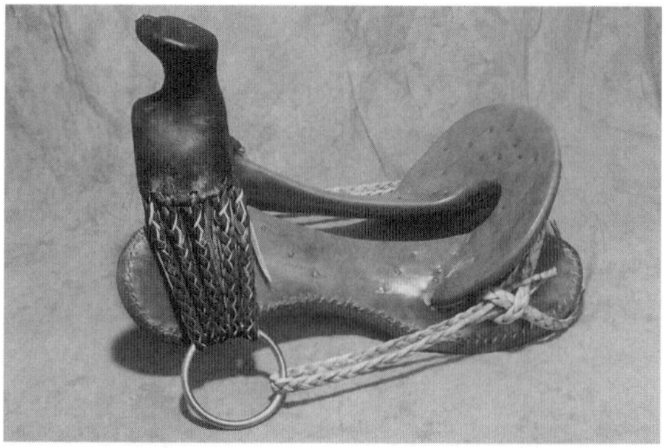

Rawhide-covered saddletree with ʻaweʻawe *(rigging) complete, including* ʻili hope *looped around rigging ring and wrapping around behind the cantle. Robin Dutson photo; W. C. Bergin Collection.*

below the saddle horn. The *paniolo* now added a rear cinch *(kaula keko)*, forming what is referred to as full double rigging. Over the last hundred years, the Parker Ranch horse has become a more heavily muscled bullock-type mount, and the prominent withers of the Thoroughbred type horse were compromised by the round-back effect that resulted from Morgan and quarter horse infusion. For purposes of heavy roping, this necessitated a more heavily and broadly rear-cinched saddle. Heavy roping could include wild bulls *(pipi ʻāhiu, buiso* or *orejano)* or domestic cattle, for horn trimming *(hao komo)* or lancing abscesses due to lumpy jaw *(ʻauwae puʻu* or *boludo)*. A rear cinch became a necessity.

The same cowboy, when in need of a stripped-down saddle for New Year's or Fourth of July races, merely lifted off the rear cinch apparatus and tightened up the *ʻili hope* to return to "three-quarter single" rigging. *Paniolo* often returned to the single rigging option when training bronco colts. The shipping saddles used to *hōʻau pipi* at Kawaihae were rigged three-quarter single routinely.

By 1960, most *Paka paniolo* (Parker Ranch cowboys) had completely shifted over to the three-quarter double rig, as the Thoroughbred and standard bred influence on horse conformation and attitude was waning significantly.

Aside from developing a stout and prominent dally horn *(ʻōkumu)*, the Hawaiian saddle maker crafted an ingenious, attractive, and strong front rigging called the *ʻaweʻawe*, which means "tentacles."

The building of the saddletree, its rawhide covering, and the *ʻaweʻawe* rigging are amply covered in a paper entitled "The Making of the Hawaiian Tree Saddle," written by my wife, Pat Bergin, and I quote her here:

> The role of the saddle maker was an important one in early Hawaiian communities whose economies were dependent on ranching. Frequently, two or more individuals employed by a larger ranch would service the entire area. It was not unusual for retired or disabled cowboys to become saddle makers later in their careers. Then, of course, there were those who took an interest in the skill very early in their youth.
>
> Construction of a saddle from start to finish could easily take 200 hours or more, not including the drying, seasoning and curing of the hide and wood, which normally took about six months. A good saddle maker might turn out about 15 to 20 completed saddles each year.
>
> Hawaiian saddles were basically patterned after Spanish saddles and traditionally had taller, wider horns, slightly higher cantles, braided front rigging straps (ʻaweʻawe) and a rounded skirt and tailpiece. Hawaiian saddle makers stayed away from cured leather for most of the saddle as rawhide was not only hardier and more readily available, but withstood moisture much better. It is said that the ʻaweʻawe is the strongest saddle rigging made, and it was adapted as part of the Hawaiian saddle because the cured rawhide could withstand the high moisture in Hawaiʻi and not rot.
>
> There were certain materials needed and steps to be followed to construct a

saddle from start to completion. Basically, these raw materials were: a tree-form or mold; wood and dowels for tree construction; neneleau (sumac) or hau for the fork and horn; rawhide to cover the tree; cured goat skin for lacing; a strip of rawhide approximately 5 feet long and 14 inches wide for the ʻawe-ʻawe; four brass rings for rigging; a pair of wooden stirrups; cured leather; and a pair of tapaderos.

The first step was selecting and harvesting wood for the saddletree. There were many varieties used by Hawaiian saddle makers. Some of these were: neneleau, hau, guava, milo, mango, ʻōhiʻa, kiawe, naio, māmane, pride of India (inia), Portuguese cypress, and monkey pod. Typically, koa was not used because of its heavy weight and the dowels had a tendency to squeak. Older or dead trees were usually selected, as there were fewer tendencies to warp when being seasoned.

The two bars (iama) and cantle (poe) could be made from any of these woods. The fork (hāmana) and horn (ʻōkumu), however, were usually carved from a stronger, forked branch of neneleau or hau, which were difficult to obtain as they usually grew in deep, dangerous gulches and took two men to harvest: one man to descend and cut it and another to haul the pieces up by rope.

The bars and cantle were then carved to fit the tree mold while the fork and horn followed the natural contour of the forked branch and were shaped in one continuous piece. The finished tree was then allowed to dry for about six months after which any warping or flaws that could hurt the backs of the horses were corrected.

The second step involved covering the seasoned tree with water-softened rawhide pieces laced into place with goatskin. Covering the tree was an important step. One large piece was used to cover the pommel and seat first, then underside pieces were added. The goat skin used for lacing was usually obtained from Kona, stripped into approximately one inch widths, soaked in water and stretched before it was used for lacing. Upon completion, the rawhide-covered tree was sun dried and allowed to contract.

Creation of the ʻaweʻawe was next and followed essentially the same procedure. A wet strip of rawhide approximate 5 feet by 14 inches was divided longitudinally into two pieces and wrapped around the horn down the fork and stripped into 12 parts on either side of the fork where it was braided into 4 strands that went through and under the 4-inch brass rings to join itself at its point of origin. This too was allowed to dry thoroughly. Wooden bars were threaded under the bars through both rings to provide tension during drying.

The final step was the addition of cured leather pieces. Using patterns, the skirting, seat with rear rigging rings, stirrup leathers, fenders, tailpiece and tapaderos were cut and fit to the completed tree and wooden stirrups. A lala was optional. The front cinch (kaula ʻōpū) was usually made of cotton, spun yarn (kaula lopiana) or manila (malina) cording and hand woven between the two brass cinch rings. The rear cinch (kaula keko) was made of heavy skirting leather between two brass rings. Both cinches were attached to the saddle rigging by rawhide latigo ('ili huki).

This then was the basic saddle. Decorative markings were added as desired.

Mochilla or *Lala*

Other vestiges of old Mexico saddles were the *mochilla* or *lala* and the tailpiece *(lala 'ihope, anquera)*. The cowboys occasionally referred to the *lala* or *mochilla* as a Mother Hubbard for its extensive apron effect on the saddle. It is usually made of a single large 3-by-3-foot piece of leather, or a laced pair of the same total dimensions, that is draped over the completed saddle. It is held in place by a hole in the front that accommodates the saddle horn and in back by a slot that accommodates the cantle. Sometimes it is tied in place by saddle strings and leather rosettes or metal *conchas*. The Hawaiian cowboy preferred a round-bordered *lala*, while his mainland counterpart enjoyed the square-cornered *mochilla*.

The *lala* added comfort to the saddle seat, protection from the elements for the under parts, and a guard against rope pressure when folded back over the thigh while the cowboy dragged a calf to the fire for branding. It was routinely used by the Parker *paniolo* until the period shortly after World War II. Today it is used only by traditional native Hawaiian cowboys such as Sonny Ke'akealani, formerly of Parker Ranch, and Kimo Hoopai Sr. and Mark Conley of neighboring Ponoholo Ranch. For the *vaquero*, the *mochilla* often draped low enough on the sides to protect his legs from horse sweat. In these cases, stirrup fenders *(fenderas, rosaderos)* were not needed. With the smaller *lala*, the Hawaiian cowboy needed the protection of stirrup fenders *(pale ke'ehi)*.

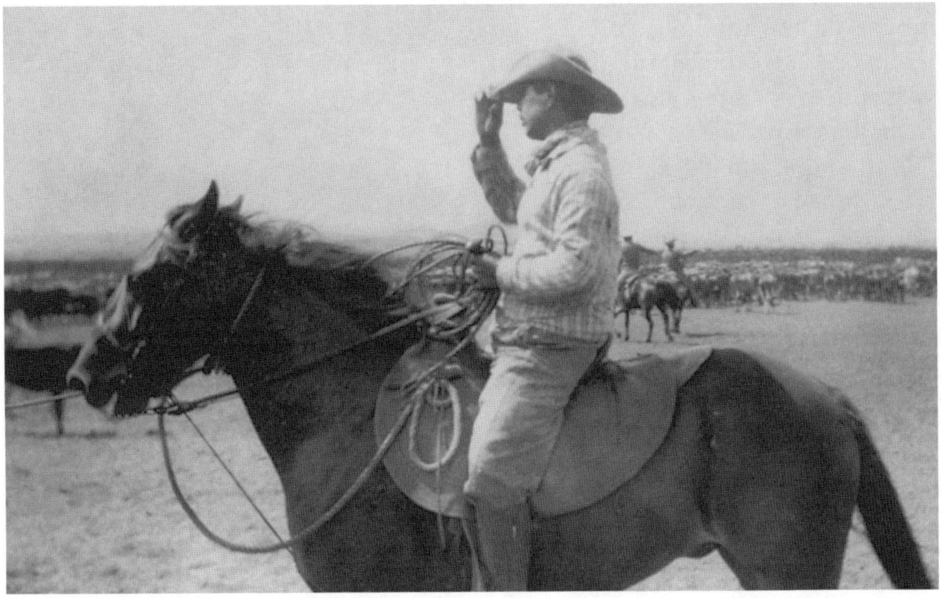

A mount illustrating the lala *and* lala 'ihope *at Pu'uhihale Corral cattle branding. A. W. Carter Collection.*

La Cincha or *Kaula ʻŌpū*

The cinch *(kaula ʻōpū)* of the early Hawaiian cowboy was modeled after the mohair *cincha* of the *vaquero.* Later, strands of cow hair (tail switch), horsehair (mane and tail), cotton, wool, or a twine called spun yarn were used to weave the *kaula ʻōpū* in a pattern learned from the Mexicans. Spun yarn is a twine made from Manila or sisal fibers. The cording would be fastened to the two end rings *(argolla* or *pōʻai)* by "pig knots" *(los nudos de puercos).*

A cinch may measure from 2.5 to 3 feet in length, depending on the size and type of horse being used. The nature of the work also is a factor, with wild bullock roping requiring a broader, more heavily stranded cinching than a *nānāʻāina (campiando)* man's casual riding. Original cinch rings were of black iron. Eventually brass, bronze, copper, and Monel were found superior due to their rust resistance, especially on the coastal plains *(kahakai)* with their salt spray *(ʻehu kai).* Today stainless steel rings are common, but the discriminating *paniolo* prefers the *nā lina keleawe* (brass rings), as did his forebears. Cinch rings are ideal at the 3.5- to 4-inch diameter, as pressure under tight cinching and strain is distributed evenly over a broader surface. Narrower cinch rings mean narrower cinches, causing pressure to be borne over a narrow band of a horse's girth. With a wild bullock lunging against the lariat, critical pressure and even traumatic chafing *(pōʻai* sores) of the horse's thin-skinned girth can occur. Chronic *pōʻai* sores put a good horse out of commission for months, and only time and a little ointment *(ʻaila pipi,* aloe sap or Bickmore salve) brings healing *(hoʻōla).* The *vaquero* relied on the advice of the *curan-*

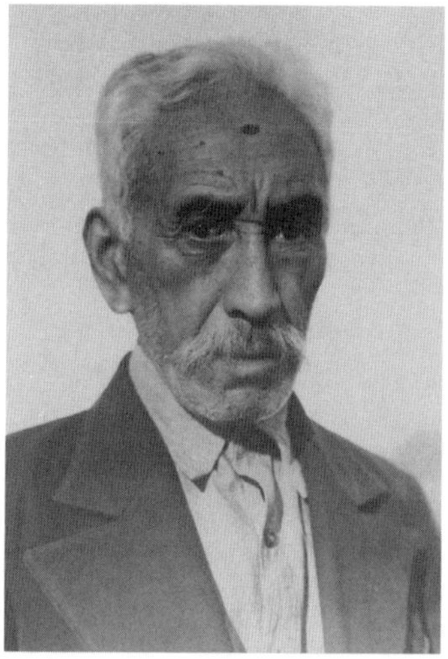

Samuel Puna Kauwē, noted ranch saddle maker and native Hawaiian confidant of A. W. Carter. A. W. Carter Collection.

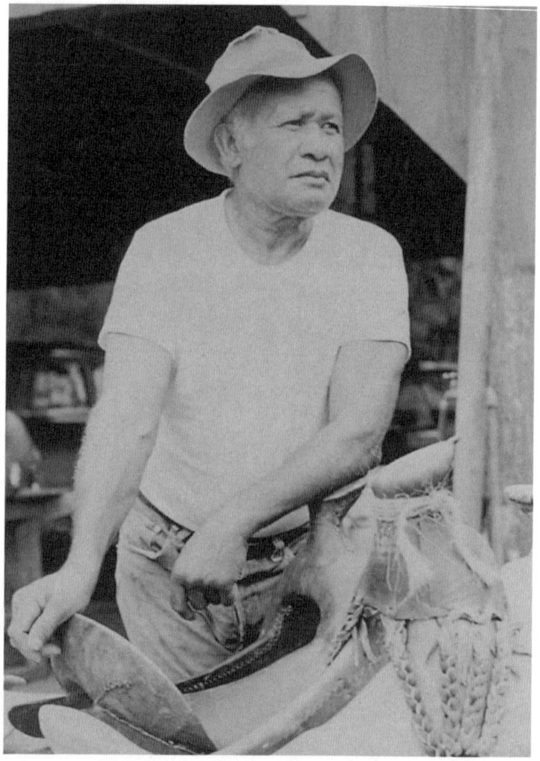

John Kianiani Kauwē, nephew of Puna and Hogan Kauwē, with uncovered, rawhide-covered, and ʻaweʻawe*-rigged saddletrees. Hartwell Carter Collection.*

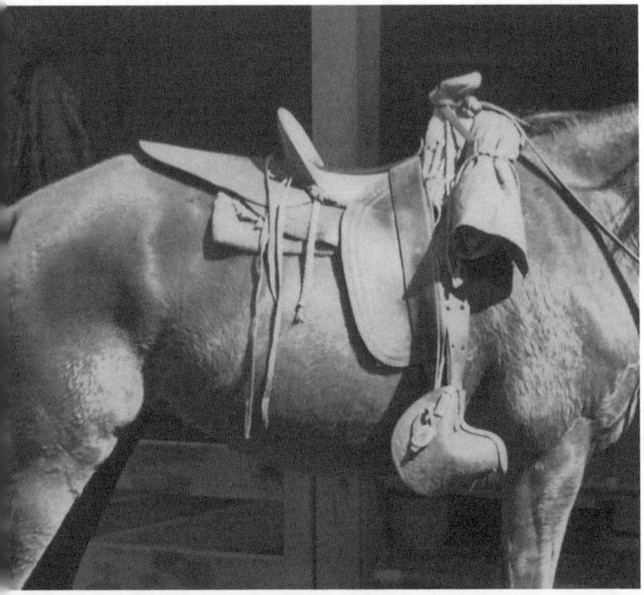

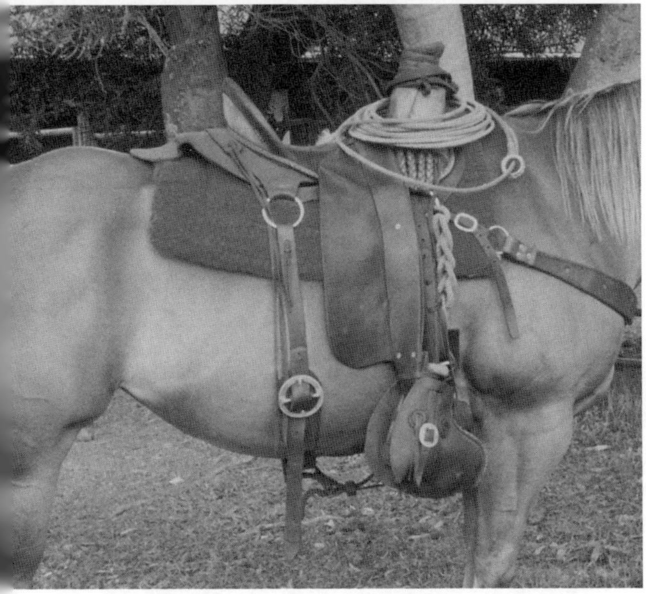

Top: *Typical Hawaiian tree cowboy saddle, 1927. The saddle blanket is of military origin. A. W. Carter Collection.*
Above: *Typical Hawaiian tree cowboy saddle, 2002. The saddle pad is of orthopedic fabric. The saddle is equipped with a* hapa lala *for rear cinch rigging.*

dero (healer) for such treatment, whereas the *paniolo* had an arsenal of Hawaiian medicines—notably aloe.

As the early days of Parker Ranch horse training became more centered on the Breaking Pen, aspiring cowboys would be assigned as roughriders to "green-break" the horses that would become their work string; there would be other assigned mounts and tasks. The technique of cinch making was one of high priority in terms of training. Once assigned to a cowboy gang or station, the *paniolo* used this skill in perpetuity, making new *kaula ʻōpū* as necessary, with the ranch providing the cording. The ranch discontinued supplying cording in 1991 when the warehouse closed. Cowhands now use commercially made cinches, and few of the younger *paniolo* have learned the skill.

The Rear Cinch: *Kaula Keko*

A unique fixture to create a rear cinch rigging evolved because the *paniolo* wanted as an option a back cinch for heavy roping, branding, weaning, and shipping. In place of the permanently mounted rear rigging, a sort of hybrid developed with a *hapa lala* (half *lala*) that sat over the rear saddle seat with a broad slit to accommodate the cantle. Riveted on each side were brass rigging rings to which the rear cinch could be rigged, either by rawhide cinch straps or billets for a buckle-on rear cinch. In all cases, the rear cinch had a keeper strap *(ʻili waena)* to keep it tacked forward and avoid the backward shifting that creates the flank strap effect, which makes any horse want to buck. The *hapa lala* could be laced into place via saddle strings or left to its own security; once mounted and cinched up, it can go nowhere. When not needed, it could simply be lifted off.

Tailpiece: *Lala 'Ihope*

The tailpiece *(lala 'ihope)* of a Hawaiian tree has a *vaquero* saddle counterpart called an *anquera* or *anquitera* that traces its origin to the armor plates, usually of leather, that graced the saddle of the conquistador of the sixteenth century. Functionally, it can barely be justified for accommodating a tied-on slicker or bedroll. It works well when another cowboy hitches a ride on the rumble seat to regroup with his runaway mount. Some cowboys prefer shorter, more practical versions. The *lala 'ihope* (tailpiece, flanker, *anquera)* is semicircular in shape, fastened behind the cantle by saddle strings *(ma'awe),* and extends over the horse's rump and partly down the flanks. The conquistador version dipped far down over the flanks to protect from sword and lance warfare.

Stirrups: *Arzones* or *'Ili Ke'ehi*

The term "stirrup" can be inclusive of both the stirrup leathers (straps) and the receptacles for the rider's feet, which are usually made of wood. The hoodlike, protective leather coverings over the wooden stirrups are called *tapaderos* or taps.

Joining the saddletree to the *tapaderos* are the stirrups, made up of a doubled leather portion *('ili ke'ehi* and *pale ke'ehi; arzone* and *fenderas, rosaderos)* and a wooden portion usually housed within the *pālulu* or *'o'oma (tapadero).*

For the Hawaiian cowboy, the integrity of the stirrup leathers was vital to his safety. The stirrup fenders *(pale ke'ehi)* merely warded off the sweat and grime of his well-worked horse. Given the high humidity and rainfall of the Islands, cured leather is not durable in strength, and a broken *'ili ke'ehi*—especially under intense working conditions—usually spells disaster. Frequent oiling of the leather coupled with frequent replacement of stirrup strings, usually of rawhide, was necessary. While leather was dear from the early days of Parker Ranch, the company also fostered safety for the workingman, and stirrup leathers and bridle reins were readily replaced when worn out. Stirrup leathers then, as today, were a full 3 inches in diameter and made of heavy skirting leather. The stirrup strapping—paired straps about 5 feet in length—is riveted to the aligned fender at both the top and bottom. At the top, the stirrup leathers are folded inward over the bar of the saddle and threaded down between the saddle skirt and tree, usually accommodated by a 3- to 4-inch wide depression on the under surface of the *iama* (bars or *tablas).* This inner strap of the stirrup leather is drawn farther down through the *'awe'awe* rings to the midway point.

The lower portion of the outer stirrup strapping continues downward below the fender and enters the *tapadero* or stirrup cross bar from the back. It then loops forward and upward to meet its descending counterpart at midpoint, where the two ends are overlapped by at least three parallel lace holes and snugly laced together. Parker Ranch tradition calls for the lower stirrup leathers to be rotated clockwise before lacing to their upper counterpart. This rotation causes the stirrup to remain facing forward, which makes for easy mounting and dismounting—but mostly serves to provide lower limb comfort for the rider.

As with other leather portions of the saddle, the fenders typically are stamped in a pattern unique to that cowboy's saddle. The inner and outer stirrup leathers are rarely stamped, as the wear and tear makes such visual enhancement impractical.

The *Tapadero*

"Taps" to the Anglo-American cowboy, stirrup guards or *tapaderos (pālulu* or *'o'oma)* remain a popular vestige of the Mexican saddle owing to the protection provided from brush, trauma, cold wind, and rains. The bulldog type *(ihu keko),* pancake type, and the pointed-nose *(ihu pua'a)* and rabbit *(lāpaki)* or mule-eared *(hoki)* versions are all found throughout ranch history, and the use of *tapaderos* remains a popular accessory. Beyond protection from the elements, the rider is more secure on an unruly or bucking horse with a lowered center of gravity. Spur action is also accentuated by the added centrifugal weight. Rabbit-eared *tapaderos* have been used at parting gates to slap an errant cow away from the gate. Finally, the *tapadero* adds character to the striking pose of a working cowboy in any venue, while still being a remnant of old Mexico.

The Parker Ranch Standard

All of the leather components of the ranch saddle—especially the actual stirrup leathers —were built to high standards. The "Parker Ranch Standard" meant using quality leather of heavy weight and high cure. It meant neat and tidy use of standard patterns and techniques. It meant beveling all cut edges for safety and appearance sake. But the most characteristic standard was that of individualized patterns of stamping *(lau kuni)* the borders of the saddle parts with designs unique to each *paniolo*. This tradition can be traced a century back, when the individual stamps were hewn from teeth and bone of various species, such as whale ivory *(niho palaoa),* and a variety of metals including steel, copper, brass, nickel, and Monel. The saddle makers of old had patterns that were *tapa (kapa 'āpana)* or quiltlike in appearance. When commercial production of steel leather stamps began, more typical Western designs prevailed. Floral carving was done until the late 1950s by Tony Mahi, the only known saddle maker to use such a design. Frank Hess, of recent Parker Ranch saddle making history, estimated that five hammer blows to each stamp die were required to achieve acceptable consistency in depth and appearance. The skirt piece alone, he calculated, needed some 1,980 blows on average. A completed saddle, including the *tapaderos,* would often require more than 6,000 hammer blows.

The Saw Knife: *Pahi Olo*

Accessory saddle accoutrements include a saw knife with case, a hammer and nail pouch for reshoeing in the field or fence repair, and knife cases, the latter of which may be packed by the rider himself, usually on the lower right leg or belt.

The saw knife *(pahi olo)* is a traditional instrument of the *paniolo,* handmade from a discarded carpenter saw and cut into a 2.5-by-12-inch blade, with the offside carefully honed into a long knife-edge. The saw edge is used for cutting tree branches and cow horns, while the knife-edge is there to be used as a machete, skinning blade, or razor. Legendary bush-ranging cowboy Ioane Ha'a, who reportedly was a rascal cattle rustler, is said to have used his *pahi olo*

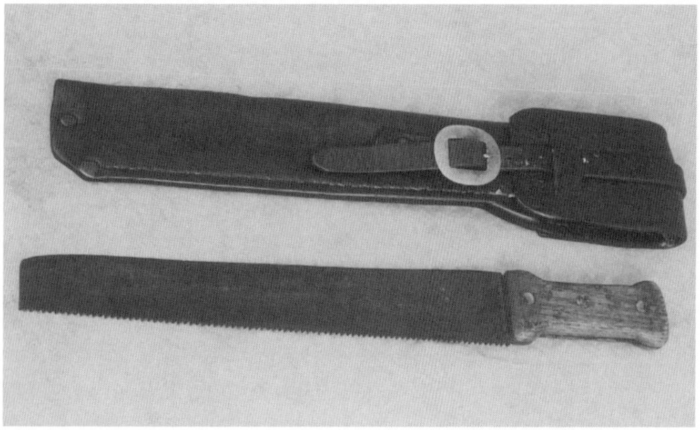

Saw knife (pahi olo) *and case. A saw knife is serrated on one edge, with a cutting blade on the other.*

to shear his flowing white beard to bluff his way past Parker Ranch personnel searching for him. The saw knife is kept in a sheath *(wāhi pahi)* usually tied over the left *'awe'awe* along with the *kaula 'ā'ī,* while the lariat or *kaula 'ili* is secured over the right rigging.

Ke Kaula Hele

The *kaula hele* is a *paniolo* saddle accessory with multiple uses, including as a "piggin" string to hog-tie calves, wild pigs, sheep, goats, or, if need be, *pipi 'āhiu.* Bullock tie ropes *(po'o waiū* or *kaula 'ā'ī),* however, are typically made of heavier rope (five-eighths-inch to an inch in diameter) and carried specifically for that purpose. In both cases, Manila is the rope most used.

When a cowboy refers to his *kaula hele,* he is describing a smaller, three-eighths-inch diameter rope about a fathom *(anana* or *brazada)* and a half long (9 feet). A loop or *kīpuka* is spliced into one end so that when threaded, a mini-lariat effect is created as in a typical "piggin" string. The *kaula hele* is most commonly used to tether the stirrups to each other beneath the girth of the horse. The tether is formed by looping the rope around the right stirrup leather *('ili ke'ehi)* immediately above the *tapadero,* drawing the balance *(koena)* with mild tension along the girth to the left stirrup leather, and tying with two or three half hitches.

When it is not being used in this manner, the *kaula hele* is usually tied in a small neat coil on the right foreskirt or *lala* of the Hawaiian tree. *Paniolo* who use this accessory frequently leave the loop tethered to the right stirrup and coil the detached balance from the left side, then neatly tie it on the right foreskirt. When saddling an unruly mount, the *kaula hele* is then readily available to retie to the left stirrup after cinching up.

The *kaula hele* serves as a safety device as the rider secures his position on a bucking horse by rotating his lower legs and turning his boots to a toe-out position while secured in the stirrups. This position causes the *kaula hele* to be tightly drawn, creating a locked-in limb posture for the cowboy. Furthermore, the toe-out position turns the spur toward the horse, adding more control to the limb grip of the rider. Such security in limb posture gives the cowboy significant advantage against a pitching and bucking bronc. The *kaula hele* also offers the rider limb security when riding in rough terrain, where the horse needs to lunge over a log *(piha'ā),* a crack in the lava *(awāwa),* or sinkholes *(nā pohā).*

The experienced cowboy developed the reflex to turn his toes out in response to the threat of a buck or lunge of his mount. In typical riding posture the more relaxed toe-forward position prevails, leaving the *kaula hele* in a slack position.

Use of the *kaula hele* as a locking-in device has its critics. Mainland cowboys look with disdain at a horseman who hobbles his stirrups. They call the tethered stirrups "cheaters" because the man is given significant advantage over the bucking horse. It is claimed that the *kaula hele* also restricts a rider's freedom for full spurring action, as lower limb movement is restricted.

Several prominent *paniolo* roughriders also disdain the *kaula hele.* Harry Kawai, a noted horseman of early Parker Ranch, Godfrey Kainoa of the last twenty-eight years, and Wayne Tachera of current times have pitted themselves against any bronc without benefit of the *kaula hele.* These men are exceptionally tough horsemen with considerable skill, balance, and athleticism. Other outstanding cowboys do not necessarily enjoy putting themselves at risk against unruly horses on a daily basis. Kawai, Kainoa, and Tachera savor those moments in a sportsmanlike manner. As age steals away the zeal, balance, and athleticism, even someone like Kainoa, a seasoned reinsman, might be found using a *kaula hele* on his green-broken colt or ill-mannered cow horse—notably common in the offspring of a King Ranch–bred stallion named San Lauro. Getting bucked off hard enough and often enough eventually causes a doubting cowboy to rethink this noble attitude.

Other detractors claim that the binding effect of the *kaula hele* can leave a rider pinned or saddle bound in a horse fall *(hina pū).* For the experienced *paniolo,* this risk is minimal, as he will loosen his stirrup grip at the first indication of an impending fall. A good horseman, in fact, knows when to "step off" at the strategic moment in the onset of a fall.

The *kaula hele* is also commonly used as a "piggin" string to gather and tie the limbs of livestock. In the course of the workday or weekend hunt, the *paniolo* may encounter and rope a fat hog, lamb, or goat for home use. The term *"kūpe'e"* is the Hawaiian word for hog-tying, the process used to restrain livestock after capture. Smaller game like pigs, goats, and sheep are loaded onto the cow horse for the ride back to headquarters. The sign of a well-seasoned ranch horse is when he can "pack pig." If the cowboy wants to hide the hog for *pau hana* (after work) retrieval, the hog-tied animal is placed in a shady *(malumalu),* moist *(pulu),* and cool *(ma'ū)* hollow *(hālua)* in the interim. This is caregiving.

Tying only the hind legs of an animal around the base of a tree or shrub is one variation of hog-tying used on the ranch. The trunk is positioned between both feet. This loose tie of the hind feet allows the animal to rise on its forelimbs and move about the tree in a circular manner to enable it to move with the shade as the day wears on. This technique creates less stress on the circulatory and physiologic makeup of the animal that ordinarily comes with being hog-tied and left on its side. A less stressed hog dresses out tastier pork for roasting *(ho'omo'a)* in the wood stove *(kapuahi)* or smoking *(ho'ouahi)* in the smokehouse *(hale uahi).*

The policy allowing cowboys the chance apprehension of a fat *laho'ole* while riding the paddocks and pipelines *(nānā'āina)* varied with each management regime. This was grounds for dismissal under A. W. To the seasoned livestock manager like Charlie Kimura, who came up through the ranks knowing the importance of the *i'a pua'a* for the family table, discretion was the key factor for both the cowhand and management. For the ranch cowboy, hunting is the last remaining benefit, and either overly harsh management or overzealous hunting puts this privilege at risk. Furthermore, commercial hunting is a genuine opportunity for diversification on ranchlands across America, including Hawai'i. Discretion in this venture is also the key in providing balance between the cowboy's privilege and the obligation of a ranch to utilize all resources to generate income through diverse ventures. Abuse by a cowhand or manager/trustee can be a threat to this traditional value of *paniolo* life. A well-informed employee base, coupled with considerate management and trustee personnel committed to land and animal stewardship, are the basis of discretion.

Saddle Blanket: *Pale Lio*

The saddle blanket *(pale lio)* is a universally important component of horse tack, as it provides the protective medium between the saddle and rider and the horse's back. The quality of care given to the back and limbs of a horse has great bearing on the animal's functional longevity and soundness. Preparation of the back to receive the blanket and saddle is vitally important. Brushing *(ho'opalaki)* firmly with the flow of the hair pattern sets the skin tone for receiving the *pale lio* that provides the protective medium between the saddle *(noho lio)* and the horse's back *(kua lio)*. Frequent readjustment of the saddle and blanket during a day's work is vital. Maintenance of a saddle blanket requires periodic washing, brushing, and drying.

For the early Hawaiian horseman, saddle padding had an English character since the horse came with early British settlers. Wool, cotton, or combinations thereof were the fabrics used. The less affluent horseman often used folded burlap bags as a practical and affordable substitute for saddle padding. This practice still exists today in remote valleys where packing of gear is done with the mule or donkey.

With the advent of the *vaquero* came the *carleita,* made of densely woven wool. For the early *paniolo,* the *pale lio* was likely made of cotton or a cotton/wool combination. These saddle blankets were converted from bedcovers handed down for a secondary use. When folded first in half, then refolded into three equal parts, a typical-sized bed blanket provides adequate horse padding and the proper dimensions of about 30 by 30 inches and 1.5 to 2 inches in thickness.

By 1900, commercially made horse blankets became available in the Islands, but wool remained the favored fabric for this purpose. Mohair (goat hair) blankets as well as the woolen products were in use on Parker Ranch. The popular issue was a woolen blanket, grey in color and usually sporting a pair of red longitudinal stripes. The brand name was Iron Clad. These continued in popularity for several decades.

Army blankets began to appear beneath the saddles of Parker Ranch men shortly after the end of World War I. The military issue was an excellent wool fabric of superior standards. A specific folding procedure yielded a durable horse blanket of consistent thickness. The army blankets were issued preshrunk from an 82-inch length down to 62 inches. Width was a consistent 30 inches and the blanket weighed 3.5 pounds. The military sold a grey blanket of similar weight and dimensions as surplus, but the original olive drab army issue was the most popular during the 1920–1950 period of Parker Ranch history. Toward the end of that era, army blankets were difficult to find—and what could be found was costly. The horse equipment industry thus entered the Hawai'i market with enthusiasm.

This caused a shift from the use of a blanket to that of a pad. Quilted hair pads became popular and were readily available through mail-order suppliers such as Hamley and Company of Pendleton, Oregon—the noted makers of the Parker Ranch Hawaiian saddletree—who entered the scene when local craftsmen retired from that skilled profession. These pads were hardy, durable, and about 30 by 30 inches square. They also produced an oval configuration. These were made of 90 percent pure curled animal hair pressed into a felt pad and lined on the top surface facing the saddle with heavy army duck canvas, which gave it added durability in Waimea's rainy conditions.

The horse equipment industry took a quantum leap in technical improvements to horse padding when an orthopedic fabric pad was produced around 1970. It was sold under the brand name of The Cool Pad. The product got its name from the fabric used in its construction, which was borrowed from hospital bedding and padding whose weave allowed ventilation that prevented bedsores and improved circulation in bed-ridden patients or those having casts of plaster of Paris. Improved air and vascular circulation bode well for the backs of working cow horses. The orthopedic saddle pad became the mandatory *pale lio* for the Parker Ranch horse in 1972. Saddle pads used today are further improved versions of the original orthopad.

Another influence left by the vaqueros was music with a Hispanic touch—particularly involving the *'ukulele* or guitar. It is a well-known fact that the outgrowth of this influence gave birth to the slack key guitar tuning *(kī hō 'alu)* and its relationship to *paniolo* folklore, which is deep-seated and genuine. Much like the steel guitar, it is said that yodeling was born in thecradle of Hawaiian music. Both aspects are basic to country and western music, called "cowboy music" by *paniolo.* The advent of the *'ukulele* via the Portuguese is significant, as its role matured in the skilled hands of notable Hawaiian cowboy musicians throughout the Hawaiian territory.

7

THE FOUNDATION OF PARKER RANCH 1832–1899

John Palmer Parker's development into an independent rancher in his own right is the focus of this chapter. But to leave the impression that he was the sole player in the early livestock business would not serve history. The roles of his peers, sons, and grandsons are vital components of this rich history.

By the late 1830s and early 1840s, the Hawaiian government had assumed a dominant role in cattle ownership because unbranded cattle on crown lands were included as an asset. Other operators included William French, R. C. Janion, W. H. Green, William Hughes, and Jack Purdy (also known as Harry Purdy or W. W. Purdy). Of these counterparts to John Palmer Parker, the influence of William French is most notable. Industrious in nature, French started his ranching operations at the same time as Parker and quickly diversified his operation by establishing a tannery and providing saddle and shoe-making services. R. C. Janion also became a neighbor of substance when he was appointed to depopulate cattle on government lands. At this juncture, some overlap occurred when unbranded cattle belonging to John Palmer Parker were absconded. In a strict sense, this constituted cattle rustling. The influence of Green and Hughes faded over the years, but the Purdy relationship to ranching grew to become one of great interest.

The Hawaiian government continued to wield an influence on the emerging beef business and, by ranching cattle on crown lands, became an actual owner/operator about 1850. In 1841, Kuakini (Governor Adams) had issued a *kapu* on the slaughter of cattle solely for hides and tallow, encouraging the consumption of beef beyond the shores of the Big Island—notably the metropolitan markets of Honolulu and, by 1850, the holds of whaling ships. Three years earlier the actual branding of cattle became a government function—often a self-serving one, since it established ownership to wild cattle on government lands. The official brand inspector was one William C. Beckley, whose role was broad and sweeping regarding government claims of cattle ownership. The eventual leasing or selling of pasturelands to private ranches made Beckley a figurehead, but the livestock industry in the Hawaiian Islands in 1847 had established the branding of hide as a legal testimony of ownership. The Great Mahele of 1858, permitting individual ownership of land, encouraged the development of privately owned ranches and plantations.

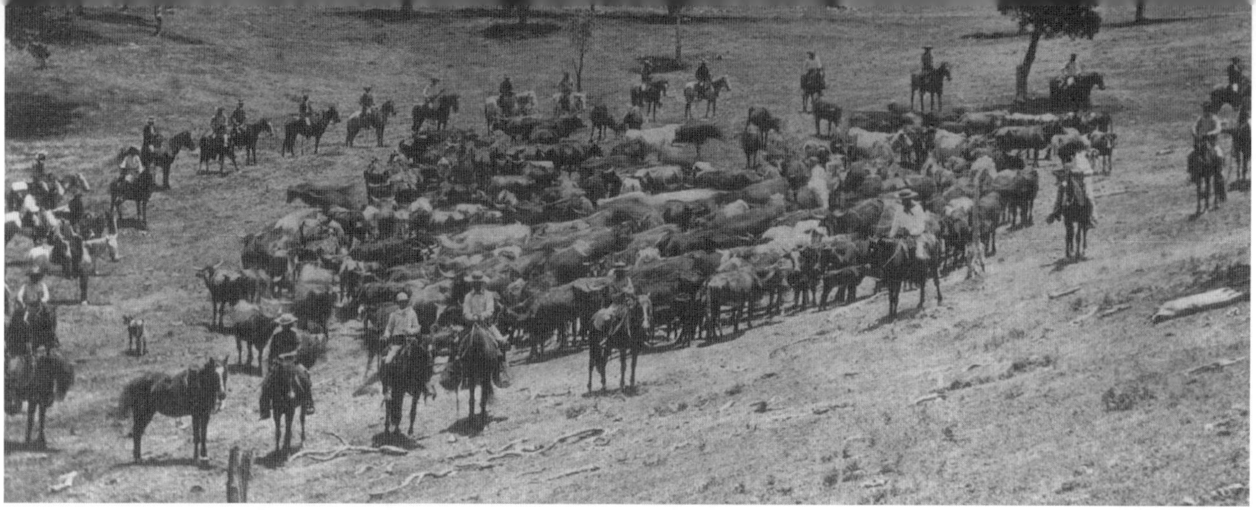

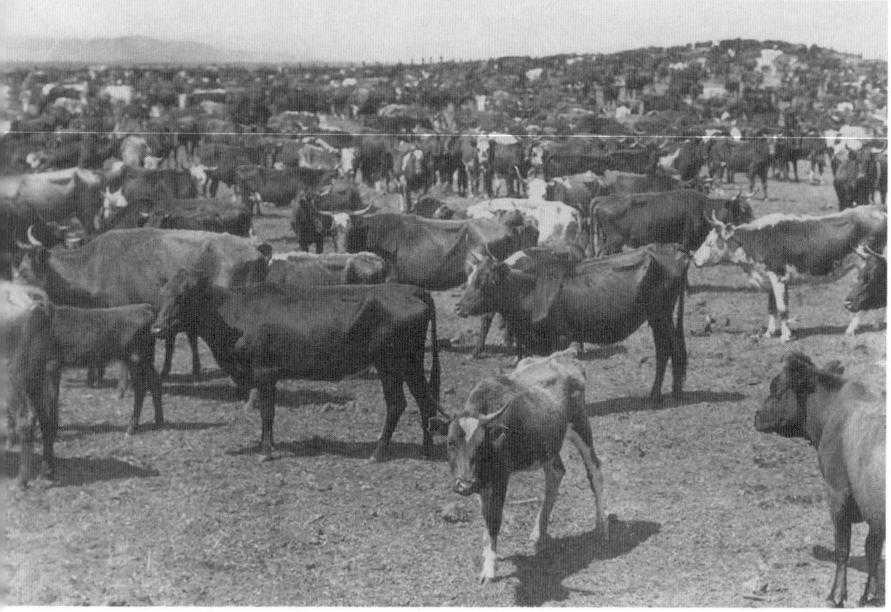

Top: *Herd encircled in hills of the Kohala Mountains, 1898. Thirty-two cowboys gathered this longhorn herd. Riderless horse belonged to the unknown photographer. Bishop Museum photo CP 27948.*

Left: *Longhorn herd gathered in a holding pen. The cows on the left have "P" brands. Bishop Museum photo CP Robert K. Bonine, BM CF 28505.*

Bottom: *Cowherd crowded into Pu'uhihale Corral for branding. Thirty-two* paniolo *are in the corral. Rider in foreground has long-eared* tapaderos, *leggings,* lala, *and* lala i hope. *Rider alongside the gate is using a turned-back* lala *and long leggings with pancake* tapaderos. *Both sport hat leis. Bishop Museum photo CP 102, 745.*

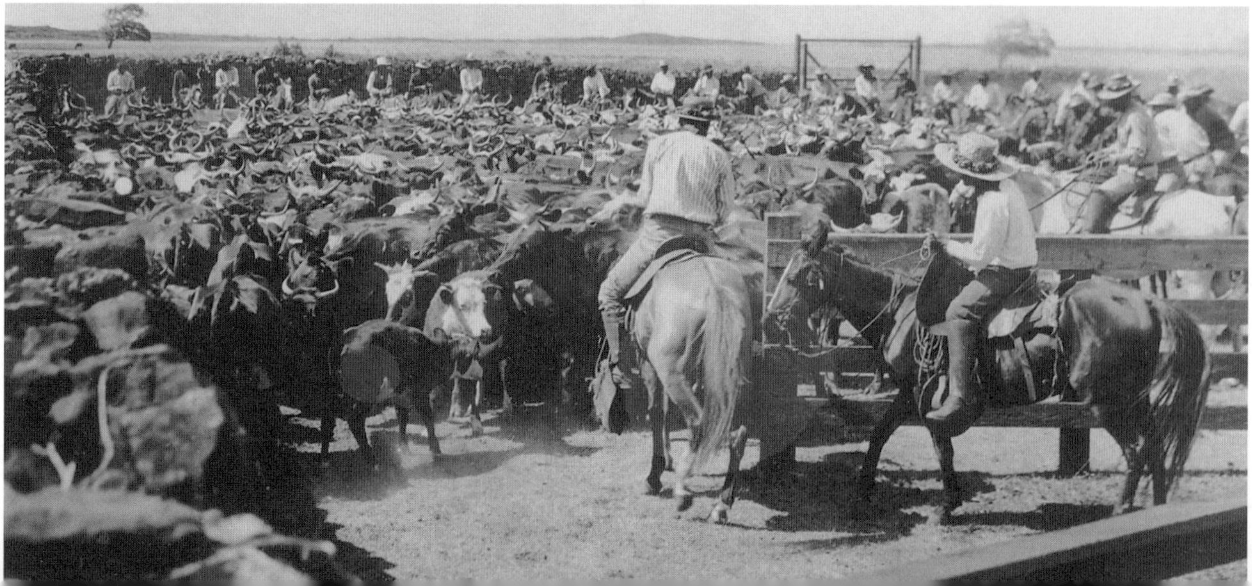

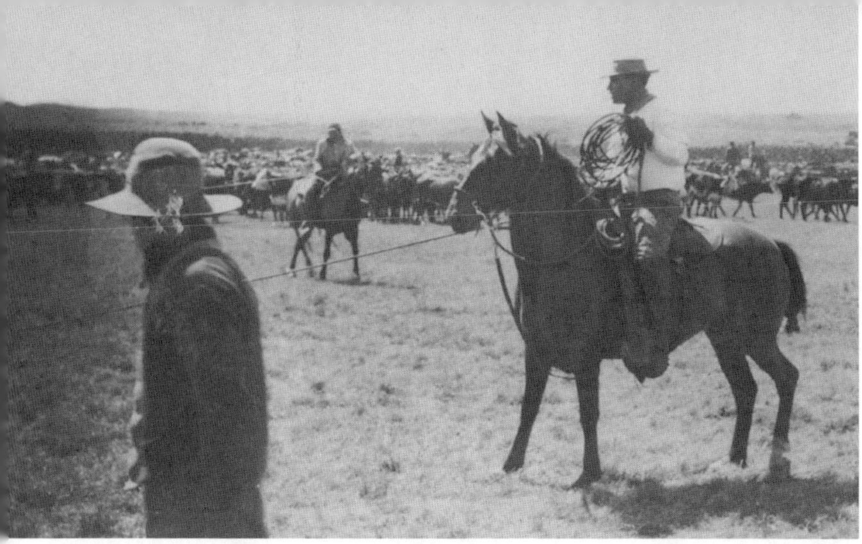

Left: *Branding begins. The cow horse has a neatly tied horsetail. Cattle in the background are of mixed longhorn/Hereford breeding. A. W. Carter Collection.*

Bottom: *Branding at Pu'uhihale Corral were a family/community affair. Pukalani* kauhale *(enclave) is in the immediate background. A. W. Carter Collection.*

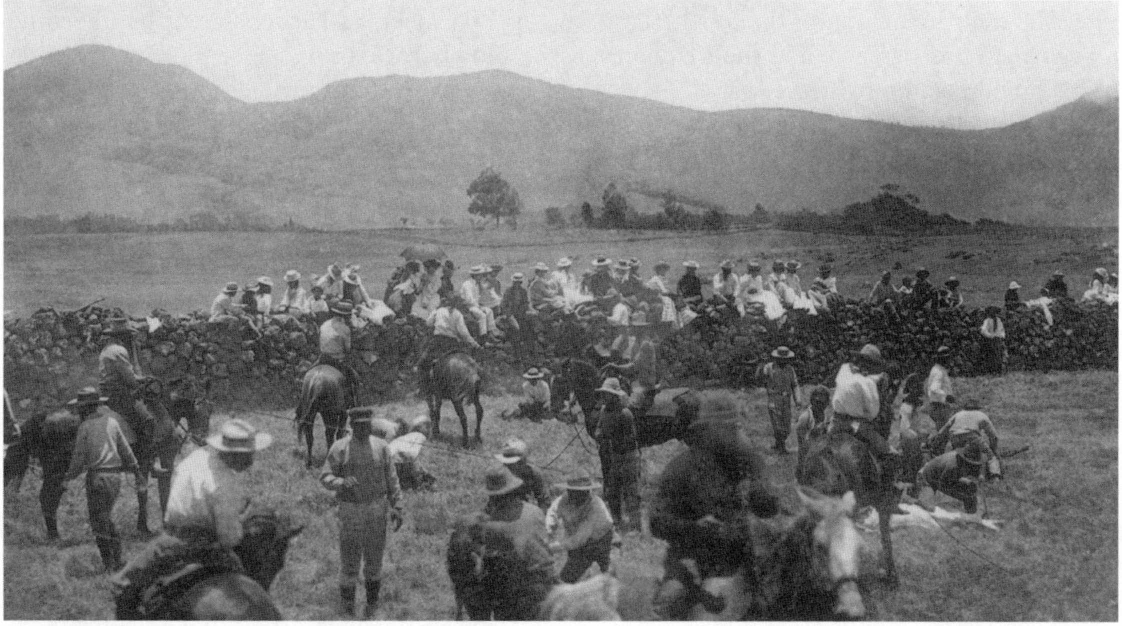

Parker Ranch Cattle Brands

The use of hot brands to sear symbols of ownership onto cattle hides, like many livestock handling practices, has a Hispanic origin. Hernando Cortes brought his own brand to New Spain (Mexico)—three Latin crosses placed on the rib cage of cattle. By 1530, the Mesta, a stockman's organization, was formed and produced the first brand book of the Western Hemisphere. The use of such brand books as a public record of all registered brands spread throughout the Americas, and particularly the American West. In the ensuing centuries, the proliferation of ranches along the coast of old California brought forward brands representing specific *haciendados (*ranchers*)* and Catholic missions. Quite likely, the longhorn cattle arriving in Hawai'i as gifts for its king in 1793 were branded according to their mission of origin.

By the 1850s the Hawaiian cattle industry was calling for formal brand registration, as misappropriation of livestock became widespread. Hawai'i Island governor Kuakini imposed the first brand laws, and Beckley became the first brand inspector. This was a critical event for John Palmer Parker I, as his own landholdings were being subjected to cattle rustling at an epidemic rate. His traditional brand—the block "P"—was formally recognized, and the frequency of his brandings intensified.

As Parker acquired adjacent lands, in many cases he assumed ownership of the brands of the purchased cattle. Some brands were discarded once inventory and boundaries were clearly established. In ensuing decades, properties such as the Humu'ula Sheep Station Company (HSSC) were acquired and their brands came with the deal. Some of these brands were kept in use by the ranch through 1965, including the "flag" brand and the block "T," "U," and "V" (see brand charts). These date back to the era of the Haneberg brothers' tenure at HSSC in 1880. The block "V" brand has even extended into the new millennium, used as a birth-year brand when rotated from an inverted V (designating 2000 as the birth year), a lazy V when pointed to 3 o'clock (2003), a proper V when pointed to 6 o'clock, (2006), and a lazy V when pointed to 9 o'clock (2009).

Historically, many a horseman cringed at the thought of marking a mount's hide in any way, given the fact that each and every horse, like a human, is a distinct individual. But for the purposes of theft prevention and inventory control, branding of horses has survived as a broadly accepted stamp of ownership up to the present day. The early Hawai'i livestock industry and Parker Ranch of old were no exception.

The registry of brands for Hawaiian livestock was done on an island (county) basis until 1955, when the responsibility was shifted from a county to territorial basis. Under the island/county system, the "P" brand could technically be used by a neighbor island rancher. This was an issue with A. W. Carter, as Parker Ranch had become a blood stock provider on a territory-wide basis. Even within the county jurisdiction, regulated by the sheriff's department or police chief, the similarity among brands became an issue, and A. W. reflected his discomfort in a June 18, 1928, memo to the Honorable J. W. Russell of Hilo. In this case, a small rancher in Hāmākua, Jacinth Paiva, registered a block "P" with a low tail similar to Parker Ranch's high-tailed "P" brand (note chart). Carter's memo was clear:

> I want to get our P brand re-registered, and could you arrange to have Paiva's brand disallowed as it is altogether too close to our brand, and our brand could be easily added to, to make it his brand.
>
> When I purchased Humuula, I purchased their brand which is a flag, and all of the numerals. I want to get these re-registered. The exact drawing of this brand will be sent you from our office in Waimea.
>
> Please represent me in these matters.

Delbert E. Metzger answered Carter's request on June 20 in two memos, providing an interesting background:

HORSE & CATTLE BRANDS OF PARKER RANCH

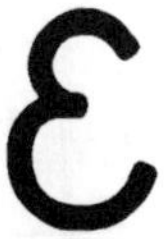

"BACKWARD 3"
Originally a horse brand of Humu'ula Sheep Station company (HSSC); became a code brand for Parker Ranch (PR) thoroughbred stallion Ihi'ihi

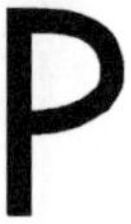

"BLOCK P"
General ownership brand of PR since 1855

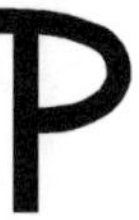

"BLOCK P" brand with a "high tail"
PR ownership brand

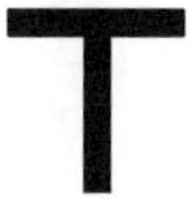

"BLOCK T"
Acquired in HSSC purchase in 1914
Code brand for polled Hereford herd (1920–1965)
Code brand for Junior (Morgan) progeny (1920–1965)
Junior colts were light mouthed and of medium build
Used by Territorial, state, and federal government to brand TB reactors (1908–present)

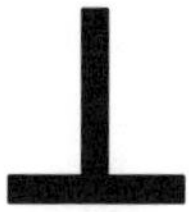

"INVERTED BLOCK T"
Acquired in HSSC purchase
Code brand for Loa (Thoroughbred) progeny

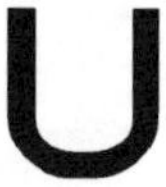

"BLOCK U"
Acquired from HSSC purchase
Used as a miscellaneous code brand (1920–1950)
Sometimes inverted

"BLOCK V"
Acquired from HSSC purchase
Used as a code brand for stallions: Eastertide (TB) (1914–1935), a noted racing and remount stud; later Otto (Morgan) (1945–1965), a chestnut that sired dependable workhorses
Used by territorial, state, and federal government to indicated cattle vaccinated for brucellosis

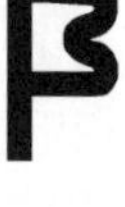

"FLAG"
Acquired from HSSC purchase
Used as a code brand for registered horned Hereford cattle (1920–1965)
Occasionally applied backward

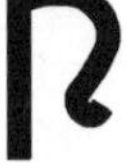

"BLOCK P OPENED"
Used under Penhallow management (1960)
Code brand for Angus cattle

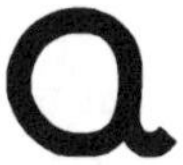

"SMALL CASE, SCRIPT A"
Acquired from HSSC purchase
Miscellaneous code brand, rarely used

Brands are described in practical terms. A "block" brand refers to a typical number or numbers as well as capital or lowercase letters of the alphabet, presented in typical upright position. The "Block P" is an example.

A "tumbling" brand indicates the rotation of its figure by 45 degrees as in the "Tumbling H," applied as ⟁, indicating an unregistered color sired of PR stallion Leo Honcho.

A "lazy" brand indicates rotation of the figure by 90 degrees as in the "Lazy H," applied as I, indicating a colt sired by PR stallion Gallant Bear.

An "inverted" or "upside down" brand indicates the rotation of the figure by 180 degrees as in the "Upside Down T," applied as ⊥, indicating a colt sired by PR stallion (TB) Loa.

A "reversed" or "backward" brand is applied as indicated. A "Backward 3" is applied, as Ɛ, indicating a colt sired by PR stallion (TB) Ihi'ihi.

In Hawai'i state laws (HRS Part II, Sec 142-148), brands are registered as letter and/or numeral combinations, as well as open and closed figures of reasonable design. Single block letter or numeral brands have long been disallowed, except for those registered prior to 1955.

Background information on brands.

During Mr. J. W. Russell's absence I open his business mail and give to it any attentions I consider immediately necessary.

Upon reading your letter of the 18th I called at the sheriff's office but found that Mr. Pua was visiting in Kona and no one here knows when he intends to return.

I looked through the new brand registration files and found that under Application No. 327, dated April 24, 1928, Jacinth Paiva of Honokaa applied for and there was issued to him the same day a registration certificate for the exclusive use of the brand P. The clerk who issued the certificate explained to me that he had no discretion under the new law, Act 57, S.L. 1927, and besides, that he did not at the time know of the similarity to Parker Ranch brand.

W. J. Rickard deputy sheriff of Hamakua was in town and I got hold of him to inquire who Jacinth Paiva is, how many cattle he has, what brand he has been usine [sic] etc., and to see if Paiva could not be urged to give up the P brand. I told Rickard that you were much disturbed over the matter as it opened a clear road and invitation to cattle stealing, and I asked him to get hold of Paiva and explain to him that he was only borrowing trouble by trying to cut in on your long established brand. Rickard told me to have you call on him upon your return to this island, that he felt quite sure he could get Paiva to release the brand and adopt some other. This would probably be the best way to get the matter cleared up. I had intended to urge Pua to recall the certificate under claim of irregularity and misunderstanding at the time of its issuance, and upon the ground that it is approximately the same as that carried by hundreds of your stock, but no one here knows his exact whereabouts or when he will return, and no one here will move in the matter without his instructions.

Since writing to you earlier today I thought it best to go back to the sheriff's office and register at once the brand P in your name, so accordingly did this and send certificate herewith. Up to this date it has stood on their record as belonging to Samuel Parker.

In the second memo, it is of interest that until 1928 the "P" brand had remained under the name of Samuel Parker when he, in fact, was bought out in 1906. Technically, the branded livestock remained his property. This oversight was unusual for A. W. In a 1949 memo to his father, Hartwell Carter reaffirmed the ranch's brand registration status.

Placement of the "P" brand on cattle varied. In the pre–A. W. Carter era, the brand was applied high on the left rump in clear view of the observer. As hide values appreciated, A. W. mandated that the brands be placed on peripheral sites such as the gaskin. The gaskin is a muscular bulge on the outside of the lower hind limb between the stifle (knee) and hock (heel). In cured cowhides, these edges are of minimal value and often trimmed for discard. Marking of the periphery of the hide preserves its central body for wide, unblemished use.

Carter also imposed a year brand to indicate the birth year of the cattle. This was a significant management advance, as Carter pegged the desired market age of grass-fattened cattle at three years as opposed to the five to six years of previous times. He was able to do this with improved pasture management and cattle genetics, and year branding made a clear target date. Furthermore, Carter mandated that mother cows be retired after a nine- to ten-year tenure, along with other causes such as lameness, cancer eye, mastitis, lumpy jaw, and orneriness. Placed immediately below the "P" brand, the year brand gave field personnel a critical yardstick for measuring performance relative to age. In later management regimes (Rubel/Lent 1970), the "P" and year brandings were placed together back up on the rump with the reasoning that readily visible ownership marking was more important than hide discounts.

Parker Ranch Horse Brands

The branding of horses on the ranch has a similar history, but it includes an emphasis on sire brands on each horse to indicate paternal parentage. Early records and photographs show that the right shoulder was branded "P" and the left shoulder or hip was reserved for the sire's mark. Year brands did not become prevalent until the advent of A. W. Carter.

The Haneberg brothers improved the branding of horses when they managed the Humu'ula Sheep Station Company in pre-Parker times. They ran captured feral and imported horses as well as locally purchased quality stallions. The latter sires were of Thoroughbred, Morgan, and draft breeding. Each sire had a special brand that was imposed on all his offspring. Of the feral stallions, there was an 'Āinahou (captured in the *kipuka* 'Āinahou) and a Laumaia sire (captured in Laumaia). The Hanebergs brought in King Thomas of Sydney, Australia, and Johnny Morgan from William Rice of Kaua'i. These were Thoroughbred and Morgan stallions, respectively.

A typical workhorse of HSSC would have a year brand on the shoulder or hip as well as the brand of origin if purchased. The "P" brand of Parker and the "frying pan" brand of James Horner's Kūka'iau ('Umikoa) Ranch were prevalent on both horses and mules.

For individual horse combination brands, the Hanebergs would rotate the numeral brands "0" through "9," as well as the three letter brands, "T," "U," and "V." The "flag" brand could be reversed so that it "waved" to the right or left of the viewer. Consequently, each horse might have three to four brands combined to make its unique identification number. Along with the horse's name, these were logged in German in the Haneberg operation ledgers.

As Samuel Parker Sr. and Jr. had their tenures at the helm of HSSC, horse quality was further improved with key imports and purchases, and the branding program was retained, albeit simplified. When A. W. took ownership of HSSC for Thelma Parker, he continued the practice (but not the registration) of the "0" through "9" numeral brands as well as the block "T," "U," "V," the "flag" brand notwithstanding.

With the advent of A. W. Carter's penchant for accountability of inventory, he developed the Parker Ranch Standard for horse identification (see horse brand chart). The rules were clear in the 1902 execution:

- Excluding Thoroughbreds, all ranch horses were branded with the block "P"; geldings on the right, mares on the left gaskin
- Excluding Thoroughbreds, all ranch horses were marked with a birth-year brand; geldings on the left, mares on the right gaskin
- Excluding Thoroughbreds, all ranch horses were sire branded according to paternal parentage beneath the mane; geldings on the right, mares on the left neck
- All registered Thoroughbreds were serially branded on the right neck, numbered in concert with their respective numbered registry with the American Jockey Club

Carter's regimented system of permanent identification of horses bode well with the coming of the military remount program. While many of the ranch remounts were predominantly Thoroughbreds, the military standard included grade horses also. In all cases, the remount system included a military brand.

Parker's Perseverance

As government lands became available for lease or purchase, John Palmer Parker seized the opportunity for expansion. But he was

HORSE SIRE BRANDS OF PARKER RANCH

0 Originally a sire brand for Frolic, a highly regarded Thoroughbred (TB) stallion of Sam Parker Sr. that was acquired in the HSSC purchase. Later became the sire brand of Diablo, an imported Palomino/TB stallion. Frolic sired polo & speed horses (racing); Diablo sired decent workhorses.

1 Sire brand of Opua, a brown native son of Ormesby, an imported remount TB stallion. Opua produced colts with the speed, stamina, and intelligence needed for cow horse work. He sired Alulu, a stallion that was also a great cow horse sire.

2 Sire brand of Lindsley, an imported chestnut Morgan stallion. Progeny were noted for broncky attitudes but were capable of dependable work careers given their stamina and surefootedness. Lindsley was built more as a draft-type horse.

3 Sire brand of Querido, an imported by Morgan stallion of noteworthy progeny. Likely the most successful workinghorse sire in early Parker Ranch (PR) history. Querido was built more as a Thoroughbred-type horse. His colts were light mouthed and light footed with great stamina.

4 Sire brand of ContraCosta, sorrel son of Sappho out of a grade TB mare used to produce "heavy" horses for branding and shipping chores. His colts were some of the rankest to break but were strong backed and good looking.

5 Sire brand of Sappho, PR's first registered quarter horse stallion in Hawai'i, a palomino with prepotent speed and color characteristics but poor footing. This brand later went to another quarter horse import, Peppy Deedee, a brown bay by Poco Bueno. Excellent cow horse sire. Progeny gentle but not noted for stamina. They were round-backed horses with little wither.

6 Sire brand of Naniloa, a palomino son of Diablo out of a Frolic mare. Another popular sire of good work-horses. His colts were well framed but coarse headed. They were gentle in nature and strong in stamina.

7 Sire brand of Heine, an imported brown Standardbred stallion. Progeny noted for being mean spirited but very good cow horses. Strangely, this trotting horse sired colts that enjoyed cattle work. They were broncky and hard riding.

8 Sire brand of Mauna Kea, a black native son of a feral stallion from Humu'ula out of a Percheron mare. Stood at the HSSC under the care of Willie Kaniho. Progeny were gentle, strong, and sure footed; described by some as "flint hooved."

9 Sire brand of Lucky Tom, a chestnut stallion imported by R. L. Hind I of Pu'uwa'awa'a Ranch; sold to PR. The colts were easy to train, gentle but strong, and long winded. He rivaled Querido as one of the greatest cow horse sires on PR.

10 Sire brand of NicolaB, an imported bay Standardbred stallion. Progeny were hard to break, but made strong, sure-footed cow horses. Pronounced "Nickel Bee" by ranch hands.

11 Sire brand of Moanalua, a brown son of Rodgers, a remount TB. Produced dependable cow horse offspring. Also sired speed horses.

12 Sire brand of an unnamed son of Diablo;[6] had limited use.

X Brand used on ranch colts and fillies of unknown parentage (sire).

faced with the problem of how to improve his stock at the same time. He was aware that growth in cattle quality should ideally parallel growth in numbers—but his task was formidable. A massive body of water, the Pacific Ocean, hindered access to improved seed stock. While Britain was a likely resource, the great distance separating it from Hawai'i made this impractical. Thus he chose to seek help from Down Under—Australia and New Zealand. The Royal Agricultural Society of New Zealand and Australia, having pioneered the same quest, was generous and provided dependable advice and breeding stock resources, mainly for O'ahu ranchers. Soon the Royal Hawaiian Agricultural Society (1850) emerged as the conduit for such services, recommending as a common goal the improvement of cattle breeding to the point of reaching finished market weight (a 500-pound carcass) at three years of age as opposed to the six or seven years required by the highly inbred feral beef animals typical of Hawaiian ranges.

Four men—John Meek, Charles G. Hopkins, Henry Sea, and R. Moffatt—were appointed to a council of the Royal Hawaiian Agricultural Society, the Committee on Neat Cattle. Moffatt, a rancher from Kahuku, O'ahu, was the spokesperson for the council and is credited with bringing the first Hereford cattle to Hawai'i—a move that around 1900 would benefit the Parker Ranch operation immensely.

The Royal Hawaiian Agricultural Society broadened its role by advising ranchers to be bold in culling deep and in castrating all males but the most exceptional. It encouraged ranchers to be equally firm in retaining as replacements only the above-average heifers and sending the culls to finishing paddocks for marketing. The society became downright adamant about keeping the replacement heifers secure from wild bulls, going so far as to recommend the fining of owners of ranch bulls that failed to consider boundary fence security as a major responsibility.

Establishment of a local territorial stockmen's association was not unique to Hawai'i. In fact, it has its origins in the influence of old and New Spain. In 1529, Mesta, the Mexican Cattlemen's Association, was born of the common interest of several cattle raisers in Mexico City. In the Western Hemisphere, Mesta became the granddaddy of all organized cattlemen's associations, including today's National Cattlemen's Beef Association.

Like the advice of modern-day university extension beef specialists, however, organizational agendas and idealistic recommendations had to be put into a practical context by ranchers. John Palmer Parker basically sought to improve a feral cattle operation, but improved genetics could do little until the wild herds were replaced with manageable livestock. To accomplish this meant recurrent frustration.

The Royal Hawaiian Agricultural Society, for whatever reason, deplored the Spanish method of livestock marketing (Hawaiian version) as brutal and inhumane. But Parker nevertheless forged ahead vigorously by gathering the wild bullocks by rope or trap and leading them live to a central processing area in Waimea, a shipment point at Kawaihae, or a series of strategically located corrals for domestication. Not only did Parker develop a marketing strategy, he also greatly reduced further forest degradation by stocking available pastures with an increasing number of manageable cattle. To these domesticated cowherds, bulls of better selection could be added, and there was constant monitoring for the reentry of wild bulls that could set back progress in a heartbeat. The crippling or killing of an imported purebred bull by mustang bulls *(laho 'āhiu)* was a serious threat, and care-giving cowhands were quick to relieve these rogue bulls of their testicles—and occasionally their lives. By 1859, the wild cattle estimate of 12,000 head was nearly matched by domesticated cattle, estimated at 8,000. By 1900, Parker Ranch wild cattle numbers were down to 5,000.

There were subsequent management challenges for Parker. For his first crop of half-breds, pasture would have to be set aside and secured with stone fencing *(pā pōhaku)* and spring or stream water provided. While incorporating advice from the Royal Hawaiian Agricultural Society for upgrading the ranch herd, he methodically opened up new increments to controlled grazing, making a secure venue for launching a genetic upsurge in the quality of his bulls.

While seed stock sources included U.S. mainland territories, Australia, and New Zealand, one of the early dependable bull sources was Van Dieman's Land, known today as Tasmania. Van Dieman's Land is credited with being a superior source for Hereford and Angus breeding stock and purveyors of quality Thoroughbred horses, along with the Durham (shorthorn) bulls. During this period, England and Scotland entered the Hawaiian purebred market, adding Red Devon to the list of breeds.

Given the constraints and challenges that Parker faced, the periodic importation of purebred cattle gradually altered the genetic makeup of the large, unruly herds of feral cattle. His priority would remain the continued conversion of feral cattle to domestication. For the ensuing decades, he had his work cut out for him.

It was an era passing; the days of the wild cattle were numbered when, on January 8, 1847, 2 acres of land at a section called Mānā were acquired by the Parkers from the Land Commission, forming the nucleus of Parker Ranch. The era had witnessed the hunting of wild cattle for hides, tallow, and beef for the whaling ships, but that was coming to an end. The year that land at Mānā was acquired, John Palmer Parker was fifty-seven years old and had three children with Kipikāne; the children were now young adults. Mary was twenty-eight that year; John Palmer Parker II was twenty, and Ebenezer was eighteen.

Improving his material welfare via the development of a productive ranching operation was directly in line with the fundamental Christian principles by which John Palmer Parker lived, worked, and conducted business. This sense of familial responsibility combined with innate livestock management principles guided his first mission as a bona fide rancher: the development of a ranch headquarters. Underscoring this mission was his awareness that without a dependable water supply, adequate range forage, and an increasingly domesticated cowherd, the need for corrals, outbuildings, and a ranch house would be moot.

Hale Mānā, the centerpiece of ranch operations, was constructed from native *koa, 'ōhi'a,* and *māmane* milled at the Hanaipoe sawmill nearby. Later a Paliho'oūkapapa sawmill was built. Typically rectangular in its earliest stage, Mānā's ground floor accommodated a large central hall, four bedrooms, a dining room, and a kitchen with an enclosed veranda. The upstairs contained three more rooms for utility needs. The construction of the entire home was done using *kui lā'au,* or dowel pegs of native woods. Square-paned windows placed appropriately throughout gave a New England effect to the architecture of Hale Mānā, reflecting Parker's eastern seaboard upbringing and his excellent skills in carpentry. The home was surrounded by well-kept gardens of native flowers and staple vegetables.

For ranch buildings, Parker and his sons built a saddle house/blacksmith shop, a barn for bullock wagons and plows, and a stone-enclosed meat house adjacent to a small household dairy. An outdoor cooking *hale* combined *imu* and smokehouse features. Two soundly constructed stone and mortar cisterns covered with the same roofing of dark grey slates to match Hale Mānā served all of the above. With periodic whitewashing, the Cape Cod effect was picturesque.

When every detail was completed, Parker—steeped as well in Hawaiian religious tradition—called on Makua Laiana, the Reverend Lorenzo Lyons of 'Imiola Church, his dear friend and parson, to invoke the word of the Lord in both English and Hawaiian while consecrating the cornerstone of Hale Mānā. In true spiritual tradition, Parker placed beneath this cornerstone a loaf of bread, some coins, an ear of corn, and the *Paipala Hemolele*—his Holy Bible. Hale Mānā was now blessed for occupation by a close-knit, strongly disciplined Christian family of old Hawai'i.

Mindful of his growing cowherd and landholdings, Parker paralleled his home-building mission with one of corral *(pā'eke)* construction, using native help and materials. Hawaiian people made up his entire ranch staff, as well as the household help for his lady, Kipikāne. Building *pā pōhaku* for his corral at headquarters gave it permanence in function and design. (Remnants of these walls and cisterns can be seen at the original Mānā home site.)

The stone corrals typically included a *pā loa* or wing fence to ease the crowding of the cowherd into the interior pens, usually two in number. Gates of *māmane* were hinged via the upright on one end spanning from its base in a wooden or stone receiver to the uppermost overhead crossbar that was countersunk to receive the upper end of the same upright. *Māmane* bars would span horizontally from the hinged upright to the opposite vertical upright that swung in, opening the gate *(puka pā).*

Central to each pen within the *pā 'eke* was a stout *'amana,* or forked post, with the bottom of the *Y* about 4 feet from the ground. The top of the *Y* was often 8 feet from the ground. With the bullock roped, the cowboy passed his lasso between the *Y,* dallied up to the saddle horn and, together with his horse, snubbed the bull (or milk cow) up tight for neck tethering—hence the rodeo event named *po'o waiū. Po'o waiū* (*po'o* means "head," *waiū* means "milk") got its name from early Hawaiian ranching families who secured household milk by roping a common range cow, tying her head to a tree or post, and milking her. The name was later applied to the process of tying cattle to a tree or post and, eventually, to the timed rodeo event—omitting, of course, milking the animal. The process evolved from an earlier version used by the *vaquero* and *paniolo* in which the fork of a tree served as the *'amana.*

An Expansive Mood

With the forthcoming land acquisitions, Parker showed foresight. The growth in his cowherd came from adjacent lands and intensified the use of this solidly built *pā'eke,* which initially served a section (640 acres) of new land and, very quickly, a few thousand acres. As it was, his ranch facility had already delivered an order from purveyor William French for a hundred barrels of salted beef to the port of Lāhainā (now Lahaina).

In 1850, laws were passed that simplified land acquisition for Parker. As a result of these laws, he was allowed to purchase the 640 acres surrounding the Mānā headquarters. Further, on January 29, 1851, the Interior Department of the Kingdom granted him permission to purchase another thousand acres at 75 cents per acre. A year later, Kamehameha III granted Parker a lease on the land of Waikoloa, which was eventually deeded to the ranch by outright purchase. By this time, W. W. "Jack" Purdy had acquired a 640-acre section of land neighboring Parker's that would also eventually become part of Parker Ranch.

Parker broadened his livestock foundation by diversification. By 1853 he had developed a peach orchard and expanded the dairy operation to include butter and cheese via a milk cowherd of nearly 150 producers. While not the most productive dairy cows by any standard, they yielded 50 cents a pound for butter, giving Parker an annual income of about $2,300. The expanded dairy operation was located at Paliho'oūkapapa, about 2 miles east of Hale Mānā.

The surplus milk was consumed by a drove of domestic swine whose forebears were feral—yet they yielded robust hams and shoulders for curing and smoking. Trimmings of both pork and beef slaughter also were routinely smoked for commercial and home uses. Parker was adept at sausage making and used native animal casings to pack beef and pork with appropriate herbs grown from the gardens of Hale Mānā. Garden herbs were used in the preparation of pickled pigs' feet and headcheese as well. These delicacies were limited to household use or to gifts for friends and guests. Guests at Hale Mānā often included dignitaries such as the French consul, Charles Victor Crosnier, in 1857.

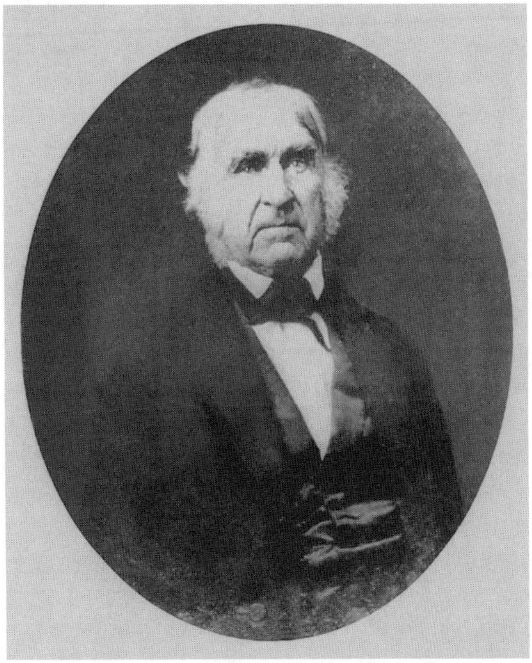

John Palmer Parker I. A. W. Carter Collection.

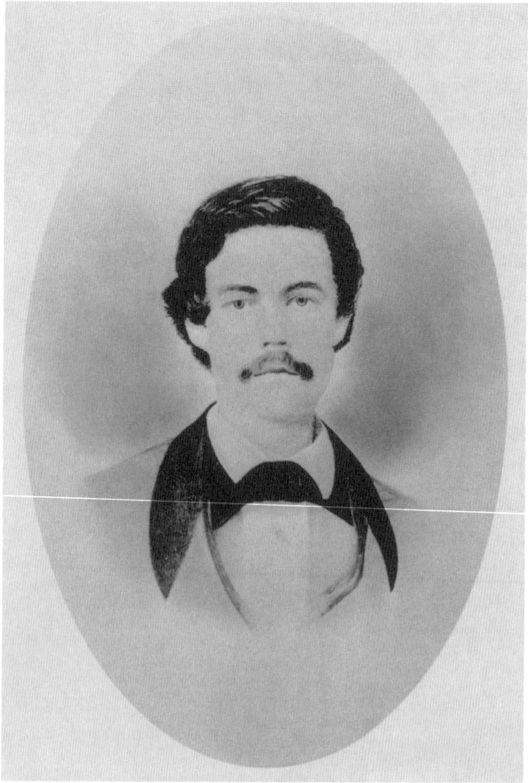
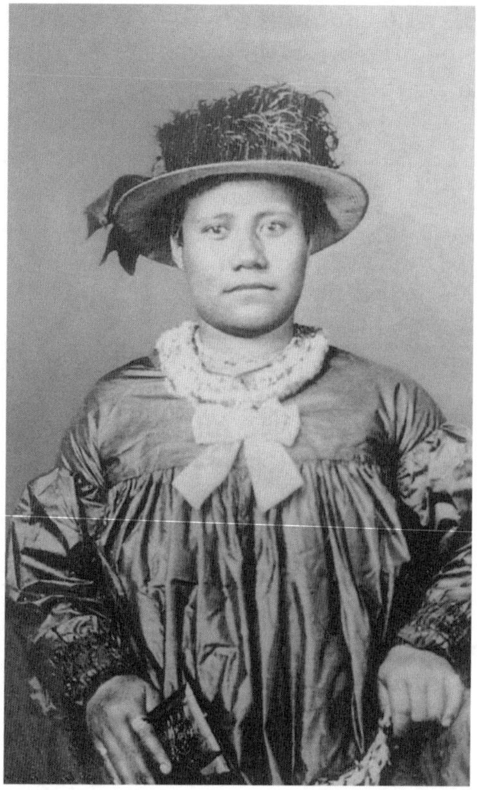

Clockwise from top left: *Ebenezer Parker. A. W. Carter Collection.; Kilia Nahulanui Parker. A. W. Carter Collection.; John Palmer Parker III. A. W. Carter Collection.; Samuel Parker Sr. A. W. Carter Collection.; John Palmer Parker II. A. W. Carter Collection.*

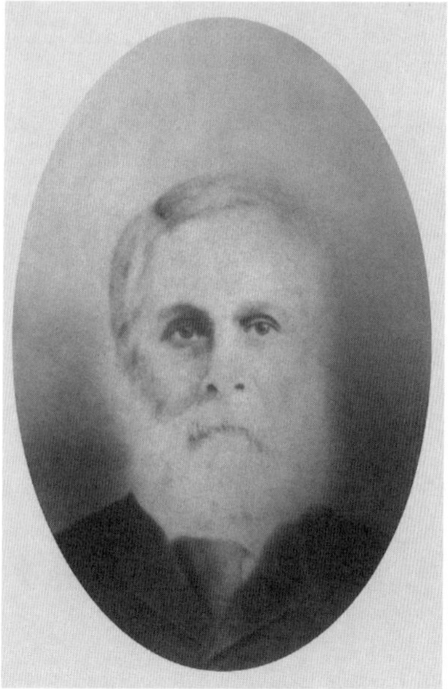
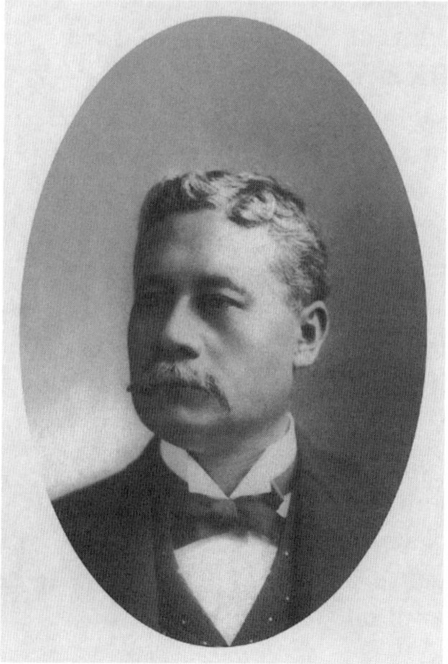
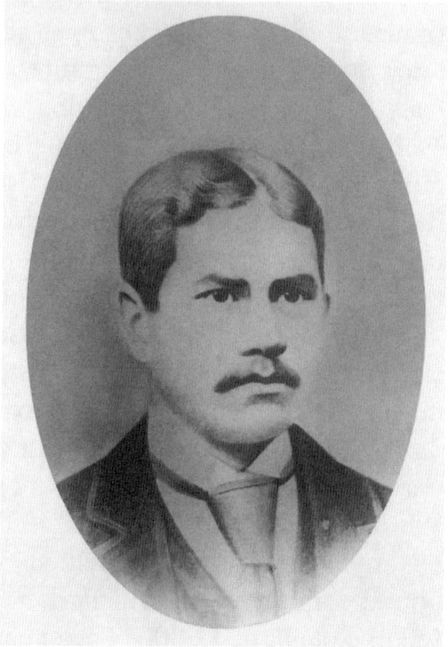

Chickens and turkeys scurried about the ranch headquarters paddock, providing an adequate supply of eggs. The turkeys were converted to roasters and eventually went wild, now abounding in the countryside of Waimea in full pomp and gobble.

All ranch efforts toward growth and productivity were reinforced by the support of Parker's two sons, John Palmer Parker II and Ebenezer. Sadly, in 1855 Ebenezer's life was cut short by an overwhelming digestive tract episode—probably peritonitis—brought on by an apparent foreign body (a plover bone) penetrating the intestine. The younger of the two boys, he showed natural ability as a horseman, had a keen eye for cattle, and possessed a great love of the good life. John Palmer Parker II now had to assume nearly all physical, emotional, and managerial support for his father, who, in 1856, saw the New Year open with a severe drought that left the district of Waimea a veritable dust bowl.

In this era, Mary Ann Kaulalani Parker—the only daughter—went on to marry a pure Hawaiian named Waipa after a failed marriage to Moses Brown Fuller. This union produced William Ulumaheihei Waipa—first of eleven children—who became the first prominent cowboy *luna* (foreman). The previous marriage to Fuller, however, had produced a daughter, Martha, who became the matriarch of the Low, Hind, and Ka'au'a clans. John Palmer Parker II married Hanai, who produced one son, Samuel, who died at age one. Parker and Hanai adopted a nephew, Samuel, who was one of four children by Ebenezer, Parker's brother, and his wife Kilia. The adoption took place after Ebenezer's untimely death.

Sam was off to a good start with his adoptive parents, who adored the "Little Prince Cupid," as he was affectionately known. Sam grew up somewhat pampered, particularly by his adoptive mother, Hanai, who saw to it that he was duly educated at O'ahu College. Parker and Hanai later adopted Sam's son, who became John Palmer Parker III.

Despite the tragedy of Ebenezer, John Palmer Parker recovered from a yearlong drought and was still in an expansive mood. He acquired the upper lands of Pā'auhau (37,888 acres) and later Pā'auhau Makai (8,165 acres). His mood was reinforced by his desire to secure the bulk of the unbranded cattle that wandered beyond immediate ranch holdings around Mānā. A government agent, E. P. Adams, somewhat overzealous in his role, was freely appropriating large numbers of Pā'auhau cattle, many of which were unbranded offspring of Parker cows. Further, Adams' activities grew closer to the Mānā headquarters. After initial complaints to the government brought only mild restriction to Adams' activities, Parker was successfully awarded upper Pā'auhau on June 11, 1861, via government Grant 2769. About a year later, Parker bought Pā'auhau Makai from C. R. Bishop in a highly convoluted transaction that originated from an aborted sale of this parcel by the parents of David Kalākaua. These land acquisitions brought needed control over Parker cattle grazing on adjacent unassigned parcels.

Parker garnered the loose parcels of Pā'auhau by effectively buying 400 acres from George Hardy in 1860, 150 acres from Phillip Ryan in 1861, and the final 100-acre remnant from Lewis Cockrill in 1866. While complete acquisition of the lands of Pā'auhau did not relieve all rustling concerns, John Palmer Parker had made a significant improvement.

In addition to confusion over cattle ownership with the government, a consortium of livestock owners—namely the Waimea Grazing and Agricultural Company (WGAC)—caused further questionable cattle misappropriations and business deals.

Waimea Grazing and Agricultural Company

Among early businessmen in the Waimea area, William French was one of the first entrepreneurs. In addition to keeping cattle and sheep on a significant basis, he further developed accessory businesses to include tallow works and a blacksmith/carpenter shop. His original home in Waimea in 1840 is now the historic Spencer House.

A traveler named Olmstead toured some of the activities of William French and others and made the following observations:

> Tuesday [June 29, 1840] I accompanied Mr. French on a walk to a place about two miles distance where the business of tanning is carried on under the direction of Chinamen. The establishment is extensive and the leather exhibited to me was of very superior quality as far as I was capable of judging. Besides a saddle maker close by the tan works, Mr. French has a shoemaker and carpenter in his employ.
>
> Our principal object in taking the walk was to witness the marking of a lot of cattle that had been driven down from the mountains not long since. Great numbers of wild bullocks are caught in the mountains every year by the hunters. The lasso, the principal instrument in their capture is made of braided thong upon one end of which is a ring forming a slip noose which is thrown with astonishing precision around any part of the animal. Even while at a full gallop in pursuit, the hunter grasps his lasso and giving it two or three twirls around his head with his right hand, throws it unerringly and entangles his victim by the horns or limbs. . . . For their capture a mode frequently resorted to by the hunters was to dig deep pits and cover them with underbrush and dirt. . . . The bullocks to be marked were driven into a pen towards which we directed our steps. They were noble animals and had been tamed by tying them singly with tame cattle for a time. . . . There were not far from 40 bullocks marked on this occasion intended for the Clementine in her trip down to Honolulu. They are then put into pasture to be fattened for the supply of ships visiting Honolulu in the fall season.
>
> This brig Clementine had upon its deck about 40 head of bullocks arranged closely together with their heads turned inwards. They were tied down by the horns to a strong framework of spars so that there was no danger of their getting loose.

French's primary business came from the whaling industry, to which he supplied beef—fresh, salted, or dried. He depended heavily on John Palmer Parker and Jack Purdy to maintain his market for beef as well as hides. He owned Līhu'e Livestock Farm and had a store at Pu'uloa. His livestock trade was taken over by Frank Spencer and James Louzada upon French's death in 1851.

Robert Chesire Janion came to the Islands in 1845 and quickly moved into mercantile, agriculture, and shipping enterprises. He took as partner W. H. Green, a trusted friend and attorney. Employed as comptroller was T. H. Davies. In 1861, Janion and Green signed the deed and charter of the Waimea Grazing and Agricultural Company (WGAC), with offices in Honolulu and Waimea. A year later, Janion acquired considerable strategic parcels around Waimea and, through WGAC, he and his partners set out to monopolize the livestock industry. Parker's concern for rustling was not over.

Janion's men immediately began to harvest any and all unbranded cattle on any lands in the area, disrupting an earlier arrangement that John Palmer Parker and Jack Purdy had with William French that divided the feral harvest equally three ways. Janion went further by securing the right from the government for such harvesting of cattle on its lands. His agent, Joseph H. Mallett, however, proceeded to kill unbranded animals on private land as well. Mallett went a step too far by writing an imprudent letter to Parker stating that he would continue harvesting cattle on Parker's ranch regardless. This position, if left unchallenged, could eventually misappropriate any or all of John Palmer Parker's unbranded cattle that he had accumulated over a dozen or more years.

Rather than directly challenge Janion and his agent, Parker chose to go public. On July 5, 1860, he had the *Pacific Commercial Advertiser* publish the irresponsible Mallett letter. After a few weeks of simmering community controversy over the conduct of Mallett and his boss Janion, Parker wrote a tactful and diplomatic memo to King Kamehameha IV, stating that he attempted to resolve the current crisis by practical and diplomatic means. He went on to share with the king his reluctance to take the route of litigation except as a last resort. Having sensitized both the royalty and the public to his dilemma, Parker secured from the Department of the Interior the unquestioned right to any unbranded cattle on his land.

Frank Spencer, through WGAC, launched other attacks on Parker, mainly contesting his claim to more than 17,800 acres in the lands of Ka'ohe and Kemole. This took years to resolve.

Still undaunted in his mission to monopolize, Janion bought out competing operators Frank Spencer and James Louzada, making them stockholders in WGAC. This partnership did not last long; Janion bought them out and they became employees.

In its heyday, WGAC killed fifteen head of cattle daily for the whaling market—but the market began to wane. Whale prices collapsed in 1864 and the need for sperm oil died two years later. There was now little or no demand for the barreled tons of beef, pork, mutton, and tallow that sustained WGAC. By 1865, WGAC had changed to the hide and wool market to keep its Waimea operation intact, but its final demise came shortly thereafter.

Following his shrewd handling of the Janion/Mallett letter in the printed medium, Parker calmly watched his thorny rival slip into oblivion while he remained focused on what he knew he could do very well. By this time the numbers of wild cattle had decreased dramatically, and management of the WGAC had become disconnected and inept. Even T. H. Davies, the comptroller, became alarmed and over time bought out all of the holdings, which by then had grown to include an iron works, a cooling station (ice house), and a number of plantations. This move formed the nucleus of an enterprise, Theo H. Davies, which became one of the "Big Five" when sugar and shipping commerce was at its peak in the 1950s. That was a full century after Janion first set out to build his monopoly of the livestock and mercantile trade in the Hawaiian Islands.

The Era of John Palmer Parker II: 1850–1891

With the ranch soundly up and running, John Palmer Parker finally turned his attention to himself and his failing health, leaving John Palmer Parker II to assume most of the day-to-day operational management of the ranch. With the death of Kipikāne some years earlier, Parker had married a Hawaiian woman from North Kohala named Leiakaula, who provided warm companionship at Hale Mānā. His son John shouldered the ongoing pressures of the livestock industry, allowing Parker to enjoy his leisure with his grandchildren, friends, and neighbors while maintaining steadfast devotion to his faith.

Some of the business pressures facing John Palmer Parker II were not new ones. Efforts to improve the quality of the cowherds were limited because importing improved breeds was out of the question financially. Meanwhile, since 1851 other ranchers had been importing bull stock, and it must have made the Parkers envious. In that year, John Meek of the Royal Hawaiian Agricultural Society announced that a Scottish-bred Aberdeen Angus bull was coming to Hawai'i via Tasmania. In 1853, W. L. Lee, president of the society, announced that Gordon Hopkins was bringing Hereford cattle, and Robert Moffatt was importing Angus cattle; both men were located on O'ahu. In 1855 the society imported Saxon cattle, with Dr. R. W. Wood of Pu'uhue Ranch on the Big Island holding one-fourth ownership.

Undaunted by these parallel and historic events in the livestock industry, John Palmer Parker continued to concentrate on critical bull selection, cowherd domestication, and feral cattle extinction through his final years. His son would be confronted with an array of crises over the ensuing decade, while cow quality was barely holding its own; there were many more immediate priorities.

One of these priorities had to do with the ailing senior Parker's continued stress over the contesting of his claim to 17,800 acres of the lands of Ka'ohe and Kemole by Frank Spencer of the WGAC. Physically weak and with impaired vision, John Palmer Parker would not live to enjoy the decision in his favor regarding the claim. A death vigil began in 1867, with Makua Laiana and Reverend Elias Bond, a North Kohala missionary, serving communion to the Parker patriarch. (On July 7 of that same year, in Waterford County, Ireland, William Carthage Bergin was born. His namesake grandson would graduate a century later as a Doctor of Veterinary Medicine and prepare to move home to the Big Island as a general practitioner.)

While on O'ahu for further treatment,

John Palmer Parker died. It was August 20, 1868. He was interred at the cemetery at Mānā back on the Big Island on April 11, 1869, alongside his beloved Kipikāne and second son, Ebenezer.

As John Palmer Parker II gathered the reins of the ranch in the decade leading up to his father's passing, other crises occurred. They included the April storm of 1868 that hit Kawaihae Harbor in the form of a series of waves that demolished the wharf and its support facilities. Four years later, a storm of greater vengeance returned to this port and plundered the recently restored facilities. Livestock commerce was hampered over a period of about six years as rebuilding in the wake of both storms took time and money from a struggling government. This occurred despite the existence of alternative points of embarkation at Kīholo in North Kona and Māhukona in North Kohala. These were not convenient alternatives, however, considering the long distances across arid waterless lands.

Through this period of hardship, John Palmer Parker II and his adopted son and nephew, Sam—who lived away on O'ahu leading a comfortable, even elegant social life—were left a fortune in land holdings that required diligent management. Quite likely, it was the intent of John Palmer Parker that his son and grandson jointly manage these holdings, but in reality Sam Parker was more at home in the social graces than in sharing the serious mission of ranch management. Still, John Palmer Parker had divided his holdings between these two men in a manner that was reasonable and equitable.

John and Sam received nearly 47,000 acres of deeded land between them. While some of this land was in the form of large parcels (the lands of Pā'auhau), others included taro farms in Waipi'o Valley that were used to produce the staple poi for ranch purposes. Property around the Mānā headquarters was also divided, but with a practical tone, giving Hale Mānā to Sam and to John the lower part of the 7-acre homestead including the original ranch cabin, along with other cottages and outbuildings—stables, barns, and corrals.

Equally divided between them were the watering devices such as tanks and cisterns, the dairy, sheep station, and the servants' quarters along with the livestock, rolling stock, weapons, instruments, storage trunks, tack, harness, and saddlery.

In fairness to Sam Parker, it is important to note that he was only fifteen at the time of his grandfather's passing. He was attending O'ahu College (later known as Punahou School) when he learned of his inheritance in detail. He did not find it necessary to disengage from his educational goals to return to a ranch that had been ably managed by his adoptive father John, whose conservative ways could be counted on for preservation of the joint Parker holdings. Besides, Sam was enjoying his early adulthood in the presence of nobility, people of high social countenance, with fine music and drink and a splendid social whirl. But this love for the good life would likely follow him home to Mānā, so home he went.

Samuel Parker Sr.

Sam married Pānānā Napela, a charming Honolulu socialite, on August 23, 1871, in Wailuku, with the Reverend Mr. Alexander officiating. Not long thereafter, the newlyweds moved to Mānā, where Sam proceeded expeditiously to upgrade the premises to match his social status. Much of the extravagance in Sam's style left John Palmer Parker II discomforted. The renovation of Hale Mānā and adjacent improvements was the beginning of a rift between the two men. All support facilities were distanced from Hale Mānā, separating the horse pen and barns from the immediate premises to the lower portion of the 7-acre homestead. Sam quickly moved to strengthen his presence through the purchase of a 640-acre section from Jack Purdy at Po'o Kanaka. Typical of Sam's nature, he generously left the 5-acre homestead and home for the Purdy family. Sam was an accommodating person, his style marked by congeniality that typified aloha in the true sense of the word. His imposing stature of 6 feet 2 inches and 235 pounds was distinct and his spirit kindly. He expressed these qualities through social interactions, entertaining nobility at Hale Mānā with a true Hawaiian flair.

Sam and Pānānā became the parents of nine children. The middle child was a son christened John Palmer Parker III, who was later adopted by his adoptive grandfather and true granduncle, John Palmer Parker II. Later, this became a key factor in how the ranch was inherited by John Palmer Parker II's adoptive granddaughter and natural granddaughter of Sam, Thelma Kahilu'onāpua'api'ilani Parker. Two other sons earned recognition, Sam Jr. for eventually acquiring the Humu'ula Sheep Station and the youngest, James, who enjoyed painting.

A logical and significant land purchase by Sam, given his love of the ocean and fishing, was the property at 'Anaeho'omalu Bay in 1906. Accessible until the 1950s only by sea, the long sand beach was fringed with palms, and an enchanting lagoon was well stocked with mullet *(ana'e)* to complete the scene. The components of the name, *'anae ho'o malu,* translate as "the gentling of the milkfish or mullet." This beach retreat first served a social purpose for dignitaries who canoed there as part of Sam's entourage, but more importantly, in the new century it became a camping site for ranch employees and their families. Eventually a beach house was built, together with quarters for a caretaker who lived there permanently. Helen Lindsey Roman describes her 'Anaeho'omalu memories in the following excerpt from the Lindsey Family reunion document (July 1999):

> For the luaus we always went on the sampan from Kawaihae to Anaehoomalu to gather the mullet in the Loko Kai. This was always done a few days before the designated event. On these trips, Barbara, the hostess, would always include her house guest or two and Thelma, Lily, Emma, John Rickard, Joe Pacheco, Piko and Alana Bell, Theodore Saffrey and me. (There were others but their names fail me at the moment.) These were people who loved music too and we had a grand time with all the Hawaiian songs; some current English ones and the hapa haole tunes. Emma and Barbara played the ukulele and Saffrey on the guitar. After the anchor was dropped at Anaehoomalu, we all took turns in the dinghy that took us to the beach where we waited until the waves receded before we got out. Then everyone gathered at the Loko to help the Beach Caretaker prepare the net for the hukilau. With one

end of the net being held by the men on land, the Caretaker took the other end in his rowboat and laid it in a semicircle in the Loko and brought his end back to us to pull in. The fishes were gathered and put in boxes of ice and taken to the sampan, then placed in a larger container filled with more ice. After this work was done we shared the picnic lunch which always came from the Carter kitchen. Some of us swam and got papa'a and scoured the beach from one end to the other. All that and trying to learn to swim! The boat ride back to Kawaihae was always uneventful—tired, sunburnt, cold and wet as a result of the waves splashing over the bow. The musical instruments were silent, as they had to be stowed in the cabin below the deck.

As the second millennium approached, the current trustees chose to sell this parcel. This brought deep-seated resentment from ranch employees and members of the Waimea community, who were further aggrieved by the understanding that the price received came on the first day of public offering and instantly rolled in escrow, bringing a far greater value to the interim buyer. Despite its sad ending, the initial purchase and longstanding stewardship of this property, with its immense recreational value, speaks well of the foresight of Sam Parker Sr., the business integrity of A. W. Carter, and the vision of Richard Kaleioku Smart.

Sam's Water Venture

Leading up to the June 11, 1906, christening of the Kohala Ditch Company, John Hind of Hāwī Plantation struggled to find a partner in this venture to bring irrigation water out of Honokāne Valley to the rich but dry plains of Kohala Iwaho.

Already owning some water rights in North Kohala, Sam Parker was a likely candidate for partnership. J. T. McCrosson and M. M. O'Shaughnessy, experts in the field of irrigation, joined forces with Hind and Sam Parker and formed the Kohala Ditch Company. Hind was president, Sam Parker vice president, and A. Lewis Jr. served as secretary.

Through their foresight and efforts, the Kohala Ditch Company put forth over 20 million gallons daily in its opening period. This volume reached 70 million gallons daily, moving water 21 miles from the water head as far southwest as Puakea Plantation's 6,000-acre estate. Later, Sam would partner in the Kehena Ditch Company and again be a pivotal figure in the development of water resources in North Kohala, probably unaware that Parker Ranch would in the near future acquire the Kohala Ranch Company (Pu'uhue and Puakea) and savor the natural resources made possible by his efforts.

John Palmer Parker II Moves Forward

Meanwhile, back at the ranch, John Palmer Parker II remained focused on livestock production, a conservative stance created by a struggling cattle market that generated scant income, which often went to cover expenses generated by his nephew's extravagance. At times, cattle prices were so low that except for the hides, whole carcasses were rendered. The drought of 1877 did not help matters. Cattle crowded the southeastern foothills of the Kohala Mountains in search of water, leaving the good grazing areas of waterless Waiki'i. Despite all efforts, cattle losses were severe.

Other cattle operations such as the WGAC consortium were in even worse straits and sold out their herds to John Palmer Parker II. This was a bold move in the midst of a drought, but this transaction brought with it the lands of Līhu'e, not far from present-day Pu'u'ōpelu, which created greater access to the well-watered foothills of the Kohala Mountains. Without a doubt, the piping of water resources from this mountain range to the foothills of Mauna Kea would have enhanced grazing and carrying capacity, but given the frugal nature of John Palmer Parker II, such a thought would not appear realistic until the ranch neared total financial collapse.

Parker missed opportunities for enhanced grazing efficiency and carrying capacity by not adding more fencing to keep herds intact in suitable pastures. By this time, the Western United States was facing the harsh realization that the days of free-ranging cattle were over and fences had become a realistic tool of cattle management. As ranch management control moved out of the direct hands of the Parker family, both water development and fencing would be seen as serious steps in modern livestock and range stewardship.

To be sure, John Palmer Parker II was highly regarded in the general community and livestock industry. He was well mannered and considered wealthy. In addition to buying lands at Līhu'e, he made other strategic acquisitions in the immediate area of Waimea village. He had been appointed a Noble of the Kingdom by King Liholiho as a measure of his place in royal circles, probably with some nudging from Sam. He was also serving as fence commissioner for South Kohala and inspector of stallions for North and South Kohala.

In keeping with his role as a community leader, Parker moved ranch headquarters to Pu'u'ōpelu, in part to escape the constant social bustle of Sam and Pānānā. In 1879, Parker and his wife Hānai moved into their stately home at the 7-acre Pu'u'ōpelu homestead, and from this vantage point the ranch operations would be guided as best they could be, considering the free-wheeling nature of Sam Parker.

Sam's Sugar Venture

Sugar lands flanked Parker Ranch in North Kohala and Hāmākua, except for the lands of lower Pā'auhau, but none of the ranch lands had the annual moisture, moderate terrain, or tillable soils that lent themselves to sugar production. Through the mutual friendship of King David Kalākaua and Sam Parker—who was the king's frequent host at Hale Mānā—and sugar tycoon Claus Spreckels, the idea to convert the lower Pā'auhau acreage to sugar was conceived in 1879.

In that year, Sam, with the backing of Spreckels and William G. Irwin, established the Pā'auhau Sugar Plantation Company on 2,000 acres. Although a difficult plantation to work, the soils were sticky and claylike, qualities that made irrigation by furrowing a possibility. Furthermore, the Pā'auhau Landing provided a convenient point for shipment of sugar.

Spreckels' proposal was clear: Sam would begin planting immediately and the sugar magnate would advance money to build the mill in time for harvest. The proposal was not well received by John Palmer Parker II, but Sam forged ahead with planting, using up ranch grazing and financial resources. (A 2000 forestation project in Pā'auhau raised similar concerns for prudent resource management, as viewed by middle-level livestock managers.)

The sugar venture was a disaster. The mill was not constructed in time and the delayed harvest had to be milled at adjacent mills. Claus Spreckels would not relieve Sam of the mill construction debt, which forced the mortgage of Parker Ranch to William G. Irwin and Company, Ltd. By 1887 the financial picture was degenerating rapidly and, for a moderate fee of $200,000, the entire Parker Ranch was entrusted to Charles R. Bishop and Company.

The management of this already great empire would thus fall into the hands of trustees. The year 1887 marked the end of an epic era of four decades of direct and sole control of the ranch by heirs of the Parker and Kamehameha bloodlines.

Trusteeship Begins

The move toward trusteeship had benefits beyond fiscal ones, particularly with regard to John Palmer Parker II's well-being. The ranch was severely compromised in the sugar fiasco brought about by Sam Parker, plus his other dealings, which created expense instead of income for the entire operation. With Parker's health failing, it was a great relief when the trust in 1888 hired a young and competent part-Hawaiian fellow named Paul Lapa'ula Jarrett.

Jarrett came with a background that would give him an understanding of the historic complexities of the ranch operations. He also was a friend of Sam's from O'ahu College and an employee at the failed Parker/Pā'auhau sugar operation; but most importantly, he had a background in agriculture and he had a progressive outlook.

Paul Jarrett stepped into an almost winless maze of dealing with the affairs of a ranch that had a pair of antagonistic owners, a group of creditors that financed a bailout, and trustees who mandated policy. Jarrett was fortunate, however, to have a few years' counsel from John Palmer Parker II. Jarrett would have to live with some on-ranch resentment from men such as Ulumaheihei Waipa, the prominent lead foreman and cousin of the Parkers (his mother was Mary Ann Waipa, daughter of John Palmer Parker I), who felt that he was being groomed for management. Furthermore, the unpleasant reality of day-to-day management meant juggling the wishes of dual and disparate ownership and absentee trusteeship while accommodating the discomfort of a *kanaka makua* (one with natural leadership qualities and respect of peers) like Waipa, whose huge frame towered over Jarrett's. Jarrett was fortunate to have as head cowboy Kaho'ohanohano, who eventually followed Waipa as head cowboy foreman when the latter moved to O'ahu.

While Parker's counsel was welcome, the flamboyant and robust ways of Sam did not appeal to Jarrett. The costs of Sam's social calendar were severe when one considers the drain on ranch resources. It was not uncommon for Sam to appropriate a fat steer to be dressed by ranch hands and prepared as a barbecued welcome for his friends and dignitaries; or to order numerous mounts from the horse strings of the cowboys for his guests to ride the countryside in pursuit of wild cattle, pigs, and sheep.

Guided trail rides of his entourage to the *kuleana* (homestead) in Waipi'o not only took its toll on the hard-working cow horses but also on the loss of cowboy time and labor in running the large struggling ranch. Probably as serious an offense as any was the cowboy's loss of control over the health and welfare of his mounts. To a cowboy, such misuse means total disregard for his mount's well-being. This was particularly stressful since ranch horse strings may have consisted of only two or three mounts per man. Such disregard is demoralizing and insulting in every culture that embraces the bond of man and horse.

Some of Sam's antics approached unlawfulness and lacked ethics from a business point of view. One example of Sam's actions, despite the ranch being in trusteeship, was harvesting wild cattle from the outer reaches of the ranch with the help of a pair of unemployed native caballeros, converting the cattle to dried beef, and selling it in Honolulu. The income would be divided among the three of them. One of these *paniolo* was a young James Stevens. His partner was Tom Lindsey. Jarrett put a halt to this activity by offering the *paniolo* ranch employment. The day of such gentle wrist slapping disappeared within a few years when A. W. Carter became sole trustee.

Parker's counsel provided needed support for Jarrett, but there were moments when Parker himself challenged the new manager's ideas. Fencing was the most prominent example of Parker's resistance to progressive ways. Jarrett also met comparable resistance from the trust company when he proposed to pipe water to the outback of the ranch. He was able to complete a pipeline from the water intake in the Kohala Mountains to the plains of Holoholokū and Keamuku, but in the overall scheme this was not adequate. During Jarrett's tenure, the major water project was never funded in time. Many years later, in the fall of 1898, the trustees finally relented and allowed such improvements.

The Passing of John Palmer Parker II

The time was approaching when the counsel of John Palmer Parker II would no longer be available. "Softening of the brain" was the diagnosis then of what today is called Alzheimer's disease. Fortunately, before his passing in 1891, Parker entrusted his will to Jarrett, who clearly understood his wishes. Parker left his half of the ranch to his *hānai* son, John Palmer Parker III, Sam's son, effective at the age of majority. In the interim, Sam was named sole owner of the ranch as the result of an earlier probated will that made no mention of John Palmer Parker III. This issue was resolved and ownership was acknowledged to be equal.

With the death of John Palmer Parker II, Charles R. Bishop resigned as trustee. In 1892, the ranch creditors replaced him with Godfrey Brown and P. C. Jones. Concentrating on continued conservative management of the ranch by Paul Jarrett, the trustees were oblivious to the fact that John Palmer Parker II's final will gave his half interest to John Palmer Parker III at the age of majority. John Palmer Parker III spent many summers in the interim cowboying out of Mānā, but he remained focused on his education on O'ahu as he entered early adulthood.

Ranch management continued through the adjustment period that followed the loss of John Palmer Parker II. For the dozen years that Jarrett ran the ranch, he deserves credit. Despite severe administrative obstacles, he continued with vigor to advocate for more water and fence lines. While he was quite frugal, he made heroic efforts to maximize income, to reduce debt, and to relieve the ranch of its creditors.

Many of the qualities found in Jarrett were sound human values. His work ethic called for early rising with the cowboys, and *pau hana* (end of the workday) was commonly after dusk, at which time he would turn to his bookkeeping chores. Jarrett kept sales, time, and expense books himself. It was not uncommon for him to rise at 2 A.M. to accompany the cowboys with the fattened cattle to Kawaihae for shipment. Not an avid horseman, Jarrett did not consider himself a cowboy, but his presence with the men on long and arduous days gained him their deep respect. He was a genuine caregiver of animals and made sure cattle being shipped always had access to drinking water. Jarrett was intolerant of animal abuse in any form. Even in dire financial times, he developed plans for larger horse strings for the men, who struggled with only two or three mounts that had inadequate feed and rest.

The Tragedy of John Palmer Parker III

Young John Palmer Parker III found a bride in Elizabeth Jane "Tootsie" Lanakila Dowsett, daughter of prominent O'ahu rancher James I. Dowsett. They were wed on April 18, 1893. Parker was still unaware that by Hawaiian probate law, he was entitled to half interest in Parker Ranch, left by the death of his *hānai* father less than two years prior to his wedding day. This entitlement would occur upon his reaching majority. A daughter, Annie Thelma Kahilu'onāpua'api'ilani, was born on March 17, 1894, within a year of their marriage. Tragically, John Palmer Parker III died less than two months later and, in effect, left his half interest in Parker Ranch to his baby daughter. Being only nineteen years old, he had not reached the age of majority. Hawaiian probate law passed his half-interest share of the ranch directly to his newborn offspring upon his death. A kindly gesture of his widow, Tootsie (Elizabeth), affected this transaction. In *The Parker Ranch of Hawai'i*, Joe Brennen clearly describes in detail what transpired:

> Young John Palmer Parker III died in Honolulu at his father's house on May 8, 1894. He died at the age of only nineteen years, two months, and two days. His funeral services, too, were held at the Sam Parker home in Honolulu, and the body was removed to Mana for burial. Having not attained his majority, J. P. Parker III had not inherited his uncle's estate. Instead, it went directly to his baby daughter, Annie Thelma Parker.
>
> The reason for the inheritance might seem a little vague without some explanation. John Parker II had adopted his grandnephew (Sam Parker's son), John III, who automatically became John Palmer Parker II's natural heir. Then, with John Parker III still a minor at the time of his death, in addition to the fact that he made no will, one-third of the inheritance would, under ordinary circumstances, go to his widow, Elizabeth Dowsett Parker (Tootsie), and two-thirds to the child. Tootsie, however, relinquished her dower right in favor of little Thelma. Consequently, in 1894,

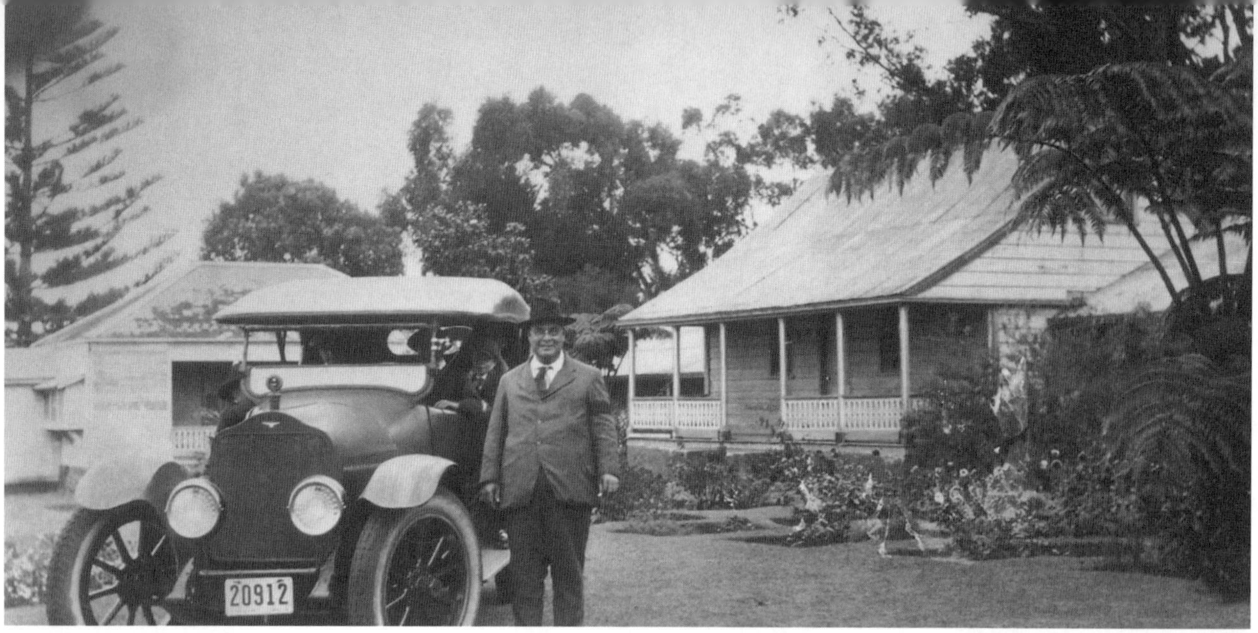

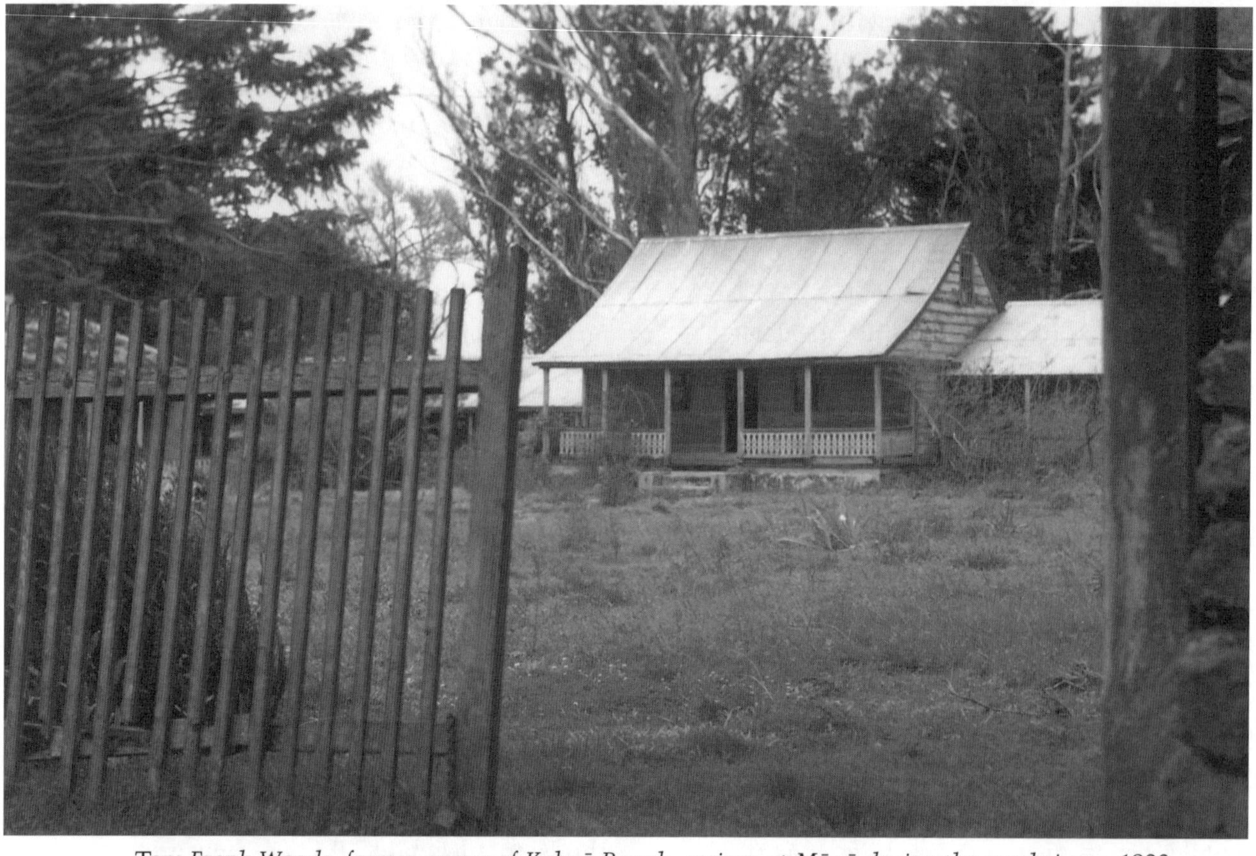

Top: *Frank Woods, former owner of Kahuā Ranch, arrives at Mānā during the good times. 1923. Bishop Museum photo CP 115,306.*
Above: *Old Parker homestead in disrepair. Hartwell Carter Collection.*

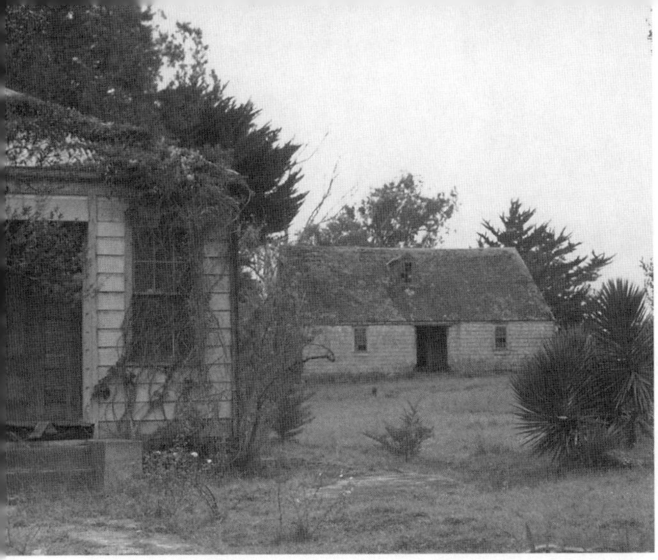

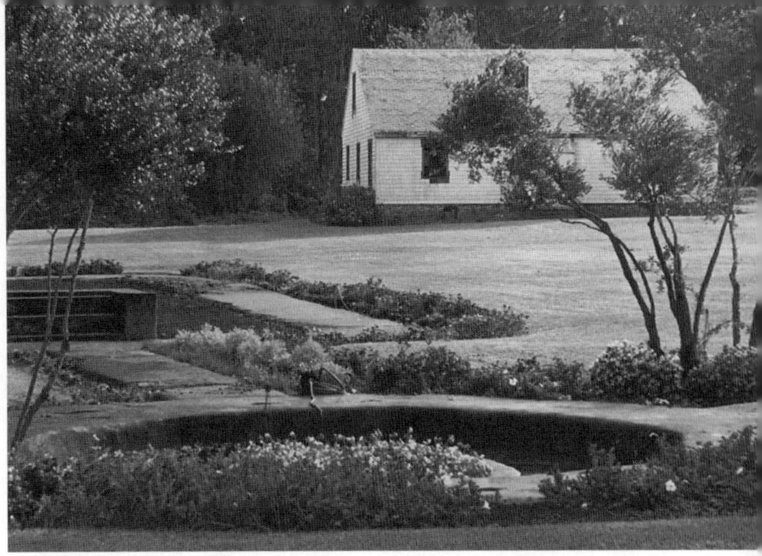

Left: *Original Mānā house in background. 1945. Hartwell Carter Collection.*
Right: *Original Mānā house, postrestoration. 1955. Hartwell Carter Collection.*

the little girl came into full ownership of her father's half-share of the estate.

James I. Dowsett, father-in-law of John Palmer Parker III, who was appointed interim guardian for baby Thelma, discovered her inheritance-in-fact when he reviewed legal documents. Shortly before Dowsett's death in 1898, he impressed upon Tootsie the impact of Thelma's half inheritance of Parker Ranch. Armed with this information and determined to do her daughter right, Elizabeth Jane Lanakila Dowsett Parker, on September 25, 1899, chose A. W. Carter to be Thelma's guardian and trustee manager of her property. Shortly thereafter, and in concert with Sam's agreement for management of his half interest, A. W. Carter assumed responsibility for running the entire Parker Ranch.

Paul Jarrett would soon be set aside as manager as trustee changes came about, but both he and John Palmer Parker II deserve credit for keeping intact a massive land and cattle empire despite weather, financial, health, and family crises. It could be said that the first John Palmer Parker was truly a builder, while his son John Palmer Parker II and Paul Jarrett were caretaker managers. Then a man who became an icon, Alfred Wellington Carter, ushered in a new dawn of Parker Ranch history.

8

A. W. CARTER

Biography and Human Resource Development, 1867–1949

It was the call of the sea that brought A. W.'s grandfather, Captain Joseph Oliver Carter, from Charleston, Massachusetts, to sail around Cape Horn and on to the Hawaiian Islands. A captain by age twenty-eight, he met and married Miss Hannah Lord of Maine. They bore six children, one of whom was Samuel Morriss Carter, who as a young adult launched a seafaring career in the whaling industry. Later in life, he was involved in the hay and grain business.

Samuel married a young widow whose maiden name was Harriet Layman. She brought with her into the marriage a daughter named Harriet (Hattie) whose father, Captain John Hempstead, had been a whaler. The Samuel-Hattie marriage produced several children, and among them was Alfred Wellington Carter, born on April 22, 1867.

A. W. was educated on O'ahu—first at the Royal School and then Punahou School between 1879 and 1883. Over his school years he worked at the wharves with his father's hay and grain business, during which time his appreciation of the Native Hawaiian workingman *(pa'ahana)* became deeply ingrained. He described them as "men of wonderful physique and the best of labor."

A. W.'s love for horses began in his boyhood years, enjoying the Kamehameha Day horse races at Kapi'olani Park on O'ahu when he should have been attending a Sunday school picnic at church on Fort Street. While attending bush track races

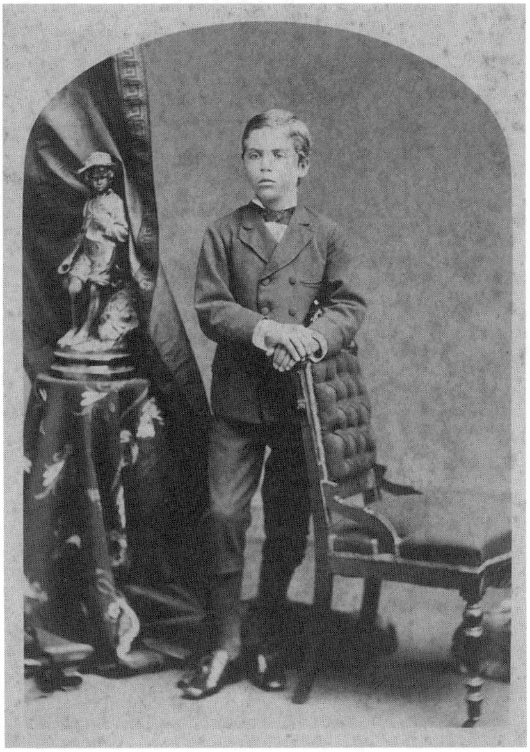

A. W. Carter as a nine-year-old boy. Photo by A. A. Montano, Honolulu. 1876.

near Punchbowl, he admired a racehorse named "Carry the News to Mary." Later, in his Parker Ranch career, A. W. christened a polo prospect "Carry the News," and this mount gained international fame in polo circles.

After completing his education, A. W. went to work for Whitney and Robertson, a printing concern. After a brief stint with Alexander Cartwright, an O'ahu insurance executive, the call of the sea beckoned him. His chores on board ship varied. Puget Sound, Washington, and San Pedro, California, were ports of call. For a short time he was fireman on the tug *Tai,* cruising Cape Flattery, picking up ships to tow into Puget Sound. Even after returning home shortly thereafter, A. W.'s love for the maritime industry never waned. In fact, his youthful seafaring experiences gave him an important background for the development of improved interisland livestock shipment systems.

Returning from the sea, A. W. was employed by Bishop Bank, but he moved on to the money order department of the U.S. Postal Service for a better salary. During these years, his yen for the ocean moved him to become one of the founding officers of the Myrtle Boat Club.

A. W.'s involvement in politically tumultuous events, including the "Bayonet Constitution" of 1887 and the insurrection of 1889, is a matter of record, but these matters are beyond the scope of this history of Parker Ranch.

A. W. left the postal service to become secretary of the Board of Immigration and the Land Office, experiences that continued to be useful in later life. His first introduction to the legal field came with his appointment as a deputy clerk in the Hawai'i Supreme Court. Inspired by this experience, A. W. left in 1891 to enter Yale University law school, graduating with honors in 1893.

After a brief period as an independent lawyer, A. W. became a law partner with his cousin Charles Carter, the older brother of George R. Carter, who later became governor of the Territory of Hawai'i. Other partnerships included Lorrin A. Thurston and Will Kinney, until an appointment as deputy attorney general of the Territory lured him back to government service. A short time later he became First Circuit Court judge, agreeing to serve on the bench for no more than two years.

On October 14, 1895, A. W. wed Elizabeth Hartwell, daughter of Judge and Mrs. A. S. Hartwell, at the home of the bride's parents in Honolulu. This marriage produced four children: Alfred "Hartwell," Edith "Edie" Millicent (later Mrs. H. Ernest Podmore), Dorothy, who died at age sixteen, and Barbara Juliette.

The year 1895 was an active one, as A. W. was named captain of the Mounted Reserve of the Honolulu Police Force. In the next few years leading up to his Parker Ranch appointment, a variety of business ventures, one of which was Moloka'i Ranch, would have an impact on his early years in Waimea. In 1896 he was involved in the organization of a coffee plantation venture in Puna on the Big Island. The Kamaili Coffee Company did not last; the lands, along with improvements, were acquired by the 'Ōla'a Plantation, a company A. W. founded earlier with Lorrin A. Thurston and Benjamin F. Dillingham.

While he maintained a law office in Honolulu, it became increasingly evident that A. W. yearned to be a rancher. On

December 20, 1897, he wrote to Louis Von Tempsky, a prominent Maui landholder and rancher: "You may not be aware that I have contemplated becoming a ranchman myself."

On January 28, 1898, A. W. realized his dream by leading a group of business partners in the purchase of Moloka'i Ranch. The group was incorporated as the American Sugar Company, with an interest in continuing ranching in part while building a sugar plantation. Water testing indicated high salinity, which eventually precluded the cultivation of sugar, but A. W.'s enthusiasm for ranching was just warming up. He immediately improved water transport for cattle and established grass seeding and fencing projects. American Sugar Company imported Devon breeding stock from California that included twenty-five bulls. Ten cows also came with the shipment. A. W. immediately plunged into reduction of herds of wild cattle and feral deer.

A. W.'s experiences on Moloka'i Ranch were invaluable for assuming trusteeship of the Parker Ranch. Besides water, wild cattle, feral deer, and fencing challenges, there was the positive experience of importing bloodstock. He also grew to appreciate three young ranch hands whom he would eventually bring to Parker Ranch. Manuel Vierra, George Keola, and a young *paniolo* whom Carter christened James K. Brighter (born James K. Edmonds in Waikapū, Maui, in 1871) were soon to join the Parker ranks in Waimea when A. W. assumed the reins of the great ranch.

In early 1900, A. W. was appointed trustee of the Bernice Pauahi Bishop Estate, also serving in various director and officer positions for government and private institutions on an intermittent basis. He gradually relinquished these responsibilities, allowing his unbridled devotion to the guardianship of Annie Thelma Kahilu'onāpua'a-pi'ilani Parker for the next fifty years.

A. W. Carter's Early Ranch Years

On September 25, 1899, A. W. Carter was appointed guardian and trustee of Thelma Parker, a role to which he brought dedication and devotion. His instincts were those of a caregiver coupled with compassion. He was farsighted to the point of being visionary, and he had the bulldog tenacity that would be required.

Carter successfully used his traits and talents to maintain his ward's estate and take it to heights of a world-renowned entity. He grasped the sensitivity of his mission, and he never shirked his duty. In the end, he brought the ranch to the realization of its promise.

In accepting the appointment as guardian and trustee, A. W. relinquished several opportunities that other men would have found difficult to abandon: a lucrative law career in the burgeoning city of Honolulu, president of Charles R. Bishop Trust, an executive career with Bishop Bank, deputy attorney general, judge of the First Circuit Court, and trusteeship of Bernice Pauahi Bishop's estate. Periodically, A. W. would continue to be of service to these institutions on an as-needed basis, but for all intents and purposes he dedicated the remainder of his life to guarding his ward and building her assets.

A. W.'s First Observations and Changes

The 1899 appointment of A. W. as guardian of the Thelma Parker Trust was already in effect, but his move to the Big Island was delayed by a rampant plague on O'ahu, as Carter had regulatory enforcement duties that could not be set aside.

In the late winter of 1900, A. W. arrived in Kawaihae and made his way to Waimea on horseback for a comprehensive survey that took nearly ten days. Although transitional manager Paul Jarrett had given him some earlier updates on ranch conditions, what unfolded before Carter's eyes was not an encouraging sight. Cattle conditions reflected inadequate feed, water, and progressive breeding. Of the 25,000 cattle grazing on the ranch, more than 5,000 remained wild. The unmanageable presence of young mustang or jack bulls was routine. Pasture conditions reflected severe overgrazing, and grass varieties were of minimal nutritional value. These factors led to market conditions that were undesirable for the cattle's lack of fat cover, minimal flesh, and overall poor health. Following is Carter's assessment of the ranch when he took over, from Lucille Brundage's *Alfred W. Carter: Hawaii's Dean of Cattlemen and Notes on Hawaiian Livestock:*

> On my first visit to the Ranch, I found the place undeveloped in every way. The stock consisted in part of about five thousand head of wild cattle, which had been inbreeding from the time Vancouver landed a few head over a hundred years previously and of which these animals on the Ranch were descendants. Aside from the wild cattle, the balance were inbred, long-legged, long horned slab sided individuals. The steers at five years would not average over five hundred pounds dressed weight.
>
> I handled about 13,000 head of wild cattle in the years right after I went to the Parker Ranch, the number having increased right along. In the early days, we would start out at three o'clock in the morning from Waimea and sometimes do nothing for a whole week but rope wild cattle in rough country. The Hawaiians became the most expert ropers in the world because their business alone was the roping of wild cattle.
>
> A wild bull or cow will run away when it smells a man. However, once they turn back, nothing will stop them from charging. I have seen them lift a horse and rider on their horns.
>
> Much of the place was covered with o-i, no effort had been made to introduce good range grasses from foreign countries.
>
> There was not a drop of outside blood bred into the cattle for years before I took charge. The cattle ran indiscriminately without separation of cows or any other method so far as I could see. For breeding bulls, young calves that looked in fair condition were allowed to run continuously with their dams and started breeding at twelve months or younger. Heifers were run without separation and became pregnant at about the same age.
>
> There had been no segregation of horses. The first year I drove and then separated them; I castrated about one hundred and twenty stallions. Some of the early horses were beautiful animals.
>
> Aside from half a dozen small paddocks, there was no fencing, no segregation and, but for a short line of pipe from the Kohala Mountains, there was

no distribution of water. Some cattle had to travel fourteen miles to get a drink and return fourteen miles to their grazing ground. The Waikii Paddock and the fattening paddock had no water. Much of the land was unproductive from a grass standpoint, having either been overstocked or never having had any grass on it.

Of the changes A. W. made immediately, two stand out: marketing policy and internal management. First, cattle shipments were suspended until fattening paddocks were restored with adequate grass and available water, so market cattle would reflect improved conditions. Secondly, A. W. had earlier brought in his brother, Fred Carter, to work alongside Jarrett to assist in ranch management while A. W. resolved the O'ahu plague issues. While Jarrett continued to work for the ranch, Fred Carter formally replaced him as manager on January 8, 1901. It was Fred who tightened the reins on the outfit until A. W. could move to Waimea and assume full control. A. W. had originally offered this interim post to Louis von Tempsky of Maui, but it was declined. Fred Carter, while lacking true ranch experience, was an appropriate alternate, however, since he was known for getting things done.

A. W. was concerned with the abundance of strange horses being pastured on the ranch. One of Fred's first missions was an exact inventory of private horses and mules grazing on company grass. The inventory was conducted on January 30, 1901, and it was discovered that various people in the community privately owned an alarming total of 888 of these animals. The outside horse inventory included 3 stallions, 1 jack, 294 mares, 110 colts, 17 mules, and 463 geldings. While ownership was spread over 140 individuals, some of whom were ranch employees, five people (or neighboring businesses) owned a third of these animals. The Kōkī consortium (hotel owners) had 77, W. Akona (general merchandiser) owned 73, Frank Johnson, who ran the Keamuku Section, had 50, Frank Spencer owned 43, and 40 head were the property of Mahuka Estate. All 888 horses were gathered for removal from the ranch's already sparse grasslands.

Fred left a favorable, lasting impression on the rank and file of the ranch—especially the cowboys. Many employees appreciated his consideration, particularly when compared with Jarrett's more austere approach. Jarrett, for example, never arranged for lunch meals no matter how remote the cowboys' mission. Keoni Li'ili'i told Will Kinney in a 1904 interview: "In Jarrett's time the men would go out and work all day on the Ranch with nothing to eat until they came home at night." In the era of the Carters, the ranch made available a *pa'i 'ai* man who met the cowboys at midday with saddlebags of dried beef *(pipi kaula'i)*,[1] *'ai pa'a* (hard-cooked taro), and drinking water. Occasionally he brought salt beef, but *'ai pa'a* always was the staple.

Community leader Sam Spencer told A. W. on March 7, 1906, as Fred's term was nearing its end, that "Waimea people are going to give Fred a send-off on Wednesday night, and judging by its preparations it is going to eclipse anything of the kind given heretofore."

A. W.'s observations of the ranch were converted to quick action. While he might have lacked confidence in Jarrett's ability to adjust quickly to a new, progressive management style, he had no doubts about his brother Fred. Fred would be unencumbered by the minimal standards to which Jarrett had been held under the regime of John Palmer Parker II. For firsthand looks, A. W. made periodic trips to the ranch as he gradually withdrew from his many important activities on O'ahu.

It did not take long for the native Hawaiian people both on the ranch and in the community (nearly one and the same) to realize that A. W.'s caregiving nature extended beyond his ward Thelma to the *maka'āinana* (common folk).

Makua

The ranch people gave A. W. the honorary name "Makua," which means "father" or "parent." Some of the most basic needs of the employees were important to him. Lucille Brundage provides comments in her book, *Alfred W. Carter: Hawaii's Dean of Cattlemen and Notes on Hawaiian Livestock,* and quotes A. W.:

> As early as 1909, Mr. Carter was considering building homes for native Hawaiians on the Ranch.
>
>> "I think the thing to ensure the happiness and contentment of the people living in Waimea, more than anything else, is to get them comfortably located in homes of their own. The situation is this: none of them has accumulated any funds and a great many of them have obligations to meet. The Ranch is going to advance the money for their homes."
>
> Later that year, material was ordered for the first two homes, one for Keau and one for Kaliko.
>
> On March 21, 1911, he wrote:
>
>> "The Ranch will have to expend considerable amount of money in the construction of homes. I believe this to be a sound business policy and the Parker Ranch should take a lead in it."
>
> Later, in 1911 and early 1912, Mr. Carter helped many of the employees purchase three acre lots in a large tract owned and subdivided by the Government. He saw to it that the lots were fenced and homes built on them for his employees. He charged no interest on the money advanced and the men paid back as they could afford it, out of their monthly wages. Some paid in full in seven or eight years and a few took as long as twelve years but all eventually paid for their homes. In this regard, Mr. Carter said:
>
>> "I felt that the Ranch was perfectly justified in doing this as a man owning his home is generally far better satisfied and makes a better worker."

A.W. went further by providing interest-free loans with wage deductions to service them, so that employees could purchase fee simple lots and build homes on them.

The educational needs of employee children were provided for in a trust deed of his young charge Thelma. It included sums set aside for welfare, education, and medical benefits for ranch workers and their families. For the girls, educational opportunities were provided through the Normal School on O'ahu and Kohala Girls School (Kohala Seminary) at the Bond Estate in North Kohala. For the young men, there was Hilo Boarding School on the Big Island and St. Louis College and Kamehameha Schools on O'ahu for those who showed desire and potential.

A. W.'s commitment to the education of Waimea's youth went beyond employee ranks, and a partial roster from 1915 is revealing:

Ernest Campbell		
	Kamehameha	1903
Kennedy Hewitt		
	Kamehameha	1903–1904
Alice Johnson		
	St. Andrew's Priory	1907–1913
Harry Johnson		
	Kamehameha	1907–1912
L. Kaapana		
	Kamehameha	1904–1906
Ben Kaaupuni		
	Kamehameha	1910
Sam Kaehalio		
	Hilo Boarding School	1908–1912
Samuel Kaimuloa (14 years)		
	Kamehameha	1903
Lazaro, Peter and Thomas Kaimuloa		
	Kamehameha	1915
David Kaumaoha		
	Kamehameha	1902
Albert Uiha Lindsey		
	Kamehameha	1908–1911
Annie Lindsey		
	Kamehameha	1908–1910
Charles Lindsey		
	St. Louis College	1913
*Anna Lindsey (daughter of WMS Lindsey)		
	Kamehameha	1911
*Willie Lindsey (son of WMS Lindsey)		
	Kamehameha	1911
James Ka'aluea Lindsey		
	Kamehameha	1907–1910
John Kuakini Lindsey		
	Kamehameha	1903–1910
*Thomas Nahiwa		
	Kamehameha	1902
*Tyson Francis Norgaard (son of Dr. Victor Norgaard)		
	Punahou	1915
Francis Spencer		
	Kamehameha	1915
Pio Family Children		
	girls to Kohala Seminary, boys to Kamehameha	
Marie Vredenburg (daughter of Wilmot Sr.)		
	College of Notre Dame after Sacred Hearts Convent in Honolulu	
Vredenburg brothers		
	St. Louis College	1910

*These were not children of ranch employees.

A. W. saw to it that the ranch covered costs, up to a limit, beyond tuition for the education of the youth. While Kohala Seminary and Kamehameha Schools were popular institutions, others included Hilo Boarding School and St. Louis College. Students benefited uniformly, as expressed in a February 22, 1918 memo to G. A. Young, administrator of the Hilo Boarding School:

> I have an arrangement with the Kamehameha Schools and I wish you to adopt the same, it is as follows: I will pay the bills for supplies of all boys, but will

> charge them at this end for all expenditures other than Tuition, Uniforms, Laundry and School Supplies. They will be obliged to work out any cash advanced and all expenditures such as shoes and clothing and items of like nature.
>
> Please inform the boys that they will be charged with the money they [have] already drawn also the shoes and pennants. At the same time I want you to keep very close account of any money they may wish to draw in the future.

Summer ranch work provided the means by which the additional school expenses were worked off. Part-time vacation help especially from the boys was the basis for the formation of the 'Ōpae (shrimp) Gang, whereby specific group assignments were executed under adult supervision. Examples of these tasks were tank, trough, and pipe cleaning and weed control. The 'Ōpae Gang (named for their small stature) continued as an effective means of converting summer help to the productive achievement of important but mundane tasks. Ranch girls were later included in the program. The quality and degree of adult supervision was the key to success of this program, which was active on the ranch until the mid-1980s. The tradition of summer employment for ranch youth ended under the trusteeship of Warren Gunderson in 1985, largely owing to the lack of quality supervision. Over the seventy-five-year existence of the 'Ōpae Gang, many future ranch and community leaders received their basic training in the work ethic, an added value of the system.

By 1924, Kohala Seminary had a routine enrollment of over a dozen Waimea girls. A memo of thanks from the administrator, Mrs. Bird, acknowledged the $1,575 annual tuition payment for fourteen ranch girls and noted that the school's gratitude had resulted in naming one of its dormitories Carter Cottage in honor of Carter's enduring support.

As he got older, Carter reflected on the satisfaction he received from supporting the youth in this way, because he knew that in the majority of cases these young men and women would go out into the world better equipped for life's challenges.

Education was not Carter's only cause. There were many and varied projects that held his interest and in which he became involved. The following letter to A. W. from Alice Spencer, written on August 31, 1911, demonstrates the degree to which the community appreciated Carter's role as Makua.

> Dear Mr. Carter:
>
> Pardon me if I am imposing on your generosity. Knowing that you are the "Makua" of the district, I naturally felt in time of trouble (if you will allow me) I ought to come to you first, although Sam repeatedly told me not to bother you as you had too much to do to give to us any of your time, I cannot see any other way and I need and must have your advise first. It is this Henry Paakiki had advised (or as the old lady, Mrs. Paakiki puts it) forced her to sign a complaint asking us to set aside some deeds she made out to me a year ago. I have written out the facts and am enclosing it under separate cover and if you could advise me before you leave for the coast

I would be ever so much obliged to you. I must again beg a thousand pardon for imposing on you with the state of mind I am in. I am at a loss what and how to think. Anything I dislike is to face public and court, the old lady has been made to believe wrongly. I think Mr. Kinney has been engaged as their lawyer I don't know. I have always thought Mr. Kinney ought to represent us in cases like this.

I am rattled so will close with many thanks and regret,

Gratefully yours,
Alice K. Spencer

Just as A. W. campaigned for veterinary service for North and South Kohala, he also pushed to get a resident medical doctor for the Waimea community. In 1873, Dr. L. A. Thompson had located in North Kohala and provided minimal service to the people of Waimea over several years. Aware of this, Carter launched a campaign for a Waimea-based doctor. Although he was not a doctor, A. W. often had to personally attend to the medical needs of the employees and their families. He was also available to help resolve family and neighbor disputes, especially when they affected an employee's performance.

A Brief Medical History of Waimea

North Kohala was a burgeoning agricultural community. Sugarcane came to Kohala in the 1800s and remained for more than a century. Dr. James Wight, North Kohala's first physician, was coincidentally one of the founding stockholders of the original Kohala Sugar Company. By 1897 there were fourteen plantations, starting in the south at Puakō Plantation (Hind) and ending at Niuli'i Mill and Plantation Company in the north. Just as large congregations of horses and mules required resident veterinary services, so did the plantations demand resident medical services for their people.

It is no wonder that Waimea, with its sparse population, had to live with limited periodic medical services. Undoubtedly the Native Hawaiian healing practitioners *(kahuna lapa'au)* provided devoted services that often went beyond the spiritual into true medical care, as most of the herbal medicines in their armamentarium have proven factual pharmacological bases. Unfortunately, overzealous missionary leaders condemned such practice as quackery and voodooism, relegating it to covert administration. Fortunately, on the other hand, the skill and knowledge of the Hawaiian medical practitioner survived into the new millennium despite the passing of its notable healers, such as Papa Henry 'Auwae, originally a Parker Ranch hand, who died in 2001. This field of medicine has enjoyed a renaissance that is now irreversible.

It is not recorded that Dr. Wight served the Waimea community in the decade leading up to the arrival of Dr. Thompson, but it is reasonable to assume some contact was made with the Parker family, as the family's North Kohala ties were strong. Following Dr. Thompson was Dr. H. L. Ross, who began treating Parker Ranch employees and the Waimea community in the early 1900s. He was followed by Dr. Claude Buffett, who remained a few years.

Waimea received its first resident physician in 1931. It is not unusual to find some

Waimea residents who were delivered by Dr. William Dutcher Whitman, a Nova Scotian. His grandson, James "Jim" MacKinnon Whitman, later became the business manager of Parker Ranch. A graduate of Stanford and the Wharton School of Business, Jim rose to the Parker Ranch post after a brief career with First Hawaiian Bank. He retired recently as executive director of the Damon Estate and set up residence in Waikoloa. In 1989 Jim's son, great-grandson of Dr. Whitman, John Mark Whitman, returned from college and training to his Waimea home and serves the community as an emergency medical technician with the county fire department.

After Dr. Whitman left Waimea in 1939, a longtime North Kohala physician, Dr. R. T. Treadwell, served Dr. Whitman's patients on a limited basis. A new associate of Dr. Treadwell's, Dr. A. H. Harrell, was assigned to fill the needs of Waimea until 1941, when a skilled physician and surgeon, Dr. David T. Woo, moved to Honoka'a and also practiced in Kamuela.

Born in Hawai'i, Dr. Woo received his medical training in China. Well ahead of his time, he recognized the role of alternative therapy such as acupuncture and herbal medication, while establishing a reputation for excellence in Western medicine and surgery. During his Hāmākua, Waipi'o, and Waimea practice years, Woo developed a genuine interest in the history and preservation of the Hawaiian horse, a casually recognized breed of mustang horses that were feral descendants of imported horses. While Waipi'o Valley still abounds with these horses, the Mauna Kea slopes were the last home to many. Willie Kaniho, foreman at Humu'ula Sheep Station during these times, contributed toward their preservation by keeping a stud named Mauna Kea at service—the offspring of a feral stallion and half Percheron dam. Kaniho and Woo were close acquaintances, based on their common interest.

During the years of Woo's service to Waimea, my father, W. N. Bergin, M.D., provided overlapping medical and surgical support when he moved to Laupāhoehoe in 1937 after a time spent in North Kona (1933–1935) and at the Kula Sanatorium (1935–1937).

Woo and Bergin dramatically enhanced radiological and surgical services to a broad geographic community that included Parker Ranch employees and their families. These facilities included Kohala Hospital in Kapa'au, Parker Ranch Dispensary in Waimea, Honoka'a Hospital, Pa'auilo and O'ōkala Dispensaries, and Laupāhoehoe Hospital. While ranch employees with the Pu'uhue and Puakea Sections were treated at Kohala Hospital, cowboys from Keanakolu, which was part of the Humu'ula Sheep Station, often were treated at Laupāhoehoe Hospital. One such employee was Gabriel Malani, who rode home from Keanakolu on weekends to his *kuleana* at Laupāhoehoe Point. Big Joe Ignacio worked at Humu'ula Sheep Station but lived in Pāpa'aloa, near Laupāhoehoe Hospital. Honoka'a Hospital was available to ranch employees who lived on adjacent sections such as Pā'auhau and First Gate. The Waimea Dispensary, built and owned by Parker Ranch, served the needs of the Waimea, Waiki'i, and Keamuku employees.

In the mid-1940s, both Woo and Bergin moved on to pursue other medical career options. Both local boys, they developed a deep sense of mutual respect and appreciation for one another. In 1949 the Hilo Medical Group was born, and they became partners with other noted physicians such as Archibald Orenstein.

After their departure, Dr. Gardner Black moved to Waimea and provided caring medical services on a consistent day-to-day basis. From this moment forward, the ranch and community of Waimea was never without resident medical services. In fact, prior to the new millennium, Waimea emerged as the health and fitness center of the Aloha State, with North Hawai'i Community Hospital as its crown jewel. This facility and its reputation would never have been realized but for the foresight and benevolence of Richard K. Smart and the philanthropy of Earl Bakken, the renowned inventor of the pacemaker and relatively new resident of the Big Island.

Makua Continues

When the Great Depression hit and wages had to be cut, A. W. Carter did it across the board, with no one spared—including himself. He described the dismal situation as follows (from Brundage's *Alfred W. Carter*):

> Since the large reduction we were compelled to make in the price of beef, the situation demands that we cut our expenses. In order to operate, I have been forced for the first time in thirty-one years, to cut all salaries and wages from the top to the bottom and we are eliminating all permanent improvements and cutting down the repair and upkeep on the Ranch.
>
> Of course, this condition will not last forever but on the other hand, in my opinion, this year will be a bad one for us and I really don't look for any large betterment of conditions until the Fall of 1933.
>
> The cattle industry has been pretty well hit. We are sending in 8 to 11 months old calves to the market, something I have never done before. The price of beef had to be reduced still further today, for our cattle, we are getting about half of what we got a year and a half ago. Everybody in the beef industry is losing money.

Restoration of wages followed when relief and recovery were in sight.

Always the caregiver, A. W. was resolute in following the intent of the Thelma Kahilu'onāpua'api'ilani Parker Trust, especially as it related to the treatment of employees. He was held in high regard among the working class largely because he treated them with respect. He firmly believed in maintaining their dignity, no matter how low their role in the ranch hierarchy.

There had been incidents in A. W.'s pre-ranch career that clearly demonstrated the character of the man. One of the first cases in his budding law career came from Hawai'i Supreme Court Chief Justice A. F. Judd, who ordered the foreclosure of a mortgage on property owned by an elderly Hawaiian. The man pleaded with A. W. to allow him a week's delay to gather the funds to cover the arrears prior to auction. A. W. was ordered to deny the request and proceed with foreclosure. Instead, he approached the court and offered his resignation from the case in light of the harsh treatment of the *kanaka maoli* (Native Hawaiian). The judge relented and withdrew the order. Carter collected the arrears, satisfied the mortgage lien, and the Native Hawaiian retained his land.

His upbringing and early years stevedoring in his father's firm could help explain A. W.'s stature in the Hawaiian community. He took the time to learn the workingman's culture. He understood their plight early in his manhood, and he deftly used this knowledge for the betterment of ranch relations in the years to come.

He respected Hawaiian ways in his ability to speak the language so eloquently. *"Mākou, mākou ka pilikia,"* he would lament when things went wrong, accepting responsibility with a "mea culpa, mea culpa, mea maxima culpa." He appreciated the Hawaiian reverence for life and respect for the dead. The draping of his own coat over tragically lost steersman William Lono in a shipping accident in Kawaihae was open testimony to the Hawaiian community as to how deeply he embraced his people.

There were unpleasant moments when A. W. imposed his firm authority in the dismissal of a wayward employee, but he maintained dignity in the process. A touching example of the degree to which he went was the firing of Charles Akiu Jr. from Parker Ranch's Kahuku Ranch. Charlie Akiu Jr. is the same *pini* man noted earlier, reported to have fatally injured a horse during a shoeing altercation. Rather than dismiss Charlie through a phone call to his boss, Martin "Makini" Martinson, which would have been more convenient, Carter made the day's drive to Kahuku and sat with Charlie and Martin to deliver his discharge order and present him with his final pay. I had the opportunity to work with Charlie Akiu at Hawaiian Ranch, owned by C. Brewer in the early 1960s. He was a quiet old gentleman charged with feeding orphaned calves at Na'alehu Dairy. Charlie shared with me his dismissal from Parker Ranch and the circumstances under which it occurred. He acknowledged his altercation with a Kahuku horse unwilling to be shod, and he was repentant. But despite his dismissal, he spoke very kindly of the dignity that A. W. showed him in this most discomforting situation.

The Humility of Christmas

The naming of paddocks, corrals, and buildings is typically the prerogative of managers or owners, but in the following incident it was the cowboys who chose the name.

Fencing off open range was an early initiative of A. W.'s in order to create herd control and improve grazing conditions. He had already surveyed and staked out a large *pāhu'a* (open flat) southeast of ranch headquarters when the posts and wire arrived by mule-drawn wagons. On an early and chilly morning in mid-December, he gathered the crew and began the fencing with diligence. After ten hard days of working alongside the cowboys bearing shovels and *'ō'ō* (spade bars), the end was in sight but a day away from finishing. A. W. asked that they join him the following morning at 4 o'clock for the final thrust of completion. To the man, they all joined him and toiled a full day, stretching the final wires by 5 P.M. A. W. expressed his appreciation for their diligence, and home they all went.

After about a week, Carter overheard the cowboy foreman Old Tom Lindsey issuing the orders of the day in flowing *'ōlelo kanaka* (Hawaiian language), asking the men to *"hele iloko pā kalikimaka"* (go on to

Christmas Paddock with the cattle). Miffed, Carter wondered how the paddock got such a name—and by whose authority! He pressed Lindsey for an explanation. A. W. sheepishly bowed his head when Old Tom told him that the final day of fence completion was actually Christmas Day, when all of them should have been home with their families, but instead they honored and embraced the boss' wish to finish the job. A. W. had completely lost track of time in his determination to complete one of his first initiatives; he humbly acknowledged his oversight to the men, and the name for the paddock remained Christmas, probably as much in honor of their hard work as A. W.'s admission of having erred. For nearly a century this paddock has kept its name, with its size reduced by half in 1997 since part of the area was a Department of Hawaiian Homes Lands lease that expired.

A. W.'s respect for others extended beyond ranch boundaries. During the Depression years, financial problems were compounded by severe drought, which seriously affected neighboring ranchers. Despite Parker Ranch being in comparable straits, Carter extended the availability of water to the homesteaders, and their herds ultimately were salvaged. Comparable acts of kindness were extended to ranch neighbors to the north, Kahuā Ranch, when he made available heavily forested pasture reserves, thus averting catastrophic losses. His description of the situation, an excerpt from a letter from Carter to Atherton Richards, follows: "About three months ago, Ronald von Holt telephoned to me and told me he was in trouble and asked if he could put his cattle in a private forest reserve which is owned by us. I consented and they put a large number of cattle in there where there was a good amount of water and considerable feed, which I think saved him from losing his cattle. When a man appeals to you on the ground that he is in trouble, one cannot do otherwise than to help."

This action prompted a letter to Carter from Atherton Richards eight months later: "Last week, while at Kahua, I had occasion to ride back into our forest section and learned of what actually had occurred last fall in connection with the emergency rising as a result of the drought and the consequent distress of our herd. Your kind helpfulness and generosity in allowing us the use of your pasturage and water are greatly appreciated. Please be assured that the failure to acknowledge your courtesy sooner was due to my ignorance of the situation and to no lack of appreciation of how much it meant to us."

Carter the Strategist

A. W. made many early, strategic decisions that guaranteed the continued growth of the ranch. Among them, in addition to acquiring land, was the procuring of superior seed stock in dairy and beef cattle, horses, and sheep, while initiating an early and permanent preventive health program for the stock. The following is from an editorial in the *Honolulu Advertiser* in 1932: "In the history of these Islands, no greater economic development of a single property through the efforts of one man has ever been accomplished, and that, not as husbandman of this own property, but for the benefit of a succession of minor heir[s] whose welfare was entrusted in his guardianship by the courts."

While the copious accolades were

rightfully deserved by A. W. for his benevolence, his business acumen when dealing in behalf of his ward Thelma K. Parker was rigid. In the area of land acquisition especially, A. W. dealt aggressively in favor of her material welfare, regardless of who else may have wanted or needed the property. In the course of his Parker Ranch career, he ran roughshod over the property designs of notable and respected businessmen, including E. H. Wodehouse, Henry Holmes, Robert Hind, John S. Low, Alfred Magoon, Fred Wunderberg, Evelyn (Eben) Low, and Atherton Richards. In A. W.'s defense, his quest was directed toward the betterment of little Thelma. But others were also entitled to reasonable acquisition of ranchlands—especially when the subject property abutted their own.

Carter's Spiritual, Social, and Recreational Initiatives

A. W. had deep convictions regarding fundamental human values reflecting honesty, diligence of duty, loyalty, and sound moral standards. Hard work coupled with one's devotion to family and faith were Carter's own values, and he labored to create a social environment that would develop those standards within the hearts and minds of the Native Hawaiian people of Waimea.

He financed the building of churches across all denominations, just as was done for a variety of educational institutions. He attacked saloons with equal vigor, developing a nearly prohibitionist posture.

His brother Fred was no less adamant about fighting the evils of alcohol after witnessing the social strife in the village of Waimea when he arrived in 1901. In a memo to A. W., Fred lamented: "Waimea is at present a wide open town, drunkenness is rife at Akona's (store) and Koki's (hostel); Suemoto and others run open house. I think the Ranch should put in a store, restaurant and hack (livery stable) and run these people out of business. Drunken deputy sheriff and officers, no arrests for drunken brawls, rows or anything else."

Not unlike the Wild West of the continental United States, little Waimea town in early Carter times boasted two saloons and an open dance hall. There were two blacksmith shops and five general stores.

The general stores were owned by Chinese merchants Lum Gee, Chock In, and Akona; the latter also housed one of the saloons. Akona Store functioned as a one-stop shop in that a restaurant, barbershop, and boarding rooms were featured. The Chock Hoo Store and Restaurant in Kawaihae was owned by the Chock In family and was a favorite stop for the cowboys after a long day of shipping. Chock Hoo was noted for delicious fresh-baked pies. Later a Japanese merchant, I. Oda, along with the Suemoto family, opened a general store in Waimea that featured a saloon and general merchandise. As the Carter era began, it was the presence of open saloons as well as the bootlegging of alcohol that raised ire.

It wasn't long before the sale of alcoholic beverages was curtailed, but not without incident. In March 1903, Fred responded with gusto to the proposed reopening of a nearby saloon. He is quoted from a memorandum he forwarded to L. A. Andrews, sheriff of Hawai'i in Hilo.

> The deputy sheriff Sam Spencer has spoken to me in regard to the saloon at Kawaihae and I told him I would write to you. I presume there is no one as well qualified to air his or her views on this

subject or, do I believe that this question interests or concerns anyone more than it does myself and the interest I am representing. Being in charge here before there was any saloon, then during the time the saloon was open and also since it was closed I consider I am in a position to know and judge intelligently in regards the good or evil result of such a place. The saloon is decidedly a detriment to the property I am in charge of viz: The Parker Ranch.

Neither before the saloon was open or since it closed had I any trouble with the men but during the time it was open I on several occasions had to order men out of the shipping pen, who were under the influence of liquor and not only endangered their own lives but of those sober men who were working with them. I know personally that this saloon did not tend to stop illicit selling but rather to increase it as the illicit sellers could say and did say, he got his liquor at the saloon.

It is a notorious fact that people laid drunk around Kawaihae after the saloon was open. Fishing by the Hawaiians stopped. I think the Court records will bear me out that there were many more arrests for drunkenness and crimes, and misdemeanors that can be directly traced to the saloon and its influence during the time the saloon was open than either before or since there was a saloon.

Nestled between these strong moral postures was a commitment to create social, spiritual, and recreational endeavors for the ranch people and the community at large. In the ensuing four decades, A. W. Carter introduced the sporting events of baseball, billiards, football, soccer, golf, and boxing. He introduced opportunities to develop an appreciation of music and the Native Hawaiian heritage while fostering the establishment of scouting and home guard troops. He created a community hall for all to enjoy.

Ranch-related events were equally nurtured, including horse racing, polo, and roping contests. A large community *lūʻau* was always the centerpiece of these ranch activities. In highlighting these activities, one must reflect on Carter's sincere belief that one's leisure time is best spent in wholesome physical, social, and spiritual—as well as recreational—endeavors.

Horse racing was hardly new to the village, but this sport was profoundly enhanced by A. W.'s personal enthusiasm. Neighboring ranchers were often invited to participate—notably the Horner Ranch (Umikoa-Kūkaʻiau) on the eastern boundary. Not yet referred to as "stations," the different sections of Parker Ranch competed as three-horse relay teams against one another and guest outfits.

On December 22, 1903, Robert Horner wrote A. W. regarding Umikoa's entry for the approaching New Year's Day race meet: "The boys are starting over with four animals. Now I will be frank with you; we are going to Waimea for a good days racing and not to take over a lot of money, so the boys there need not think they can retire from business just yet. We are putting up our best race mounts and just enough money to make it interesting." Often the races would be combined with roping and baseball and, later, silent movies.

By 1905, Eben Low was campaigning for the better ropers of the ranch to enter the Cowboy Contest at Kapiʻolani Park on

O'ahu. A. W. had the cowboys' horses shipped on the SS *Kinau* along with the contestants: Sam and Ikua Purdy, Sam Mahuka, Martin Makini Campbell, Po Ma'ina'aupo, and Ka'aihue Pua'a. Low later wrote Carter expressing his gratitude for sending the ranch team, noting that "the matter was promoted by me and Mr. Robert Shingle and it turned out to be a great success, the actual count at Kapi'olani Park was over 4,500. No larger crowd was ever witnessed at Kapi'olani Park before; our boys did splendidly. Ikua Purdy was the grand champion."

Low was certainly a promoter, and he did much good for Hawai'i's *paniolo* image and history. It was his 1907 trip to observe the Cheyenne Frontier Days Rodeo that convinced him that Hawai'i's own could outdo their mainland counterparts—hence the 1908 campaign when he took Ikua Purdy, Archie Ka'au'a, and Jack Low.

It is important to note that many of the matched ropings on the Big Island during that era were won by either William Akau of Kawaihae or Keoni Li'ili'i, both Parker Ranch cowboys who often beat Ikua, Ka'au'a, and Low.

After the Frontier Days event, the ranch sponsored an epic matched steer-roping contest at Waimea in the arena that became known as Minukeole (less than a minute) after Keoni Li'ili'i won with a record time of under sixty seconds. Entries were limited to ranch employees and included support personnel such as fence and watermen. There were more than fifty entries, including a Kamakawiwo'ole, James Harbottle, and Manuel Vierra, along with a host of eight Lindseys. Of the several Purdys entered, Ikua was a known favorite. The Kauwe, Spencer, Chesebro, Akau, Kawai, and Lanakila families were also represented.

In the 1912 employees-only New Year's race meet, there were eight heats, all a quarter mile in length. Prizes for the first two races were for colts by Eastertide and Royal F, respectively. George Mahalopiha Lindsey won the first on Black Beauty, while Kamaki Lindsey on Mokihana won the second heat.

There were other races of interest. A. W.'s daughter Barbara, riding Moonlight, entered the sixth race, for the prize of Shetland ponies. She competed against Charley Boy Lindsey (Kale Kuli II), Keoki Opio (young George Mahalopiha Lindsey), and a youngster named Kepo'o, who won the race. The Shetland pony race was also an event at the following Fourth of July meet, won by a boy named Matsuichi (Matsu) Yamaguchi, who as an adult was nicknamed Mack and became assistant foreman of the cowboy gang. History was repeated about fifty years later when his twin grandsons, Mark and Mack Yamaguchi, match-raced a pair of Shetland ponies at the 1966 Fourth of July races at Parker Ranch. Mark grew up to become foreman of the Keamuku Division twenty years after his victory.

Judges and other officials at the races frequently included A. W.'s assistant ranch manager, O. L. Ollie Sorenson, along with James Wight and Sam Parker Jr., the latter doubling as a starter. Makahālau supervisor Alex Sandy Napier was a timekeeper, along with William A. White, Waiki'i supervisor, while the finish-line judge was Robert Victor Ka'ilianu, the teamster.

By 1909, A. W. Carter was planning for a new polo field between the ranch office (where the present day banks are located) and a *pahu'a* (flat) called Makahikilua (where North Hawai'i Community Hospital and Lucy Henriques Medical Center are now located.) Polo mallets, balls, and leggings were amply stocked, along with colored jerseys for three teams. No doubt A. W. was laying the groundwork in preparation for supplying the polo horse market of O'ahu, including U.S. Army personnel. On-ranch polo games were likely a version of cowboy polo.

Fields were built for other recreational projects. In 1913, on the advice of Sorenson, A. W. had a field landscaped for baseball. As Sorenson wrote on June 27, "The boys want to play baseball and I am having the grass cut for the diamond near the stables and will have the back stop put up by tomorrow. I trust you will approve."

Sorenson later wrote Carter on November 15 regarding billiards: "The pool table is getting very popular up here. Wouldn't it be a good idea to put in a bowling alley just makai of the social hall? I know the game and believe that if it was put in, the Akona Store would seldom have Ranch boys loafing there." Eventually billiards and bowling were put in place in the social hall, soon to be named Barbara Hall in honor of the eldest daughter of the Carters.

Music was appreciated at the social hall. An orthophonic Victrola record player was provided with an appropriate repertoire, including classical, contemporary, and even Japanese renderings. Other musical instruments were available for use, including string and steel guitars, Kumalae ukuleles, and bass violas, along with violins and bows.

The original social hall and its amenities were buildings constructed by the ranch on government property. The first building, about 40 by 50 feet, was located south of the present-day post office, with the adjacent bowling hall of 30 by 80 feet. Sorenson petitioned A. W. for an addition of some 20 by 50 feet to accommodate a cavalry armory that would include lockers for uniforms, weapons, and related equipment. Carter, somewhat of a military enthusiast, wholeheartedly supported the accommodation of a home guard.

By 1915, with small wars already flaring up all over the Asian mainland, the Cavalry Troop of the ranch was nearly a complete unit of its own, but the question of adequate mounts arose in a meeting of the men. Archie Ka'au'a inquired: "Will the Ranch supply all those joining with mounts? There are already 58 boys who have signed up and we can easily get some more from the Ranch to make the requisite number 65." With affirmative support of Carter, a fully manned and mounted troop was well trained and equipped by the end of 1919. Carter's deep patriotism was reflected by his unequivocal support for the home guard.

A. W. fervently embraced the scouting movement and established a ranch-sponsored troop (27) by 1917. The troop was started with only eight members under Scoutmaster Waiau Lima. While not all of the scouts were employee children, all were given the opportunity of part-time jobs on the ranch, developing early on the work ethic that Carter embraced. The scouting movement flourished, with Hisao Kimura eventually becoming the scoutmaster. The troop made significant contributions through its various projects, both in community efforts and in ranch programs such as seed harvesting, grass planting, and noxious weed eradication. In a 1923 letter from Dr.

H. B. Elliot, Carter was commended for his islandwide support of the scouting movement. At that time, A. W. served as district representative of South Kohala for the Kīlauea Council of the Boy Scouts of America.

By 1919, social hall activities included movies, which were silent until 1932. In a gesture of wholesome censorship, A. W. reviewed the film lists supplied by Consolidated Amusement Company for appropriateness. Emphasis was on good clean comedies mixed with travelogues and human-interest features. Eventually, Westerns became popular, with the likes of Hoot Gibson, Tom Mix, and local boy Buster Crabbe. A nominal admission fee was charged for nonranch people. Sugiyama Brothers, a Kohala movie company, eventually assumed the film service, but A. W. retained his role as film critic.

By 1935 films were shown as part of the Fourth of July racing festivities, which included baseball and a hospitable *lūʻau* for the whole town. By World War II, with the Marine divisions housed at Camp Tarawa, movies were also shown in the building that today houses Kamuela Liquor Store.

Golf was introduced in 1920 and enjoyed an early surge of interest. Often twenty to twenty-five cowboys would be out on Sundays enjoying the game on a nine-hole course. Interest waned when mowing the grass was relegated to the golfers on a voluntary basis.

Boxing was also introduced in the early 1920s, but not without some reservations on Carter's part. Yutaka Kimura had written Carter that "by listening to the boys, they are very anxious to get a set of boxing gloves." Carter replied, "I will have a set of boxing gloves sent up, but I don't want the boys to get into scraps with them. If possible, I will get someone over a little later on to give them instructions in boxing and it is better for them to have this before they use the gloves." Carter was able to send a U.S. Army boxing instructor named Stewart to provide this training, and the sport proceeded as a genuine outlet for many an anxious young ranch hand. Young Harry Kawai was a formidable boxer.

Of all the conventional sports, baseball has enjoyed the most enduring role in ranch recreation. In preparing for the 1936 Fourth of July festivities, Carter fielded two ranch teams, ordering two dozen baseball pants, jerseys, caps, and stockings.

The relay race teams fared just as well for uniforms. Carter ordered jerseys (sweatshirts) in eight colors, representative of the eight islands, to field eight relay teams from all sections of the ranch, including the Breaking Pen and dairy crews. Jersey sizes ranged from 36 to 42.

For the *lūʻau* that year, a temporary pavilion was built on the grounds of Puʻuōpelu, seating about 700 people, with separate accommodations for 300 children. More than 5,000 *laulau* (leaf-wrapped food packages) were made for the occasion.

In typical Carter style, ample expressions of gratitude were extended to the people who prepared the *lūʻau*, notably Mrs. Lioe Kaʻanaʻana and Piko Bell. In a 1936 memo to Theodore Vredenburg, then head superintendent of ranch operations, Carter had this to say:

> Everything went fine on the ranch. I do not believe there is another place in the world where they could construct a building, prepare food, and give such an

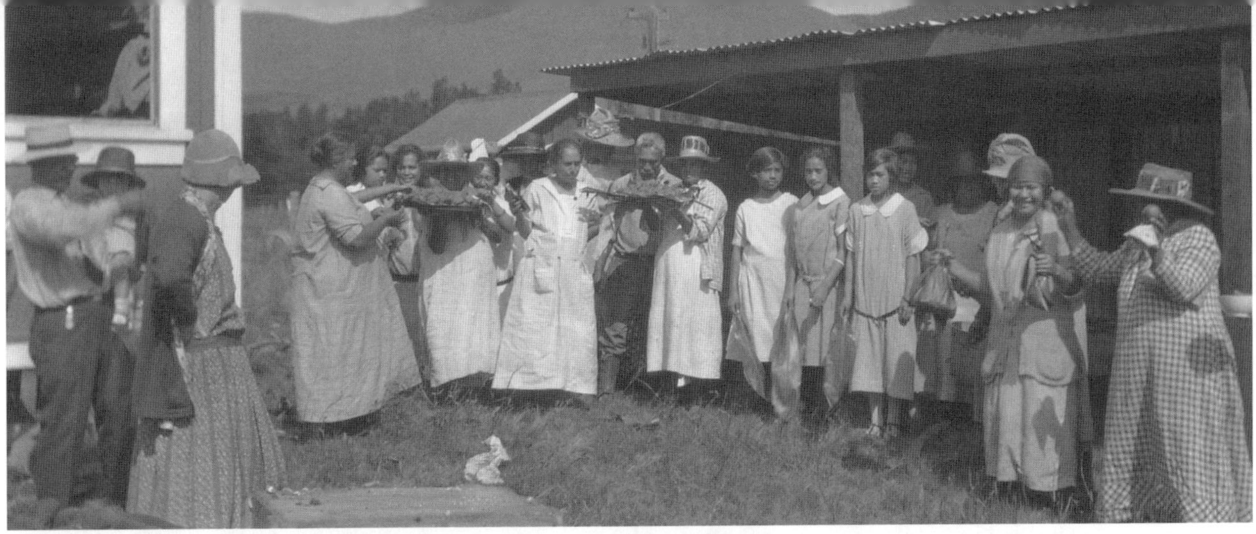

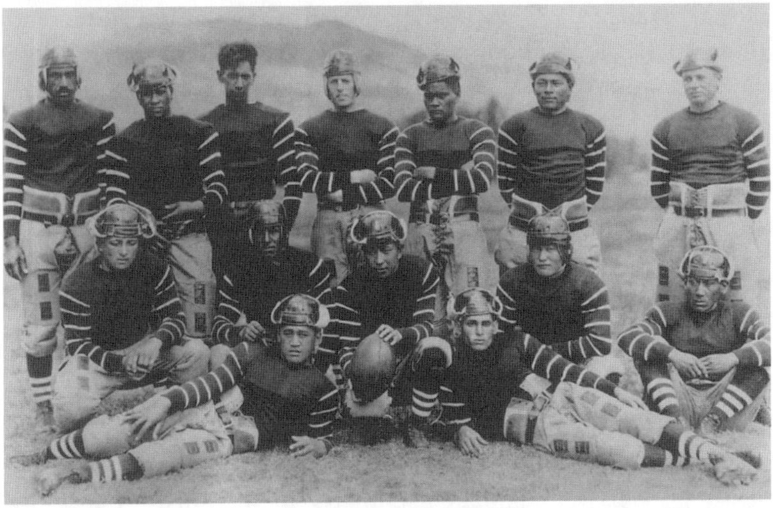

Above: Lū'au *Gang at Barbara Hall. A. W. Carter Collection.*

Left: *Parker Ranch football team, 1937. Back row: Hogan Kauwē, Joe Pacheco, Theodore Saffrey, unidentified, Robert Puakea, Agapito Tolentino, and Charles Kalaniali'i Lindsey. Front row: Billy Boy Ma'ina'aupō, Alex Arthur, coach George Ing, Saburo Kuroyama, and John Liana Jr. Lying on the ground: Eluwini Ka'aekuahiwi and Bill Breitha. Hartwell Carter Collection.*

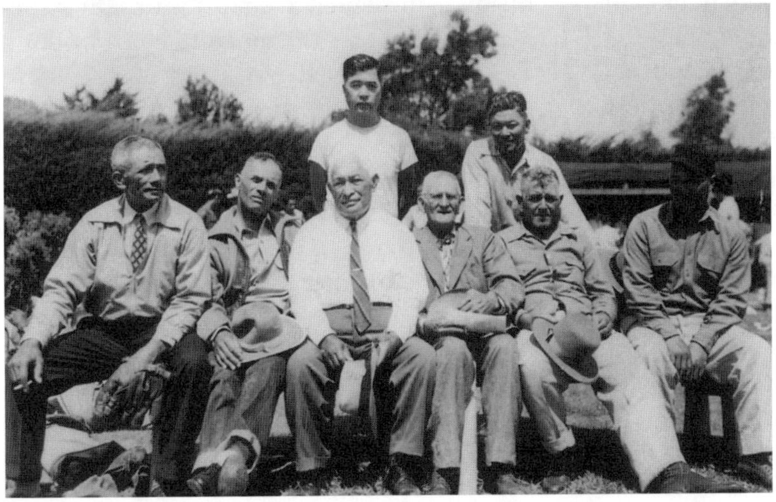

Parker Ranch Makule League baseball team, 1932. Standing in back, left to right: Masao Yoshimatsu, Yutaka Kimura. Sitting: John Purdy, John Kuakini Lindsey, Dick Kawai, Joseph Snow Maertens Sr., Alex Bell Sr., and Henry AhFong. Hartwell Carter Collection.

> entertainment as was done in Waimea in the short period of time that was at our disposal.
>
> I am sending to each of the women who worked on the luau some material for a dress. It [is] simply a small present in acknowledgement of the great work they did.
>
> I enclose herewith a list of the men who worked, and I should like to have the ranch contribute some small present to each of them. Please tell Henry Pang to give to each of these men, as they wish, either a hat, sweater, or something of about the same value.

Frequently on these occasions, A. W. would coordinate the layover of Captain Willie Punahoa and the crew of the *Humu'ula* so they could take in the three-day festivities celebrating Independence Day.

Support for the Religious

A. W.'s reverence for faith extended far beyond his denomination and community boundaries of North and South Kohala. His generosity went beyond financial help to include the donation of lumber, building materials, beef, and pork—even the gift of a replacement mule for the Catholic priest, Father Charles Windels of Hōlualoa, whose old donkey gave out while returning home from several days of ministering to the people of Waimea.

The allotting of property to different denominations was handled equivocally. In 1910, applications for church sites for both the Buddhist and Mormon faiths were granted.

In 1923, A. W. received a request for a large hog for the dedication of a new church on Manono Street in Hilo. The church was of the Kalawina faith (Congregational), and its founding pastor was the Reverend William Ellsworth Edmonds (Ekemana), recently of Maui. Still present at its founding location, the name of the predominantly Native Hawaiian denomination was Ho'omāna Na'auao o Hawai'i. Edmonds and his associate pastor, James Kekela, had planned the large ecclesiastical celebration for April 15, and both clerics wrote Carter a sincere memorandum of gratitude. At this juncture, A. W. may or may not have known that Reverend Edmonds was the full brother of long-lost James Kaleima'ema'e Edmonds (Palaika), who disappeared from his childhood home in Waikapū, Maui, nearly forty years earlier. James kept his real name a well-guarded secret when he stole away to Moloka'i in 1885 and launched a *paniolo* career, eventually coming to work for Parker Ranch. For all anyone knows, however, Carter may have been aware of the relationship and spared an extra fat hog for the true brother of his loyal *paniolo,* Palaika.

As further testimony to Carter's even-handed generosity to all faiths, including his own, he denied a request from Reverend Henry P. Judd for the ranch to contribute to a new Waimea pastor's salary. In his memorandum to Judd, Carter stated, "I have chosen not to favor any church over the other. We will help them all as long as they do good work. I would not feel that I had any right, as Trustee of the Parker Ranch to emphasize my religious views in any way."

A very special spiritual initiative of A. W.'s was his involvement in the reinterment of the remains of Protestant missionary Reverend Lorenzo Lyons to 'Imiola Church. Upon the cleric's death in 1886, he was buried in the family plot adjacent to their home. Over the course of several decades, the property fell into serious disrepair. The house was eventually demolished, leaving Lyons' grave and marble spire monument deserted and neglected.

A. W. was sensitive to the reverence Native Hawaiian people have for human remains and was touched by their expressed desire to move Lyons' grave to the grounds of the beloved 'Imiola Church. Carter picked up on the initiative in 1939 through the current pastor, the Reverend Moses Ina-'ina, to make necessary arrangements. Reverend Lyons' granddaughter, Emma Lyons Doyle, acknowledged that the gesture was made possible through the "generosity, interest and direction of Mr. Carter." The exhuming and reinterment at the church graveyard was executed, followed by resetting the headstone and marble spire monument and culminating with an inspiring spiritual service. Those of us today who revere the spiritual prominence of the marble spire seen in the distance may quietly thank not only the reverend for his missionary contributions but Carter for his reverent preservation of the remains of the icon Lyons. Both men possessed a reaffirming faith in God.

Closing this section celebrating Carter's humanity is a poignant vignette recognizing his love for the Hawaiian people. It is a demonstration of his deep-seated respect for the nationality and heritage of the people of Hawai'i nei.

This memorable incident began in 1940 when Clifford Bowman, a prominent territorial Board of Health figure from Hilo, who had established the first Hawaiian civic club on the Big Island, contacted Carter. Bowman had also launched Hawaiian civic clubs in Kona and Laupāhoehoe, both of which received generous support from Parker Ranch through Carter. Bowman wanted Waimea to have its own club, and Carter's reaction was one of warm enthusiasm. Their exchanges of memoranda speak to the occasion. Bowman wrote: "As the scholarship fund we are creating is for the benefit of Hawaiians from the Island of Hawaii, it is our intention to organize Hawaiian Civic Clubs in the various communities of the Island. The object of the organization is to promote the welfare of the Hawaiian people."

In his response, Carter wrote: "I think it would be a very nice thing to have a club in Waimea and the next time I go up, I will get some people together and arrange such a club. My son and I cannot become members as I notice it is only for persons of native Hawaiian ancestry, but that in no way will affect our interest in your organization, and the purposes for which it is created." Hence followed the birth of the Waimea Hawaiian Civic Club, which endures to the present day.

Horse Program Leadership: 1900–1949

Between the twilight years of the nineteenth century and the advent of the A. W. Carter era, the breeding, weaning, halter breaking, and training of ranch horses was led by William Akau, grandson of the family patriarch, William Paul Mahinauli Akau. Care of the breeding stallions was a major responsibility.

A. W. had bigger plans for William Akau, however, given his businesslike nature coupled with his highly skilled seamanship. William was commissioned by Carter to move back home to Kawaihae and assume the combined role of harbormaster and postmaster. This move gave Carter a deep sense of security regarding freight and the receipt and shipment of mail. James Palaika, retaining the staff of Pōai Nahale'a, Samuel Awa'a Sr., George Keola, and Charlie Duncan, assumed Akau's horse responsibilities.

Under the long stewardship of James Palaika, the Breaking Pen responsibilities grew, as did the horse numbers, and eventually the breeding program was assigned to others, such as Albert Uiha Lindsey and James Merseberg. These men operated these components through the period leading up to Hartwell Carter's assumption of ranch management. In his twilight years, Palaika did cowboy and *nānā'āina* work. Albert Uiha Lindsey followed Palaika as Breaking Pen foreman. Under both Palaika and Lindsey were roughriders (horse breakers) including the following:

Willie Palaika
Harry Kawai
Jordan "Red" Quintal
William Campbell
Thomas Milimili "Bear" Lindsey
Joe Maertens
Francis Spencer
Harry "Kahuku" Purdy
George Purdy
John Stevens
Frank Vierra

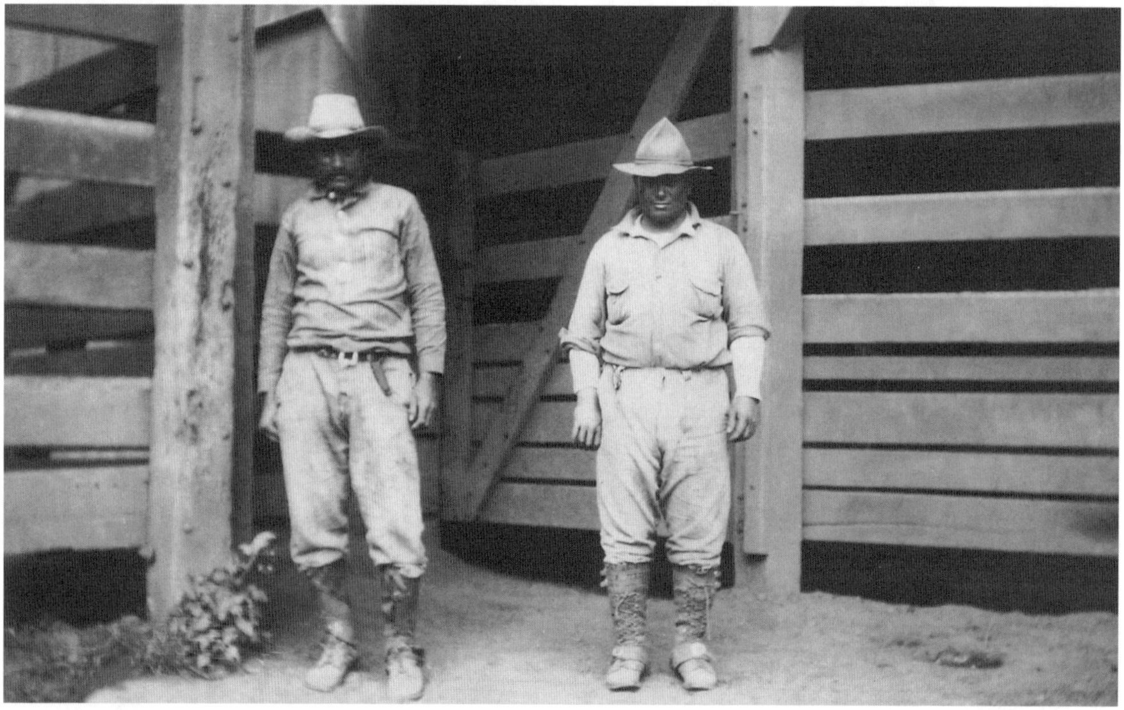

James Palaika Brighter, foreman of the Horse Program, including the Breaking Pen, with assistant Pō'ai Nahale'ā. A. W. Carter Collection.

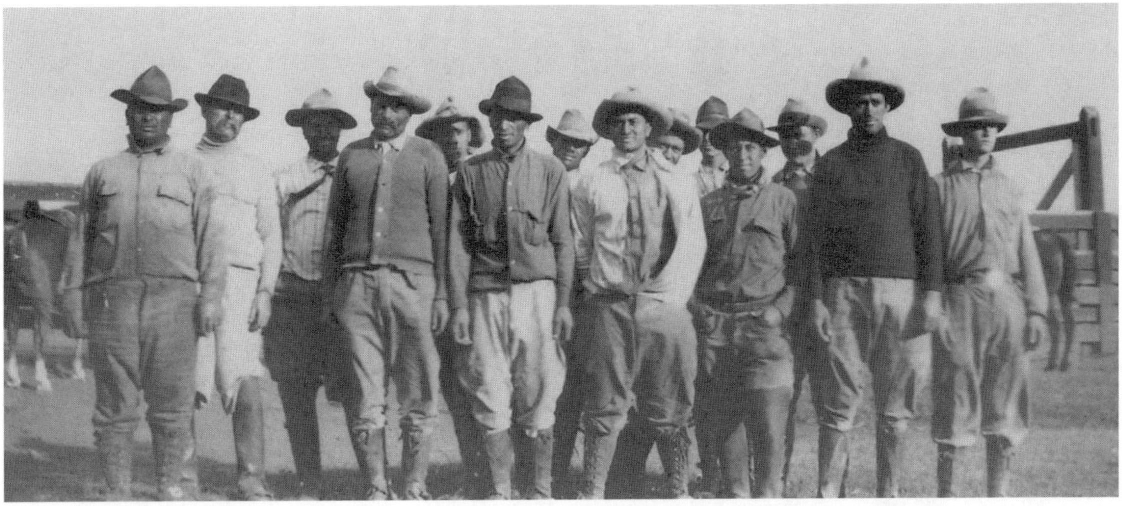

Palaika with the early Breaking Pen crew. Left to right: Pōʻai Nahaleʻā, Francis Pakana Spencer, unidentified, Palaika, unidentified, unidentified, Kuakini Lindsey, Theodore Vredenbury, unidentified, unidentified, Alex Arthur, unidentified, Bill Breitha, and unidentified. A. W. Carter Collection.

By 1940, Harry Kawai rose to become Breaking Pen foreman for five years, until he was chosen to lead the Cowboy Gang as *paniolo nui*.

In 1945, George Purdy assumed the reins of the Breaking Pen and, like Palaika and Kawai, developed a distinguished career as leader of the Roughrider Gang. All three men were role models for the young horse trainers, fostering a high level of horsemanship, confidence, and work ethic. They were calm, firm, and effective leaders, exemplary in grooming leadership prospects among the roughriders.

During this period, Merseberg continued his role of overseer of the horse-breeding program, including registered stock of several breeds. Nicknamed "Dictionary," Merseberg had a phenomenal memory for the pedigree of any horse, and the Carters enjoyed listening to his soft chanting of the genealogies that reached back as far as four generations of steeds. For a brief interim period, Joe Pacheco served as foreman of the Breaking Pen.

While the breeding program was designed and coordinated by A. W. and later Hartwell Carter, it was the field personnel who delivered the breeding stallions to the appropriate bands of mares that were kept uniform as to breed, registry, and function. The registered Thoroughbred dams had a higher priority compared to the unregistered Thoroughbred and grade mares, whose offspring were intended as workhorses. These mare bands were closely monitored by *nānāʻāina* men to oversee breeding, gestation, and foaling. The stallions would be returned to Pukalani Stables after a four-month breeding season. Palaika and Merseberg followed up with foal descriptions relative to these dams to effect proper registration with such associations as the American Jockey Club, the American Palomino Horse Club,

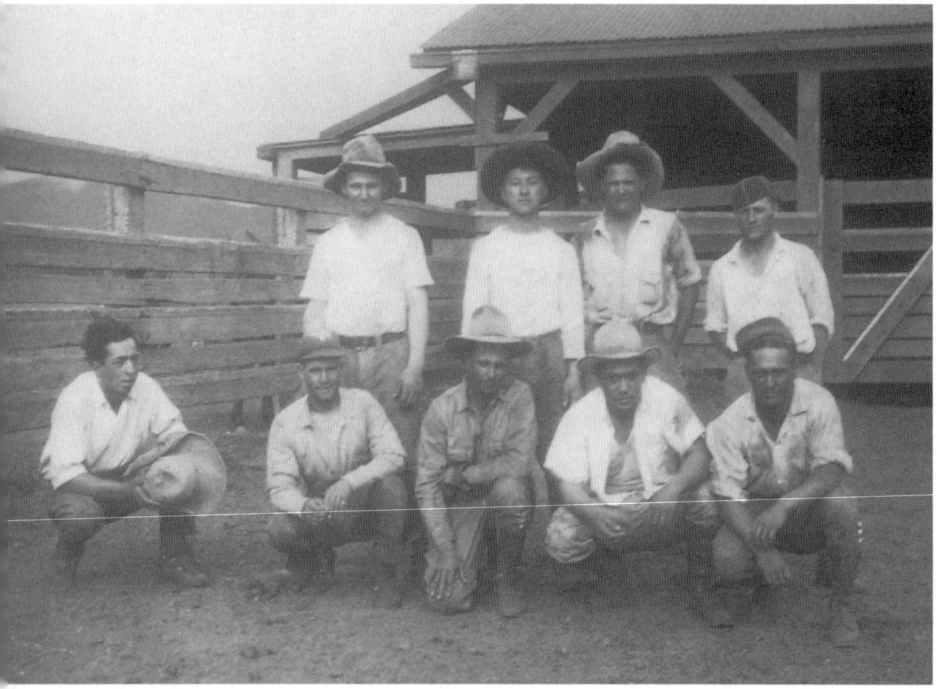

Left: *Albert Uiha Lindsey with Breaking Pen crew. Lower row left to right: William Campbell, Adam Quintal, Uiha Lindsey, Sam Liana, and Willie Palaika. Standing: Joe Maertens, Masaru Mussolini Okura, David Dango Lindsey, and Jordan "Red" Quintal. A. W. Carter Collection.*

Below: *Uiha Lindsey and Breaking Pen crew. Thomas "Bear" Lindsey, Harry Purdy, Uiha, William Kealoha, Joe Maertens, and Francis Pakana Spencer Jr. A. W. Carter Collection.*

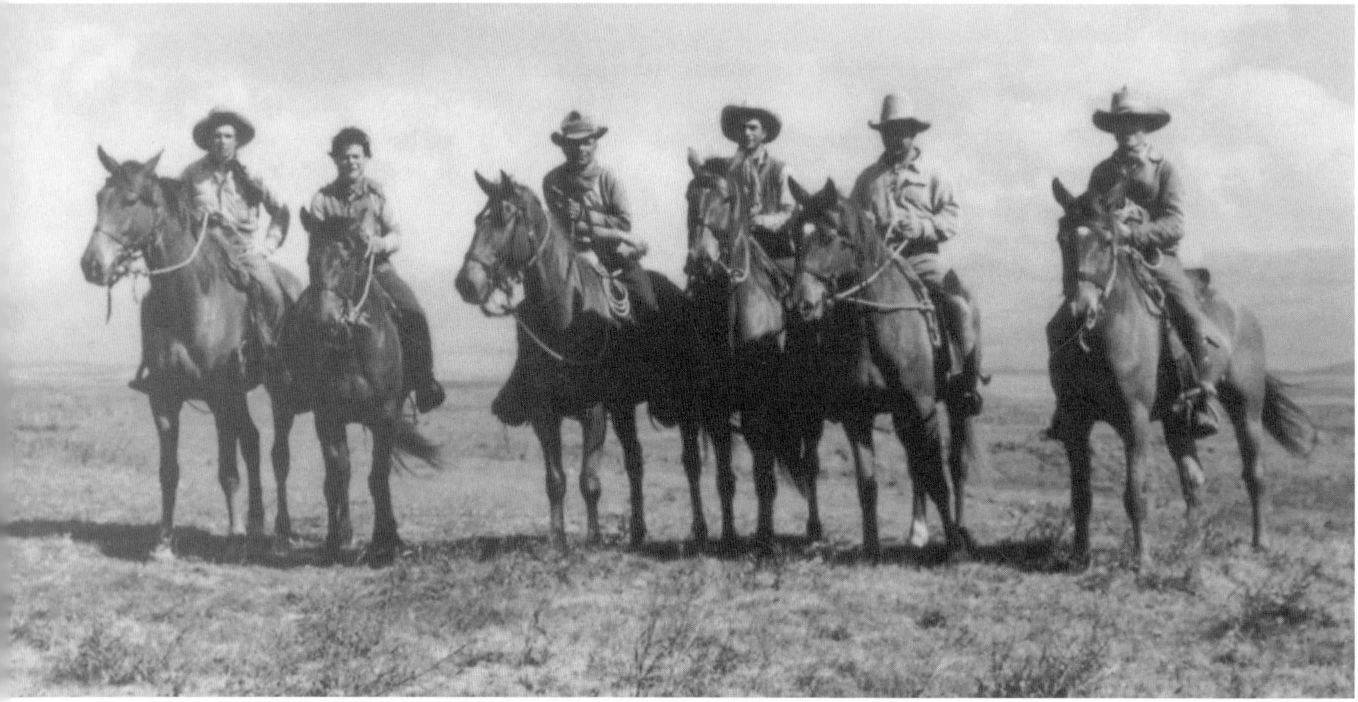

and the Morgan Horse Club. Much later (1948), the American Quarter Horse Association became an additional registry. Teddy Bell later assumed Merseberg's role during the transition between the Hartwell Carter and Penhallow periods. Bell had left the ranch and Rally Greenwell appointed Walter Stevens to manage the horse breeding in 1969. Alex Penovaroff, Tommy Goto, and finally Donnie DeSilva were subsequently placed in charge of the broodmares, in that order.

At weaning time, the foal crops were separated from their dams at secure corrals such as Wahine Kea and Pu'u Mahoelua. With the dams remanded to their respective permanent pastures, the foal crops would be driven to the Breaking Pen for halter breaking. Once gentled, the weanlings would be sorted as to breed and gender, then turned out in premium, high-elevation grazing paddocks. The stud colts would go to Kemole I and II and the fillies to Pu'u'anu'anu for growing out. As yearlings they would return to the Breaking Pen for branding, castration, and further identification as to potential function.

At this point, the young horses were returned to their respective growing areas for as long as two to four years. Their final trip to the Breaking Pen meant being sacked out (prepared for saddling) and assigned to be ridden by any one of the eight to twelve roughriders. In the period between the 1920s and 1940s, it was not uncommon that these men rode five- to six-year-old stout and snorty horses that had reached total maturity. Each roughrider rode eight to ten colts a day. After a six- to eight-week green-breaking, which included saddling and riding out into paddocks independently, the horses were picked by the cowboys and station men for new additions to their work strings. By this time, the green-broken horses had progressed in reining from the *palaka* line to the single rein of the snaffle bit. All four hooves and limbs would be gentled for handling, although shoeing of horses as a broad ranch practice did not arrive until the 1970s. By this time George Purdy had retired, and Louis Akuna, a veteran cowboy and horseman, became Breaking Pen foreman, with Donnie DeSilva as his assistant.

The Aura of the Cowboy Gang

The charm of Parker Ranch centered on its icon—the cowboy. Whether it was in the form of art, photography, or music, the *paniolo* was the celebrated centerpiece in ranch history, and A. W., more than anyone known, appreciated the colorful history of the Parker Ranch cowboy.

The concept of a roving cowboy grew simply out of necessity, given the expanse of the lands and the magnitude of the cowherds. The leader was always a charismatic and capable native cowboy with exemplary talent in all aspects of horsemanship and cattle handling. With only one exception (Yutaka Kimura), the head cowboy foreman was always of Hawaiian or predominantly Hawaiian descent. Physical stature was irrelevant, but leadership skills were paramount. Keoni Li'ili'i and Willie Kaniho were both of small stature, but their reputations were immense.

Ulumaheihei Waipa—son of Mary Ann Kaulalani Parker (daughter of John Palmer Parker I) and Kame'e'iamoku Waipa—was the first recorded cowboy foreman, a position referred to sometimes as *luna nui* or *paniolo nui.* Called Lumaheihei, he came with a huge frame and handsome demeanor. He was groomed by his uncle John Palmer Parker II to assume the ranch management post, in recognition of his respected role as a fine horseman and broadly experienced cattleman. His intricate knowledge of the grasslands, plains, and forests was handed down through the royal lineage from which he descended. Lumaheihei was articulate in both English and his native tongue.

To his credit, Lumaheihei built several wild horse traps in strategic locations on the southeast slope of Mauna Kea. Vestiges of one of these corral traps can be found in the flats above Pu'u Mali in the Hānaipoe Mauka area. At lower elevations, he built corral traps in both Kapa'akea and Wahine Kea hollows. Today, purebred cattle graze on paddocks of Wahine Kea I, II, III, and IV, where the former wild horse trap is located. Lumaheihei's efforts effectively reduced the bands of feral horses that played havoc in the Hānaipoe area.

The original headquarters of the Cowboy Gang were at Mānā, where the ranch began. As the Parkers moved to Pu'u'ōpelu and many of the working families followed, Waimea became the center of operations. By 1895, a very large working corral made of high and wide stone walls was built. It was named Pu'uhihale Corral and was developed later into smaller pens. Many of the early cowboys kept their tack at home and rode spare horses to overnight house paddocks at their homes. Many of these pens can still be seen today.

But the true saddle house/cowboy stables became Pu'uhihale Corral, where eventually saddle and tack lockers and a long shed barn were built. One must view Pu'uhihale Corral to appreciate why it has been called the "cowboy's *heiau*" (place of worship), with its walls still in place, 6 feet wide and 6 feet high. It is hoped that the ranch will retain this awesome monument to the *paniolo.* Since the disbanding of the divisional organization that came with the January 14, 2002, Voluntary Separation Program, Pu'uhihale has fallen into disuse.

With the ranch under trusteeship in the late 1890s, Paul Jarrett was appointed ranch manager instead. Lumaheihei took the decision very hard. Although he remained on the ranch during the transition, he lost heart for the outfit and left for Honolulu shortly thereafter. He became a sheriff and was politically active in the Hawaiian government. He tragically met his death in a severe horse fall on the corner of King and Kalākaua Streets on O'ahu.

Kaho'ohanohano succeeded Lumaheihei and was a quiet, spiritual leader. By this time the crew had grown to nearly a dozen men. Some of the more prominent cowboys included James Stevens, Thomas Lindsey, Keoni Li'ili'i, and "Po" Maina'aupō. Eventually, Thomas Lindsey moved up to the leadership post when Kaho'ohanohano was appointed *nānā'āina* man, beginning an important tradition of routinely riding assigned paddocks. Kaho'ohanohano spent the twilight of his *paniolo* years harvesting the hides of deal range cattle, yielding 25 cents per hide delivered. At this time, men in the Cowboy Gang included:

David "Hogan" Kauwe

Ka'inapau Kailikini

William "Bill" Lindsey

"Po" Maina'aupō Sr.

George Lindsey

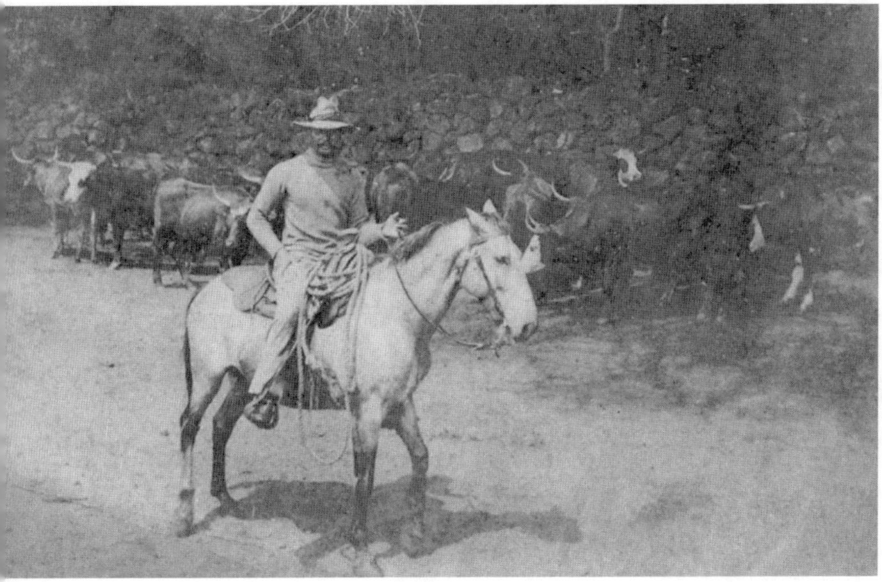

Kaho'ohanohano, an early ranch paniolo nui *or* luna nui *preparing to* hō'au pipi *(swim cattle) at Kawaihae. Longhorn market cattle are in the background. A. W. Carter Collection.*

Sam Purdy

Archie Ka'au'a

Ka'aihue Pua'a

John Chesebro

Iakopa Kahai'kupuna

Keoni Li'ili'i replaced Tom Lindsey when A. W. Carter came to the ranch. Lindsey became *nānā'āina* man, as did his predecessor, Kaho'ohanohano. The *nānā'āina* man had an important role but a lonesome existence. The degree of responsibility was reflected by the person they reported to—the livestock manager. The timing of major events such as branding and weaning were scheduled largely on the observations of the *nānā'āina* man. This tradition lasted more than seventy years, fading out as the Rubel/Lent team compartmentalized the ranch into four geographic divisions.

Another tradition that paralleled the Cowboy Gang was the employment of a large string of workhorses—usually twenty to twenty-four. On a given workday, a cowboy might ride one mount and lead a second horse to the designated paddock for a cattle drive. The fresh mount would be saddled for the drive, while the traveling horse was tied in the shade of a tree or turned loose in a holding pen adjacent to the corral. At the end of the drive and corral work, the traveling horse was remounted for the ride home. Quite commonly, however, a wrangler would drive the selected cowboy horse band to the designated corral a day before the cattle drive, requiring only that a traveling horse reach the work site and return home *pau hana* (at the end of the workday). As an example, for a gang riding to Big He'ewai for a day's work and back, the distance covered from the Pu'uhihale Corral headquarters to the area of Keamuku was more than 25 miles.

After Keoni Li'ili'i, cowboy foremanship went to Willie Kaniho, who was brought

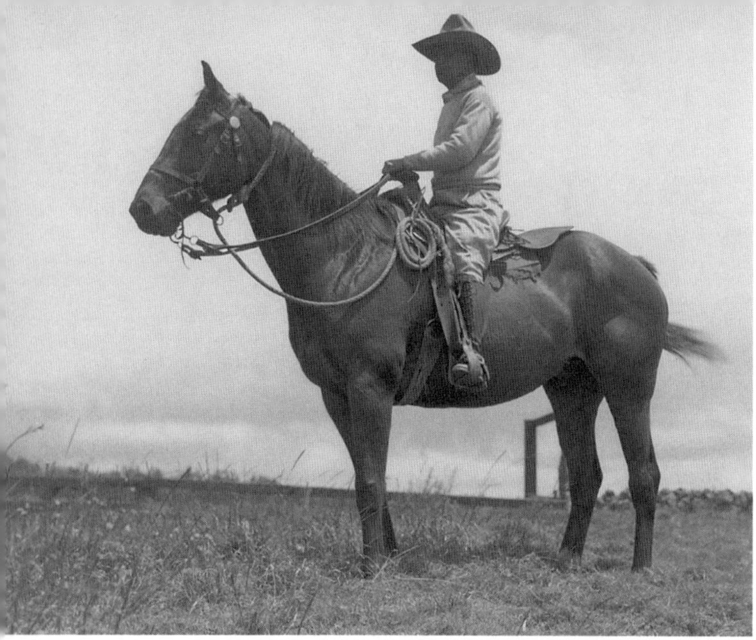

Above: *John K. Keoni Li'ili'i Lindsey mounted near Pukalani Stables. Hartwell Carter Collection.* Below: *Willie Kaniho, legendary ranch cowboy foreman. Hartwell Carter Collection.*

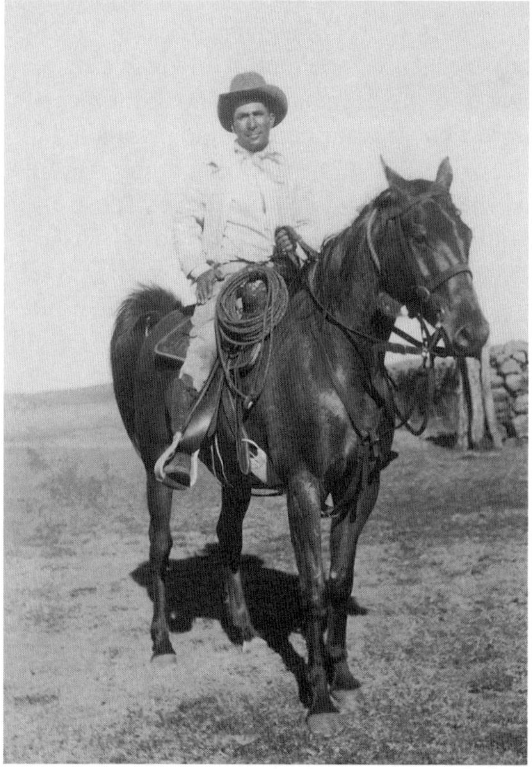

down from Humu'ula. Willie's role became legendary and he was, of course, highly regarded by A. W. Carter. By 1935, the Cowboy Gang reached its peak in numbers and talent. The men, numbering fifteen, included:

William Kaniho, foreman
Alex Akau*
Matsu Yamaguchi, assistant foreman
Awili Lanakila
David Kauwe (Hogan)
Harry Koa
Harry Kawai
George Purdy
Joseph Pacheco
George Keola
Frank Vierra
John Lindsey Jr. (Kuakini)
Henry Ah Fong Sr.
Kaliko Ma'ina'aupō
William Kawai Jr.

There were also miscellaneous cowboys available to help in the big drives, shipping, branding, and weaning. These men included:

Ah Sing Akim Seu, *nānā'āina*
Manuel Ferreira
Sam Kaehalio, *nānā'āina*
Moses Keola
Charles Lindsey
John Lindsey Sr.
Small Johnny Lindsey
Sam Lindsey Jr.
James Merseberg, Horse Department

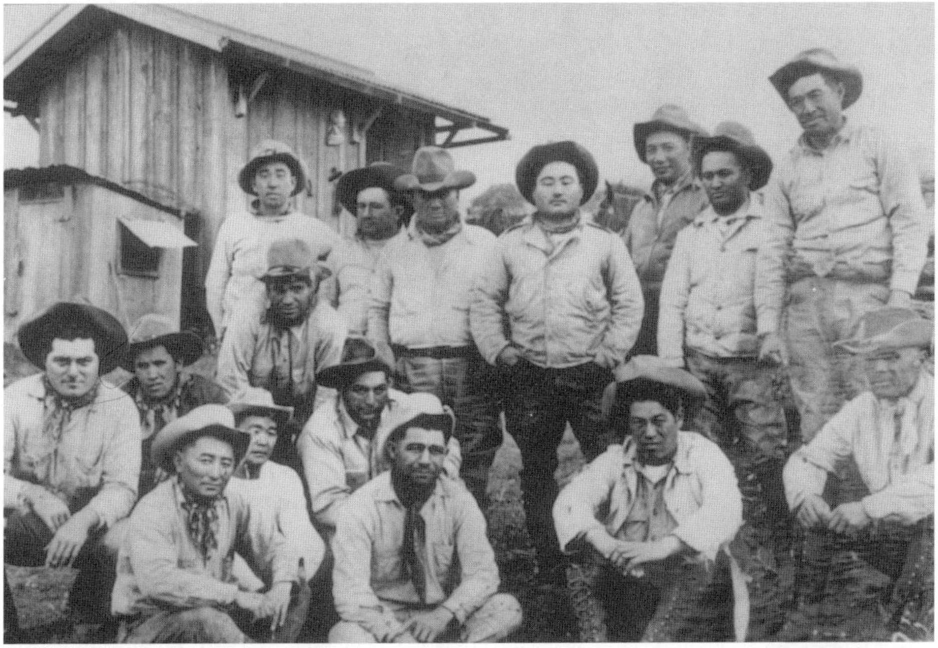

Cowboy Gang at Hanaipoe House. Standing left to right: Shiro Fujitani, Charlie Lindsey (Kale Keiki), Kaliko Ma'ina'aupō, Gilbert Shinko Hokama, Henry AhFong, Willie Palaika, and John Samoa Lekelesa. Kneeling front row: Francis Pakana Spencer, Herbert Dead Lindsey, Bob Kui Sakado, Saichi Morifuji, Foreman Harry Kawai, Willie Kawai, Jimmy Merseberg, and John Kuakini Lindsey. Hartwell Carter Collection.

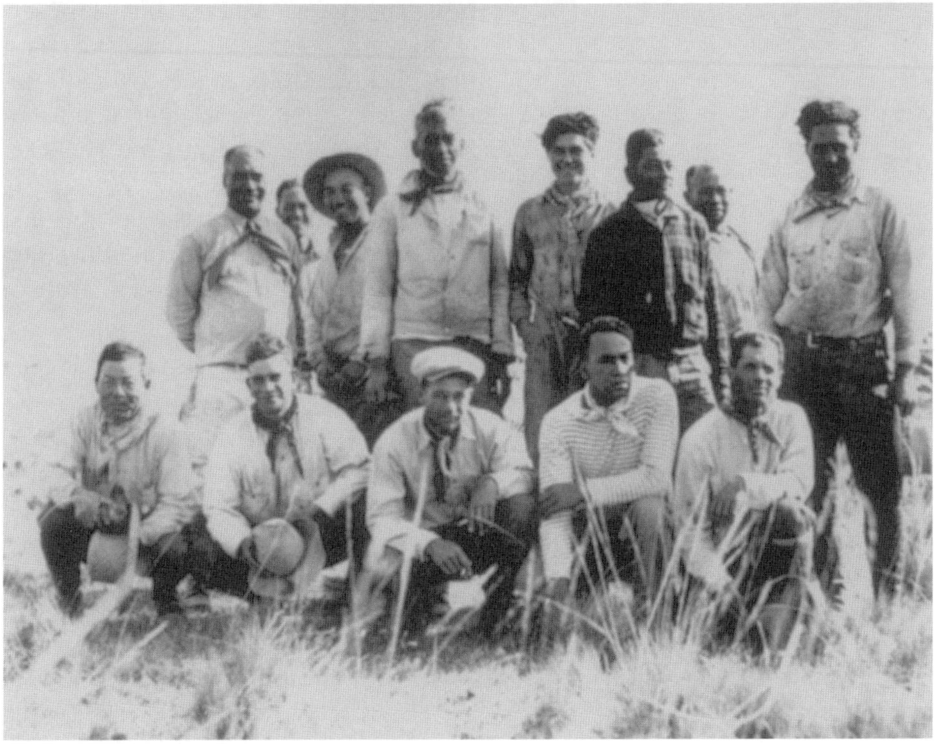

Cowboys at 1935 Humu'ula flow. Front row left to right: Kiyoshi Tagawa, Frank Vierra, Willie Kawai, unidentified, and Kuakini Lindsey. Back row: Awili Lanakila, Lawrence Kea, William Campbell, Hogan Kuawē, Francis Pakana Spencer, Agapito Tolentino, John Kauwē, and Robert "Red" Spencer. Hartwell Carter Collection.

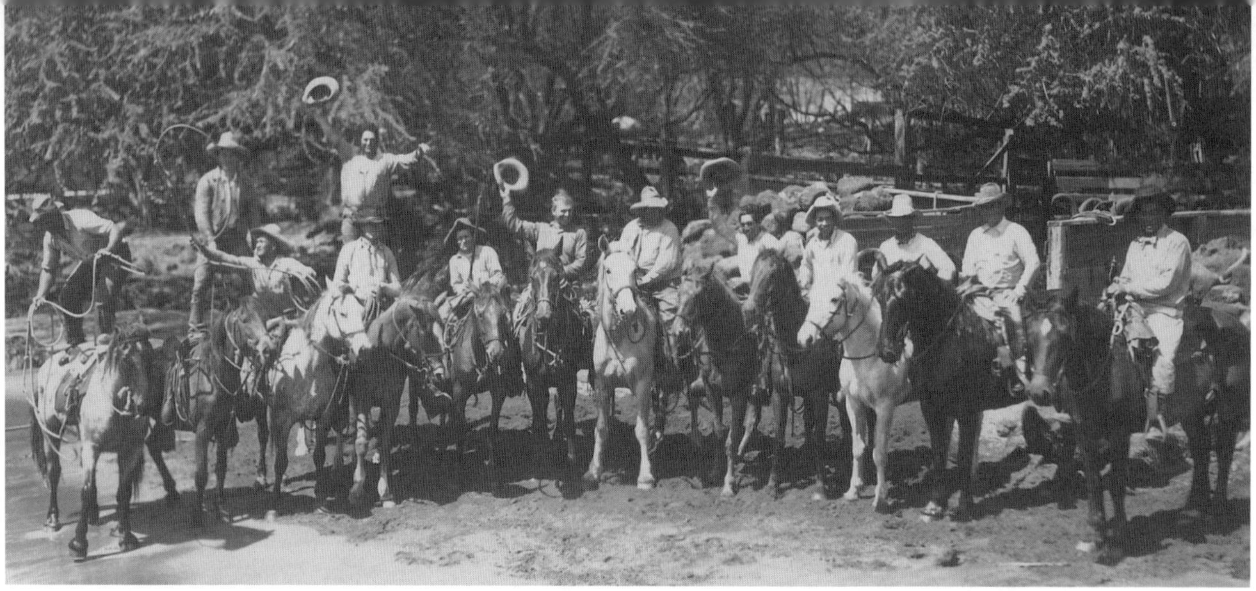

Cowboys at Kawaihae hō'au pipi. *Left to right: Alex Akau, James Mersebery standing on two horses, and Willie Palaika. Standing on horse: Harry Kawai, Prince Ross, Red Quintal, Joe Maertens, Joe Pacheco, Sonny Kaniho, William Campbell, Kuakini Lindsey, Frank Vierra, and Harry Koa. Hartwell Carter Collection.*

Frank Paiva

Jacinthe Paiva

John Purdy

Francis Spencer

Frank Spencer

Sonny Kaniho**

Frank Kawai**

*Based out of Humu'ula

**Schoolboys who worked weekends and summers

Over the ensuing years, several prominent men rotated through this leadership position, including Albert Uiha Lindsey, Harry Kawai, Yutaka Kimura, and John "Samoa" Lekelesa. Assistant cowboy foremen included Matsu Yamaguchi, Bob Sakado, Henry Ah Fong Ah Sam, Joe Pacheco, and John Kuakini Lindsey.

Charles "Kale" Stevens was the last Cowboy Gang foreman before this elite group was disbanded by the Rubel/Lent team in 1971.

NOTE

1. *"Pipi Kaula"* is the common term for jerked beef. Usually sun dried, it can also be smoked. *"Kaula"* is the mongrelized form of the word *"kaula'i,"* meaning "sun dried" or "jerked." The latter term is the anglicized form of the Spanish word *"charqui"*—hence the name of the meat product, "jerky."

9

IMPROVEMENTS IN ALL CLASSES OF LIVESTOCK

1900–1949

Through the early years of the ranch and the eras of John Palmer Parker and John Palmer Parker II, improvements came by selection within existing longhorn herds. There was no outside genetic influence, probably due to the cost. In his early observation, A. W. noted that of the 25,000 cattle on the ranch, some 5,000 were wild and unbranded.

The clear progress from cowherds of immediate feral origin and horse bands of slightly improved stock to the exemplary status of the ranch livestock demonstrates that Carter established some practical goals and set out methodically to achieve them.

Although the Royal Hawaiian Agricultural Society had been bringing in Aberdeen Angus, Durham shorthorn, Devon, and Hereford bulls since 1850, it was late 1899 before any of those British breeds stepped onto Parker Ranch soil. During that year, interim manager Paul Jarrett secured a Red Devon bull. Not long before the appointment of A. W. as trustee, Jarrett also brought in ten purebred shorthorn bulls of reasonable quality that were distributed only to select and controlled cowherds. Shortly thereafter, he brought eighteen half-bred shorthorn (half longhorn) bulls. Before they could be spread among any cowherds, A. W. rejected them as inferior, lacking the phenotypic and genetic substance needed to make a significant contribution to improvement. These bulls were castrated and sent to fattening paddocks for market.

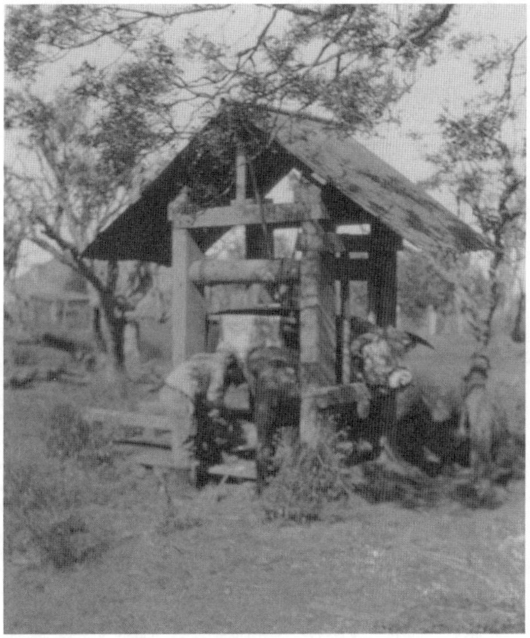

Parker Ranch's first restraint chute for doctoring cattle, especially bulls as shown, typically referred to as a stock. Such restraint chutes provide safety and security for both man and animal. Makahalau, 1924. A. W. Carter Collection.

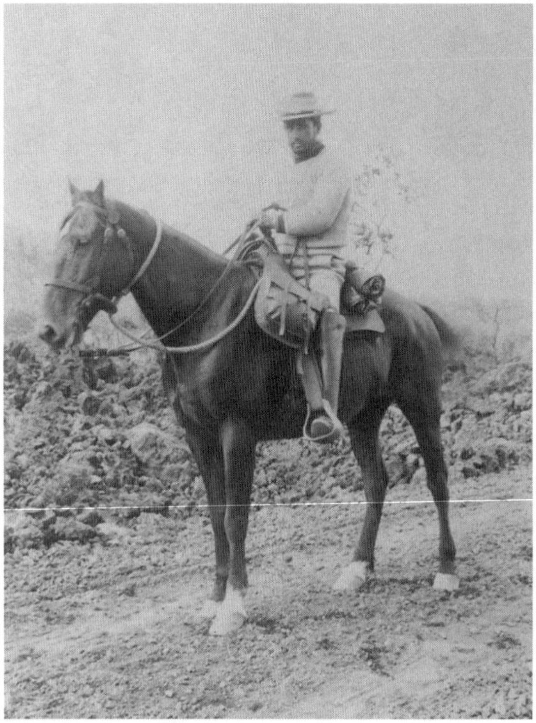

Keoni Li'ili'i outfitted for bringing home the bull Oddfellow from Hilo Wharf, 1901. A. W. Carter Collection.

As for the cattle improvement goals, a standard was set above and beyond the level of the Royal Hawaiian Agricultural Society. A. W. elevated the early ranch standard from producing five- or six-year-old butcher cattle yielding less than a 500-pound carcass to a two- to three-year-old steer dressing at 650 pounds. Beyond high carcass quality in a relatively youthful steer as an end product, A. W. demanded that the purchased breeding stock be free of disease. With the ranch aggressively expanding into dairy operations, it was vital to prevent diseases such as brucellosis and tuberculosis by restricting the purchase of seed stock to those tested negative via veterinary certification. A. W. opened his improvement campaign based on preventive veterinary medicine.

During his first year on the ranch, A. W.'s brother, Fred Carter, purchased more than forty head of Hereford, Durham (shorthorn), and Holstein bulls from the U.S. mainland and local sources. Parker Ranch records indicate that the first registered Hereford bull imported was named Oddfellow.

When Parker Ranch took delivery of Oddfellow at Hilo Harbor, there began an epic tale of caregiving centered on Johnny Lindsey (Keoni Li'ili'i), who was asked by A. W. Carter to lead the bull home on an eight-day, 60-mile trek. Since he came from the livestock show in Pendleton, Oregon, Oddfellow was accustomed to a pampered diet and routine, including bedded stalls. Lindsey received a bull whose diet was suddenly grass of minimal nutritional value—plus occasional grain feeding. The terrain to be covered varied from mud, grass, and gravel to some hard roads as they trekked through the villages of North Hilo and entered the tall cliffs and valleys of Maulua, Laupāhoehoe, and Ka'awali'i. Lindsey packed a minimal grain supplement that he divided between the bull and his horse. As these resources were depleted, he bartered from village and plantation sources for animal feed and meals for himself.

En route, Oddfellow's hooves suffered considerable wear and were in need of protection. With due compassion, Lindsey trimmed his saddle of its skirting leather and fashioned shoes for the comfort of his charge. They climbed from the coastline of Hāmākua to the village of 'Umikoa, headquarters of Kūka'iau Ranch, where this neighboring outfit provided feed and recu-

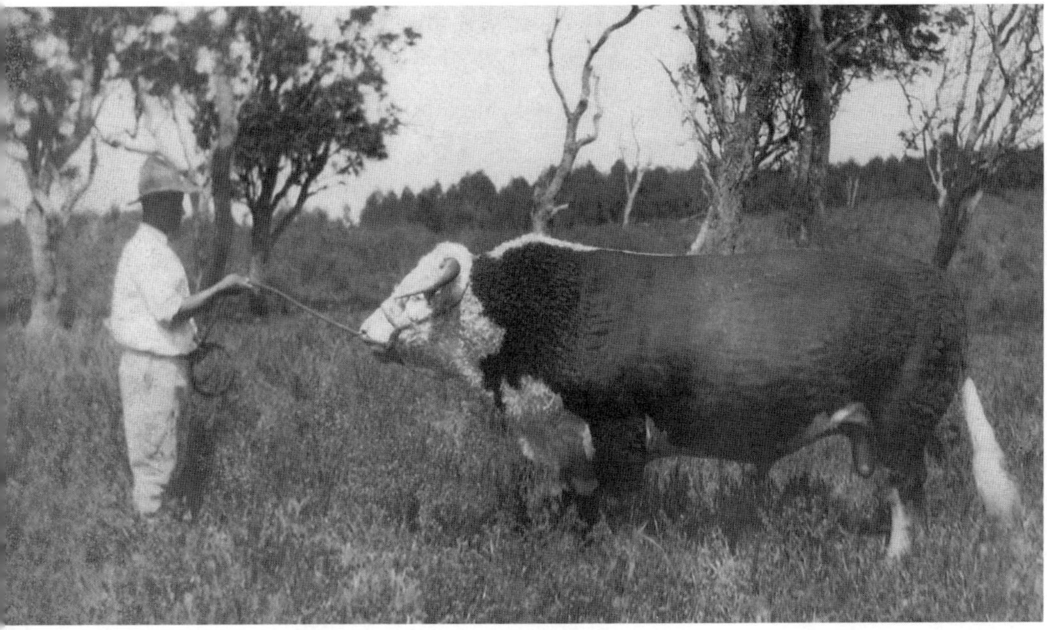

An Oddfellow-vintage bull. A. W. Carter Collection.

peration for Oddfellow. John rode home through Hānaipoe and received a quiet hero's welcome from A. W. Carter. Carter later wrote of the event:

> Fortunately I started the importation of purebred cattle in the early part of the century. Our breeding cows are all purebred Herefords now [1940], descendents of imported registered heifers, and of course we have used no bulls except purebred animals.
>
> The first purebred Hereford I ordered was "Oddfellow." This was purchased on November 9, 1901. John Lindsey took eight days to bring him from Hilo (where he was unloaded) to Paauhau. John cut up pieces of leather from his saddle to make shoes for the bull. I knew then that John was a good man.
>
> "Oddfellow" was raised on Mr. Gudgel's farm in Kansas, as was "Kamiakin's" sire "Beau Donald 31st". "Kamiakin" was the second purebred Hereford bull bought and shipped to Parker Ranch and came from Washington State. Other cattle purchased in Washington in 1908 were also of "Beau Donald 31st" get. "Anxiety 4th" blood was strong in the pedigrees of these early imports.

By 1906, A. W. purchased registered Hereford bulls and heifers by the carload from prominent Midwestern and Southern breeders, among them J. M. and W. H. Curtis and Gudgel and Sunprow of Missouri and Kansas, the Giltner Brothers of Kentucky, and the Cudahy Farm of Kansas City, Missouri. Many of these bulls carried the genes of Prince Rupert, another cornerstone sire of the Hereford breed. In the same year, a carload of Hereford heifers came from Senator A. J. Splawn of North Yakima, Washington, and Governor John T. Sparks of Reno, Nevada. The bulls Columbia, grandson of Lord Pretty Face, and Kamiakin, by Beau Donald 31st, were used on these heifers.

After the first purebred imports arrived in Hilo, Kawaihae then became the usual port of entry. Still, the 12-mile walk from Kawaihae to headquarters was grueling, with frequent water and shaded rest stops necessary in light of the pampered state of these imports. By this time, Makahālau was designated the purebred breeding station and, after a few days of rest in Waimea, the imports trekked for 9 miles along a grassy meadow to reach their permanent home. Full acclimatization to the environment, soils, grasses, and water and minerals took an entire year after arrival.

In 1907 Fred Carter was sent on a bull-buying trip, bringing home thirty-seven more registered Hereford bulls. In 1910 A. W. exercised his skill at brinkmanship on a buying trip of his own. At the Kansas Hereford farm of H. M. Curtis, he was shown a set of bulls and heifers from which he selected five males and twenty females for purchase. As Curtis was leaving the Farm with Carter, A. W. passed a discrete meadow and noticed some typey heifers. "Typey" is a ranching term that refers to an animal with the ideal frame, size, posture, musculature, and proportionate dimensions, and A. W. had a consistent eye for such livestock. Upon inquiring of Curtis about these choice females, he was brushed off by his host, who appeared to have squirreled away these superior individuals to retain for himself. Curtis relented when Carter gruffly threatened to cancel his purchases altogether. The carload destined for the ranch included these topped-off heifers as well.

During this same period, A. W. Carter hired a fine herdsman named James Wight, who accompanied a shipment of Hereford stock back to Hawai'i. Makahālau became his new home. A Hereford senior sire named Bonnie Lad 24th was also purchased from J. C. Robinson and Son in Evansville, Wisconsin, on a 1912 buying trip. Alex Napier, a young cattleman, was the stock tender accompanying this shipment, which included heifers from J. M. Curtis and the Giltner Brothers. Like James Wight, Napier also settled in Makahālau. Later, when Wight was made head overseer of Parker Ranch, Napier replaced him as herdsman. Napier, who was called Sandy by his coworkers, went on to establish a family that continues to hold a place in Hawai'i's ranching history.

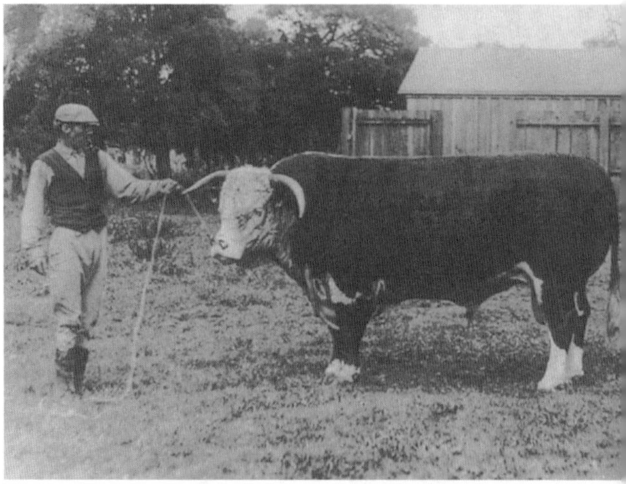

Alexander "Sandy" Napier holding the imported bull Bonnie Lad 24th. Napier Family Collections.

Because the Red Devon and shorthorn herds proved less adaptable than the traditional Hereford, A. W. moved toward a goal of an entirely purebred ranch composed of the latter breed. Importation of registered Herefords continued on a large scale until about 1913. By that time the ranch had developed special purebred herds to generate bull and heifer replacements; further importations were limited to select individual sires of the highest order. In 1918, Parker Ranch purchased another noted sire, Paragon 12th, for the princely sum of $6,000. A. W.'s purebred herd began to show results. Between 1912 and 1918, market carcass weights increased from 440 to 663 pounds.

R. R. Giltner of Eminence, Kentucky, an early supplier of bulls to the Parker Ranch, paid a visit in 1923 and was impressed with the entirety of A. W.'s import purebred program that had been developed over two decades. Giltner, when comparing other large ranch operations throughout the West that used his bulls, described Parker Ranch very favorably. By 1929, the ranch had 400 purebred bulls and about 2,500 purebred females. The commercial herd numbered 9,000 cows.

Not settling for anything less than perfection, A. W. moved toward the Domino strain of Herefords. In 1929 he purchased two sons of Dandy Domino 2nd, the noted register-of-merit sire and son of the great Prince Domino from the Banning-Lewis herd of Colorado Springs, Colorado. In 1930, he brought home Prince Domino 7th, a direct son of Prince Domino, which he purchased from Fulscher and Kepler of Holyoke, Colorado, for $4,000. The Domino lines added even greater carcass traits and passed on a gentle nature in their offspring, a quality appreciated by Carter. The even-tempered purebred ranch calves also matured earlier, and they developed deep, fleshy hindquarters.

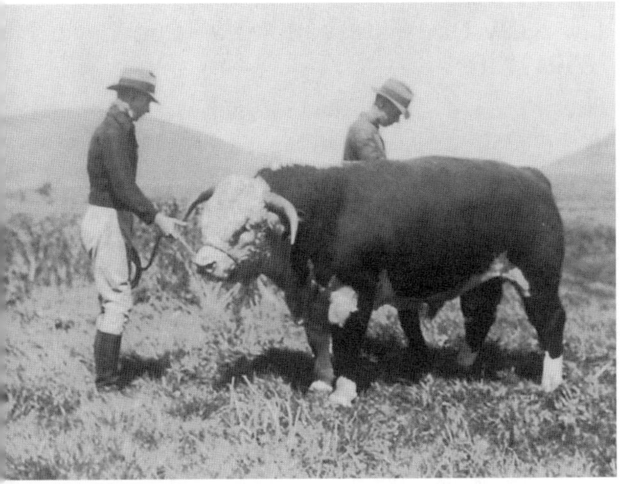

Visiting mainland judge critiquing Parker Ranch home-raised, single-herd sire. A. W. Carter Collection.

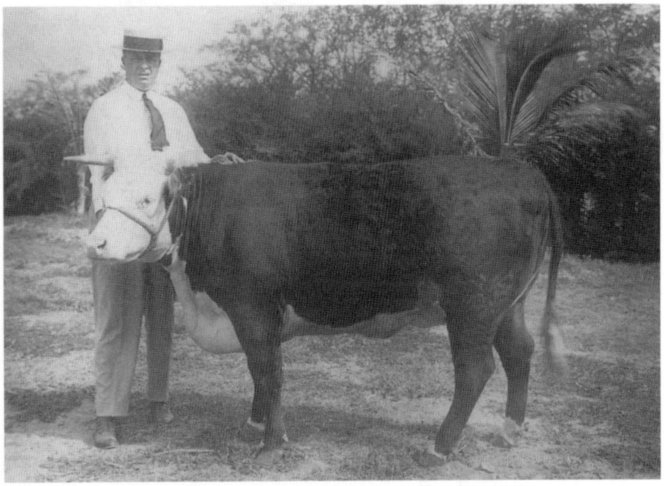

Ranch purebred herdsman James Wight at the halter of an imported heifer. A. W. Carter Collection.

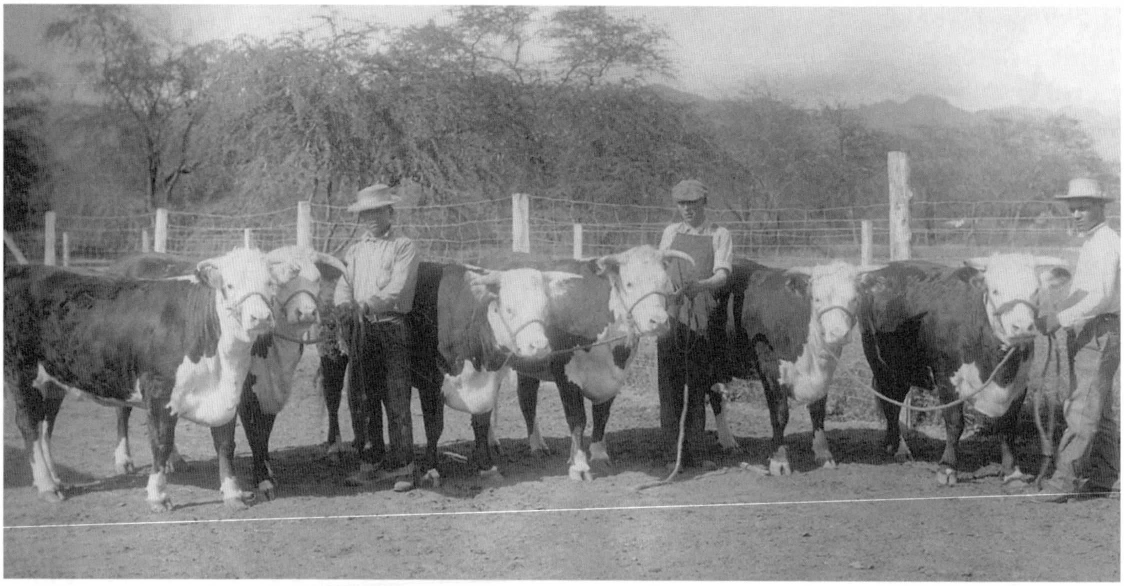

A battery of imported, registered heifers. A. W. Carter Collection.

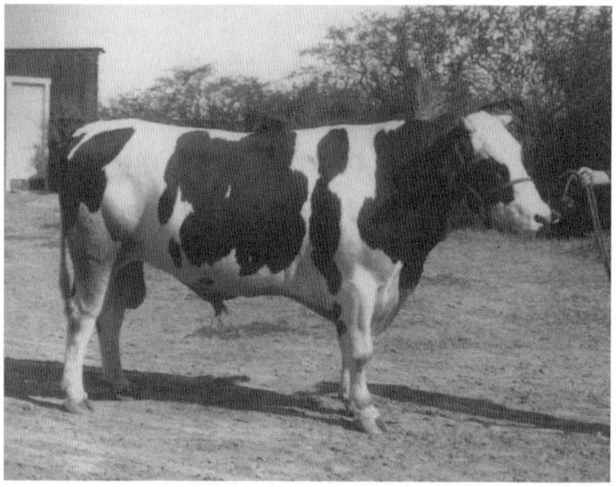

Imported premium Holstein bull. A. W. Carter Collection.

In 1937, at the Wyoming Hereford Ranch auction in Cheyenne, Carter commissioned for purchase the third top bull, Super Superior 3rd. Shortly thereafter, J. B. Wight, an Oregon State graduate and son of former herdsman James Wight, attended the bull on his trip home. Young Wight later went on to Cornell University, where he received a degree in veterinary medicine.

Parallel to beef cattle improvements, A. W. launched a comparably designed program in dairy production.

The Early Dairy Story of Parker Ranch: 1850–1920

The original dairy of Parker Ranch was built in the 1850s at Paliho'oūkapapa (Old Dairy) as an expansion of the home-use milking cowherd in Mānā. The need for an expanded and commercial dairy came with a larger employee base at Mānā as well as a growing community. In Waimea, milk and butter were delivered by horse-drawn wagon. The rudimentary milking barn was framed with *māmane* posts and beams and had walls of *koa* boards and battens. Windows were sparse and were more for lighting than ventilation. Paliho'oūkapapa was situated at about 5,000 feet elevation, 2 miles east of the Mānā headquarters and more than 12 miles from Waimea village.

The milk cow was typically a crossbred, having some Durham (shorthorn) or Devon breeding with a mixture of feral stock. Production of 2 gallons a day, with morning and night milking, was considered median performance, sparing some for nursing calves.

As the end of the century approached, little had been done to upgrade the Paliho'oūkapapa facility or the quality and productivity of the milk cowherd. This became an early challenge to A. W., who saw the need for dairy products throughout Hawai'i, coupled with the potential for great financial returns.

A. W. made a direct move by securing better-grade milking shorthorn bulls from the mainland and Holstein bulls from New Zealand. Milk cow numbers grew to about 750 head of unregistered crossbred cows; marketing Parker Ranch dairy products was expanded to include Hāmākua. By 1908, Parker Ranch was an exemplary dairy producer using only purebred bulls.

The milking parlor was upgraded and a cooler room added to accommodate butter and cream storage. Four men were kept busy processing dairy products. In 1916 a silo was built to accommodate corn silage at the Old Dairy and another at Makahālau, where there was also some corn farming. Later, these silos were sealed in an attempt to hold water for storage. The silos failed in both the silage and water storage functions.

Complementary programs such as pork and chicken production made the Paliho'oūkapapa facility a well-integrated force in Parker Ranch's overall production scheme, despite the disadvantage of the long wagon haul over many miles to town. A further complication was an outbreak of brucellosis in the dairy herd in 1919. This disease is caused by a bacterium called *Brucella abortus* and is nicknamed Bang's disease after the Danish veterinarian who first described it. It is an abortive disease that is highly contagious and transmissible to man in the form of undulant fever, so named by its variable fever that spikes and drops in clinical cases. It can be fatal in humans, but rarely in cattle, which become carriers and can spread it via breeding, calving, and so on.

Also in 1919, a fourteen-year-old hand named Yutaka Kimura became a milker under the foremanship of Iwataro Ishii. From humble beginnings, Yutaka Kimura emerged as a major player in the growth and productivity of both the dairy and beef operations of Parker Ranch. Yutaka shares many of his experiences in *Parker Ranch Paniolo: Yutaka Kimura,* a book authored by Jiro Nakano, M.D.

The epidemic of Bang's disease at Parker Ranch caused A. W. to shift from the periodic use of territorial veterinarians such as Doctors Elliot, Rowat, and Onissimenko, who were on call from Hilo and Kohala, to retaining a world expert in the field of reproductive diseases in cattle. A. W. was determined to rid the ranch of this disease and others and launched a preventative veterinary medical program to bar its reintroduction. This meant testing existing suspects and slaughtering those infected, coupled with ranch-imposed prepurchase testing of all imports. Parker Ranch's disease prevention posture was implemented nearly fifty years before mainland counterparts would take such bold steps. Integral to this program was Dr. W. L. Williams of Cornell University.

By 1920 Dr. Williams was a world-renowned veterinary scientist and author. He was a standing professor in the Department of Obstetrics and Research in the Diseases of Breeding Cattle at New York State Veterinary College in Ithaca. He was contacted on January 2, 1920, by A. W. to purchase disease-free dairy breeding animals with a history of 30-pound daily milk production in the females. For the bulls, their progeny of females also needed to have a high milk production rate. In his reply to Carter, Williams proposed launching this buying mission on June 9 of the same year.

In the interim, a flurry of letters was exchanged between them, and the role of Dr. Williams expanded to include a six-month consultative mission to Parker Ranch. During Williams' tenure at the ranch, Yutaka Kimura became his protégé. Other issues were opened for discussion, such as tuberculosis and cancer eye in cattle, as well as investigating the losses of newborn calves and cerebrospinal meningitis in horses. A. W. also asked for help in securing a good dairyman, as noted in a letter to Williams dated February 27, 1920: "I desire to obtain a good Dairyman who will be able to handle the cows and make butter. In selecting a man to bring the bulls and heifers out here, you might keep that in mind. I intend installing an up to date creamery and should like to get a man for the same. He should be a good herdsman as well as butter and cheese maker. I would pay $150.00 per month and in addition would supply cottage and free firewood."

On March 9 of that year, Dr. Williams outlined the plans for his consultative tenure on the ranch, which would coincide with the arrival of twenty purebred (registered) heifers and five bulls he had selected and forwarded to Hawai'i, based on exhaustive efforts to acquire the finest breeding of Holstein milk stock from disease-free farms in Massachusetts and British Columbia. These were purchased via a $45,000 letter of credit forwarded to him from A. W. on March 8.

In addition to setting up a registered herd for Parker Ranch, Dr. Williams thoroughly reviewed the health status and performance of the five-hundred-head cowherd located at the new dairy at Pu'ukīkoni, 8 miles from Waimea. His memo to A. W., dated March 9, 1920, reveals his astute, in-depth review:

> According to present plans, I hope upon my arrival to take up a thorough study of the pedigreed cows and the bulls used in the pedigreed herd. For this purpose I would plan to make a careful examination of the genital organs of each individual and record the findings upon another group of blanks, which I shall forward a little later but which will not be used until after my arrival.
>
> By having the dates of service and the results of breeding, I can go over the records and get at once the reproductive efficiency of the individual cow. If the bull used is also noted, I can take these services and group them and learn regarding the efficiency of each individual bull. Then, if the disposition of the calf is recorded, I can also trace its efficiency, if it has remained in the herd, and can thus arrive at the efficiency of the family including both the heifer and the

bull calves. I have understood that you have about 500 of these pedigreed animals in which case it will require a month to go over them thoroughly, working a part of the day and leaving a part of each day free to take up other work, which may demand attention. After this work is completed, I would then take up other lines of work according to conditions. My idea is that while attending to emergency cases which come up, I can take up what seems to be the most important problem at hand and get a complete record of conditions of certain groups.

Later, I shall be forwarding some instruments and drugs that I think I may need and which would probably prove difficult to secure. I shall aim to advise you when these are shipped and shall have them addressed to me in your care. Upon their arrival, it will be necessary merely to put them away where they will be safe pending my arrival.

By late March, Williams highly recommended a Mr. Mills of the senior class of the Veterinary College. Mills had a year's training in butter and cream production at the State School of Agriculture before entering veterinary school. Furthermore, he had been the attending herdsman on a previous shipment of Holsteins to Frank Baldwin of Maui. A May 31 letter from A. W. concurred with Williams on the selection of Mills as the new herdsman and dairyman for Pu'ukīkoni Dairy. In a final note before his arrival, Dr. Williams was cautioned by Carter to be careful in the use of the term "contagious abortion," or Bang's disease. In the words of A. W., "This territory is a small one and reports not only spread rapidly, but soon become exaggerated." It is obvious that A. W. wanted to contain the brucellosis outbreak at Paliho'oūkapapa—both in the cowherd and in the gossip circles.

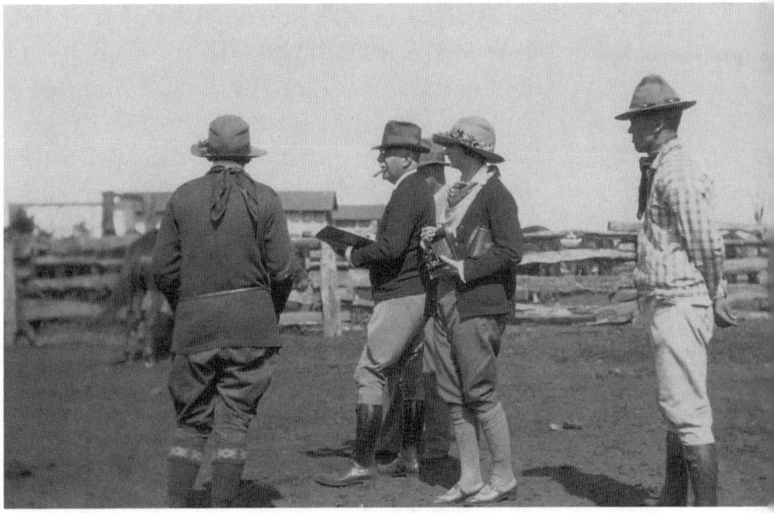

A. W. Carter, second from left, grading dairy heifers at the Paliho'oūkapapa (Old Dairy) facility with an entourage of ranch guests. PPS Collection.

Early Valuable Livestock Contacts

Although the history of veterinary medicine on the ranch is treated in a separate chapter, it is timely to state that in 1895, Hilo received the island's first veterinarian. Dr. H. B. Elliot served the plantations and ranches in East Hawai'i but remained available to Parker Ranch on an on-call basis. In 1905, Danish veterinarian Victor Norgaard was named superintendent of Animal Industry, Board of Agriculture and Forestry, Territory of Hawai'i. Dr. A. W. Rowat, who was assigned to North and South Kohala, briefly joined Norgaard and Elliot in 1908. Rowat administered rules and regulations on inspection and testing of imported livestock.

A. W. Carter undoubtedly played a major role in the promulgation of the regulations and certainly in the later appointment of a government veterinarian for North and South Kohala. With the inspection rules firmly in place, he concentrated on developing agents for premium stock from opposite ends of the globe—Tasmania and New York.

When A. W. was owner of Moloka'i Ranch, he became well acquainted with one of the managers there—George C. Munro, a native of New Zealand. In later years, Munro returned to New Zealand and became a prominent agent for the exportation of seed stock to the Hawaiian Islands. Munro served Parker Ranch very well in this capacity, providing Hereford and shorthorn bulls and ram and ewe merinos.

Horse Contacts

A. W. had an eye for phenotype and character of the ideal ranch horse: reasonably large framed, with stout legs and solid black feet, and a lengthy neck knitting a prominent wither to a well-coupled back that is driven by moderately heavy hindquarters. As for the head, beauty was ideal but not a requisite. What went on inside the head was important, however: ideally, a love for cattle and a respect for man. Speed and stamina were mandatory. Probably not unwittingly, A. W.'s standard of an ideal ranch horse fit the description of a military remount horse. He cultivated valuable horse contacts across the nation, especially in Thoroughbred and Standardbred racing and remount circles. On August 31, 1900, as an immediate gesture of horse and mule improvement, A. W. purchased a purebred grey yearling French Percheron stallion from the Minewawa Creamery of Fresno for $250. He also purchased a Maltese jack from Dr. B. F. Williams of Fresno.

Eastertide was one of the most prominent early Thoroughbred sires of ranch horses in the early Carter era. He came to the ranch via Samuel Parker, who purchased him from Prince David Kawananakoa for $1,000 in 1900. Carter described him and his colts as follows: "They are beautiful, of the Thoroughbred type, but have more substance of leg. Old EASTERTIDE was the sire of CARRY THE NEWS JR. The latter seems to get excellent mares for broodmare purposes and has done much to build up the quality of the stock as did his sire." A broodmare pasture named after this famous sire is Eastertide Paddock in the Pu'u Pueo section, Mānā Division of Parker Ranch. Eastertide Jr. was later retained to carry on this line of Thoroughbreds; quotes from military records reaffirm A. W.'s eye on the U.S. Army as a market for horses. Standardbreds were forthcoming. In 1908 A. W. imported another Percheron stallion, Bijou Blanco, who weighed over a ton, along with a German Coach stallion. A. W. set his horse goals and methodically achieved them.

Improvement of the Horse Market: 1900–1945

When Parker Ranch was born in 1847, the horse had been on the Big Island for less than forty-five years. In that span of time, horse quality undoubtedly declined from the first import of quality Thoroughbred types from California mission breeding to a smaller, coarser equine that had become feral.

This occurred despite the concomitant arrival of a carload of twenty-four horses and the *vaquero* in the early 1830s. The horses were intact males as well as geldings; the *vaquero* tradition shunned the use of female horses. An effort to upgrade local horses took place when William French (WGAC) imported a dappled grey stallion named Kalakini to Waimea. Later a racing stallion, Waimānalo, went to John Palmer Parker. Still later, another speed horse, Governor Stanford, was retired to Parker Ranch. About that time Parker Ranch acquired a buckskin stallion imported from Chile that passed on quickness, endurance, and solid conformation.

Sam Parker obtained many of the better stallions that became ranch sires. These included General Hancock in 1878 and Waterford, a grey Thoroughbred stallion. At the turn of the century, Sam had acquired Eastertide. Eastertide's descendents became famous. One of these, Carry the News, was an internationally famed polo horse, and Eastertide Jr. became a consistent sire of fine remount horses for the war effort. Eastertide Jr.'s dam, Nisa, was the first registered Thoroughbred mare imported to Parker Ranch.

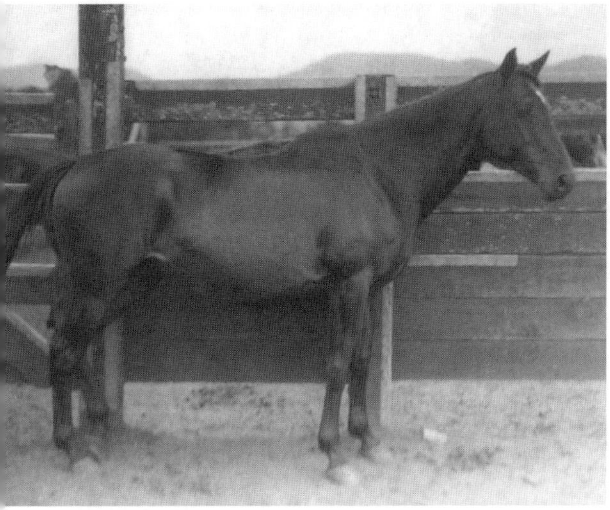

Ideal and typical ranch broodmare, daughter of Royal F, a remount Thoroughbred stallion. 1927. A. W. Carter Collection.

When he arrived on the ranch, A. W. acknowledged the presence of some good horses but was vitally concerned by their scarcity and physical condition. He described the program at that time as being the indiscriminate breeding of scrub horses. Paul Jarrett had on hand a Percheron draft stallion and two utility stallions that were brought in the early 1900s, but immediate productivity was needed from the finer stallions acquired by Sam Parker as well as those that still needed to be purchased.

In 1901, A. W. addressed the O'ahu polo market by developing an in-house training facility with readymade mounts provided by his brother Fred. A. W.'s polo horse trainers were the four Lindsey brothers: Johnny, George, Thomas, and James. They formed a polo team and received their basic training on the Fred Carter mounts. Once adept at the game, they were each issued a small string of green-broken ranch colts to train as polo horses. A. W. improved an area southwest of today's Waimea Airport for the game, and senior cowboys still refer to the area as a polo field. Later a practice polo field was developed at Makahikilua.

The polo program was designed as a marketing tool, and it functioned quite effectively, as seen in the sale of the mounts of the ranch polo team. Louis von Tempsky of Haleakalā Ranch on Maui arrived to buy fifteen of them, including the colt Carry the News, which became famous in New York polo circles. The O'ahu Polo Club and Damon Estate bought the balance of the polo

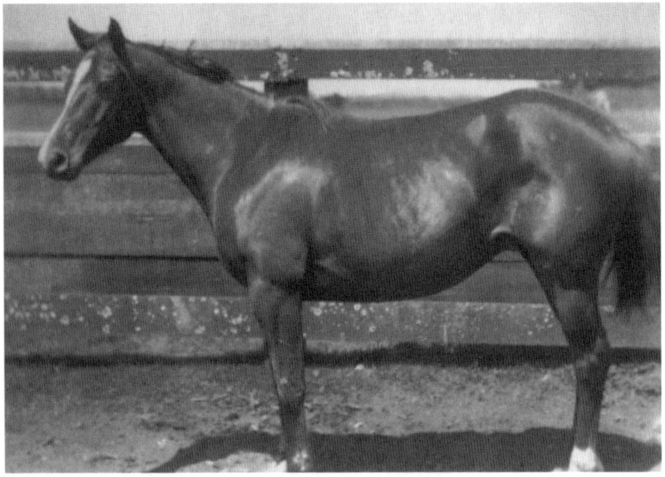

A three-year-old colt, son of Royal F: an ideal cowhorse, polo, or military prospect. A. W. Carter Collection.

mounts, although the Lindsey brothers hated to part with the colts that brought so much pride. Parker Ranch became known nationally as a resource for polo horses.

In later years A. W. reflected on the lesson he learned in watching the men part with their well-trained polo ponies. He forever changed his attitude, as is clearly evident in a December 22, 1911, memo:

> A few of the boys have got horses which might do for polo, but I would not think of taking the horses away from them under any consideration. It would take their hearts out of the game of finishing horses for their own use.
>
> After once giving a man a horse to train as a work animal and he has spent time and patience and gotten a good string together I should never feel inclined to take it away from him.
>
> I think the application of this line is one reason we have such fine cow ponies at Waimea, some of them with intelligence to do almost anything but talk. The men know that if they take care of, develop and get together a string of fine horses that I will allow them to keep the horses as long as they work for the ranch. In case I take a shine to a horse they will often offer to turn the animal over to me. This I seldom avail myself of for the reason mentioned above.

In my exposure to John Holi Ma'e of Kūka'iau, I can recall when this great horseman lamented, as did the Lindsey brothers, the loss of beloved horses he had trained and of which he had grown particularly fond. In Holi's case, remount personnel often gleaned the best of his twelve-horse cowboy string for military use annually for the war efforts. Understandably, great pride and credit go to the men who are recognized as top horse trainers. The bond between man and animal is undeniable.

In 1901, A. W. launched a lengthy, well-designed, and well-executed breeding program based on careful selection of dams and sires to generate sound cow horses, draft horses and mules, military remounts, and racing and polo horses. This was supported by broodmare bands according to breed and type, plus a stallion station at Pukalani Stables, housed with premium sires, some of which came from Sam Parker acquisitions.

One of A. W.'s first stallion buys was a Standardbred horse named Moloka'i Boy. Carter described him as a short-coupled, well-balanced, straight-legged, good-shouldered sire with a somewhat refined neck and head. Carter's description of an ideal ranch

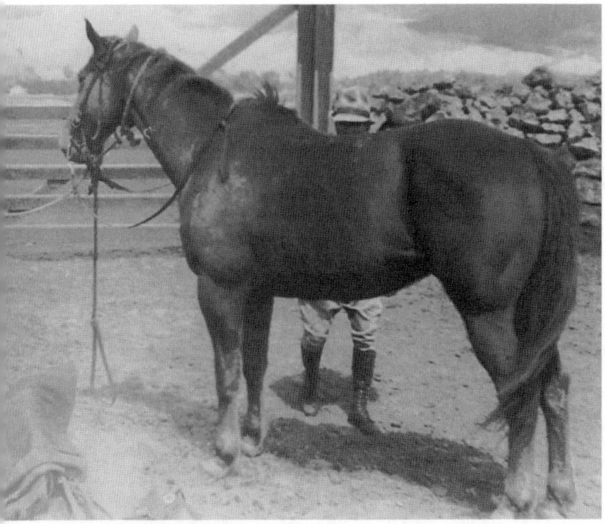

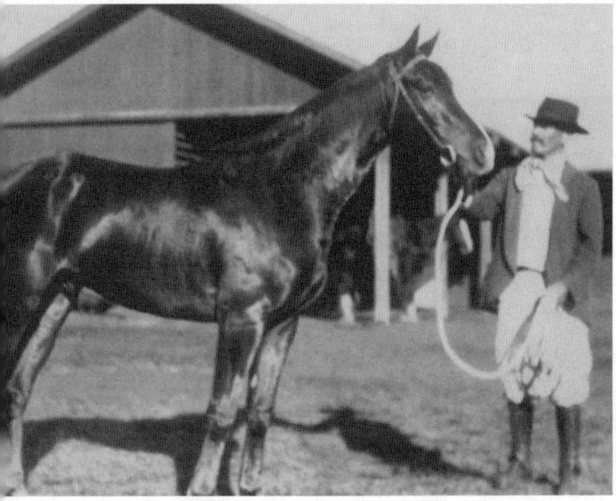

Top: *Typical Parker Ranch cow horse, July 1924. Saddling up at Pu'uhihale Corral. A. W. Carter Collection.*
Above: *Moloka'i Boy with Sam Awa'a Sr. at Pukalani Stables. A. W. Carter Collection.*

horse still fits the standard a hundred years later. Moloka'i Boy turned out to be a solid producer of cow horses when crossed on native mares. He also sired sound polo ponies.

Parker Ranch Thoroughbred and Army Remount Program: 1908–1945

Aware of the military as a market for horses, A. W. called Washington, D.C., in 1908 requesting permission to supply cavalry mounts to Schofield Barracks on O'ahu. This paralleled the needs of the U.S. Army. Within a year, the ranch was in the remount business, a market that A. W. cultivated through both world wars by producing precisely what the army purchasing officers called for in their specifications.

In 1910, Parker Ranch shipped twenty remounts on the SS *Wailele* to the Fifth Cavalry at Schofield, where Major S. C. Reynolds, a keen horseman himself, eyed them critically—and appreciatively. The ranch's remount program had been designed strictly, and its horses were readily accepted. Reynolds and A. W. developed a close relationship, and the major became an enthusiastic evaluator of the offspring of the different remounts, with Eastertide winning top praise. Reynolds was an avid polo player as well as a cavalry officer, and the army played a lot of polo on O'ahu. Reynolds' knowledge and experience made him a key consultant for the ranch remount program.

In 1913, A. W. campaigned vigorously to reinforce the Thoroughbred program. Registered mares included Sumptuary, with foal at side (nursing) and in foal to a famed Thoroughbred sire named Marse Abe. Her second foal was Gaiety, known in local circles as 'Awapuhi. As a two-year-old in 1916, this mount held the track record on O'ahu for the three-eighths-mile, winning the Silver Cup. One of 'Awapuhi's daughters garnered nearly $28,000 in winnings in California in later years. The addition of

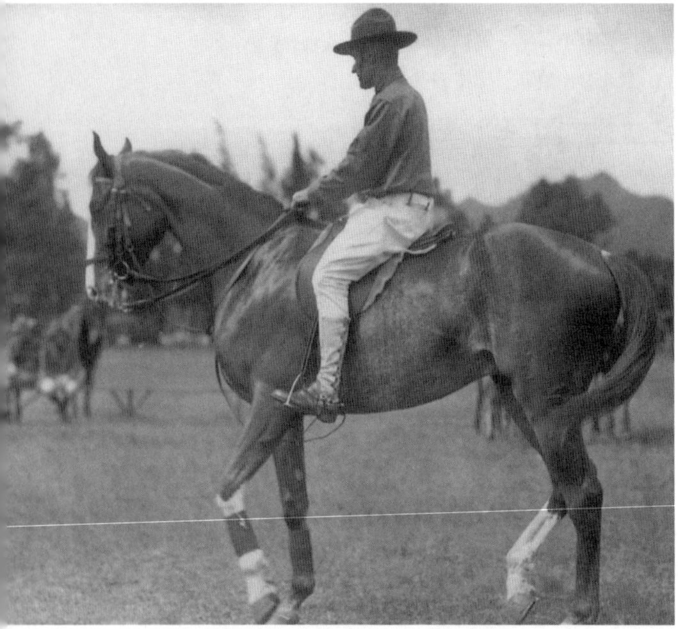

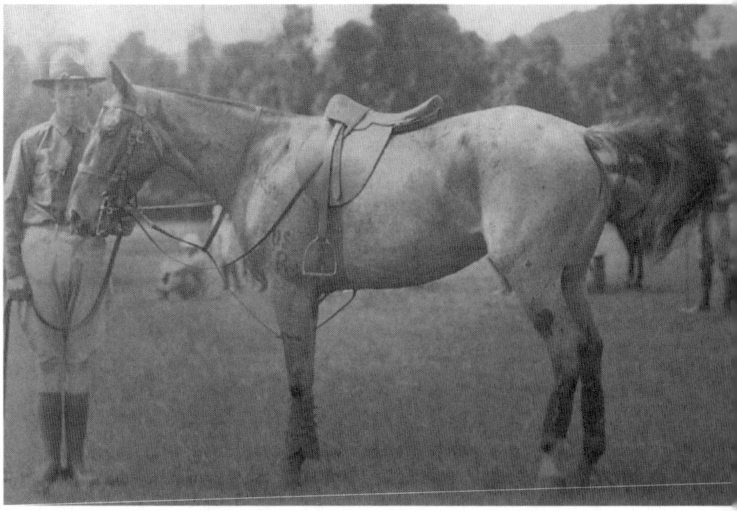

Left: *Ideal cavalry officer's mount. A. W. Carter Collection.*
Above: *A polo mount sired by Frolic, typically roan, with brands "US" and "R" on shoulder and "P" brand on gaskin. A. W. Carter Collection.*

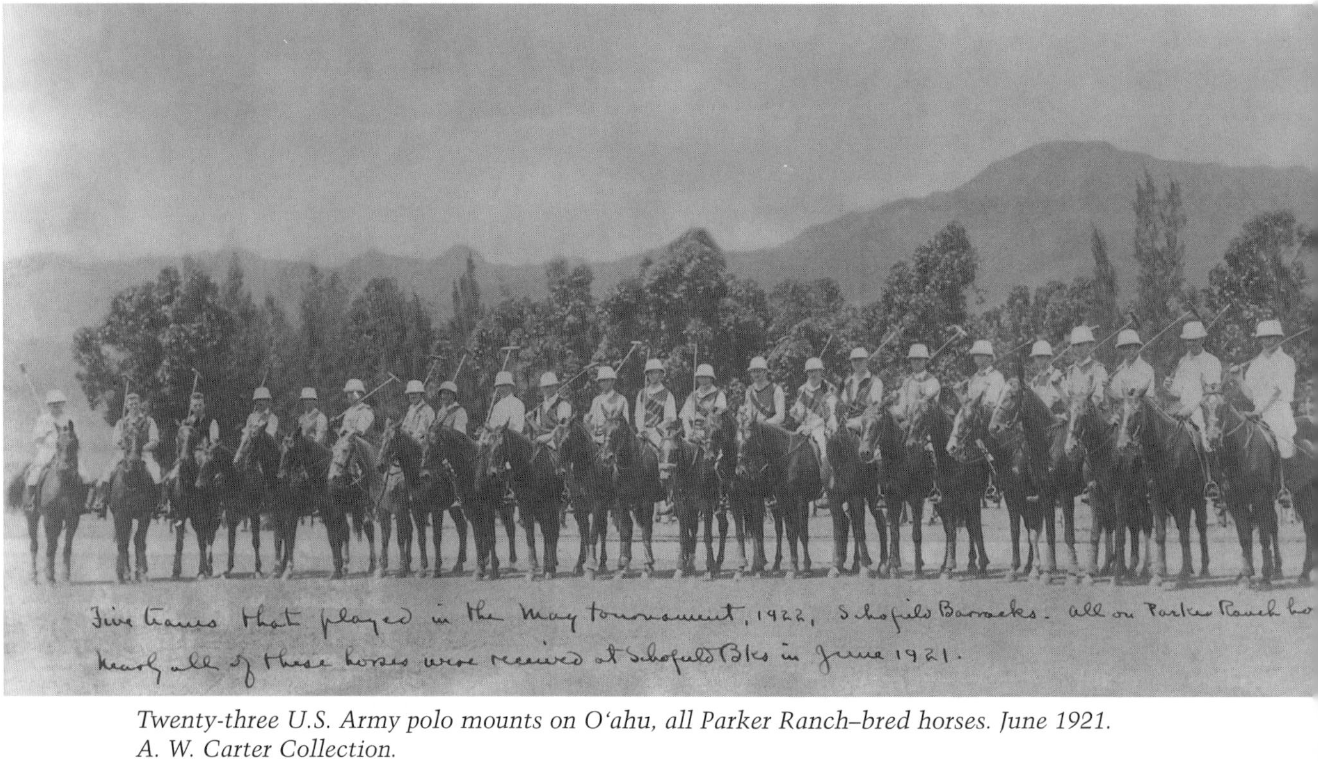

Twenty-three U.S. Army polo mounts on O'ahu, all Parker Ranch–bred horses. June 1921. A. W. Carter Collection.

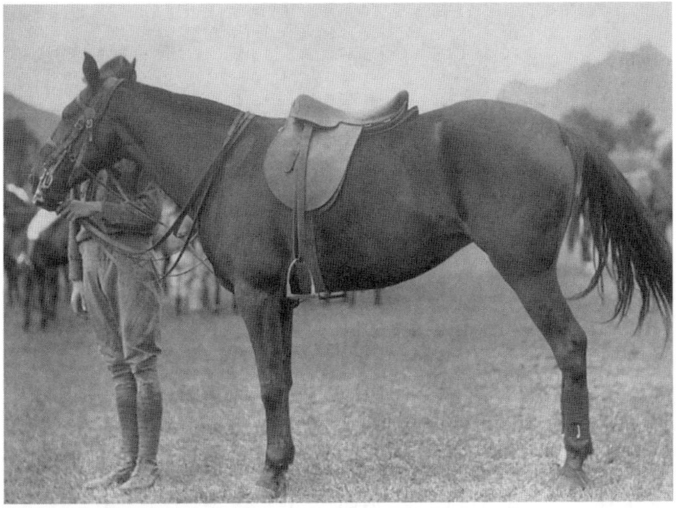

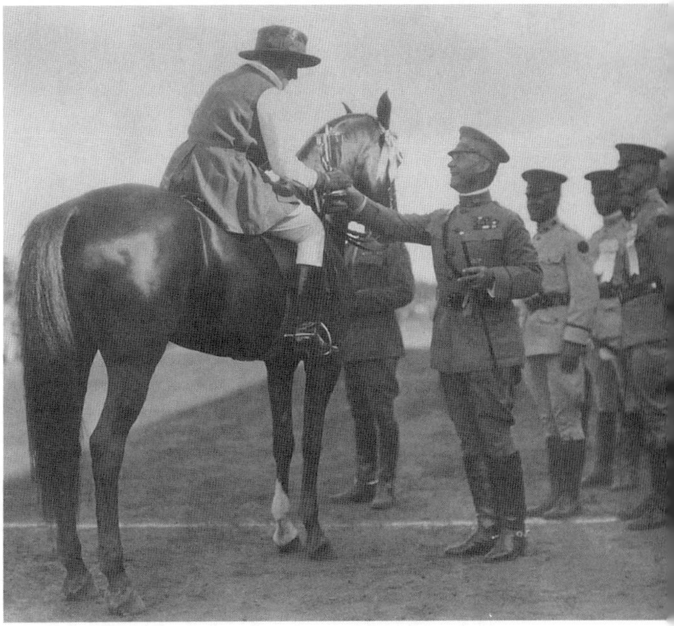

Above: *A black polo mount by Moloka'i Boy. A. W. Carter Collection.*
Right: *Military horse show, Honolulu, 1914. The horse is a Parker Ranch–bred officer's mount. A. W. Carter Collection.*

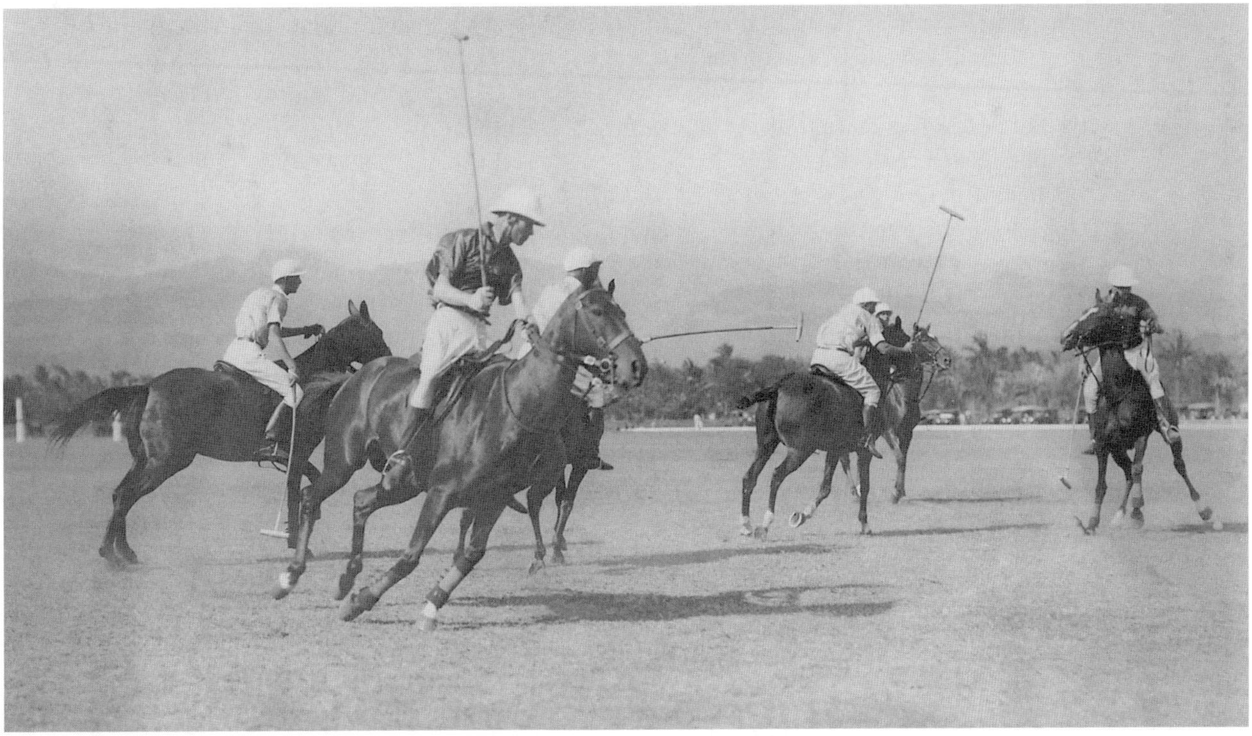

Polo action on O'ahu featuring Parker Ranch horses. A. W. Carter Collection.

Frolic, which came with the Humu'ula land purchase in 1914, was a genuine asset to the program. During the same year, A. W. acquired a pair of Irish-bred polo mares, Trilby and Little Nell. Both were unregistered but brought intelligent and dependable offspring when crossed with ranch stallions.

The year 1916 was a banner one for Thoroughbred imports. From England came Mangonel by Carbine, Donna Christine, Hidden Star, Barricade and her foal, Grafton's Barricade, and Shine. Grafton's Barricade was by Fitz Grafton, a horse that had won the Calcutta Cup. In 1917, Gold Bryan by Gold Oak and out of Miss Bryan established a half-mile record of 48.25 seconds.

Two of the imports were especially notable: Donna Christine produced a typey and speedy daughter (but quite lop-eared, a trait that reappeared in Parker Ranch's Thoroughbreds even through the 1970s), and the cavalry was especially impressed with Barricade's colt foal that grew into prominence as a remount sire.

Major Reynolds was unabashed about his love for polo, even as he reviewed cavalry remounts. One of the first remount sires, Harvester, a favorite of A. W.'s, was described by the major as being a little high-strung for polo. Gold Oak, one of the all-time finest sires of remounts, was noted for producing fine officer's charges (see the appendix for a U.S. Army polo horse-grading chart). I have had the pleasure of riding grand colts of Gold Oak, and they were consistently broncky (inclined to buck without reason or warning), quick, smart horses that never seemed to get over their snorty nature. But their superior endurance as cow horses made up for any of their shortcomings. Another prominent sire, Rodgers, a remount stallion, was sent to the ranch and produced sound colts. I also rode some of his descendents, and they were superb. Two of Rodgers' illustrious offspring were La'ikealoha and Moanalua. A solid brown, Moanalua, who was out of Waiau, became a noted sire in his own right. Mālani, one of his sons, established a track record in Hilo in 1940.

In the same year of Warshot's arrival on Parker Ranch, another large band of registered Thoroughbred mares arrived. As Carter reported, "The 1925 lot were the CULLEN BON, HOMAID, MAYSVILLE, PENWELL, MISS STRATHEM, PUNTA GORDA, NELLIE HARPER, CURRY DESHA, PAVLOVA, TEMPTRESS, BELLE TAYLOR, VON LADY, VIC, HONEST LASS and CONFETTI. A second group of five had been collected in England by Mr. Allison, of the International Horse Agency of London, and were CAUCASIA, MYRTILLA, AZCONA, SANTA JAVELINA and FREDA CONSTANCE."

In the following years, another Thoroughbred stallion, Overall, entered the stallion station at Pukalani. He came from Whitney Farms and conducted a productive breeding campaign on ranch mares, especially when offspring were crossed on daughters of a stud named Ormesby. Overall was by the famed Peter Pan, the leading sire in the United States in 1927. He produced several prominent racehorses, including Puna out of Pukalani, Molokama out of Kāhili, Kaneohe out of Maile, Akia out of Neneleau, and Hanohano out of Donna Christine. In the Fourth of July races in Hilo, Molokama established a 1-mile record time of 1:42.4.

Ormesby was the third remount stal-

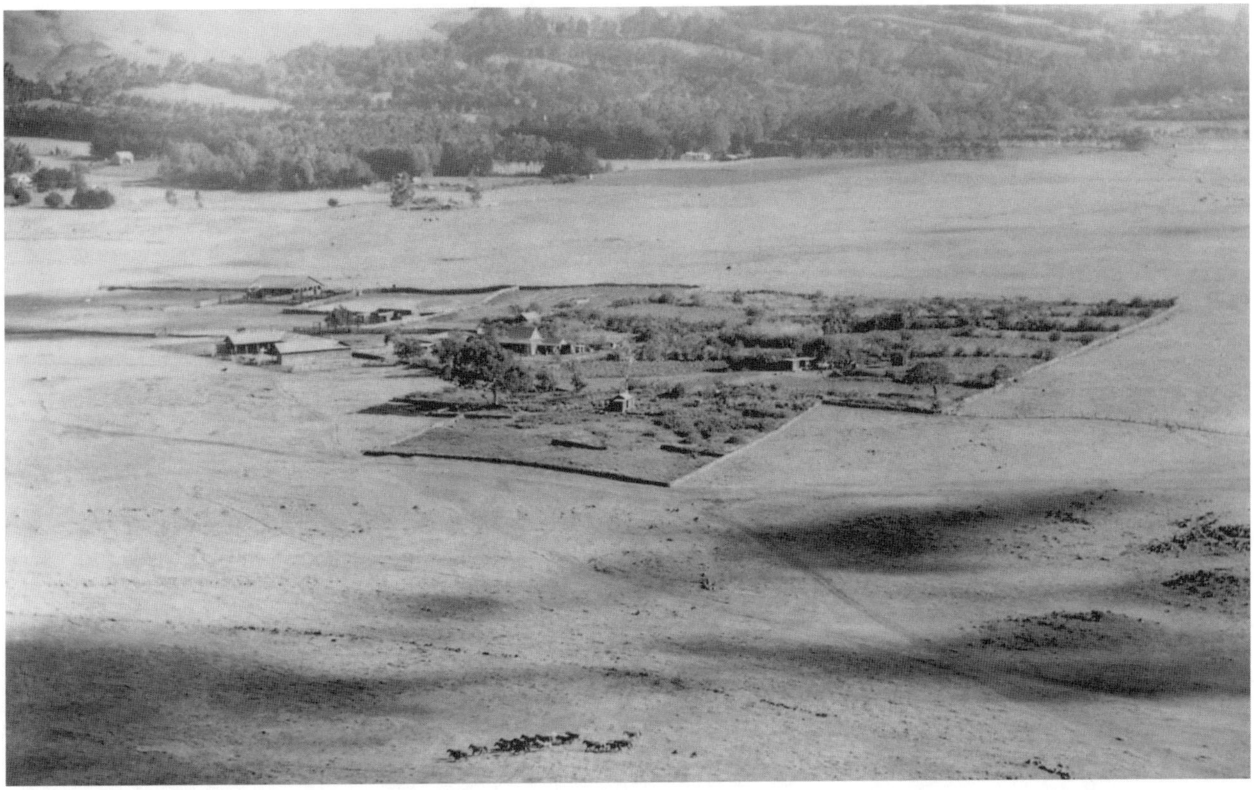

Aerial view of the Pukalani Equestrian Complex, including the stables and Breaking Pen on the left and several kuleana *(taken from 7,000-foot elevation on January 22, 1925). A herd of horses runs in the foreground. Lyman House Memorial Museum 74.30.2.68.*

lion sent to Parker Ranch. He was by Sir Galahad III and had garnered about $17,000 in races that placed him first in seven, second in four, and third in nine races before he was retired to Parker Ranch as a coming seven year old. The U.S. Army had paid $150,000 to the Breeding Association of America for this three-quarter brother of the famous Gallant Fox, winner of the 1937 Kentucky Derby and the 1938 English Derby. On his own record, Ormesby placed in the 1936 Kentucky Derby. One of his sons, Ikaika, won the Maui Derby in 1940 and 1941.

Major Reynolds' superior, General C. D. Herron, a frequent guest of A. W.'s at the ranch, filed this report in 1939 with the quartermaster general in Washington, D.C.:

> I am delighted by the Parker Ranch and by its Thoroughbreds. It is on the slope of a great quiescent volcano, rugged enough to develop wonderful feet and bone, and steep enough to give fine endurance. The grass is superb all the year around and of a kind that brings cattle to top market condition without a single ear of corn. Furthermore, the colts

Schofield Barracks - June 1922

GRADER'S NOTES

A - Good now
A' - Will be good
B - Fair now - Probably will be good
C - Playable - never be good.

CAVALRY OFFICER'S COMMENTS

	IDENTIFICATION	YEARS OF AGE	
			EASTERTIDE (sire)
C	Mare	5	Plays regularly. Slow
	Mare	6	Never played. Fine looking. Used as officer's mount. Can be made polo pony.
	Mare		Chunky. Can be made to play. Never be very good.
	Horse	6	Never played. Very large. Good officer's mount.
A'	Mare	5	Plays regularly. Will be fine.
A'	Horse	6	Not played. Looks like Oahu but bigger and coarser.
A	Roller Coaster		1920 - Fine pony. Big.
B	Mildred		First lot 18-1921. Plays very fast. Not yet properly trained.
A	Jimmie	6	1920. Very fast. Plays regularly.
A	Red	6	1920. Fairly fast. Plays regularly.
	Horse	6	Too big for polo. Pretty good.
A	Oahu	6	1921. Best pony in our string. Won troopers mount at Dept. Show. Has had lots of work.
A'	Glad Tidings		First lot of 18-1921. Played last fall. Was sick 6 months. Will be a very fine pony.
A	Chemise	6	First lot of 18-1921. One of the best.
A'	Madam E	5	Will be fine pony. Not played.
			FROLIC (sire)
A'	No. 32	5	Will be very good.
A	Mare	5	First lot of 18-1921. Plays regularly and well.
	Horse	5	Will be fair.
C	Skeeter	6	1920
C	Chicken	6	1920
	Mare	5	Good prospect.
A	Jane Wilder	6	1920
A	Louise Noel	6	1920
A	Nancy	5	Very fast
C	Gipsy		1921
A'	No. 40		First 18-1921. Very fast
C	Helen K	6	
C	James	6	
A'	Shirley		First 18-1921. Stubborn
	Gray Mare		1921. Plays a little. Will be fair.
	Snowball	6	1921. Foolish but well trained. Will be good.
A	Nita		No brand. Looks like Frolic. not fast.
	Mare		Big gray. 1921. Pretty good looking. Not much.
B	Mickey Jr.		
A	Garity	6	1920. Not fast. Very handy. Has had fine training by Swing.
B	Grey Dawn	5	1921. Officers horse. Big and plays well.
B	Queen	6	

			GOLD OAK (sire)
A'	No. 52		First lot of 18-1921. Very fast.
C	Mare	5	
	Irene		Plays. No good. Is a fool.
C	Madge		Big clumsy buckskin.
	Chestnut Oak		Very fast. Fool. Will never play. Came with Officer's horses 1921.
A'	Great Northern		1920 - Plays. Very fast.
A'	Golden Glow		Very fast. Fine disposition. Came with Officer's horses 1921.
	Reigner's horse		Officer's 1921. Very fast and fine looking. Stubborn.
A'			Why has Manna Loa 17 brand? Molokais, Harvesters, etc.
C	Molokai Mare		
B	No. 9	6	1920
A'			Big bay Brugelli. Very fine looking. 5 on right side of neck.
C	Molokai Mare	6	
	Big Bay		Good Brugelli
C	Polly		Molokai
C	Two cents		Molokai
A	Hawaii	6	Molokai. As fast as any pony in the lot will be a wonder.

Reproduction of a military polo horse-grading chart reflecting the popularity of Parker Ranch sires Eastertide, Frolic, and Gold Oak.

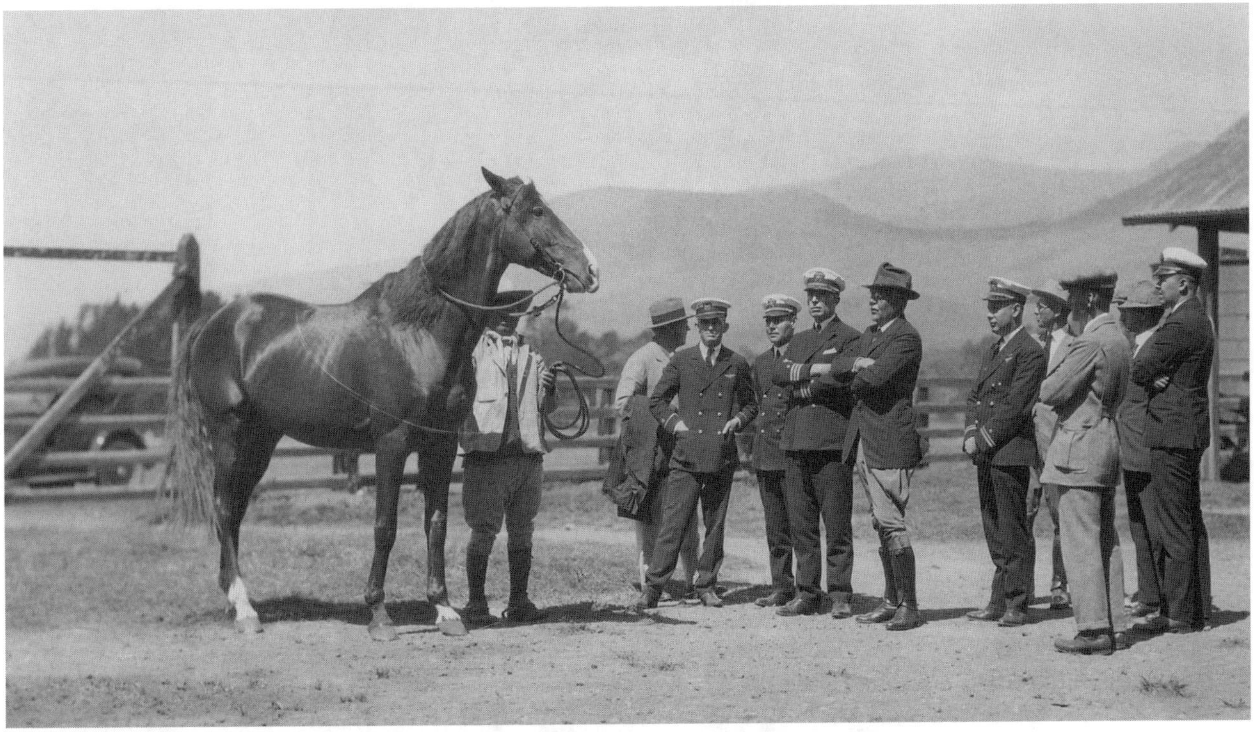

A sorrel cavalry prospect with U.S. Army Remount Grading Committee. A. W. Carter is in riding breeches with arms folded. A. W. Carter Collection.

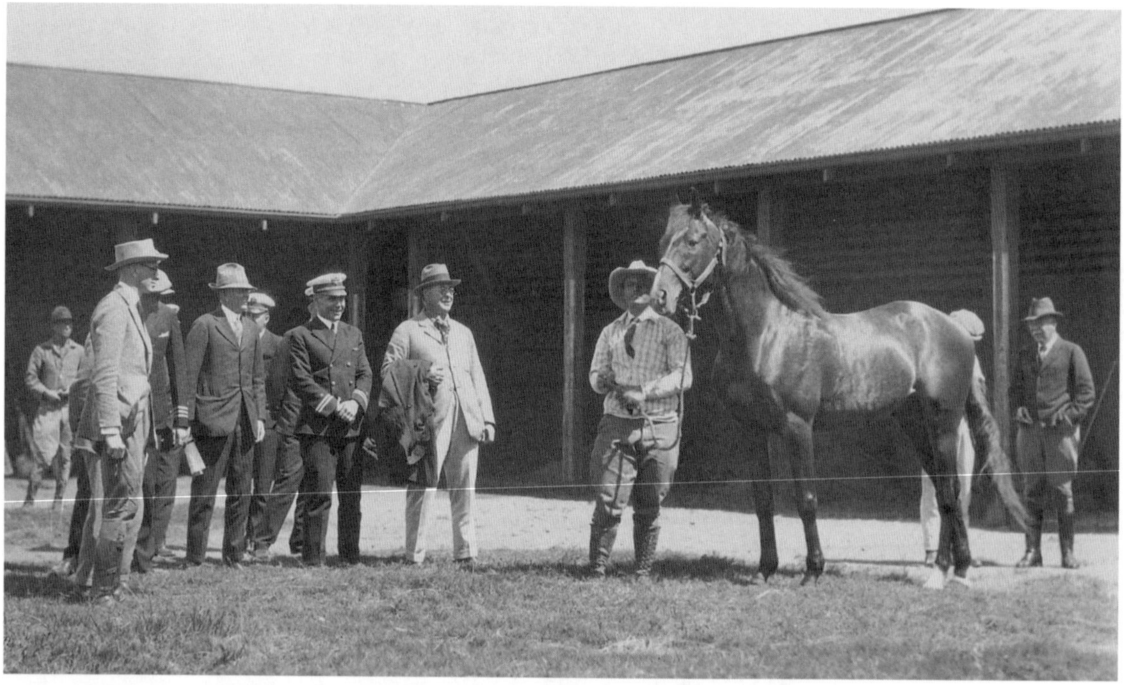

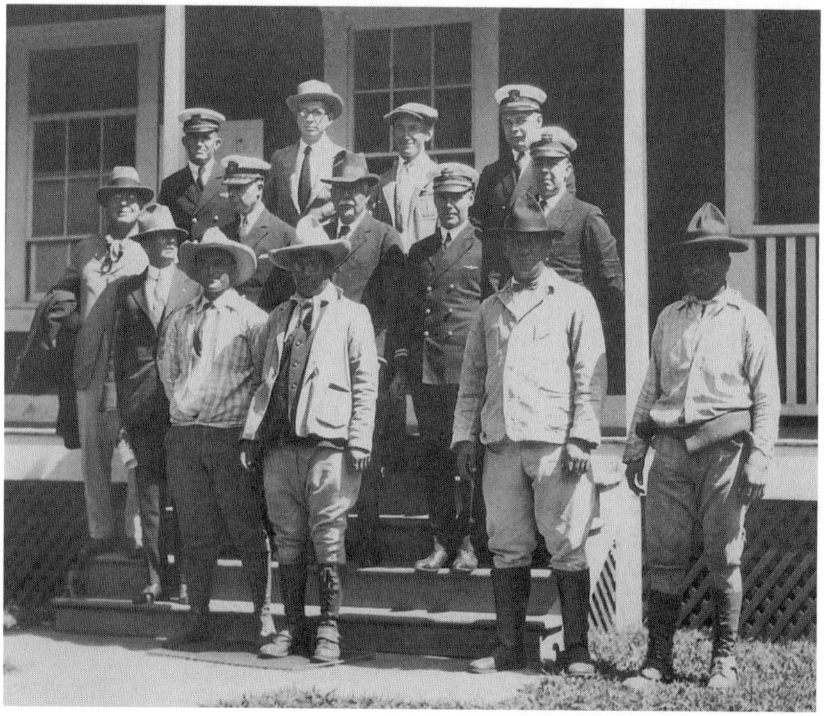

Above: *A bay cavalry prospect with Theodore Vredenburg at halter. A. W. Carter is at the rear of the horse. A. W. Carter Collection.*

Left: *Remount Committee with Parker Ranch personnel at ranch office. A. W. Carter Collection.*

are so efficiently handled that they are as gentle and affectionate as kittens.

The main interest of the Remount here in Hawaii is in this Parker Ranch, the Trustee of which is Mr. Alfred W. Carter, a keen and able horseman much interested in Thoroughbreds and in raising the standard of the 2,500 horses upon the Ranch, as is his son Mr. Hartwell Carter. The Ranch has 100 registered Thoroughbreds and about 200 grade Thoroughbred mares. It owns a Standard bred, a Morgan, a Palomino and two fine Thoroughbred stallions including Ormesby. Mr. Carter is now breeding the pick of his thoroughbred mares to him and Overall (1922) by Peter Pan out of Frillery by Broomstick.

The Standardbred stallion to which Herron referred was Heine, the Morgan mentioned was Querido, and the Palomino was Diablo, a three-quarter Thoroughbred. Another Palomino named Sappho followed from W. B. Mitchell and Sons of Marpha, Texas. Sappho was of the early quarter horse

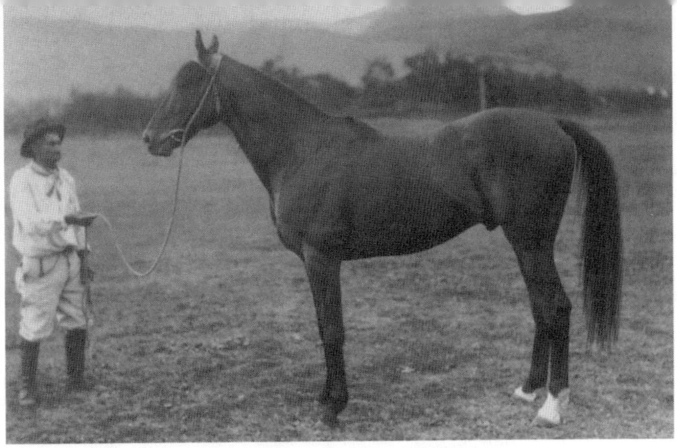

Gold Oak at Pukalani Stables, with Charles Duncan at halter. A. W. Carter Collection.

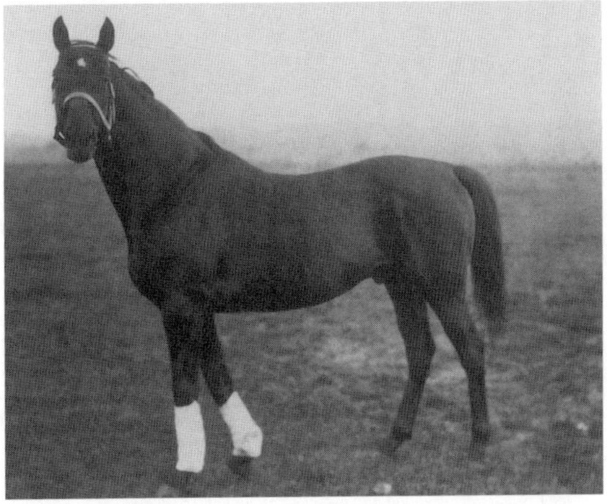

Famed Parker Ranch sire Warshot of remount breeding. A. W. Carter Collection.

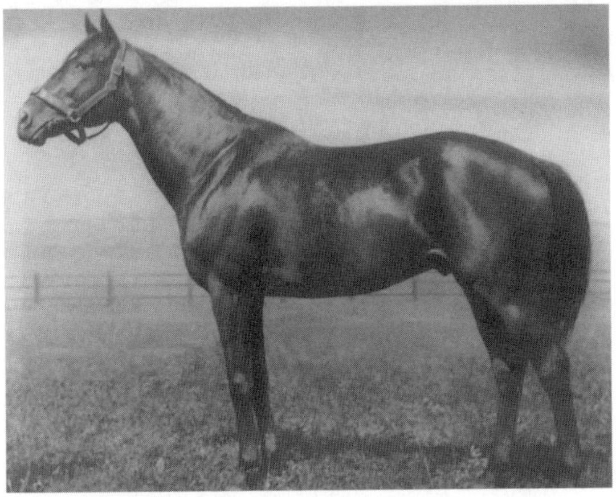

Famed Parker Ranch sire Overall of remount breeding. A. W. Carter Collection.

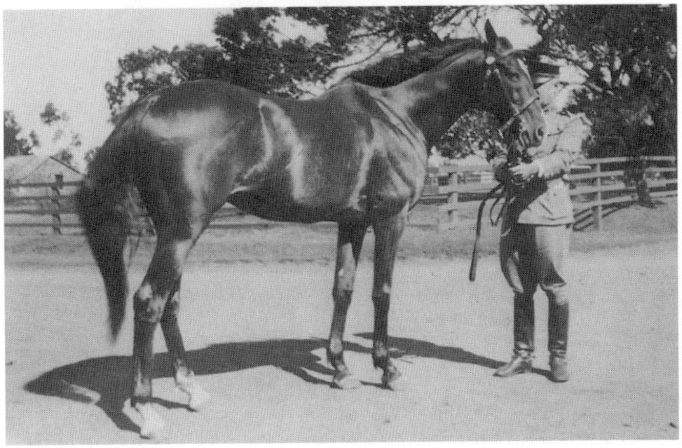

Gold Oak at Front Royal, Virginia. A. W. Carter Collection.

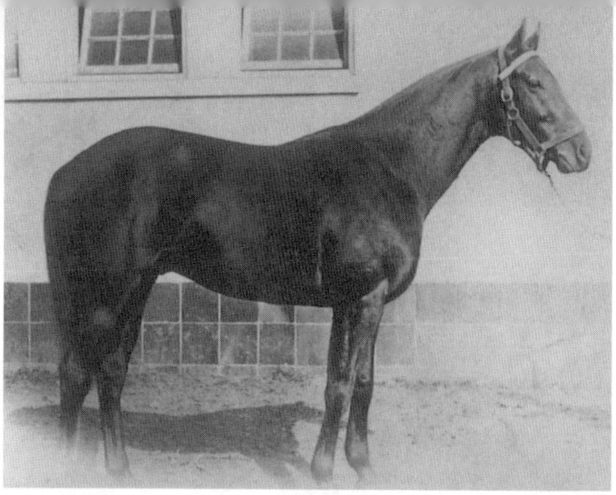

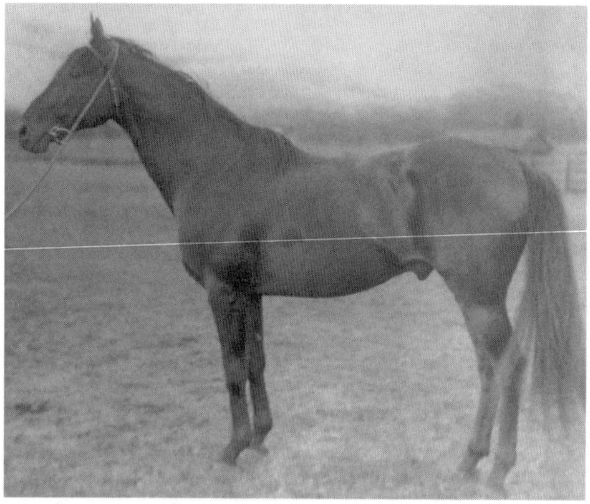

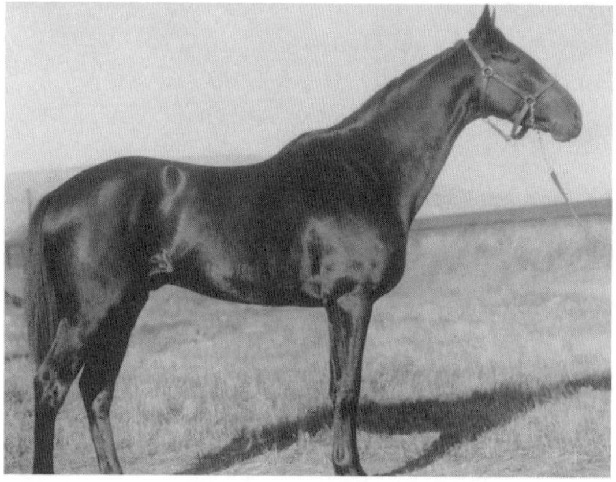

Top: *Famed Parker Ranch sire Rodgers of remount breeding. A. W. Carter Collection.*
Middle: *Famed Parker Ranch remount sire Harvester by Solitaire II out of Carnation II. A. W. Carter Collection.*
Bottom: *Famed Parker Ranch remount sire Ormesby, before leaving California for Parker Ranch. A. W. Carter Collection.*

vintage and conformation—and the first of this newly established breed to be imported into the Territory.

Continuing to strengthen the Thoroughbred broodmare bands, A. W. purchased a grey stallion in 1931 named Herodiones, sired by The Finn, which was a distinct outcross from the existing bloodlines. As in the case of Gold Oak and Rodgers, Herodiones got circulated among the ranches of the Big Island. The grey stallion matched Gold Oak for both good and bad qualities. It was never a shame to be bucked off the offspring of either stallion, but the horses had endless stamina when it came to cowboying—especially in the wild cattle operations that still existed after World War II. Despite the reputation of Herodiones' colts among the cowboys, Johnny Carroll, the noted race trainer for Parker Ranch, was partial to the grey stal-

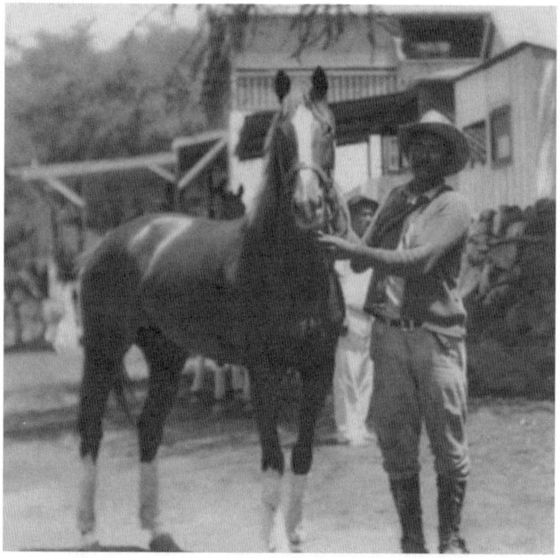

Parker Ranch horse foreman James Palaika at Kawaihae, with Niumalu at halter prior to shipment to California. Hartwell Carter Collection.

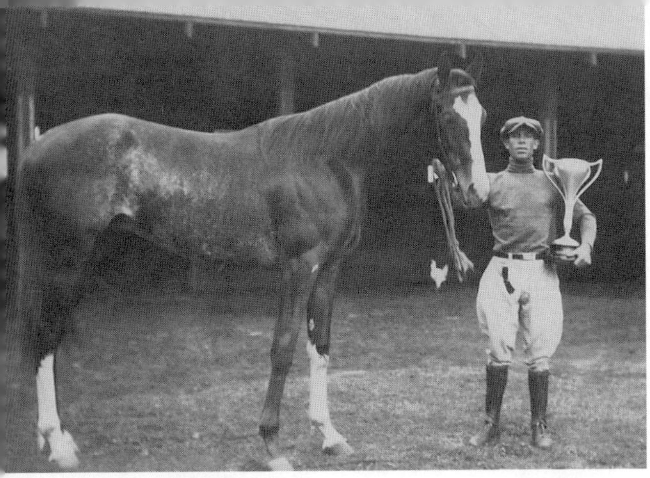

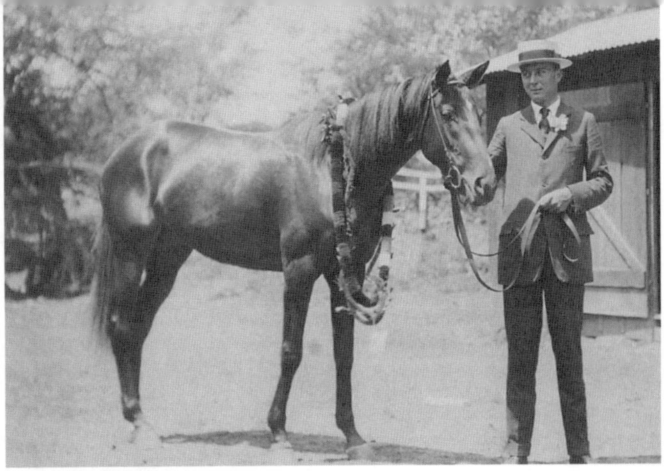

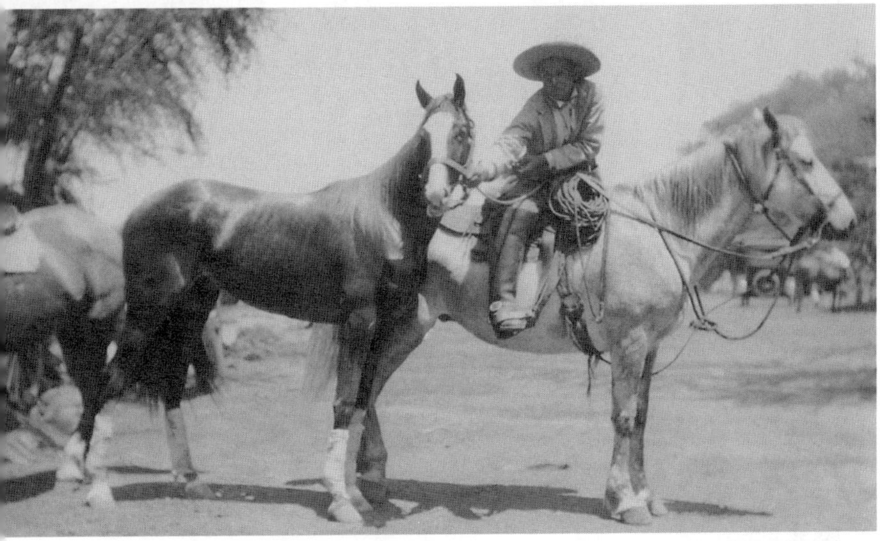

Top left: *Johnny Carroll, noted Parker Ranch Thoroughbred racing trainer, with Aukai at halter at Race Track Stables. Hartwell Carter Collection.*
Top Right: *Hartwell Carter with lei-bedecked Thoroughbred Kāpena at halter, preparing for California shipment from quarantine station in Honolulu. Hartwell Carter Collection.*
Bottom: *Hōkūao in the winner's circle. Hartwell Carter Collection.*
Left: *Charles N. Lindsey II with Akahiloa at halter, preparing for shipment out of Kawaihae. Hartwell Carter Collection.*

lion's offspring, with the familiar claim that mellow handling produced mellow horses.

In 1935, four Parker Ranch Thoroughbreds were shipped to California for training in stakes races. A California horse breeder, Neil McCarthy, received and trained Malia, Aukai, Akahiloa, and Niumalu. They were sold after successful races at Bay Meadows and Tanforan Park in 1936.

By this time, as a remount station, Parker Ranch outflanked the quality and productivity of other remount stations located in El Reno, Oklahoma, Fort Robinson, Nebraska, and Front Royal, Virginia. Major Reynolds made the following assessment:

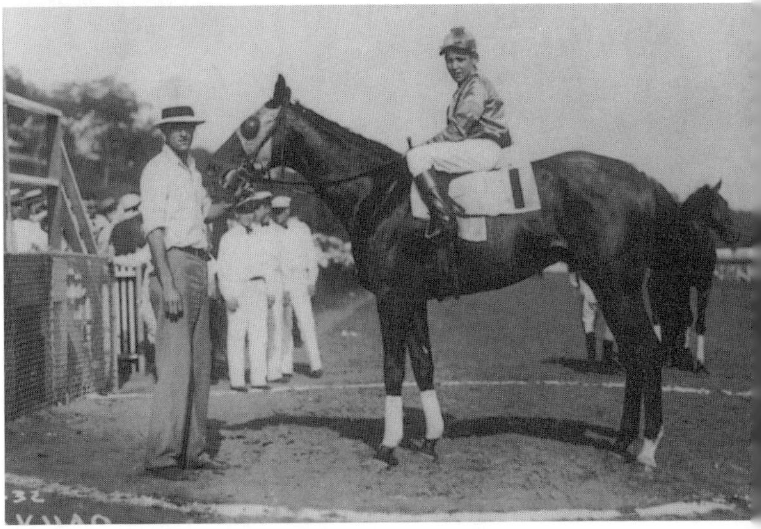

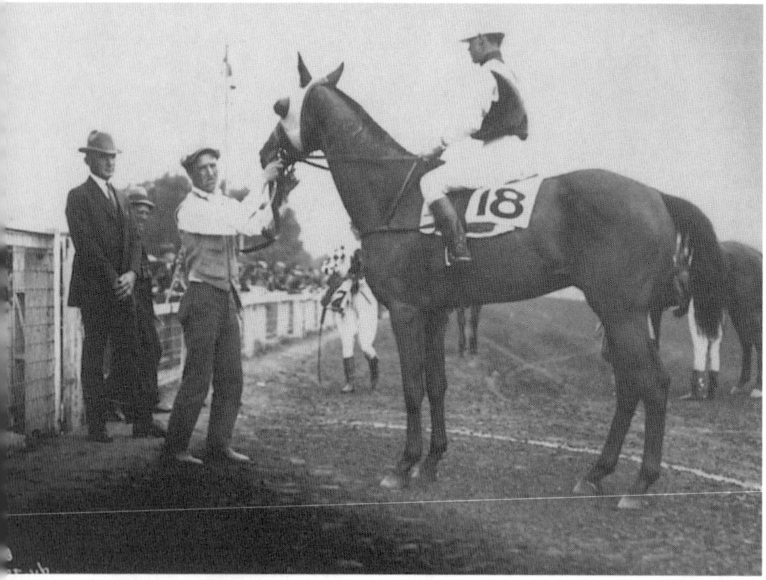

Above: *Kīlauea in the winner's circle at Tijuana. Hartwell Carter Collection.*
Below: *Kāpena in the winner's circle at Tanforan Park, San Francisco, with unidentified jockey and California owner/ Parker Ranch agent, C. A. Hartwell. Hartwell Carter Collection.*

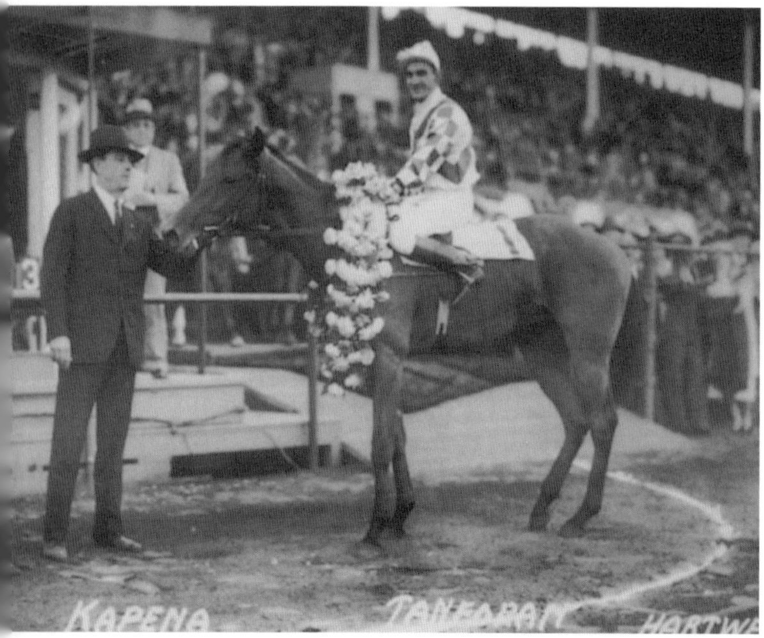

"Breeding operations at the [other] Remount Stations cannot begin to compare with those on the Parker Ranch. Mr. Carter has built up a very fine type of horse from range stock [since 1900]. His breeding has been very intelligent and I believe much better than any ranch breeding in the United

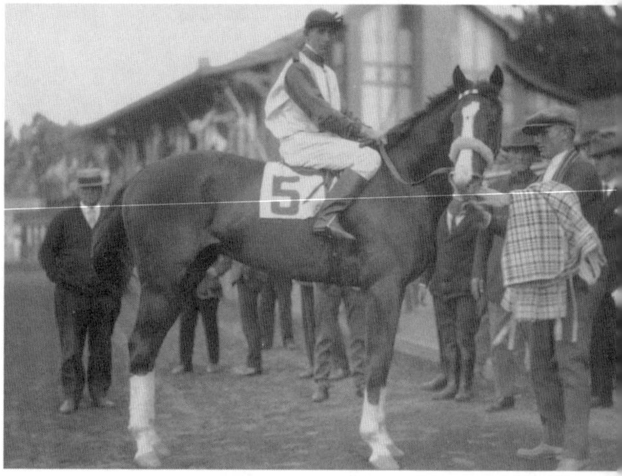

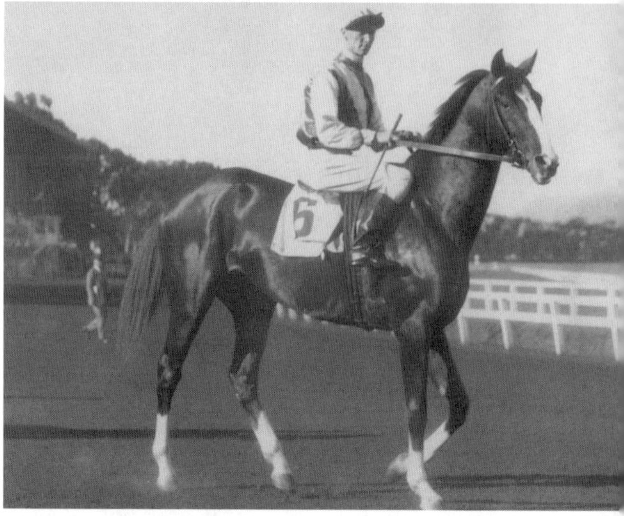

Top: *Gaiety saddled and ready for Tanforan. Hartwell Carter Collection.*
Bottom: *Gaiety entering the track at Tanforan Handicap Race. Hartwell Carter Collection.*

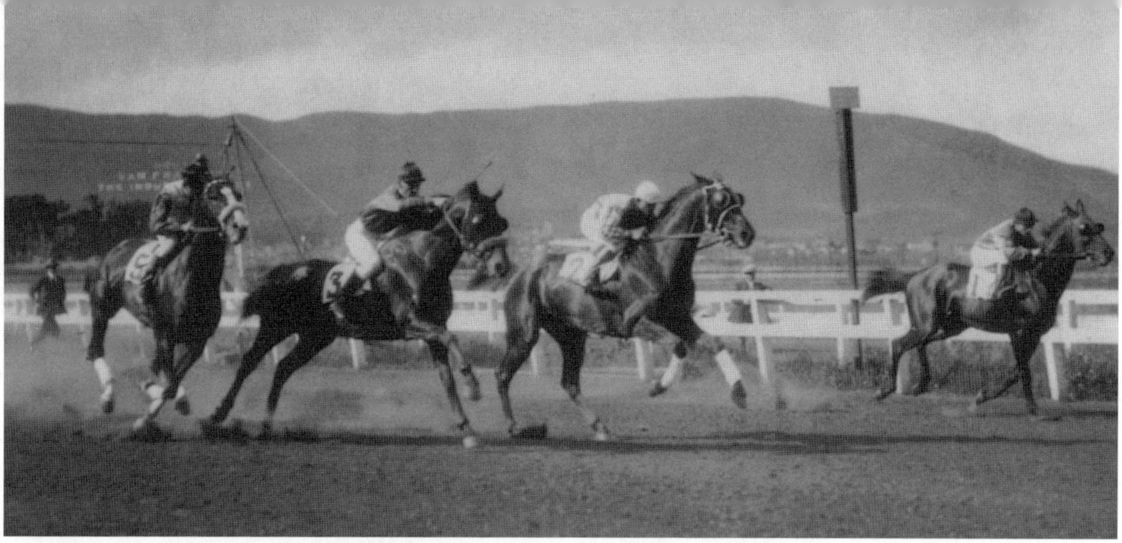

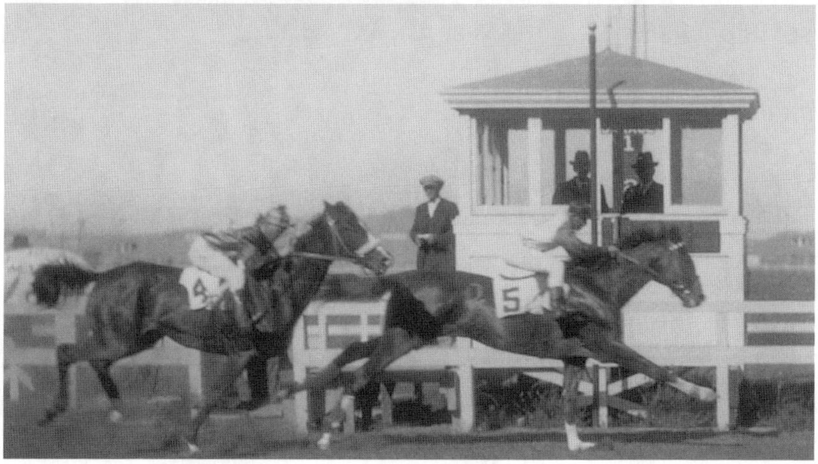

Top: *Gaiety coming from fourth place on the outside. Hartwell Carter Collection.*
Left: *Gaiety winning the Tanforan Park Handicap. Hartwell Carter Collection.*

States. As a matter of fact, he is as careful and more successful than many horse-breeding farms in the Middle West where Thoroughbred horses are raised."

Standardbred horses were also common, with Heine being one of the most prominent sires. His offspring were strong-endurance workhorses, but notably mean and broncky (also referred to as cold-backed).

A. W. had another close associate who was to become an illustrious figure as a hard-charging army general. George S. Patton, then a major, was stationed in Hawai'i, where he joined enthusiastically in local polo matches. His first contact with A. W. was a request to buy a personal mount, a hunter/jumper Thoroughbred. Patton also had considerable success as a trainer and took part in horse shows with his daughter Ruth Ellen in the saddle. Having Patton—clearly destined for greater things within the army—as a friend was an asset for A. W. and for the ranch, and Patton contributed significantly to the ranch's Thoroughbred and remount programs in the years immediately preceding the bombing of Pearl Harbor and the advent of World War II.

During the war years, Parker Ranch continued its active horse breeding directed toward remount production. In the postwar years, Hartwell directed the ranch toward a working cow horse–breeding program.

Mr. Carter never saw General Patton again—but in 1942, Patton wrote him from Morocco: "I often think of the happy times I had in Hawaii with you and your family and look forward to the day, which I think is distant, when I shall be able to return. Perhaps when I have finished this war, I may be able to."

Draft Horses and Mules: 1900–1935

Recognizing a demand by the sugar plantations for draft horses and mules, A. W. began in earnest to address this market. A good-quality work mule was valued at $250–$350 by the plantations. A pair of registered Percheron stallions, acquired earlier, when crossed on Thoroughbred mares, produced daughters that became broodmare prospects for the imported jacks; they yielded stout

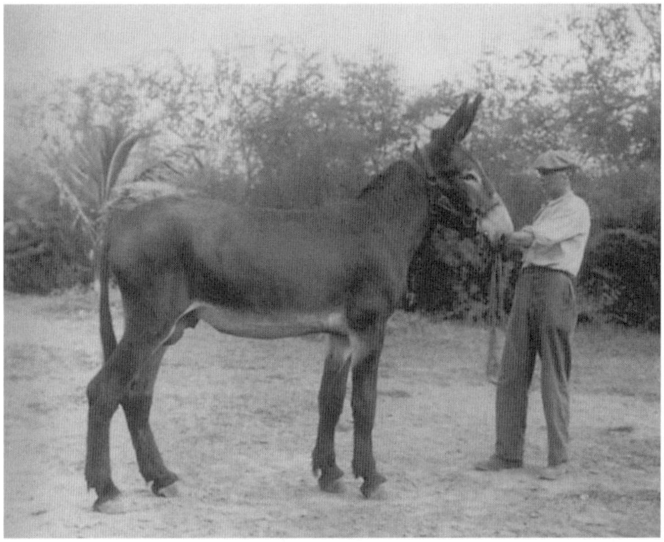

Imported Kentucky jack from Giltner Brothers. A. W. Carter Collection.

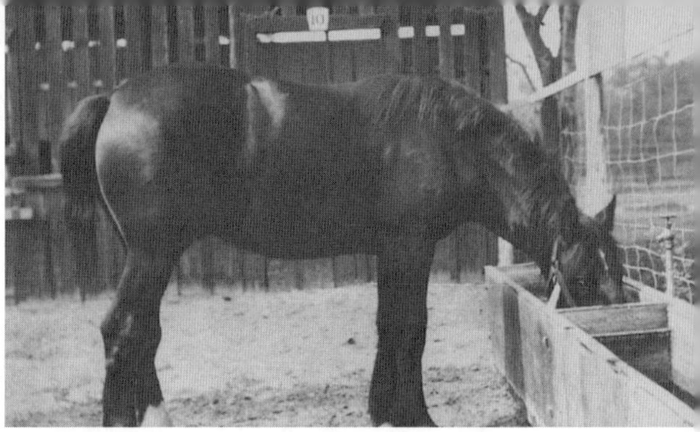

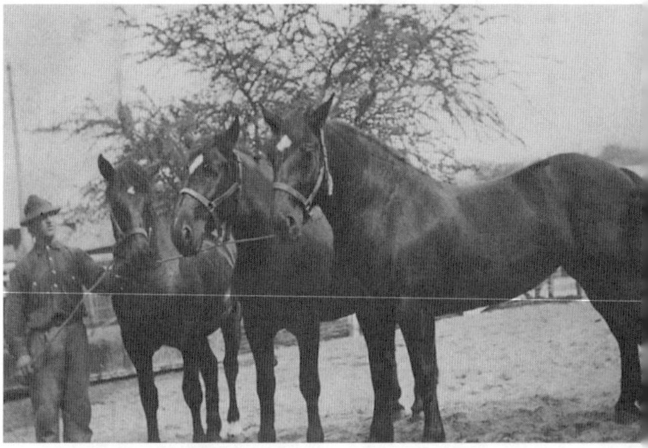

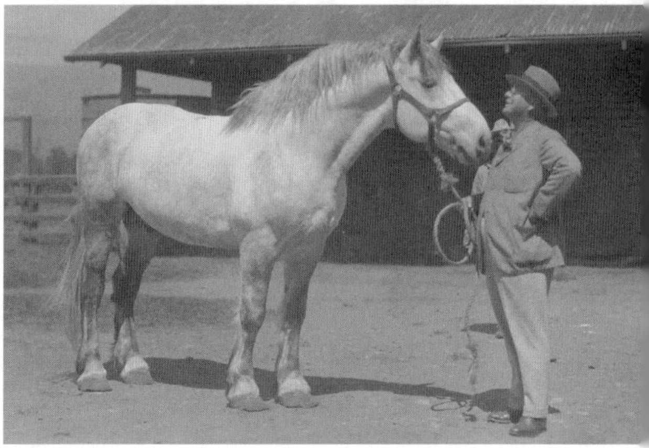

Top: *Imported three-year-old American Percheron stallion in Honolulu quarantine. A. W. Carter Collection.*
Middle: *Imported three-year-old American Percheron fillies at Honolulu Quarantine Station en route home to Parker Ranch. A. W. Carter Collection.*
Bottom: *Imported American Percheron Bijou Blanco at Pukalani Stables. A. W. Carter Collection.*

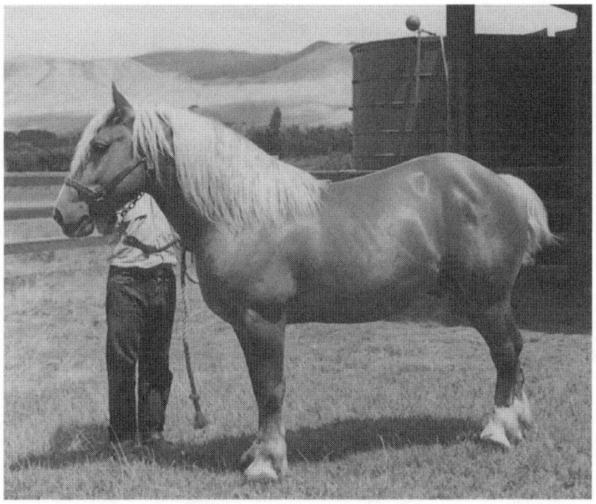

Flaxen-maned palomino stallion, an imported Belgian stud for crossing on Thoroughbred mares. Hartwell Carter Collection.

mule offspring for the plantations. Over the next thirty-five years, the ranch imported twenty-three jacks and more draft stallions, such as Bijou Blanco, as well as German coach and French coach studs (coach horses were light draft horses used for pulling a passenger wagon such as a stagecoach or buckboard).

Direct crosses of grade Thoroughbred mares also produced light mules. Some of the draft mules were retained for drayage purposes on the ranch as well. Just as was done in promoting polo horse sales by preliminary training, Parker Ranch provided added value to mules sold, in that a special stable called Pieper Stable (renamed Purdy

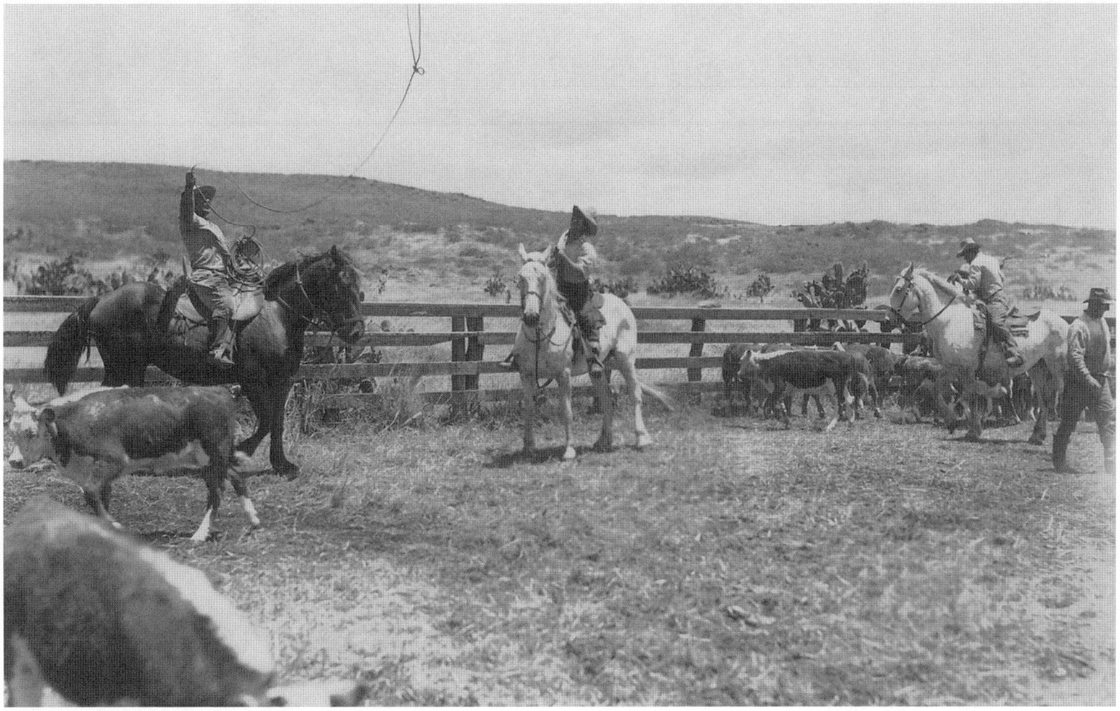

Half Percheron cow horses at a branding at Pu'upapapa Corral, Keamuku. John Purdy is at left on the black horse, Willie Palaika on the grey at center, and Alex Akau on the grey at far right. Fred Napier Maertens is walking at far right. Hartwell Carter Collection.

ADDRESS ALL
COMMUNICATIONS
TO THE MANAGER

ALEXANDER YOUNG HOTEL
MOANA HOTEL
SEASIDE HOTEL

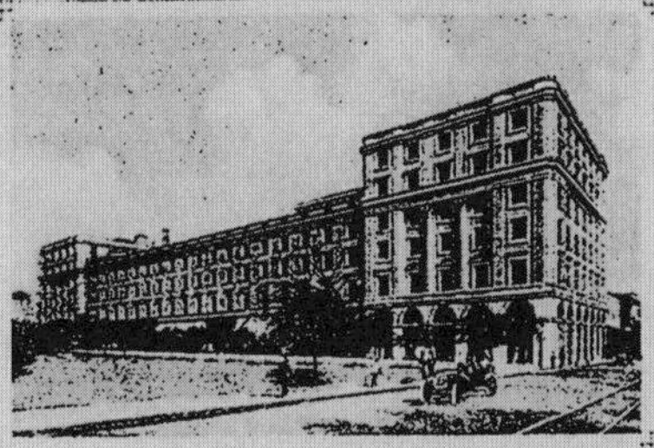

Alexander Young Hotel

CABLE ADDRESS
"YOUNGS"

HONOLULU,
HAWAII.

March 5th. 1923.

Mr. Alfred W. Carter,
Kamuela, Hawaii, T. H.

Dear Sir:

Your kind letter of recent date to hand,

In reference to the Jack which we are having Mr. Monsarratt ship to you. He decided that he wanted the last Jack which we had shipped to him for the reason that it was the smaller animal and he thought it would do better turned out on his rough range than the larger one. After looking the two Jacks over again on my way back to Hilo I decided that the Jack that we are shipping to you was the better of the two. I feel sure that you will like this Jack and if he does not prove absolutely all right you can let the boys knock him in the head. In other words I want you to be satisfied and to feel that he is all right before you pay us a cent for him.

I certainly appreciate your kindness in offering to let us ship the Jack to your place and I hope that your man will use a little patience in giving him a trial before condemning him.

My wife and I are still enthusiastic over our visit to your Island and the kindness and hospitality that was shown us, especially at the Parker Ranch. She joins in sending best wishes.

Very sincerely yours.

R. R. Giltner

Letter from R. R. Giltner, noted Kentucky bloodstock supplier, including a strong desire to please Parker Ranch at all costs.

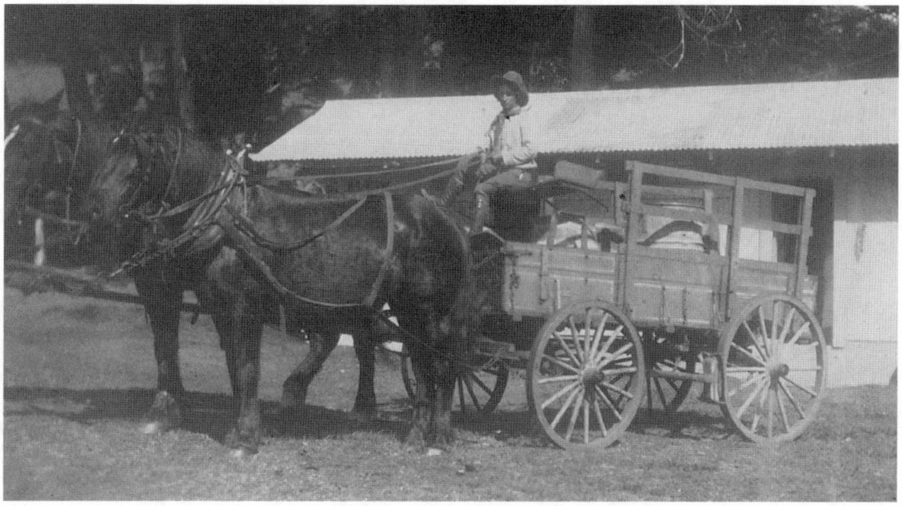

Half Percheron horses used for drayage. Teamster Louis Rodriquez. Hartwell Carter Collection.

Stable in later years) became the venue for the training of draft animals. Other draft horses also were trained at these stables, located behind the present-day Ace Hardware site across the street from the historic Spencer home in Waimea. A large, circular, abandoned water trough can be found in the brush where the stables once stood. Some of the great Parker Ranch men came from the stable, including Henry Pieper, a German teamster and father of the late Johnny Pieper, who moved on to become the patriarch foreman of Ka'alu'alu and then Hawaiian Ranch. Another was a gregarious Irishman named Tom Gallagher who sired a son named Thomas Bell. Young Thomas grew to become a prominent horseman and the eventual foreman of Pu'u'ō'ō, W. H. Shipman's ranch, moving later to Kapāpala Ranch, then owned by C. Brewer.

Another teamster/trainer was Robert Victor Kailianu, a native of Kokoiki in Kohala. He was skilled with draft horses and mules and was known for his immense patience. Kailianu had two beautiful daughters: Thelma (Mrs. George Mahalopiha) Lindsey and Mary (Mrs. David Kepa) Bell. The teamsters played a strong role in the early development of the Parker Ranch horse program. Several of note in the period around 1920 from Waiki'i were Hulihia Lewi, Kanaloa Ho'okano, and Sam Puna Kauwe; the latter eventually became a saddle maker. In the Waimea area were Sam Kaimuloa, Kapuniai, Henry Purdy, and Kane Brown. Brown and Kaimuloa became truck drivers when mechanization of the ranch began around 1930.

The beginning of mechanization and the end of World War II signaled the end of the draft horse and mule industry in Hawai'i, as it did throughout America. While the U.S. Army continued the horse cavalry until 1949, the need for remount horses declined. With A. W. aging by the late 1940s, his son began to quietly surface as the architect of a new type of cow horse program beyond crossing the Thoroughbreds and Morgans. As the last of the draft animals were sold off the ranch, Hartwell

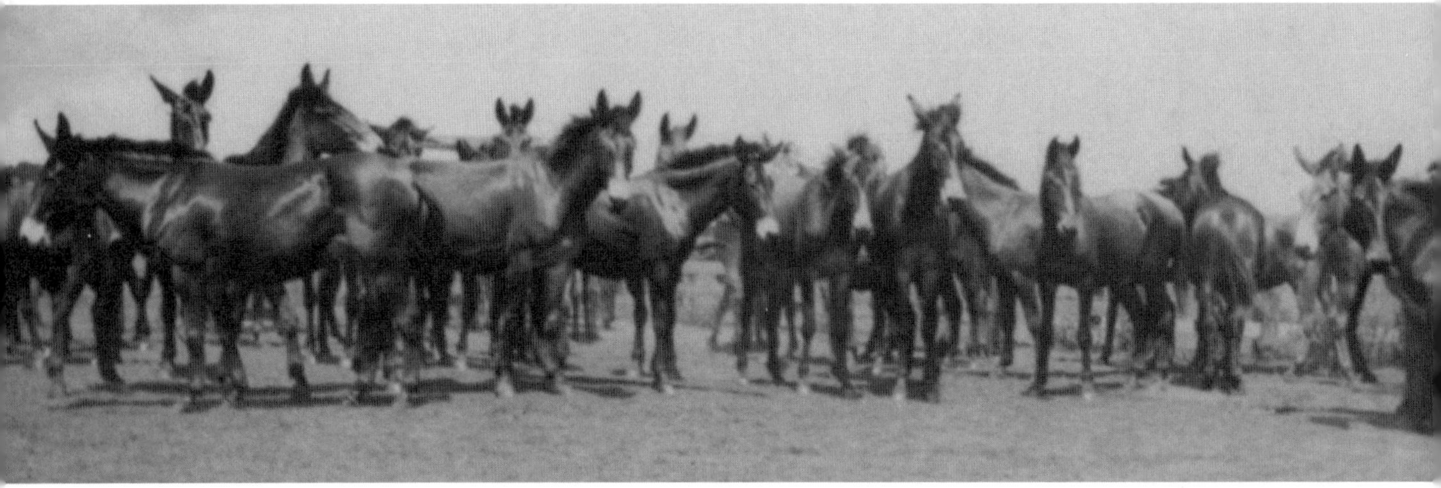

Four-year-old Parker Ranch draft mules at training time. Hartwell Carter Collection.

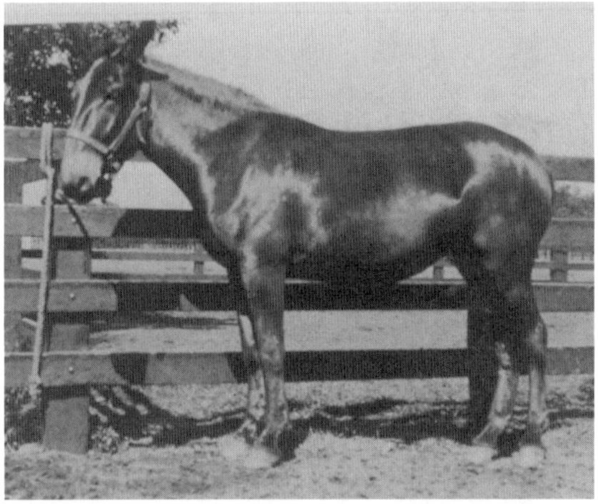

Typical Parker Ranch draft mule, illustrating his sire relative to the Pukalani Stable fence. This mule was 16 hands tall and weighed 1,300 pounds. Lyman House Memorial Museum 1379 #3.

Carter actively sought superior sires of the quarter horse type on the mainland, signaling a new direction in the horse program. Hartwell relied on the early foundation breeders of the new and fledgling breed, the quarter horse. While speed on an oval track was the single measure of Thoroughbred horse success, the quarter horse needed to excel in multiple functions: roping, cutting, short-distance racing, and showing. Hartwell saw the transitions coming and prepared the ranch for a different kind of horse.

Early Development of a Sheep Ranch: 1900–1949

Sheep were first introduced to the Hawaiian Islands in 1793 when Vancouver delivered a ram and two ewes to Kealakekua. The following year, the sea captain repeated the gesture by presenting Kamehameha I with five rams and five ewes from California.

Sheep on the Big Island proliferated in the higher, cooler elevations, in keeping with their metabolism. Some were retained as small farm flocks, but others became feral. The year 1822 found feral sheep on Mauna Kea. By 1935 there were at least forty thousand in the upper reaches of Mauna Kea; in the following year, eradication drives lowered the numbers by six thousand head.

The commercial sheep industry on the Big Island centered on Waimea. Around 1845, William French imported two merino rams for his native flock. John Palmer Parker had already developed a flock of his own by that time. French, at his ranch at Līhu'e near Waimea, was exporting wool on a small scale.

By 1853, French imported Saxon sheep and in 1858 Spencer and Louzada brought in six Saxon-merino crossed sheep from Germany, the first in history from that country. In 1860, John Palmer Parker sold 192 wethers for $550 to the Waimea Grazing and Agricultural Company. In 1862, the emperor of France, Napoleon IV, presented four Rambouillet rams from the Royal Flock to the king of Hawai'i, and Samuel Spencer of Waimea received these.

In the ensuing years, WGAC continued to diversify and expand their operation, including the sheep flocks. In addition to the Līhu'e headquarters, they eventually developed sheep stations at Kalai'ehā, Lahohinu, and Keanakolu.

A. W., in the meantime, received some early exposure to sheep ranching in his Moloka'i Ranch venture in 1898, prior to his trusteeship of Parker Ranch. Once on Parker Ranch he acquired two sheep operations as part of strategic land purchases: Pu'uloa Sheep and Stock Company and Humu'ula Sheep Station. The Pu'uloa outfit included the Keamuku Sheep Station. This was purchased from the MacFarlane interests and included 4,500 ewes, 800 lambs, 300 wethers, and an undisclosed number of rams of reasonable breed—mainly merino

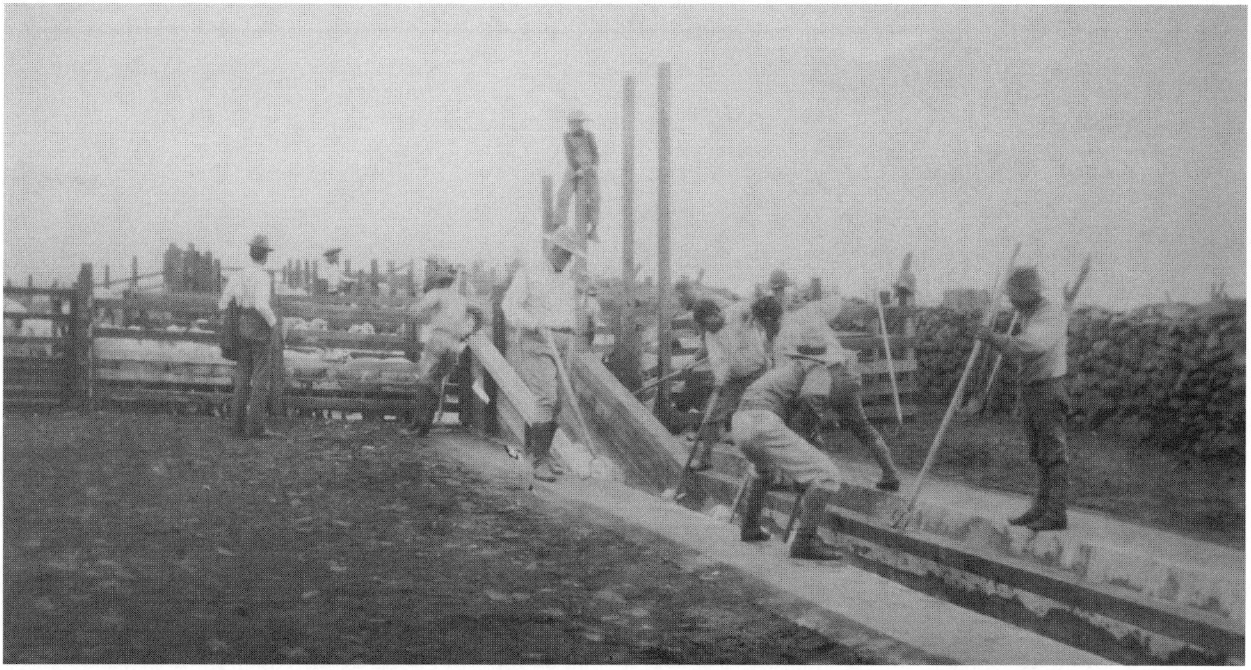

Dipping sheep at the Keamuku Sheep Station. 1906. A. W. Carter Collection.

crosses. Also undisclosed was the exact number of semiwild range sheep that grazed in adjoining Kālawamauna, but estimates were between 2,000 and 2,500. At the time of purchase there were substantial improvements, including a dwelling, men's quarters, and outbuildings that housed scales and equipment for shearing sheep and packing wool. Pens, corrals, and yards were left intact also. All told, A. W. acquired nearly 8,000 head of sheep, together with station improvements and equipment—but the centerpiece was the Līhu'e *kuleana* (property) itself adjacent to Waimea, followed by the Keamuku lands, which included Kālawamauna.

Sheep Operation Centralizes at Humu'ula

In the next decade, A. W. gradually converted Keamuku/Kālawamauna to cattle ranching and the Līhu'e lands to better use, while the sheep operation retained its status quo until Humu'ula was acquired in 1914. Shortly thereafter, Willie Kaniho moved his family from Keamuku Sheep Station to Kalai'ehā, the headquarters of Humu'ula Sheep Station, where he launched a long and colorful career as a Parker Ranch cowboy, foreman, and superintendent.

Humu'ula Sheep Station

In early Hawai'i, adze-making and bird-catching activities were common at Humu-'ula and Ka'ohe. This area was traversed by cross-island traffic of Native Hawaiian people, including warriors. The *kaula* (prophets) practiced communication with the gods in those areas above 4,500 feet. With the advent of the nineteenth century, traditional native uses of the area declined as sandalwood harvesting and wild cattle hunting became more commonplace. By 1834, two *malihini* (newcomers) cattle hunters were camped at Humu'ula butchering and processing beef (drying or *pipi kaula'i*).

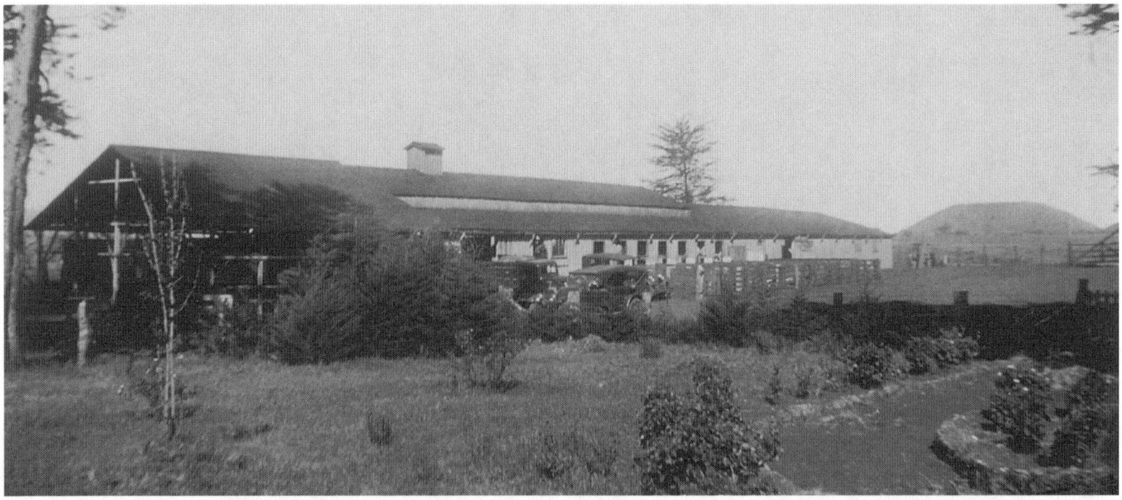

Humu'ula Sheep Station Company's shearing shed at the time of its acquistion by Parker Ranch. Pu'u Omaokoili is the hill in the background. A. W. Carter Collection.

They were James Castle and Miles G. A. Simmons at Keanakolu. Ned Gurney, who operated out of Lahohinu, paralleled them in the north.

By 1840 there were two native bullock hunters—Waiki'i Li'ili'i and Hoakimoa—also operating at Humu'ula (the latter hunter's name suggests the possible Hawaiianization of the biblical name "Joachim" or the Hispanic name "Joaquin"). Fifteen years later, the lands of Humu'ula were leased.

The original fifty-year government lease issued in 1856 was for 158,000 acres to WGAC. This was an *ahupua'a* (land division) once described as reaching from the ocean to the peak of Mauna Kea. Shortly thereafter, Francis Spencer bought a half interest in this venture. Over the years, lava flows (1935) and government appropriations for roadways, forest reserves, and military usage reduced it to about 50,000 acres, the bulk of which is administered at present by the Department of Hawaiian Home Lands of the State of Hawai'i.

About 1870 Frank Spencer and his son Ashford spun off the lease as independent contractors, further developing the three distinct sheep stations: Kalai'ehā (headquarters), Keanakolu, and Lahohinu, running about 25,000 sheep, 500 cattle, and 50 horses.

In 1876, James Gay bought the operation and incorporated it into the Humu'ula Sheep Station Company (HSSC) by 1883. A thousand shares were issued at $100 per share, with four hundred shares going to himself and four hundred to Conrad Menke, a German national. Still struggling for survival, in 1885 the company was mortgaged to a German businessman named Hackfeld, who later developed American Factors, Ltd. This became an American acquisition during World War I. F. W. Glade, president, and J. F. Hackfeld, treasurer, signed for HSSC. Both were German nationals. The Haneberg brothers, Armin and August, sugar planters from Kaua'i who originally emigrated from Germany, ably managed the Hackfeld holdings and by 1893 had become owners of HSSC.

In the meantime, the German consortium acquired the lease of Ka'ohe Four and Five, consisting of 137,200 acres. Parker Ranch had since 1876 held the lease on Ka'ohe Three, which was adjacent to Waiki'i, while the Hanebergs had already secured Ka'ohe Four and Five, adjacent to Humu'ula. In addition to sheep ranching, the Hanebergs were actively engaged in horse breeding and training. Some of the stallions used were Johnny Morgan (from Francis Spencer), Danemark, King Thomas (an Australian Thoroughbred), Willy (from William Rice of Kaua'i), Thomas Spreeden, Swan, and Makapū. They also had progeny from a pair of captured feral stallions, 'Āinahou and Laumai'a, named after the places from which they came.

The Hanebergs were meticulous in brand identification of each ranch horse and used a logbook to include descriptions, gender, age, and ability.

The German brothers were industrious and innovative. They built the Haneberg House at Humu'ula, which is still in evidence, along with an office and sheep shearing shed. They developed water for the Ka'ohe lands by tapping the spring of Houpo-o-kane, high above Pōhakuloa.

Between 1893 and 1895, many miles of stone walls were built by several Japanese

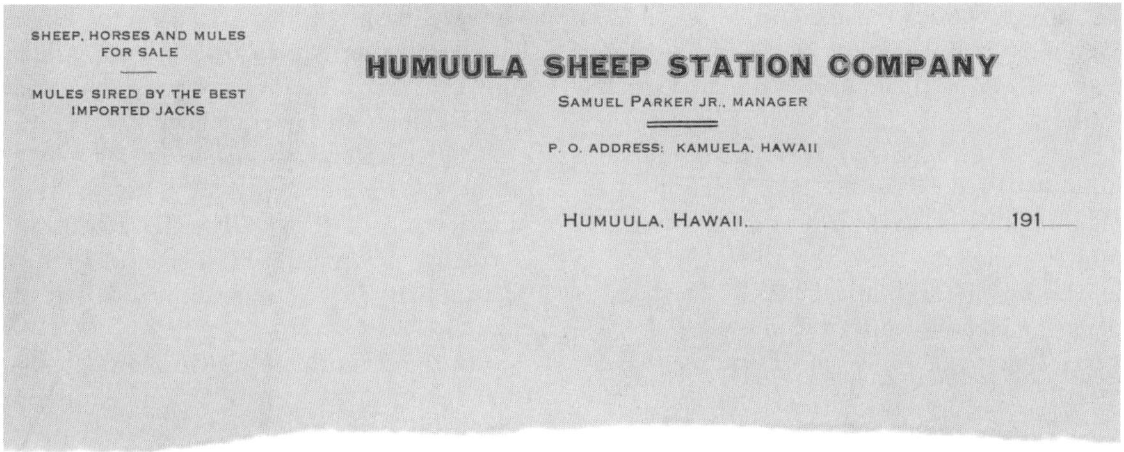

SHEEP, HORSES AND MULES
FOR SALE

MULES SIRED BY THE BEST
IMPORTED JACKS

HUMUULA SHEEP STATION COMPANY

SAMUEL PARKER JR., MANAGER

P. O. ADDRESS: KAMUELA, HAWAII

HUMUULA, HAWAII............................191___

Humu'ula Sheep Station Company letterhead.

laborers who camped at the Humu'ula headquarters. The fence along the south boundary, the *kīpuka* (oasis within a lava bed) of 'Āina hou near Pu'u Huluhulu, is still intact in portions along the Saddle Road. Parts of the wall were buried in the 1935 lava flow. The north wall can still be seen meandering steeply up the Mauna Kea slope from the Humu'ula-Mānā Road near Pu'u'ō'ō. The Haneberg brothers retired to Kaua'i, then to O'ahu.

On May 10, 1900, Samuel Parker Sr. purchased the holdings of HSSC for $75,000, creating contiguous ownership of the ranch lands all the way north to Keanakolu and his fee simple Waipunalei acreage. The Waipunalei parcel had been previously acquired in a land exchange with R. R. Hind, who gave Parker Ranch rights to this plantation parcel as he acquired the potential sugar lands of Puakō in fee. The HSSC transaction included shares of Metropolitan Meat Market in addition to the cash payment.

In 1902, when his son Samuel Parker Jr. was named manager, the total sheep flock of primarily merino breeding still numbered around 25,000. Horse numbers steadily increased, given his father's love for fine horses.

By this time there were as many as twenty residents of Humu'ula working as ranch hands and shepherds. Some of these men were carryover ranch hands from the Haneberg era. Among those employed at Humu'ula were Ernest Campbell, Johnny Norton, John Lindsey (Keoni Li'ili'i), Alfred P. Deverill, Robert Stevens, Johnny Makapa'a, Antone Fernandez, August Gramberg, Heine Meyer, and Anton Waltjen. The Japanese contingent adapted well to the tedious sheep care, including shearing and dipping under the supervision of Sam Jr. and his brother James. At this point the operation developed five distinct substations, adding Pu'u'ō'ō and Hopuwai to the original three developed by the Spencers.

In 1907, Sam Sr. was so overwhelmed by the mortgage debt to T. H. Davies and Co. that the Humu'ula property was transferred to the company in trust. E. H. Wodehouse was involved in this process. Upon relief of the debt, the property passed on to his son, Sam Jr. The senior Sam Parker in the meantime suffered a severe stroke in 1913 and was relegated to a wheelchair in his twilight years.

During Sam Jr.'s tenure in 1911, one of his cowboys named Ma'e Holi fell *(hinapū)* with his horse in Laumai'a and died. This left his family of seventeen children fatherless. One of his sons, John Holi Ma'e, who was twelve years of age at the time of his father's death, eventually worked for Parker Ranch at Humu'ula in 1919. In later years, Holi played a key role in introducing me to the *paniolo* culture and nurtured my love and appreciation for the *paniolo* lifestyle. He was also a significant influence on my life's work in veterinary medicine. John Holi Ma'e eventually returned to Parker Ranch in 1954 from Kūkai'au Ranch and remained there until his retirement.

At one point, Sam Jr. became disenchanted with the mountain ranch. Word circulated that he had offered the outfit back to T. H. Davies for $110,000 but was turned down. The same offer was made to the Baldwin family of Maui. Surreptitiously, E. H. Wodehouse, the Davies' representative, together with Henry Holmes and Robert L. Hind, approached A. W. to form a consortium *(hui)* to buy the outfit. Their position, however, was that the station was worth no more than $85,000. But A. W. had already secured an anonymous agent who approached Sam Jr. directly. A. W. was indignant about the Wodehouse scheme, because he felt it was inappropriate for a trustee of Thelma Parker to not act purely in her behalf. He made it clear to Wodehouse that Parker Ranch should secure the station directly, in the best interest of Thelma Parker. Meanwhile, Sam's offer to the Baldwin family was not taken under serious consideration, so he turned to A. W.'s agent and accepted that offer. A deal was struck in 1914 for $111,000.

Parker Ranch was now able to consolidate its sheep operation at the Humu'ula Sheep Station under the immediate supervision of Ernest "Haineki" Campbell. Campbell had worked at the Humu'ula outfit under both Haneberg and Sam Parker. Humu'ula leadership eventually went to Willie Kaniho after Ikua Purdy's interim term as foreman. A. W. himself took original inventory of all of the livestock and improvements. He immediately called for importation of superior ram stock.

The 23,000-head flock had been reasonably maintained, aside from an infestation of sheep scab. Sheep scab is highly contagious mange caused by a skin parasite, a mite named *Psoroptes ovis.* It is characterized by severe itching and inflammation and causes wool to get "taggey" and fall from the skin, leaving a dry crust that harbors the mites. Lime and sulfur dipping was the early treatment. Dipping is a two-minute total body immersion, with each sheep ducked head under at least once. The dipping vat was at the end of a chute, forcing sheep into the solution for a short swim. Parker Ranch later used a carbolic dip for this purpose.

The more than five hundred head of cattle were sorted for uniformity, age, and sex, then driven back to appropriate sections of the main ranch for disposition. The workhorses had been loosely managed and maintained, so they also were sorted for

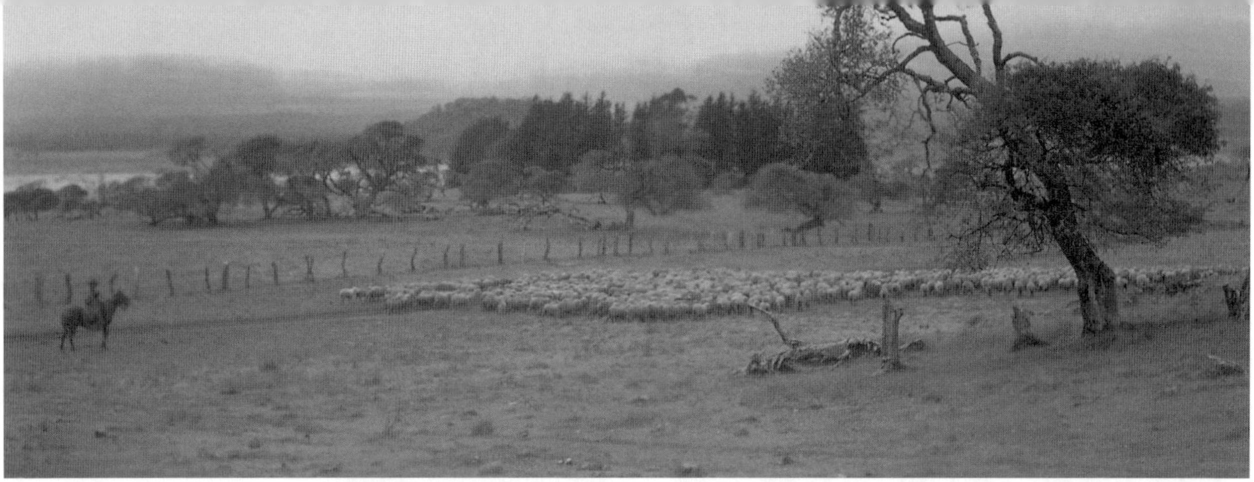

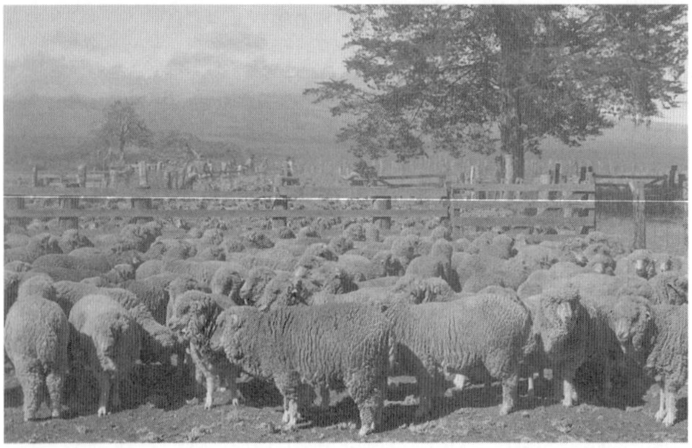

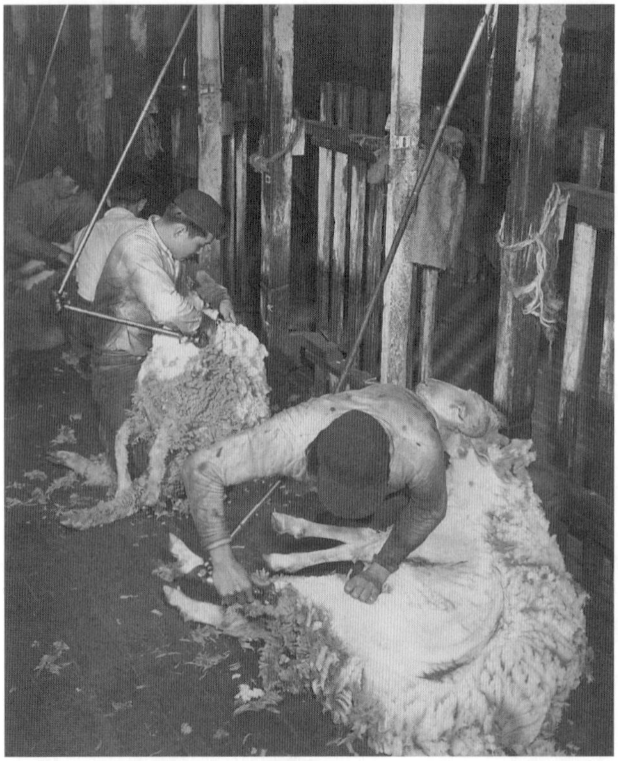

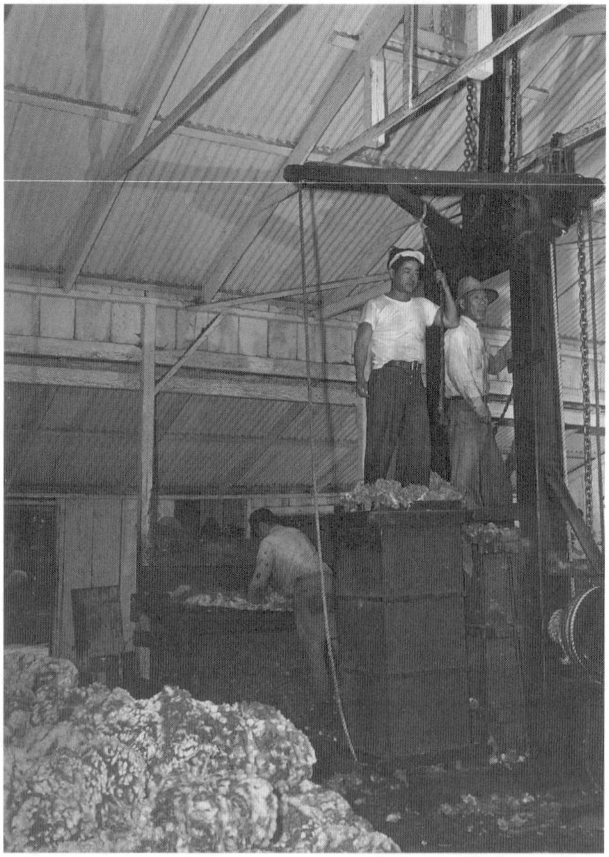

Top: *Gathering sheep near the Kalai'eha headquarters of Humu'ula Sheep Station Company. Hartwell Carter Collection.*
Middle left: *Corralled sheep ready for shearing. Pu'uhuluhulu, at the base of Mauna Loa, is in the background. Hartwell Carter Collection.*
Bottom left: *Shearing sheep of their wool. Ichiro Yamaguchi is at center. Hartwell Carter Collection.*
Above: *Baling wool with a wool press and the helpful weight of two sheep shearers. Hartwell Carter Collection.*

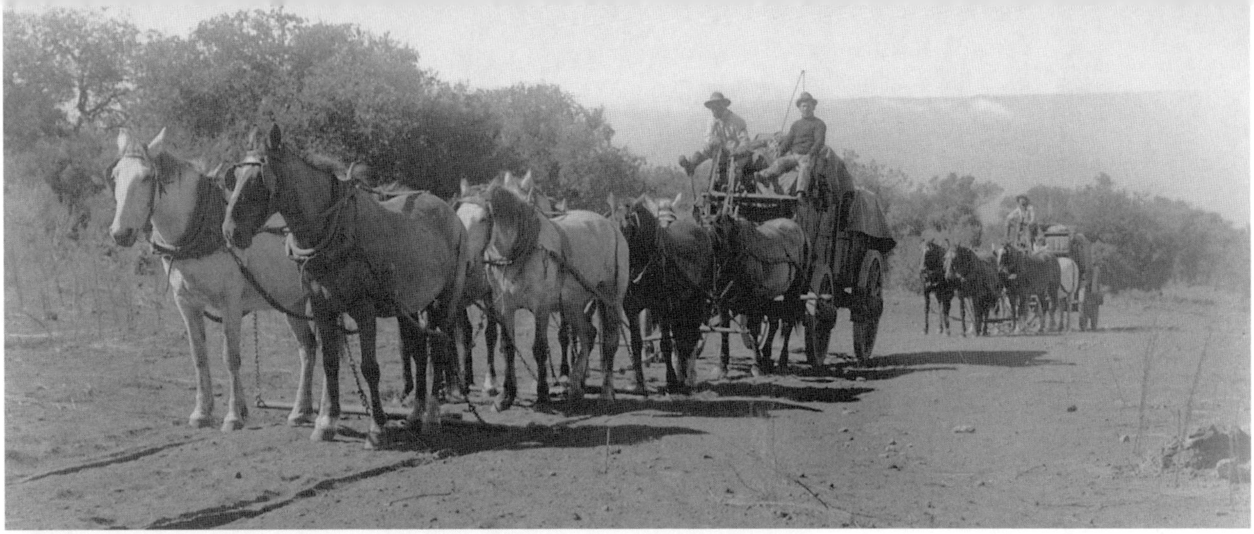

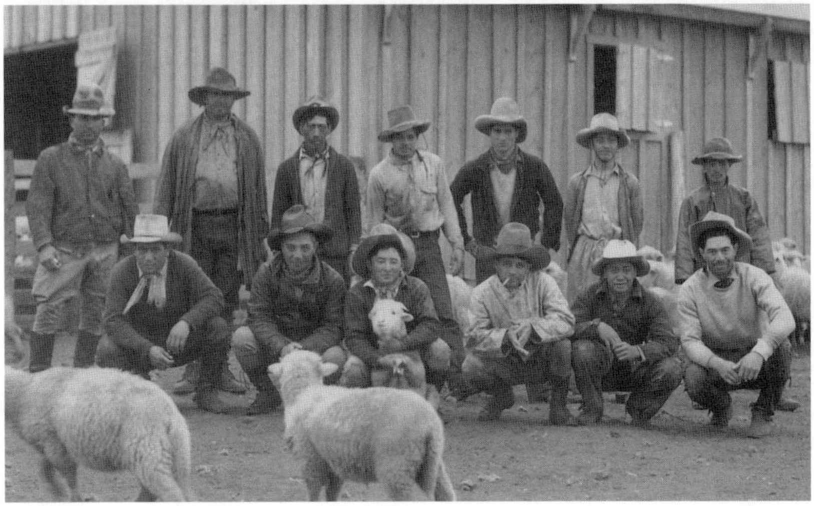

Top: *Percheron draft horses pull two wool wagons through Pōhakuloa on their way to Kawaihae. The teamsters are Kane Brown (rear) and Hulihia Lewi (foot on brake). 1910. Bishop Museum photo CF 28354.* Left: *Cowboys at Humu'ula. Front row: Sam Liana, Henry Ah Fong, Bob Sakado, Willie Kawai, Masaru Okura, and John Samoa Lekelesa. Back row: Willie Kaniho, Big Joe Ignacio, Robert Kaniho, Pasquale Paiva, Manong Paiva, unidentified, and Take Horie. Hartwell Carter Collection.*

uniformity of quality and training. The older mounts were retired; some were put down. Although Sam Jr. had an indiscriminate horse-breeding program, the crown jewel was the stallion Frolic, assigned to the stallion station at Pukalani and exposed to a special broodmare band. A. W. described Frolic as throwing "handsome colts that had a beautiful way of going, particularly for polo." The mares Frolic ran with at Humu'ula prior to his arrival at Pukalani were of decent enough quality that many of his colts coming with the Humu'ula transaction were eventually sold as army remounts. Frolic matched Eastertide's reputation as a polo horse sire.

With livestock values that included the immediate wool harvest, coupled with fencing and other improvements, A. W. valued the station at well over the price he paid on behalf of the Thelma Parker trust.

Beyond calling for imported replacement rams, A. W. ordered the construction of new dipping vats at the station for treatment of sheep scab, and he mandated full renovation of building, corrals, pens, and sheds. Within a few years, the Humu'ula Sheep Station acquisition elicited the following comments from A. W.:

> I do not profess to be a sheep man. My business has been primarily cattle. However, I have learned considerable about sheep with reference to our own peculiar conditions.
>
> The only experience that I have had with sheep has been with a special type of Merino and I have imported rams of this type from Australia. This is a wool type. By the importation of rams I have been able to raise the average wool clip to about eight pounds per animal and it is the type of wool that is desired in trade. I ship it to Boston, the headquarters of wool in the United States, and we get a very good price. For many years the wool clip continued to be profitable.
>
> The wethers and ewes were sent to the Honolulu market. The wethers dressed out at about 48 pounds. The real demand is for lamb.
>
> We are not in a position on this particular Ranch to go into the lamb business. I have raised a few lambs for our local supply on Hawaii by crossing the English Leicester, which is a large long-wool sheep, on Merino ewes and I think this is a good practice if anyone should want to follow that line of business.
>
> For many years I have used an arsenic dip to cure and prevent sheep scabs and finally eliminated it. I have continued however, to use a carbolic dip annually and I think that anyone running sheep must dip their flocks every 12 months to keep them clean.
>
> We are fortunate in not having any burrs and our wool is very clean as it comes from the sheep's back. Dirt and burrs, of course, depreciate the value [of] wool.
>
> The Merinos, which we use, are small, chunky animals, not like the large sized Merinos that you often see in Australia. We have a very rough and rugged country and we find that these do much better than heavier sheep. We run our sheep in paddocks; they are not herded. I have handled Southdowns and Shropshires but without much success.

The number of sheep gradually decreased to around twenty thousand in the next few years. The permanent fence crew that built and maintained stonewall and wire fences continued to make improvements, further subdividing into paddocks Humu'ula's vast sections of Kalai'ehā, Pu'u-'ō'ō, Laumai'a, Hopuwai, and Keanakolu. Line camps were built in Laumaia, Hopuwai, and Waipunalei, since Keanakolu as a section overlapped with it. Subsequently, the house at Waipunalei was built and maintained as a permanent home because it was located on deeded land. This became the home of Yama Horie Sr., who was the section caretaker. This family produced several children who spent their careers on the ranch. Take became a cowboy under Willie Kaniho's foremanship at Humu'ula; Yama Jr. became a saddle maker; and "Horsey" Horie retired as the motor pool foreman.

A. W. put Parker Ranch in the sheep business as a gesture secondary to the land acquisition process—namely, Pu'uloa Sheep and Stock Company, 5,600 head, and the Keamuku Ranch and Humu'ula Sheep Station Company with 23,000 head. Entering the sheep business through the back door did not diminish A. W.'s desire to excel in this aspect of the livestock industry.

A. W. had three important measures of sheep standards: cut (wool), carcass (meat yield), and constitution (health). Vast exposure to feral sheep hampered progress, as it did with cattle. A significant departure from the mainland and European sheep management practice of large roving flocks managed by shepherds, the Humu'ula Sheep Station was a compartmentalized operation with many secure paddocks accommodating individual flocks. Management of flocks was intensified under the Parker Ranch Standard, and caregiving was thus enhanced.

10

PARKER RANCH AND THE TERRITORIAL DIVISION OF THE ANIMAL INDUSTRY

1905–1920

As early as 1905, the Territory of Hawai'i (TH), under its Board of Agriculture and Forestry (BAF) and Division of Animal Industry (DAI), began formally to record progress in the livestock industry through blood stock importation as well as disease prevention and control. In annual BAF reports between 1904 and 1918, Parker Ranch is credited with advancing the betterment of livestock by importing premium breeding cattle, horses, jacks, sheep, and hogs. A. W. Carter specifically is applauded for his integral role, not only in preventative disease measures, but also in the eradication of existing animal diseases, especially on the Big Island of Hawai'i.

Victor A. Norgaard, V.S. (Copenhagen), became the first superintendent of Animal Industry, and he was correspondingly given the title of territorial veterinarian. The assistant territorial veterinarian was J. C. Fitzgerald, MRCVS (London), who later moved to Maui, providing regulatory as well as private veterinary services. A similar post designated for Hilo went to H. B. Elliot, MRCVS (Edinburgh). The V.S. following Dr. Norgaard's name designates a veterinary surgeon degree and Copenhagen designates the veterinary institution of Denmark from which he was awarded his degree. In the cases of Drs. Fitzgerald and Elliot, their degrees are designated in British fashion—MRCVS for "Member of the Royal College of Veterinary Surgeons." Elliot obtained his degree from Scotland's eminent veterinary institution in Edinburgh, and Fitzgerald—although an Irishman—obtained his degree from the English veterinary school in London. Other foreign veterinary personnel in Hawai'i during these times included Drs. A. R. Rowat and C. R. Onissimienko (Russian).

A considerable history of veterinary medicine unfolds, as the efforts of the ranch leadership and that of the Territory of Hawai'i dovetailed. In early American veterinary history, the United States lagged far behind the foreign institutions in training veterinary personnel. Formal veterinary schools in the United States opened in 1870, a century after their foreign counterparts. By this time there were only three American institutions. This number, however, doubled in 1880, and by 1890 there were more than fifteen schools of this profession. The typical American degree is DVM—Doctor of Veterinary Medicine.

While Drs. James Brodie and A. T. Baker, arriving in 1884, were the first veterinarians to locate in the kingdom, the appointment of Dr. Norgaard to head the newly formed DAI in 1905 was a reflection of his prominence in the veterinary field. The evolution of Norgaard's illustrious career as DAI superintendent was perfectly timed with the emergence of A. W. Carter as Parker Ranch's chief steward, a visionary figure for the entire livestock industry. Their parallel roles in this regard approach heroic proportions.

Once on board, Norgaard quickly imposed regulations to prevent further introduction of infectious, contagious, and communicable livestock and other animal diseases. He instituted veterinary inspections at the known ports of entry—at that time limited to Honolulu and Hilo Harbors. Norgaard went further by mandating preshipment health certification from exporters of animals to the Territory of Hawai'i. Specific negative test results were required for such shipments, which included glanders in horses and mules as well as tuberculosis in cattle. Quarantine became mandatory for imports of questionable certification and testing, as these diseases had already crept into the Islands in years prior to such regulation. The BAF, in support of Norgaard's protective efforts, went even further by appointing Dr. William F. Egan as territorial livestock inspector for San Francisco, the common port of embarkation of animals destined for Hawai'i.

The DAI was vigilant in the control of other livestock diseases such as brucellosis, anthrax, foot-and-mouth disease, and red water in cattle, as well as strangles and cerebrospinal meningitis in horses, donkeys, and mules. A. W. underscored Norgaard's efforts with enthusiasm.

Livestock Disease Control

Without a doubt, the most significant episode that brought Norgaard and Carter together in the mission of disease control was an outbreak of glanders in Waipi'o Valley in 1907.

Glanders

Glanders is a zoonotic (transferable between animal and man) disease caused by the bacterium *Psuedomonas mallei* that is highly contagious among horses, mules, and donkeys. It causes a high fever (105 degrees), ulceration of the nasal passages, thick nasal discharge, glandular swellings about the throat, and often death. Glanders has a long incubation period of two weeks with no clinical signs, making detection difficult. Passed by direct contact, the disease can also be spread among equidae (horses, mules, donkeys) through shared feed and water troughs. Public hitching posts also served as a transfer device (fomite).

Glanders wreaked havoc during the American Civil War, immobilizing cavalry forces with its highly contagious, debilitating, and fatal nature.

It is thought that glanders came to the island of Hawai'i in 1895 via a load of mules intended for the Hāmākua series of plantations. In 1898, Dr. Elliot confirmed its presence using the mallein skin test, which could provide early detection prior to the advent of clinical signs.

While Norgaard's import regulations began to stem the influx of glanders into the Territory, the disease had established itself in Waipi'o Valley. Outbreaks in the valley occurred in 1907 and 1912, posing a very serious animal health threat to Parker

Ranch. A. W. was keenly aware that the physical proximity and traffic in pack trains to and from Waimea were high risk factors.

In July 1907, Dr. Fitzgerald, then assistant territorial veterinarian, was sent to Kukuihaele Plantation to follow up on reports of glanders there as well as in the valley below. In his annual report, Norgaard noted:

> While there, Dr. Fitzgerald received information to the effect that there were a number of sick mules in Waipio Valley wherefore he proceeded to that place to investigate. He found 9 cases of acute glanders, all in mules belonging to a Chinese rice planter. All of these animals were destroyed and the remaining mules belonging to the same party were placed in quarantine in a separate pasture.
>
> Dr. Fitzgerald also ascertained that a larger number of horses and mules had been dying from glanders in Waipio Valley during the last few weeks before his arrival, and, considering that nearly all of the animals from this valley are used to carry rice and paiai all over the Hamakua districts from Kamuela to Paauhau, the situation was sufficiently serious to warrant that steps be taken without delay to prevent the further spread of the disease.
>
> A wireless message was sent to the Parker Ranch requesting that no horse

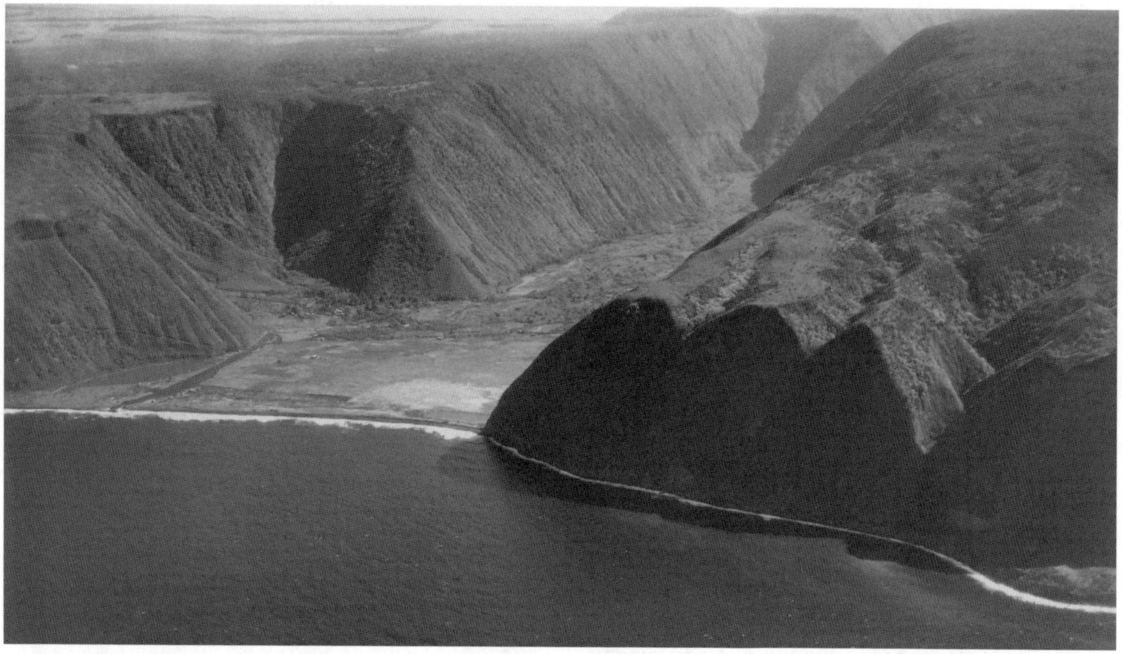

Aerial photo of Waipi'o Valley. The large square of eucalyptus trees in the upper left corner is the Baker Paddocks of the lands of Mahiki of Parker Ranch. The Lālākea Paddock of the ranch is located between Mahiki and the rim of Waipi'o Valley, indicating significant physical proximity. Bishop Museum photo CP 110,161.

or mules from Waipio Valley be allowed to pass the Hamakua gate and that the pack train which once a week brings paiai to the Parker Ranch and to the stores at Kamuela be met with wagons at said gate. At this place the provisions were transferred from the pack animals to the wagons and the pack train instructed to return immediately to Waipio Valley and remain there until further notice.

On the Parker Ranch end, at its Hāmākua Gate located at the bottom *(makai)* end of what is called Mud Lane today, Carter took further precautions with the ranch pack trains and their tenders: They were cautioned in the process of freight transfer to avoid all contact with the foreparts (head and neck) of the Waipi'o mules. Further, the rendezvous point of the Waimea and Waipi'o freight animals had a neutral zone of about 200 yards, where the actual transfer of bags of taro, *pa'i 'ai,* and rice from the Waipi'o mules to the Waimea pack trains would be borne on the backs of ranch tenders only.

Carter was clearly aware of the devastation that glanders would render on the workhorse, draft, and pack strings in terms of lost work productivity. He was also preparing vastly improved breeding bands of horses and jackasses for ranch and commercial purposes.

There were follow-up visits by Fitzgerald, as he submitted the following report:

> I left for Kukuihaele on December 11th and immediately proceeded to organize a party to accompany me into and beyond Waipio Valley. I succeeded in getting together three cowboys, one interpreter and two policemen, with whom I went into the valley on the 12th. On the 13th and 14th we rounded up all horse stock in Waipio valley, a total of 192 head. Among these was one case of acute glanders, which was destroyed, while 16 animals were slightly suspicious. These all submitted to the mallein test, but as none of them reacted they were released.
>
> On the 15th we went up the trail on the further side of the valley and rounded up the entire plateau including the 12 gulches located there. We found only 10 horses, one of which was affected with glanders and was destroyed. The 16th was spent in Waimanu valley where 30 animals were examined and one destroyed for glanders. We then returned to Kukuihaele and on the 17th and 18th rounded up and inspected all of the outside animals in the neighborhood, as well as all of the plantation stock. Among the outsiders one case of glanders was found and destroyed. After examining the plantation stock of Honokaa, which was found to be free from the disease, I returned to Honolulu.

Glanders outbreaks tapered off over the next five years but returned to Waipi'o Valley with a vengeance in 1912. During the period of quiescence, Norgaard warned that until a deputy territorial veterinarian (DTV) was stationed in North Hawai'i, eradication of glanders could not be expected. Some background history of this struggle is in order.

A. W. was well connected with the commissioners of the BAF. His cousin George R. Carter served during this period, along with other business associates such as Albert Waterhouse and P. R. Isenberg, all of whom he prevailed upon for the appointment of a permanent DTV for North Hawai'i. Reinforcing Carter's efforts, Norgaard appealed to the BAF commissioners that—with support provided between the Hawaiian Sugar Planter's Association (HSPA) and the Hawaiian Cattle Raisers Association (which later became the Hawaiian Cattle Breeders Association and ultimately the Hawai'i Cattlemen's Association)—both the plantation and ranching interests would be best served by the appointment of DTVs in Līhu'e on Kaua'i, Wailuku on Maui, and on the Island of Hawai'i, in Hilo and North Hawai'i. For the sake of the Kohala and Hāmākua Districts, A. W. immediately offered a monthly fee of $100 in support of such a DTV.

The BAF logically responded that by supporting such appointments of regional DTVs, Carter's gesture would be considered more as an insurance premium on the livestock belonging to the respective plantations and ranchers than as compensation for services rendered. Thus was born the unique territorial system of providing regional (insular) consultative veterinary appointments. By the new millennium, it would be a system that outlived its usefulness. From these first appointments up through the period following the Korean War, the DTV (later the deputy state veterinarian) system provided a unique, progressive, and useful service. The appointment of a DTV for North Hawai'i was slow in coming; the BAF began such appointments starting with Hilo.

In early 1908, Elliot was appointed DTV for the districts of North and South Hilo, Puna, and Ka'ū, with the plantations and ranches in those districts collectively providing the $100 retainer. Maui soon followed suit with the appointment of Fitzgerald as DTV, responsible for Lāna'i and Moloka'i as well. While in earlier years Drs. A. R. Rowat and C. R. Onissimienko had provided North Kohala with private services, they both moved to O'ahu to work in the veterinary laboratory, then directed by the assistant territorial veterinarian, L. N. Case, DVM (Cornell). This left no one to cover that district, which was struggling with glanders among its plantations. During this period, Elliot did extend north as needed, but only on a limited and emergency basis. In an early 1909 report, Norgaard expressed frustration:

> The Kohala plantations signified their willingness to cooperate with the Hamakua plantations and with the Parker Ranch for the purpose above mentioned, but opposition of undefined nature was met with from the Hamakua district and no deputy has been appointed up to the present time. There is a strong feeling in the district assigned to Dr. Elliot that the good to result from his appointment will, to a great extent, be offset by leaving his district open and unprotected toward the North, especially in view of the fact that glanders has prevailed in the Hamakua district for a number of years.
>
> At the same time, the managers of plantations and ranches in the Hilo district, who participate in the payment of the Hilo veterinarian's salary do not consider it reasonable that Dr. Elliot should give his time and services for the

inspection of livestock intended for plantations which have refused to contribute anything to his salary.

Norgaard and Carter shared mutual disappointment that was exacerbated over the coming years when the 1912 glanders outbreak occurred in Waipi'o—still with no resident territorial veterinary services.

Norgaard's December 31, 1912, report to the BAF described the evolution of the new outbreak in Parker Ranch's backyard, Waipi'o Valley:

> On October 30, 1912, a wireless message was received from the Manager of the Kukuihaele Plantation, Hamakua, Hawaii, stating that an epidemic had broken out among the horses and mules in Waipio Valley and requesting that the Territorial Veterinarian be notified, which was done immediately. Glanders was suspected at once, but as the message seemed to indicate that the recent heavy rains in the district might be a contributing cause and as the death of seven horses within a short period pointed to some acute infectious disease, such as epizootic laryngitis, the glanders theory seemed doubtful, even though the seriousness of the situation was realized.
>
> Waipio Valley or Gulch is an absolutely isolated district approachable only by way of a rough, steep trail, descending a thousand feet or more at an angle of 40 to 45 degrees from a point a short distance north of the village of Kukuihaele, where the plantation stables, with more than 250 head of mules and horses, are located. The transmission of the infection to the plantation animals was therefore imminent, and the same applies to the adjoining string of plantations, stretching south along the east coast of Hawaii, as well as to the Parker Ranch, all of which are visited regularly, twice a week, by pack trains from Waipio Valley carrying rice and pai'ai.

In both outbreaks, the focus of infection came from the stables of a prominent Chinese rice and taro farmer named Akaka. Akaka's operation had barely managed by purchasing available horses and mules from neighboring farmers in the valley. Apparently during the period between epidemics, glanders had been reduced to a mild incidence, and the locally raised mules had some natural immunity. This changed drastically when Akaka purchased twelve young pack mules from Parker Ranch and put them to work packing his produce up to Kukuihaele. The Parker Ranch mules were

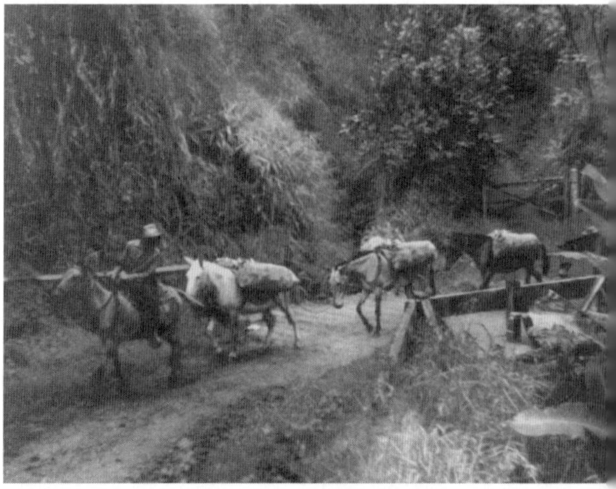

A mule train of taro produce wends its way through Waipi'o Valley. Bishop Museum photo CP 126,359.

obviously immunologically naive, for within three months all had died of glanders and an epidemic broke out among neighboring taro farmers.

On October 30, 1912, Norgaard himself left Honolulu by steamer, but not before a wire was sent to A. W. advising the reinstitution of the safeguards that had protected the ranch during the 1907 outbreak. Norgaard also ordered the Territorial Department of Health to station twenty-four-hour personnel, with support from Sheriff William M. S. Lindsey of Waimea, at the mouth of the valley to detain and segregate any suspicious animals coming out. Upon arriving himself, Norgaard aptly described the challenge ahead:

> The first day in the Valley was exceedingly disagreeable, a constant pouring rain necessitating the continuous use of oil skins and making the trail dangerous. Several times my horse slipped and nearly sent me headlong over the pali. However, we reached the bottom safely, but the mud and rain made the work slow and exceedingly disagreeable. After inspecting the isolated cases, the rest of the animals belonging to Akaka were examined, and several had to be pronounced suspicious. . . .
>
> During the following days an attempt was made to inspect all the horses, mules and donkeys in Waipio Valley, and it is believed that all the animals were seen. In rounding up both owners and animals, the Hamakua police lent valuable aid, supporting the Board of Health inspector and his assistants in securing a complete list of names of the owners and the number of animals belonging to each. A paper slip bearing my initials was attached with poi to each animal after examination, and owners who had failed to bring all their animals were promptly returned for the missing ones, whatever the excuse might be.

All suspicious cases were segregated on the valley floor, while Norgaard proceeded to monitor pack trains coming out of the valley crest adjacent to Kukuihaele.

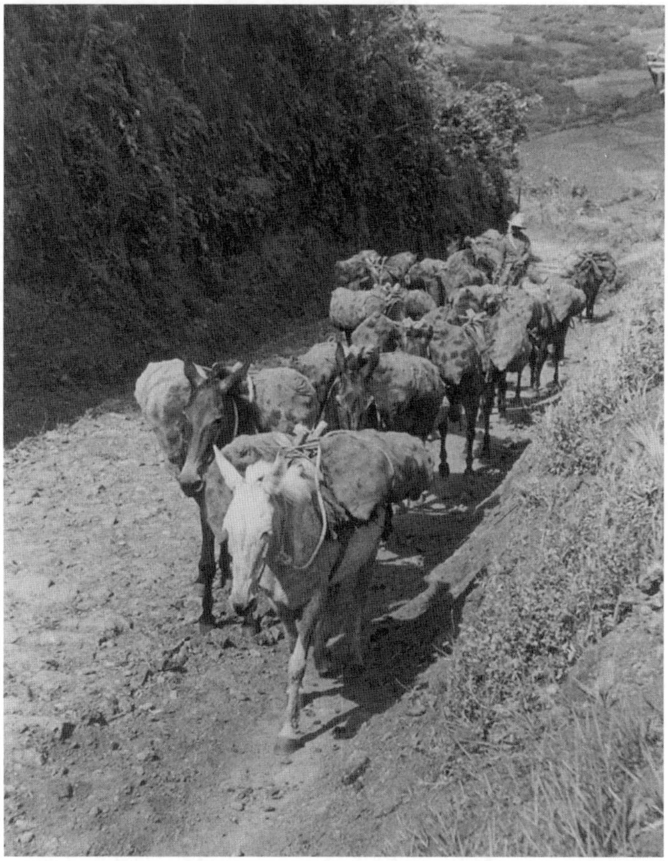

A mule train of taro climbing the steep road out of Waipi'o Valley. Bishop Museum photo CP 73308.

By 7 a.m. the following day, Akaka's pack train arrived, using twelve horses bearing two large bags of rice each. These were all Akaka had left for pack animals, and even these horses now showed early signs of glanders. They were relieved of their freight and marched back to the valley floor to join the rest of the condemned pack animals. Norgaard describes the disposing of these condemned horses and mules in tragic detail:

> The animals were, therefore, ordered sent to the beach where I shot them in such positions as to leave the bodies below the high tide line, which, the natives assured me, would cause the carcasses to be carried out to sea by the receding tide.
>
> A strange sight was presented at this place, seldom visited by anybody except for the purpose of disposing of dead or dying animals. Here bones and skulls were littering the beach, many of them still retaining traces of flesh and ligaments, indicating that they had but recently died. Four white bleached skulls were found having bullet holes in the forehead, evidently the remains of those which were killed at this place five years ago. In several cases the skulls and neck vertebrates were found tied with a rope to a large rock, without indication of injury to the skull either by bullet or axe, showing clearly that the sick or dying animals had been taken to the place and left to die without the mercy of a death blow. Considering that a majority of the bones and carcasses must have been carried out to sea, the number of those remaining in sight were a sure indication of a very high mortality among the horse stock of the Valley, which has never been know to exceed 200 head.

A. W.'s concerns were not limited to infecting ranch horses and mules; they focused also on the human famine that would result if Waipi'o's intensive taro and rice production was rendered undeliverable from the valley save for the backs of men. A severe shortage of these vital foodstuffs would certainly hamper the productivity of the cowboys, as well as risking the health and welfare of ranch families. The bulk of the villages in Hāmākua, from Kukuihaele in the north to 'Ō'ōkala in the south, would be similarly deprived.

Norgaard and Carter divided the challenge of resolving the crisis. While Norgaard immediately launched aggressive mallein testing on all equidae between Kukuihaele and Kohala in search of unexposed pack animals to replace the large Akaka string, Carter implored the territorial government to indemnify Akaka for destroyed pack animals and purchase of replacements. Norgaard gave Carter the direction for securing indemnification, but it was A. W.'s influence and perseverance that saw it through to fruition.

Akaka's woes were compassionately described in a supportive memo from Norgaard to Albert Waterhouse, then chairman of the BAF:

> I am up against a problem that in a way is new to me, and I take the liberty to lay the matter before you, and, if you deem it necessary, I should very much like to have the Attorney General's opinion thereon. The facts are simply

these: — Mr. Akaka, an aged Chinese rice planter in Waipio Valley, has glanders among his horse stock to an extent unprecedented in this Territory. What his losses have been previous to a few months ago I can only surmise, but of twelve young mules purchased at Waimea about four months ago not a single one remains; all have died from glanders. When I was here three weeks ago I destroyed 6 horses and one mule, all affected with typical glanders and now, having tested with mallein his remaining twenty-seven head of horses I have found twenty-four of them to be suffering with glanders. Now, the law, that is, the rules and regulations of our Board, is plain enough on the subject, saying that any horse or mule found by the Territorial Veterinarian to be affected with glanders or farcy shall be destroyed, the carcass disposed of, premises disinfected, and so forth at the owner's expense. . . . There are, however, conditions and circumstances, not to mention precedents, which seem to me to demand more explicit instructions or authorization before I proceed as indicated. Mr. Akaka was a comparatively rich man a few years ago, but through mismanagement of his business here, while on a visit to China, he returned to find himself very much reduced financially, in fact almost ruined. Unfavorable conditions made it difficult for him to regain what he had lost in spite of generous support from his friends, and the present difficulty which finds him indebted to the extent of $9,000.00 can result only in bankruptcy, if he is deprived of the means wherewith to harvest and market the now ripe crop of 55 to 60 tons of rice. Without any assurance of indemnification by the Government, I am told it will practically be impossible for him to borrow any more money wherewith to replace his work animals, without which the present crop and possibly the following one will be a total loss. Under these conditions you will undoubtedly understand why I am desirous of reassurance in regard to the apparently plain course outline by our rules and regulations. As an American citizen, Mr. Akaka must have some inalienable rights, covered by constitutional or other laws, guaranteeing him against the deliberate though legal destruction of his means of making a legitimate living—pro bono publico—

Now Mr. Waterhouse, if, after reading this, you have come to the conclusion that I am borrowing trouble and that I should not bother the members of the board with such details after receiving plain instructions to enforce the law, I will only ask you to imagine yourself in my position, with an old man, palsied and with tears running down his face begging me to spare at least some of his animals, explaining over and over again in his nearly unintelligible pidgin English the unfortunate position he is in—and I doubt you would fancy yourself in my place. To me there is something awfully unjust in ruining the old man for the benefit of the community, and I have done what I could to get the neighboring plantations to help him out with a view to reimbursement by the coming legislature.

Waterhouse—with nudging from A. W. —responded positively, ordering the appraisal of stock and the cost of replacing the Akaka stables and outbuildings. Norgaard then reported, "Upon my return I was pleased to meet Mr. O. L. Sorenson, Assistant Manager of the Parker Ranch, who informed me that he had been designated to act as an agent for a number of plantations, as well as for the Parker Ranch and the Hawaiian Irrigation Company, for the purpose of purchasing horses wherewith to enable Akaka to harvest his rice crop, thereby saving him from bankruptcy."

The appointment of A. W.'s assistant ranch manager, O. L. Sorenson, as the indemnification agent is testimony not only to A. W.'s influence but also to his habit of following up on all matters. The flow of taro, *pa'i'ai,* and rice out of Waipi'o was safely restored for the cupboards *(waihona)* of the Waimea and Hāmākua people, while the horses, donkeys, and mules of Parker Ranch and the Hāmākua plantations were spared another threat by the debilitating and fatal disease of glanders. By 1917 the BAF had appointed a permanently positioned DTV to North Hawai'i: Dr. Leonard N. Case moved to Waimea, making it his permanent home.

Tuberculosis

As Norgaard took office, he was well aware that tuberculosis was entrenched in Hawai'i, especially in the dairy industry. The *Mycobacterium tuberculosum* causes debilitating pulmonary and bone disease in animals and man. It is transmitted by respiratory, physical, and oral (digestive tract) means.

Bovine tuberculosis was not unique to Hawai'i in the early twentieth century, but governmental support of its control and eradication through dairy and milk inspection was dismal when compared to mainland counterparts. Norgaard opined in his December 31, 1908, report that "there is probably no state or Territory in the union which expends less money on dairy and milk inspection, or in which less efforts are being made to improve the sanitary condition of the milk supply."

It is no wonder that the process of upgrading and modernizing the dairy component of Parker Ranch relied much on retaining the expertise of Dr. Williams of Cornell University and the employment of technically trained dairy personnel (Schroeder, Mills, Wallace, and others). Parker Ranch's dairy locations were also strategically placed well within its boundaries, away from neighboring ranches, dairies, and wild cattle populations. Parker Ranch guarded its dairy herd by purchasing only stringently tested and certified replacement bulls and heifers from reputable mainland farms. The TB test in use was the federally approved tuberculin test. This was administered like the mallein test, in which an injection into the cow's skin (intradermally) of a minute and measured dose of the antigen (tuberculin) generates a measured inflammatory response. A suspicious or reactor response is indicated by the degree of swelling measured seventy-two hours after the injection.

The ranch administration was critically aware that if the dairy became infected with tuberculosis, the risks were twofold: contamination of the milk supply to the communities served and exposure of the entire beef cattle herd, which was approaching ten thousand head at the time. Exposure to this infection would have catastrophic ramifications. Granted, the risk of exposure of brucellosis from the dairy to beef herd existed at one time, but ardently applied measures to control this easily recognized disease kept it confined to the milk cows. Tuberculosis does not lend itself well to control beyond testing and subsequent slaughter of suspect and reactor cattle. It is a slow, progressive, insidious disease with belated but severe impact on humans. Considering that Thelma K. Parker Smart died in 1914 from pulmonary TB—likely acquired from infected cow's milk, as with many people of the Territory in those times—the entire dairy industry was under close scrutiny. Although Parker Ranch in its history of over a century and a half had never been considered infected, largely due to A. W. Carter's diligence, other parts of the island and Territory were being ravished.

In 1912 Dr. Elliot conducted comprehensive TB testing of dairy cows in East Hawai'i (North Hilo to Puna). Of the 1,800 animals tested, 562 were considered exposed suspects and sent to slaughter. In the same year, the Hawai'i County Department of Health recorded 1,456 childbirths, of which 352 died as infants. Of these cases of infantile mortality, more than half were diagnosed as tuberculosis from consuming contaminated, unpasteurized cow's milk. Honolulu health statistics were comparably tragic.

During this era, there were no veterinary services in West Hawai'i in terms of dairy sanitation, but A. W.'s stringent preventive health program coupled with rigid prepurchase testing safeguarded the ranch herds and, most significantly, the people of Waimea and the surrounding villages. How his ward Thelma became infected with TB will never be known, but the comfort of knowing that her home ranch was never the source of this tragic affliction underscores the wisdom and stewardship of her guardian.

The DAI persevered in its Territory-wide suppression of tuberculosis. These efforts were greatly enhanced in 1917 when the Indemnification Act was passed. This act allocated $20,000 for the depopulation and restocking of infected dairy herds, giving the farmers genuine incentive to become disease-free dairy operators. Again, the DAI in enacting this legislation appreciated A. W.'s influence. For the purposes of the ranch, eradication of TB in the dairy herds of North Kohala and Hāmākua further safeguarded Parker Ranch beef and dairy cattle from possible exposure. Just as importantly, those communities could now depend on a wholesome milk supply.

Brucellosis

Brucellosis is another contagious and infectious disease that, like TB, is transmittable to humans, causing undulant fever. In cattle, *Brucella abortus* causes abortion and infertility. It is easily spread through contact via body fluids, especially blood from a febrile cow or pig. Parker Ranch did in fact have periodic pockets of this disease; the background is discussed under the dairy section. While A. W. adopted a preventive testing policy that was heartily approved by the DAI, he also acted in support of legislation to tighten up import regulations.

Anthrax

There probably will never be a more challenging mission for an epidemiologist than tracking the point of origin of anthrax outbreaks in the Territory of Hawai'i. The first, occurring in 1917, was simultaneous with World War I, and the second outbreak immediately preceded World War II. Again, this disease is zoonotic, and its victims become rapidly febrile and it is commonly fatal if left untreated.

Hanalei, Kaua'i, was the scene of the first outbreak. Unfortunately, more than 300 head of cattle were lost before a diagnosis was made, and these animals were buried shallow without cremation. Cattle were dying at the rate of ten per day until vaccine was received and administered, by which time another 225 had died. The remaining 2,400 head were vaccinated twice before daily deaths subsided.

Within weeks, anthrax cropped up at a dairy in Honolulu. Of the hundred head in the herd of milking cows, more than a third died in forty-eight hours. A bag of beet pulp—only one out of a shipment of several hundred—proved to be laced with anthrax spores. The feed room had never been locked, allowing convenient access for someone to heavily lace the single bag of beet pulp. Vaccination was too little too late.

The final crisis came the following month on Maui, where in the course of ten days, seven distinct outbreaks occurred in widely scattered districts. All outbreaks occurred along roadways, as did the Kaua'i cases. The series of pocket epidemics ended with vaccination, cremation, and deep burials. All told, more than five hundred head of horses and cattle were lost.

In his report, Norgaard made it clear that this tragic series of anthrax outbreaks was the work of warlike Germany, a country that wanted to become a major military and economic influence in the Pacific Basin. Germany by this time had colonized Western Samoa. His December 31, 1918, report reflects his frustration and concerns:

> The work of the Division of Animal Industry has during the past two years differed from that of the preceding decade in that it had to be devoted to combating a disease hitherto unknown in the Territory, and which for a while threatened the entire local live stock industry, namely, Anthrax. For the first time in the twelve years of its existence this Division had to deal with a real epidemic of a virulent malignant disease, fatal to man and beast and the introduction of which remains a mystery to this day. The total absence of a natural source of infection and the inexplicable manner in which it jumped from Kauai to Oahu and Maui leaves, however, little doubt that the propelling force behind it was of Hun origin. . . .
>
> On both Kauai and Oahu the disease was kept confined—by stringent quarantine measures—to the premises where it first made its appearance, and there is little reason to doubt that the seven outbreaks on Maui, though widely scattered, were planted simultaneously or at least within a couple of days, and were independent of each other.

In his memoirs, A. W. reflected on the strong measures he took during this period of risk to safeguard the ranch from the dreaded anthrax:

> As soon as we entered the war, 1917, Anthrax germs were planted in the Territory. They became established in one location on Oahu and on Kauai as well.
>
> I came to Honolulu and took up the

matter of guarding the Ranch with the U.S. District Attorney. He cabled to Washington and we could not get a Federal Proclamation from the President authorizing the work which I wanted done. I then told the District Attorney that I would proceed and do it on my own responsibility. He pointed out to me that I was taking a risk as they would undoubtedly prosecute me, but he said he was in entire sympathy with the matter but could not guarantee me protection. I told him that I would go ahead anyway and I did.

At every entrance to the Parker Ranch from the outside, I placed guards. I had a list of those whom I suspected of mischief and I gave the guards instructions to bring everyone in to Waimea whom I did not know. I had them searched and then passed them on, under guard, to the edge of the Ranch lands in the direction in which they had been going. There was a tremendous howl made about it and several went to the Federal and local authorities to endeavor to obtain my arrest. These authorities were all in sympathy with my acts and for one reason or another, they threw out every complaint that came in so that I never did get prosecuted and I felt certain that this action prevented the introduction of Anthrax on the Parker Ranch. This Ranch was the primary source from which the Army and civilian population in the Islands obtained its meat at that time and of course this would have been the objective of any mischief makers.

This was a high-handed action, I realized, but it worked effectively. Anthrax, once established on a place, remains practically forever. It gets into the soil and it cannot be eliminated.

Elliot recapped the Anthrax impact on Hawai'i Island in a 1918 memo:

> The Island of Hawaii was fortunate, also, in escaping the outbreaks of anthrax which occurred in 1917 upon other islands of this group. A good deal of alarm was naturally manifested by owners of livestock, and for several months all cases of sudden death attended by suspicious circumstances were reported. Where it was possible, investigations of these cases by field post-mortem and laboratory examinations were made, but in every case with negative results so far as anthrax was concerned.
>
> During a part of the time covered by this report the writer was absent from the Territory, and the work of the office was conducted, first, by Dr. L. N. Case, Assistant Territorial Veterinarian, and for a later period by Dr. O. B. Shipman, of Hilo.

O. B. Shipman was the brother of Herbert C. Shipman, and they were the sons of W. H. Shipman, the founder of W. H. Shipman, Ltd.

A. W. was kept informed of each of these outbreaks, while measures to protect the ranch were taken in a firm and timely manner. The DAI kept A. W. firmly in the loop, a reaffirmation of the close professional relationship that had developed between him and Norgaard. This relationship between the DAI and Carter became even more significant when foot-and-mouth disease was diagnosed in California in 1924.

Foot-and-Mouth Disease

By 1924, the lead veterinarian for the Territory was Dr. P. H. Browning. Captain A. L. C. "Jack" Atkinson had assumed chairmanship of the BAF.

On February 25, 1924, it was announced that foot-and-mouth disease (FMD) had broken out in nine counties of northern California. The USDA and the University of California Agricultural Station immediately forwarded emergency information circulars on the nature of the disease, along with steps to control spreading and prevent infection.

A. W. was alerted on March 2 by Alexander and Baldwin that one of their freighters bound for the Islands was turned back by the BAF, citing quarantine regulations relative to the California FMD outbreak. The freighter was carrying a large shipment of animal feed and a significant number of mules to the Islands before it was returned mid-shipment.

In the interim, the BAF immediately imposed a total embargo of all shipments of animals, hay, and grain out of California as of February 15. A deputy territorial veterinarian was immediately assigned to trace back any shipments of hogs, the prime carrier, into the Islands within two weeks prior to the February 15 quarantine order and embargo.

The California situation was bleak. All contiguous states also placed a quarantine embargo on livestock traffic from the Golden State. During the month of March, California officials had destroyed 280 herds that in total contained nearly 24,000 head of cattle, 10,000 hogs, 4,000 sheep, and 237 goats. (Such numbers evoke memories of the tragic United Kingdom outbreak of 2000.) P. H. Browning reinforced the embargo to include all products of the soil, dairy products (unless canned), and poultry products, including eggs. Rice shipments were spared. This deeper embargo had the absolute support of Captain Atkinson and Carter.

Honolulu merchants, lacking available merchandise, joined California agricultural officials in trying to have the embargo lifted; pressure was being brought to bear. In a March 12 letter to A. W., Captain Atkinson shared some local background:

> A. L. C. Atkinson to A. W. C., who is in San Francisco, "I want to thank you for your telegram, reading, 'Yesterday outbreak in Alameda 12 miles from former cases . . .' I published this information in the morning paper and everybody was glad to get authentic information.
>
> Public feeling is divided here on this question of embargo, and merchants on whose toes we tread are beginning to holler. Some of them have got opinions from local attorneys that we have no legal right to put on the embargo. However none have dared peep very loud. We are ready for them. The Governor and the Attorney General have approved our regulations. . . .
>
> The stock men are waking up. Charlie Lucas told me that the cattle men are going to get together and call a public meeting to protest against the introduction of any animal from any port. He said they proposed to protest against it even if they had to organize a

vigilance committee. So there you are! I did not discourage him.

The whole trouble is that the public in general doesn't know the danger of the foot and mouth disease epidemic. I propose to see that our Board remains firm. . . . Because of some mutterings of merchants I asked the privilege of addressing certain committees of the Chamber of Commerce, and I laid down the law to them, and also had them told by veterinarians the danger we are fighting. Those who came to scoff went away impressed.

The Territorial Board sends its veterinarian, Dr. Browning to the Coast. It is his opinion that the Islands should exclude all products of the soil. This would include potatoes, rice, beans, etc.

Besides Browning, the BAF sent G. L. Waller Sr. of Hawai'i Meat Company to monitor San Francisco export facilities and monitor embargo compliance. A. W.'s son-in-law, Ernest Podmore, was sent to the Port of Seattle with the same mission. Alexander and Baldwin's offices in Seattle and San Francisco were supportive and cooperated with such surveillance.

By the summer, California officials were developing optimism on the containment of FMD. In a July 1 memo to Captain Atkinson, G. L. Waller reported that a USDA veterinarian from Washington, D.C., a Dr. Mohler, after touring the affected counties, was very emphatic that the territorial embargo on California products was too stringent. Mohler even went as far as attesting to the safety of smoked meats—a position that would bring clamor from modern-day federal veterinary officials. It was learned in 2001 that tainted Chinese sausage *(lup chong)* carried the FMD virus from the Orient to London, igniting the outbreak in the United Kingdom.

Between the mercantile community of Honolulu and the casual posture of the federal and California state officials, the territorial governor, Wallace Farrington, was pressured to take a position. Governor Farrington issued this statement to the BAF: "We solicit modification of Hawai'i's embargo to conform to federal regulations. Disease is now well controlled." The BAF, in a memo signed by Atkinson, politely responded that the existing embargo would remain imposed to a point significantly beyond the resolution of the FMD outbreak. The governor gradually relented when the facts were examined in detail. Further, the news was widespread in October 1924 that Texas had dealt with an FMD outbreak, slaughtering more than five thousand head of cattle within a 20-mile radius of the original infection. Dr. J. Gomez-Esparza, chief of Mexico's Department of Animal Industry, imposed an identical, sweeping embargo on Texas products.

The California epidemic took more than two years to resolve to the satisfaction of the BAF. Through the entire process, A. W. provided support for the strength demonstrated by the BAF and its chairman, Atkinson. In a follow-up memo to Atkinson, A. W. stated: "The splendid work you have done and are doing at tremendous personal sacrifice is deeply appreciated by me and I know it will be by the larger part of the community when the thing is over. Your own satisfaction, however will be simply work well done against large odds, resulting in the protection of the Territory."

While other diseases of cattle and horses did surface in ranch history—such as blackleg, red water, shipping fever, and leptospirosis—they were conditions controllable by management practices. The salient fact remains that in all cases of significant diagnostic and regulatory matters, the close relationship of Parker Ranch and the Division of Animal Industry endured a full century.

Importation of Premium Breeding Stock

Since its inception in 1905, the Division of Animal Industry became the formal agency of importation of animals, a function it serves today within the Quarantine Division.

As men responsible for the regulatory oversight of livestock importation, Norgaard, Elliot, Fitzgerald, and Leonard N. Case were in a unique position to monitor the quality of animals arriving as well as having a firm grasp of what the territorial needs were. In their annual reports can be found objective comments on the quality of imports, as well as consultative views on which direction the livestock industry had to move in its quest for excellence. In many of these observations, A. W. Carter and Parker Ranch are held up as standards of excellence. In a 1907 report on mule production, Norgaard made these comments:

> I was early impressed by the large annual importation of various classes of live stock from the Mainland, principal among which was the importation of mules. . . . $225 per head, or an aggregate of $135,000 for the two years. . . . Mules have, however, increased from 75% to 100% in value during the past five years, and the sugar plantations have refrained from buying mules unless absolutely compelled to do so. . . . There is, unfortunately, no prospect at the present time, or for several years to come, of the price of mules going down. The necessity of this Territory's producing its own mules is therefore obvious.
>
> Another point in connection with this move is the encouragement which farming or the production of feed stuffs will receive through the presence of an abundance of work animals. To produce first-class mules a certain amount of concentrated feed must be given, and as young mules must be broken to harness and their muscles developed through steadily increasing work before they can be considered finished plantation mules, this process of breaking and developing can be made profitable by making them produce their own feed.
>
> When it is further considered that it requires practically the same length of time to raise a mule as to raise a steer, and that when ready for market the former is in strong demand at five times the price at which the latter often goes begging, there can be little doubt that mule breeding in this Territory will prove profitable to the stock raisers and ultimately, in regulating the prices by a constant and abundant supply, prove beneficial to the planters and agriculturalists.

In 1908, through the Hilo port alone, 374 mules arrived, while 376 entered Honolulu. A. W. eyed the mule market of the Big Island with satisfaction. Of the 374 imports to the island of Hawai'i, 97 percent were already purchased by or consigned to the plantations around the island. The average price paid was $260 per head.

Parker Ranch was well equipped to begin a long-range breeding program aimed at the plantation market for stout, well-broken draft, pack, and riding mules. The ranch already was equipped with well-framed, stout broodmares that were bred to Percheron stallions, the female progeny of which would be bred to premium mainland jacks to produce an 1,100- to 1,300-pound draft mule. The premium jacks would also directly serve native Thoroughbred and Standardbred mares to produce pack and riding mules. Norgaard summarized A. W.'s program in a 1914 report:

> As far back as 1906 this Division began urging the breeding of larger mules here, it having been conclusively shown that such animals could be raised here, when sufficiently large mares were used. To produce the mares here first and then import pedigreed jacks for mule breeding, appeared like a very risky undertaking requiring from six to nine years before the first crop of mules would be of marketable age, and still, a careful analysis of the question satisfied Mr. Carter of the Parker Ranch that this was the only sane solution of the problem. He therefore imported a number of Percheron stallions and bred them to the heaviest mares on the ranch, the colts being broken to harness early and put to work at producing the corn necessary for their own development and that of their future offspring. The experiment proved a complete success, and while a tempting market was ready to absorb all the resulting offspring, only the male animals were disposed of. About four years ago the fully developed half-bred and three-quarter blood Percheron mares were bred to some of the best jacks imported from Kentucky, Tennessee and California, and the young mules, now maturing, promise to equal the best Missouri and California bred mules ever imported here. The foresight and courage exhibited by Mr. Carter in this undertaking will undoubtedly prove of immense importance to the Territory, entirely aside from the financial advantages to accrue from it, since the great European war is bound to deplete the mule market of the United States to an extent that will send prices soaring.

In his 1915 report, Elliot underscored A. W.'s acumen: "Six fine jacks have recently been imported, five going to the Parker Ranch, Hawaii. That mules can be raised here equal to, if not better than, any place else in the world was well demonstrated by the entries of Hawaiian breds at the Hilo County Fair. This industry is now in full swing, the Parker Ranch alone having 250 large mares in foal to imported Jacks, and in a few years the ranches here should be able to supply the entire local demand for mules."

Parker Ranch had the horsepower in a massive band of excellent, specially bred draft mares; they acquired the best in jack power; they provided superior nutrition via the cornfields; and they were endowed with a fine mule training staff headed by Henry Pieper and, later, John Purdy. A. W. put all these pieces together.

Horse Imports

One of A. W.'s early progressive mandates was to improve the quality and depth of the cowboy's horse strings—and he did this vigorously. But he also had the wisdom to blend into his vision of an ideal cow horse those characteristics that appealed to a significantly enhanced market. Carter clearly saw the remount, polo, and racing market as a target parallel to the ranch cowboy's needs. Speed, stamina, intelligence, size, and soundness of leg and back were qualities sought by both the ranch horsemen and outside buyers. Mule production notwithstanding, A. W. looked especially to the Thoroughbred as the breed of choice—and he was not alone. Dr. Fitzgerald made these observations in his 1912 annual report:

> The plantations are using better class animals as a general run than they have heretofore, and are bestowing much more care upon them. The Cavalry and Artillery Regiments of the United States stationed on Oahu create a steady demand for high class remounts and artillery horses. The increasing demand for polo ponies with breeding and speed is also a factor; lastly, the renewed interest in racing is a source from which to draw a goodly profit, if the right kind of animal is forthcoming. Taking all in all, the encouragement to the other breeders to invest in high class breeding stock was never greater than at the present time. It was feared that the innovation of the automobile would cripple the horsebreeding industry, but the reverse has proved to be the case, and horseflesh is higher in the Islands today than ever before and the demand for high class animals has increased. . . .
>
> It has been frequently demonstrated in horse breeding, especially in thoroughbreds, as among these, better facilities are offered for testing the results of breeding that sixty per cent lies in the dam. The breeding operations of many of the most successful breeders of America and Great Britain have been conducted on these lines with no meager results. It is to be hoped that the importations of the future will include more females, and that breeders will see the common sense of this policy; economy of time being one of the chief factors in modern progress.

Testimony as to the wisdom of A. W. Carter's breeding program, however, is most meaningful when it comes from the likes of J. C. Fitzgerald, regarded as one of Hawai'i's finest judges of horseflesh. In a 1914 report on horse breeding, he made some sobering points on Territorial horse production:

> Too much stress cannot be laid upon the necessity of this Territory becoming self-supporting in horse flesh, particularly draft stock, as there is little doubt that the shortage of horse flesh in the United States created by the demands of the warring governments for remounts and artillery horses, will be felt in the islands. The abnormal demand created by the tremendous orders from Europe in America can better be judged when cognizance is taken of recent statistics which show that the life of the average horse in action is five days; that the cavalry and artillery are proving the most effective branches of the service; and that agents of foreign countries are daily traveling through the horse breeding states purchasing everything available

> and at all suitable for remounts and artillery purposes.

The Remount Importation Program

The DAI veterinary personnel were in a pivotal position regarding horse import quality, largely owing to the early needs of the military for remount horses. In the first few years of importation, beginning in 1908, large numbers of military horses and mules —as many as a thousand head annually— were brought in from the West Coast to fill the army's needs. The importation gradually tapered off to a trickle in the ensuing fifteen years. This was largely a result of the remount ranches of Hawai'i filling this market from their own breeding programs— using, of course, federally supplied remount stallions.

In 1913, the U.S. War Department instituted the Remount Stallion Program to improve the light horses of the country for use in their cavalry units. Under this program, the War Department would loan stallions to ranchers as their agents. The ranchers received $10 per head for the mares they bred, plus the actual market value of selected progeny produced. Wealthy breeders of first-class Thoroughbred horses often gave these stallions to the War Department, giving the program a sound genetic basis.

Parker Ranch was, by a wide margin, the predominant remount station. In a separate report, J. C. Fitzgerald affirmed the relationship of the Thoroughbred horse to successful remount programs for the nation and Hawai'i:

> The value of this importation, with proper handling, is unquestionable and it is to be hoped that other breeders will follow the good example and make importations of thoroughbred stock while the American market for them is such that they can be purchased for about half their real value, owing to legislation prohibiting racing in the major portion of the United States. Many adverse criticisms have been made and are frequently heard in reference to thoroughbred blood, but it is the only source from which any real improvement to saddle stock can be reasonably looked for. This is clearly demonstrated by the breeding methods adopted by the Governments of the powers of the world in the breeding of remounts. In citing a few instances where most costly thoroughbred stallions are being used for this purpose, mention may be made of the acquisition of "Henry of Navarre," "Octagon," "Footprint" and "Tipsand" by the United States Breeding Bureau. These animals were presented to the nation by Mr. August Belmont, who, as a breeder of world-wide fame, could see the necessity of strengthening the country's defenses, through building up the standard of the remount. The purchase by the German Government of "Galtee More" one of England's great thoroughbreds, for $100,000, is also another instance.

Ample credit must be given to other remount ranchers throughout the Territory. Parker Ranch on both flanks had Kūka'iau and Kahuā Ranches as significant breeders. Kapāpala Ranch in Ka'ū was also well equipped to do the same.

Norgaard recapped the remount background in his 1914 report:

> The Parker Ranch Company alone has aided immensely in increasing the value

of the live stock in the Islands, although practically every ranch manager in the Territory has departed from the use of common or cross bred sires. That there still remains much unimproved stock is natural, ten years is but a short period when it comes to revolutionizing an extensive industry, and if the same progress is maintained during the next ten years, it is safe to predict that the "scrub" will become extinct, as hundreds of pure bred pedigreed sires and dams have been imported. During the past few years the stock breeders of the Territory have sold to the United States Army hundreds of island-bred horses, fully up to their stringent requirements, and what is more, the local breeders may have to supply all the animals needed by the cavalry and artillery stationed here, because it will probably be impossible for them to obtain their animals anywhere else, unless the United States, like other neutral countries, forbids the exportation of horses and mules.

Cattle and Sheep Importation

By 1907, the DAI began in earnest to report on the quality and quantity of imported purebred cattle. Of the sixty-seven head imported into Hilo, only twenty-one were dairy cattle. The balance was made up of registered beef cattle, mostly of the Hereford breed with a few shorthorns. Of the Hereford bulls, several were blue ribbon show animals and, after quarantine, Parker Ranch took delivery of them at Makahālau Station.

During 1907–1908, A. W.'s former acquaintance at Moloka'i Ranch, George C. Munro, launched a series of shipments of registered bulls from his home ranch near Auckland, New Zealand. Munro also shipped numerous breeding rams. The territorial BAF gave emphatic notice to help Munro and New Zealand Secretary of Agriculture John D. Ritchie execute these shipments, based largely on Munro's excellent reputation as a stockman, coupled with that country's disease-free status. At the time, the U.S. Bureau of Animal Industry mandated that all breeding animals be registered and that a 90-day quarantine would be imposed. The DAI accommodated these factors with emphasis, and A. W. acknowledged the doubly imposed standards with enthusiasm.

Acclimatization

In his 1914 DAI report, Norgaard provided information on sheep imports: "As in former years, the importations have been confined to two breeds, Merino and Shropshire. The Parker Ranch has recently imported one hundred and ten of the finest rams, the shipments consisting of fifty American Merinos, fifty New Zealand Merinos, and ten Shropshires."

A known factor was the slow acclimatization process in cattle and sheep coming out of New Zealand and the mainland. Parasitism and inferior forage were the reasons given by the DAI for the delayed acclimatization of all imports.

In those times, deworming agents were not commonly available. The products that were used—such as copper sulfate drenches—were quite toxic, even in mild overdoses. Some of the other importing ranchers lacked the flexibility of having isolated paddocks at higher elevations with rich forage. Parker Ranch, of course, set aside small, high-elevation paddocks that teemed with rich grasses and clovers. At these higher elevations of 4,000 feet, the annual freeze of winter would rid the forage and soils of worm eggs, making these set-aside paddocks quite hygienic in a natural way. These imports were frequently monitored for health status by qualified personnel (Napier, Vredenburg, and Wight) who, for the first few weeks, provided daily grain feedings. Such supplementing with feed was gradually tapered off to nil over three months time, as the imports gradually grew accustomed to their new home. For the blue ribbon single-herd sires, however, daily grain feeding was permanent.

In their annual report dated December 31, 1916, Norgaard and his assistant Case held Parker Ranch up as the territorial standard that other ranchers and plantations should strive to attain:

> The effect of pure-bred bulls on the beef production of the largest ranch in the Territory—the Parker Ranch—has been discussed in many of the previous reports of this Division, and is now an established fact and no longer an experiment; and with such evidence in plain sight there seems to be little excuse for the continued use of scrub bulls. The argument advanced, that the pure-bred bull cannot live under the natural conditions upon which the scrub can thrive, and the high-class bulls are hothouse plants, requiring special care and feed, good only where no effort is made at acclimating the imported animals or giving them time to accommodate their systems to new and perhaps less favorable conditions than those they were use to. Pure-bred bulls, for instance Herefords or Shorthorns, whether raised in Oregon or in New Zealand, but range-bred-and-raised, cannot be brought here and turned out in the breeding paddocks, unless the same are exceedingly rich in feed, without going to pieces. The management of the imported range bull requires considerable care and expense, at least during the first year, and unless that is provided, the animal may be a worthless wreck at the end of the first season, or before it is half over; and no amount of subsequent care will, in some cases, put it on its feet again. Such animals should, therefore, be watched very closely, should be seen at least once a week, and should be removed from the breeding paddocks at the first sign of their weakness. The same applies to imported rams, and in either case, the point should be made to have such breeding stock reach the Islands several months previous to their being needed in the breeding paddocks, or should be imported as yearlings or two-year-olds. Delays and mishaps in shipping are unavoidable under the present conditions, and even under the most favorable ones, ample time must be allowed for acclimating and familiarizing the imported animal with the new conditions under which they are to live.

The proper post-entry acclimatization standards that these men prescribed continue to be absolute mandates in handling imports to this very day, with the care of stallions included. It took a year for imported breeding animals to properly acclimatize given good care over a century ago. Imports of the new millennium demand the same care.

It was indeed high praise for A. W. and Parker Ranch, coming from the Territorial Division of Animal Industry, that the "Parker Ranch Standard" was described as the unqualified ideal.

Other Venues and Observations

Dr. Elliot was instrumental in establishing in 1915 the Annual Hawai'i County Fairs for farm produce and livestock display. Their popularity grew significantly, and Elliot submitted the following report after the 1916 livestock show at the Hilo Armory:

> The second annual Hawaii County Fair was held in the latter part of September, 1916, and proved to be a remarkably successful undertaking from every point of view. The attendance during the three days that the fair was open was estimated at over 15,000 people, many of whom were visitors from other Islands.
>
> Two of the sections which excited most remark on this occasion were those devoted to live stock and poultry. In the case of the former this was due to the astonishing exhibit of Hereford cattle and native-bred mules from the Parker Ranch of this Island. Very few people had had any previous conception of the great strides that have been made in the live stock industry in Hawaii, and the splendid appearance and general excellence of these animals made a favorable impression.

In 1917, Parker Ranch imported another hundred New Zealand merino rams. By 1918, DAI reports of mule and cattle imports were declining in number. This is evidenced by a statement made by Elliot largely crediting Parker Ranch with filling the need for such animals locally:

> The splendid development of mule breeding, of which the Parker Ranch has made such a signal success, has been dealt with in the report of the deputy territorial veterinarian for East Hawaii. His surmises that this ranch will be able in the near future to supply all the mules required by the local agricultural industries and thereby do away with the costly and unsatisfactory importation of mules from the mainland are no doubt correct.
>
> The same ranch has also taken the lead in breeding up its beef cattle and is now in a position to supply high class purebred Hereford sires to the cattle men of the Territory.

Elliot goes on in the same report to provide a status statement on the sheep industry: "The sheep industry is gradually being crowded out by the cattle and the demand for mutton also seems to be decreasing. Only the Parker Ranch continues to keep up the qualities of its wool and mutton by annual importation of high class rams from New Zealand."

In the 1919–1920 DAI reports, the impact of the automobile began to surface, but the author, Dr. Case, reinforced the role of Parker Ranch in satisfying the mule market. By this time, Case had established Waimea as his permanent residence and continued to assist Parker Ranch in striving for excellence in animal husbandry.

11

LAND ACQUISITION AND RANGE MANAGEMENT

1900–1949

A. W.'s early concerns for acquiring and converting more lands from lease to fee were matched with concerns about the lack of water delivery to some of the strongest grazing lands of the Ranch—Waiki'i. By June 1900, he retained the services of C. H. Kleugel to prepare an engineering proposal on hydraulics.

Both of these issues required the consent of Sam Parker, given his half-interest ownership at the time. Obtaining consent for the water improvements proved to be an early challenge for A. W. in his dealings with Sam, who immediately rejected the thought based on advice given to him by his own hydraulics experts. They feared that friction within the lines would keep water from ever reaching the foothills of Mauna Kea. The key water head was the Alakahi intake, located well up on the southern slope of the Kohala Mountains, from which thousands of gallons of available water could be tapped on a daily basis. A. W. had his homework done on the hydraulics to the point of neutralizing concerns for friction. He retained the services of A. W. Williamson to chart the course for the waterlines. Sam, however, remained recalcitrant, as he

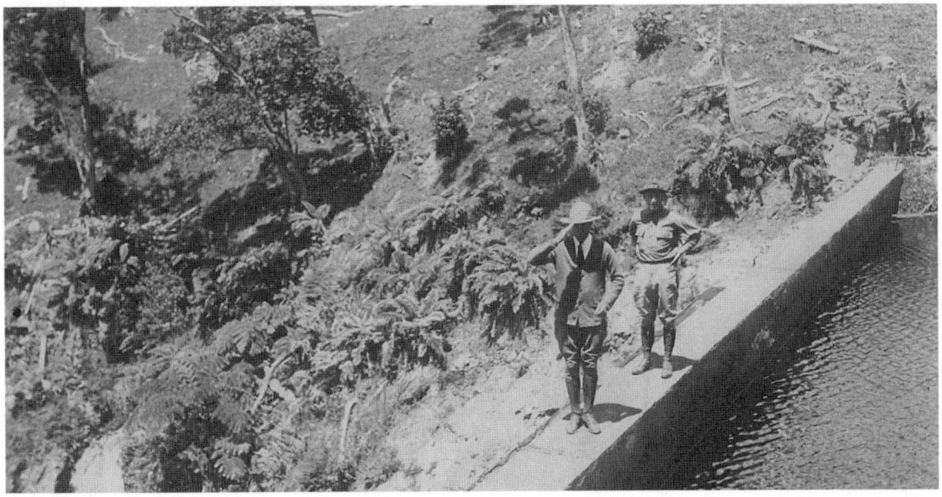

Engineers surveying the cistern/water head improvement of the Parker Ranch. A. W. Carter Collection.

wanted any available ranch income to come to him and not be plowed back into improvements. His attitude was diametrically opposed to that of A. W.

A. W. anticipated the moment when the pipeline stalemate would be broken. He patiently awaited this eventuality, although the ranch was barely recovering from arid pasture conditions and losses in beef production. Within a few months, Sam appeared before A. W. with a request for a cash advance of $5,000 for a trip to Washington, D.C. Seizing this opportunity, A. W. offered a personal loan for the amount, conditioned upon Sam's agreement to the Waiki'i pipeline proposal. Sam stormed from A. W.'s office, rejecting the proposal as a holdup proposition, despite the opportunity to allow more than 7,000 gallons of water to flow daily to the grazing fields of Waiki'i.

A. W. again waited patiently for the few days that it took for his proposal to sink in. Sam reappeared at his door to collect on the $5,000 offer—and signed off on the pipeline proposal. Materials were acquired and construction began on January 30, 1902. On the project completion date, April 7, 1902, the seventy-day pipeline mission climaxed with the delivery of nearly 16,000 gallons on its first day in use—more than twice the minimum projections. With Waiki'i now provided with water, A. W. celebrated the success of his first major challenge involving Sam Parker.

The streams of Waikoloa and Keanu-'i'omanō always traversed through Waimea, but the actual rights to the former in terms of value to the ranch were never spelled out. Although A. W. instituted proceedings for clarification in 1914, the final resolution of this issue did not come to pass until 1970, twenty years after his passing. The Kohala Mountains, from historic times to the present, offer a bountiful water head, servicing North and South Kohala and Hāmākua, including the eastern parts of Parker Ranch as far east as Pā'auhau. The history of water resources and their development on the ranch is worth a book of its own and is beyond the scope of this text. However, among the three crown rules of animal husbandry—the other two of which are adequate feed and appropriate minerals—sufficient water holds the highest order of importance in animal caregiving.

The Lands of Waikoloa: 1903

Earlier on, Parker Ranch had secured a lease on the *ahu 'ili* of Waikoloa, a land section abutting the *ahupua'a* of Waimea that contains about 95,000 acres. With a short term remaining on the lease, A. W. approached Sam Parker for support in buying this parcel in fee simple. Initially, Sam agreed and A. W. approached the owner, Ms. Lucy Peabody. Shortly thereafter, Sam withdrew his agreement of purchase and A. W. sheepishly informed Ms. Peabody that the deal was off for the lack of Parker's support. He convinced her to grant a first right of refusal should another offer come in.

Within a few months, Ms. Peabody honored the request and notified A. W. of an offer that had come in on the parcel. A. W. was aggrieved to learn that none other than Samuel Parker had formed a *hui* (partnership) to purchase Waikoloa. Thelma Parker's cousin Eben Low, along with Fred Wunderberg and Alfred Magoon, provided the financial backing for the offer. The proposal left Thelma out completely.

When confronted by A. W., who spared

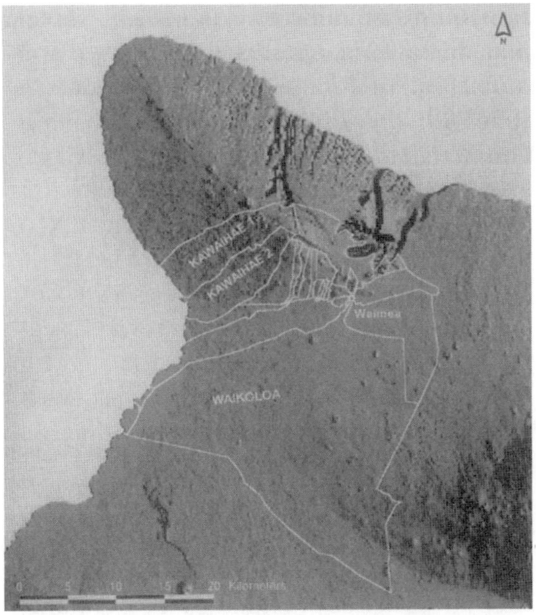

This map illustrates the extent of Waikoloa, stretching from the coast inland to Pu'u Ke'eke'e, near the present Pōhakuloa training area. Its acquisition in 1903 expanded the area of Parker Ranch tremendously.

no words of disdain for his reprehensible business conduct, Sam relented. In the presence of his former financial supporters, Sam agreed to jointly purchase the lands of Waikoloa for Parker Ranch instead. On July 13, 1903, the purchase documents were executed, with the ranch acquiring nine-tenths ownership of the lands of Waikoloa for $112,000, safeguarding the pasturelands (and surviving another Sam Parker episode). Shortly thereafter, Sam approached A. W. with an offer to be bought out of his one-half interest of the entire ranch. The next day, A. W. offered him $600,000, and Sam left to think it over.

When A. W. assumed the reins of the ranch, the combined total of fee simple land owned by Thelma K. Parker and her natural grandfather Sam Parker amounted to only some 34,000 acres, with large parcels under lease. A. W. was immediately concerned that if neighboring ranchers acquired those leases, it would be necessary to dispose of a large part of the cowherd. In the short span of a dozen or so years, the fee simple holdings of Parker Ranch grew to 314,000 acres. Strategic government leases also were acquired.

Significant acquisitions during this period were the lands of Waikoloa, Pu'uloa Sheep and Stock Co. (including Keamuku-Kālawamauna), Humu'ula, Ka'ohe, Waipunalei (purchased from Sam Sr.), and Kahuku Ranch. Later, Pu'uhue and Puakea Ranches (Kohala Ranch Co.) were acquired, bringing the total fee lands of the ranch to approximately 327,000 acres. The final purchase of the Wight Estate, Puakea Ranch, did not occur until 1946.

Sam Parker Is Finally Bought Out: 1906

There were several things in the air that caused A. W. to believe that Sam was mounting a hostile takeover of the entire ranch instead of accepting his $600,000 offer. The first salvo came in the form of a petition filed in court by John S. "Jack" Low that he be named replacement for A. W. Carter as guardian of the Thelma Parker Trust. A consortium made up of Sam Parker, Eben Low, Alfred Magoon and Fred Wunderberg, who had Sam's power of attorney, was behind this scheme.

Sam Parker further claimed the entire ranch possession based on the premise that he and his uncle, John Palmer Parker II, were actual partners, thereby blocking A. W.'s power as guardian to borrow money

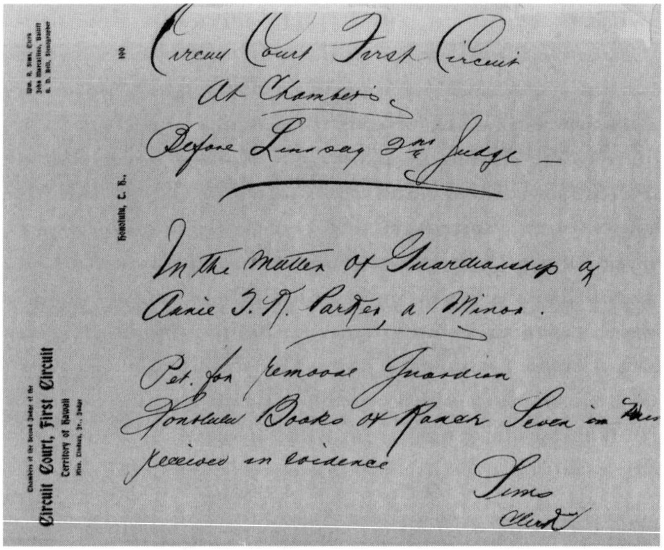

Circuit Court First Circuit
At Chambers
Before Lindsay 2nd Judge —

In the Matter of Guardianship of
Annie T. K. Parker, a Minor.

Pet. for removal Guardian
Received in evidence

Sims
Clerk

Circuit Court, First Circuit
Territory of Hawaii

This petition was intended to remove A. W. Carter as the guardian of Thelma "Annie" K. Parker. It ws entered by Jack Low, with Eben Low, Sam Parker, Alfred Magoon, and Fred Wunderberg as copetitioners.

in mortgage form to buy out the half belonging to Sam.

Things got downright mean when Eben Low met A. W. at Parker Ranch headquarters armed with a pair of six-shooters to demand full possession of the ranch in behalf of Sam Parker. In typical A. W. style, Eben Low received a brutal dressing down and left the office in a state of total disarray. Within two months, the court dismissed the case against A. W.'s role as guardian, with all court costs being borne by the plaintiff, Jack Low.

Two more years transpired in which Sam continuously battled the interests of Thelma Parker in court. In the interim, Carter brought in E. E. Conant to manage the day-to-day ranch functions while the legal battle ensued. A. W.'s perseverance paid off when a worn-down Sam Parker, with his attorney at his side, entered into serious negotiations to be bought out for $600,000, the same price offered to him prior to the drawn-out litigation. On May 8, 1906, the decree was filed in the Third Judicial Court, followed on September 20 of the same year by a partition deed filed in the First Circuit Court. Sam Parker received, for the sum of one dollar, by conveyance, certain small parcels of land, and he in turn conveyed his half interest in the balance of Parker Ranch lands for the sum agreed upon—$600,000.

Strategic Adjoining Lands Are Secured

By acquiring the *ahu ʻili* of ʻŌuli, shortly after the purchase of Waikoloa, Parker Ranch secured 4,000 adjacent acres that bridged property lines approaching Waimea town. The purchase included the former homestead of Frank Spencer and James Louzada, a well-watered 50-acre parcel that was once the headquarters of the Līhuʻe Cattle Ranch and Beef Packing Establishment.

A. W. Carter had bid successfully on the Puʻuloa Sheep and Stock Company at a land auction. The company included 5,600 head of sheep and a shearing shed complete with wool press, shearing gear, and related *ukana* (equipment/paraphernalia). On a limited basis, this sheep venture was a primer for A. W.'s purchase of the Humuʻula Sheep Station. The Puʻuloa Sheep and Stock Company added another 3,700 acres of fee land, together with two territorial leases.

Kahuku Ranch Acquisition: 1912

A. W. became aware that Kahuku Ranch, located above South Point on the opposite corner of the island, was up for sale. The land of Kahuku, originally 308,000 acres, once belonged to King Lunalilo until sold in 1860 to Charles Harris for $3,100. Shortly thereafter, a former whaler named Captain Robert Brown bought the ranch. Discouraged by the devastating effect of the 1868 lava flow, Brown sold the ranch to a consortium of ranchers named Reed, Richardson, Jones, and Kina.

About a decade later, Kahuku Ranch was sold at auction for $27,000 to a Colonel Norris who in 1910 deeded the land to Reverend MacComber for a mere $1800. Later, a Judge Cooper quietly acquired the ranch and upon inquiry by A. W. Carter offered it for sale for $150,000. Subjected to Carter's bold negotiation, the judge accepted a tender offer of $100,000 in October 1912.

A. W. felt that acquiring this 159,000-acre ranch—though it had minimal improved pastureland—was a strategic guard against government leases soon to expire. On behalf of the Thelma Parker Trust, he made the purchase, justifying space for another possible three thousand cows as a sound hedge against the risk of lease losses. Furthermore, his foresight on the appreciation of property values was keen. Having paid just $100,000 for Kahuku Ranch in 1912, he sold it in 1947 for more than $436,000. He disposed of this ranch while securing the Kohala Ranch lands of Pu'uhue and Puakea—the former Wight Estate property. This was a wise move, exchanging the remote Kahuku operation for the nearby Kohala Ranch.

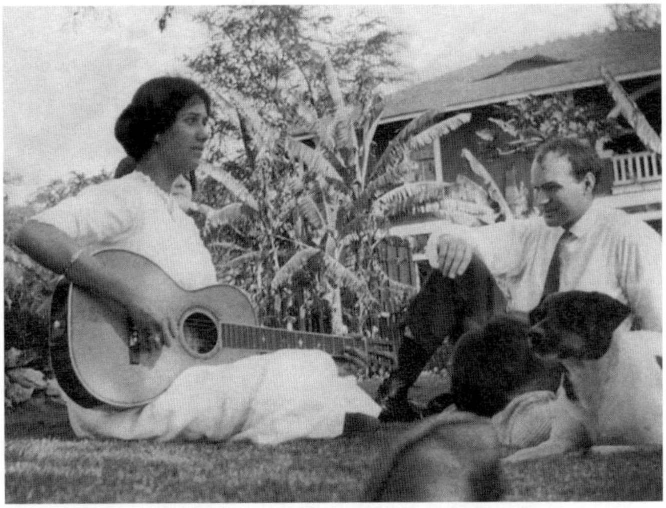

Thelma K. Parker as a young adult entertaining a malihini *guest she befriended on the cruise ship, SS* Manchuria. *A. W. Carter Collection.*

Thelma K. Parker Comes of Age: 1912

Thelma's upbringing by her mother was one of unspoiled innocence combined with reasonable discipline. She was given routine household chores, coupled with ample freedom, often enjoyed with her guitar. Many photos capture Thelma with her guitar across her lap, whether alone or with a friend or two. Although her facial expression often appeared somewhat melancholy, it took nothing away from her striking beauty. Her sensitivity and sincerity were traits of her mother; her love for music came from her father.

On one of her voyages with her mother, Thelma met a young man from Virginia who was headed to the Islands for employment with the Trent Trust Company. Henry Gaillard Smart was the product of a fine family of the old South who came with appropriate manners and credentials. It was a chance meeting that launched a storybook romance. The tall, dark, and handsome Smart wed the statuesque beauty Thelma at Pu'u'ōpelu on January 25, 1912, just prior to her eighteenth birthday.

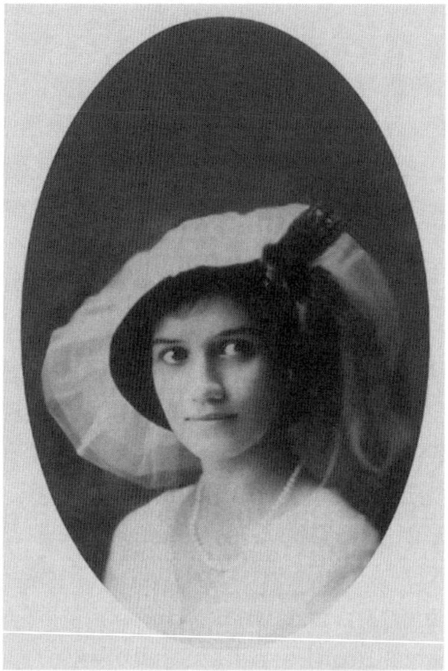

Thelma K. P. Smart as a young bride. A. W. Carter Collection.

Thelma's coming of age triggered a change in A. W.'s legal position. According to the law, A. W.'s guardianship of Thelma was terminated. The legal relationship between the two was restored shortly thereafter with the execution of the Thelma Kahilu'onāpua'api'ilani Parker Smart Trust (TKPST) deed, making Carter trustee of her property until her death or that of her mother, whichever occurred last. Her property and wishes were secure with Carter's reestablished role, but the star-struck marriage soon faced difficulties.

The Ultimate Challenges: 1913–1915

Despite her frailty, Thelma gave birth on May 21, 1913, to a healthy baby boy named Richard Palmer Kaleiokū Smart—the male heir to the John Palmer Parker legacy and Parker Ranch holdings.

Thelma's frailty of health may have been seated largely in a subclinical infection of pulmonary tuberculosis. Giving birth, nursing, and rearing a newborn naturally challenges the constitution of any new mother—and particularly a young and delicate heiress like Thelma. The less-than-pleasant relationship between her mother and husband added to Thelma's burden.

When Richard was less than a year old, the family embarked on a trip to Europe. On the sailing voyage to France, Thelma, again pregnant, began to show clinical signs of a systemic illness. She was hospitalized in Paris and gave birth shortly thereafter to a baby girl, Elizabeth Ella, a fragile newborn.

It was 1914 and the war was on, with France vulnerable to invasion. The Smart family had no recourse but to flee under the most difficult circumstances. They escaped to Le Havre within minutes of the slamming of the gates of Paris. With a nurse/nanny in tow, the Smart family of four sailed for the United States. Baby Elizabeth Ella died shortly after their arrival in New York. Smart was overwhelmed by the loss of their new baby, and what little constitution he had left was shattered upon learning that Thelma's confirmed infection was tuberculosis. Despite his tentative relationship with his mother-in-law, who had remarried and was now Mrs. Fred Knight, he sent for her. She traveled to New York from California, resolved to bringing her daughter and family back to the Islands. Thelma never reached home. She died in San Francisco at the home of her mother on November 17, 1914, surrounded by palm fronds and flowers of the Islands and soft strains of the Hawaiian guitar that she so loved.

Within less than a year, in November 1915, Richard's father, Henry Gaillard Smart succumbed to meningitis at the home of his parents in Virginia. While this left Richard parentless, it resolved the issue of being inadvertently left out of his mother's will. A. W. Carter had contested the will on behalf of the child. Working with no salary, he challenged the will in which Thelma's holdings would have reverted directly to Gaillard Smart, whose interest in the ranch was less than robust. With his passing, the issue was settled. Richard's grandmother, Mrs. Elizabeth "Aunt Tootsie" Knight, became his legal guardian, and A. W. served as vigilant trustee and protector for a second generation of Parker progeny. Mrs. Knight and A. W. had weathered unfathomable challenges over the years, sharing in grief and caregiving, but they also shared the joy and satisfaction of securing a bright and wonderful future for baby Richard Smart.

As testimony to his strength and dedication during these trying years, A. W. not only kept the ranch going, he also followed through on the Thelma-approved expansion plans to acquire the ranch lands of Humu'ula and Kahuku.

While this tortuous period took a great toll on A. W.'s emotions, it did not keep him from examining happily the recently acquired Humu'ula Sheep Station. His enthusiasm is reflected in his comments: "I am more than pleased with the deal. The sheep are superior, the cattle very good and while there are a good many scrub horses, there are a good many which are saleable this year and next to the Army. The sheep are ready to be shorn at the present time and they will average a dollar a head for wool. If Sam had come to me, I should have advised him against parting with the place but I imagine he did the right thing so far as his own interest is concerned. I felt that it would be a great mistake to let it go to other owners." A. W. had shared with Thelma the logic behind acquiring Humu'ula.

The lands of Humu'ula were mainly made up of government leases, but they were well endowed with improvements for sheep production, such as shearing sheds, barns, stables, and homes. One of these homes was somewhat ostentatious, obviously the former quarters of Samuel Parker Jr., from whom it was purchased. Coincidental with this transaction was the purchase of 23,000 sheep, 500 head of cattle, and 400 horses. As mentioned earlier, many horses were sired by the fine stallion Frolic, acquired years earlier by Sam Parker Sr. The sheep were ripe for shearing, and—at a dollar per head wool value, given the total flock of nearly 30,000, which included Pu'uloa and Keamuku—the returns were quickly garnered. At the time of the Humu'ula purchase, which included a wide band of parcels made up of Kalai'eha, Pu'u'ō'ō, Laumai'a, Hopuwai, and Keanakolu, the 7,000-acre deeded parcel of Waipunalei was bought from Sam Parker Sr. Now Parker Ranch, except for the Horner Ranch properties (Kūka'iau), completely encircled Mauna Kea with grazing animals.

Kohala Land Acquisitions: 1932

Kohala Ranch Company consisted of two components: Pu'uhue Ranch of the Woods family and Puakea Ranch, which was part of the Wight Estate. Pu'uhue Ranch rose from sea level high into the northern slope of the Kohala Mountains. Puakea Ranch was made up of lands at lower elevations that also reached the ocean. Collectively, Kohala Ranch Company comprised 12,000 acres of fee simple property and another 10,000 acres in lease lands, carrying about five

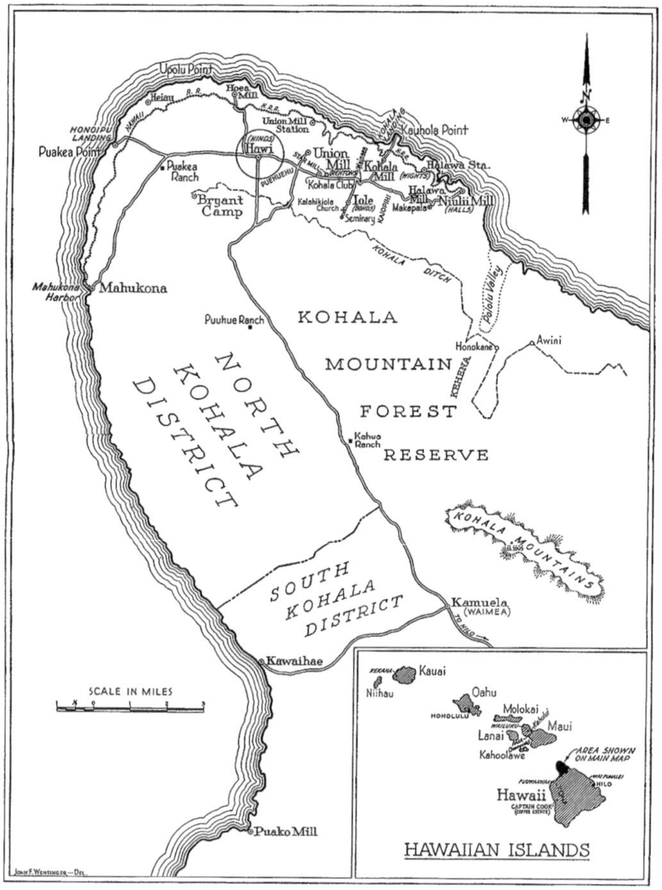

This map illustrates the relationship of Puakea Ranch and Pu'uhue Ranch to various other ranches and plantations of North Kohala, including shipping points, railways, and roadways. Map from Hind, John Hind of Hawi.

thousand head of cattle. A. W. saw as much value in additional cattle production as there was in water rights. On February 23, 1932, he closed the transaction, which, beyond pastureland, included cane and pineapple lands and a valuable lease of the right-of-way for water.

There were two problems that came with the Kohala Ranch Company. The Woods heirs carried a mortgage of nearly $200,000 on Pu'uhue. Kohala Ranch Company was also in the midst of a drought. Acquisition of this ranch by Parker more than relieved the Woods heirs of debt; it also provided relief value, as hundreds of cattle were driven over to the home ranch near Waimea to avoid death. A. W.'s posture in acquiring this land was shrewd stewardship on behalf of Parker Ranch combined with compassion, as the Woods heirs and Kohala Ranch stockholders were duly compensated. Finally, the cattle could be accommodated with adequate interim pasture until the drought ended.

Leasehold ranch lands belonging to the Wight Estate still remained as part of the Kohala Ranch Company transactions. Approximately thirteen years later, these lands—about 5,800 acres—were purchased in fee for $182,400 as an addition to Parker Ranch's Kohala land acquisitions. Under a parallel arrangement, the water rights-of-way through these lands were conveyed to Kohala Sugar Company, which in turn agreed to provide 120,000 gallons per day to Parker Ranch's Kohala lands. More than 12 miles of pipelines served these lands. Despite some guava infestation, these were well-subdivided paddocks with secure fencing that rendered a high level of productivity under A. W. Carter's management.

With these expansive moves came some immediately marketable assets, which included breeding herds with pending calf crops. Quick returns on this expenditure provided testimony to the sound decision making. The Puakea Ranch purchase yielded five breeding jacks. Water was an asset acquired along with the Kohala Ranch purchase in 1932. All paddocks were watered via existing pipelines served by five water heads on the property. These water heads *('ahawa)* were further tapped to serve adjacent (prepurchase) ranch lands, bolstering their carrying capacity.

Range Management: 1900–1949

As the twentieth century dawned and the Carter era began, progressive efforts in all aspects of Parker Ranch management were well documented. Carter's early steps in range management were practical ones. With fencing and water development plans underway, he established controlled grazing as a means of improving plant cover and for healing heavily eroded areas.

Forage Importation Campaign

In his previous ranching endeavors on Moloka'i, A. W.'s keen relationship with G. C. Munro, formerly of Moloka'i Ranch, continued when he returned to his homeland Down Under. Beyond being a resource for purebred cattle, Munro also provided ready access to grass seed—especially of Rhodes grass, which was successfully introduced and flourished on Moloka'i. Introduction of Rhodes grass to Parker Ranch was one of Carter's first moves.

Beginning in earnest in 1906, tons of imported grass and legume seeds were scattered about the ranch by the cowhands, timed carefully to follow rain patterns. Brome grass or wild oats *(Bromus uniosoides)* had already been moderately established, but the introduction of two varieties of rye grass was significant; they could withstand heavy grazing, and they were highly palatable. Orchard grass from Eurasia was also a significant admixture, as this grass thrived around the shaded tree bases infested by Hilo grass *(Paspalum conjugatum)*. Orchard grass at appropriate elevations replaced Hilo grass, as did Kikuyu grass. By its bunchgrass nature, orchard grass around tree bases allowed for coexistence with mosses, providing reasonable retention of moisture and ventilation. It is palatable and nutritious, like sweet vernal, which arrived about the same time (but more likely by accident.) A fragrant, perennial bunchgrass, sweet vernal still exists today as utility forage and is moderately drought resistant.

In 1915, Natal redtop *(Tricholaena rosea)* was introduced to the ranch. A seasonal perennial of the bent grass family with creeping underground rhizomes, redtop appears most evident in the Keamuku Division, especially after early spring rains. Its red tops cast a pleasant afterglow in the sunsets when viewed from Kamuela on clear evenings. It provides meaningful nutritional value and excellent palatability.

Weeds and Other Noxious Plants

Unlike grasses, weeds and noxious plants rarely have a person's name associated with their introduction. Some weeds survived for a relatively short period of time, were of minimal impact, and were eradicated, thereby not meriting discussion.

The alien cocklebur or *kīkānia* was an early nuisance that readily spread after rainstorms, especially in elevations around 2,500 feet. The leaves were described early as having potentially toxic characteristics when consumed by cattle and sheep. Current veterinary literature does not describe this toxic nature. While the matting of the tails of horses and cattle remains a nuisance into these times, *kīkānia* seems to be a self-limiting pest of little import. Its most significant historic contribution is as the nickname of the popular blacksmith spur—*kīkānia* spur—from its sharp, multi-pronged rowels that resemble the burr of the cocklebur.

Early ranch writings refer to *oʻī (Verbena liltoralis)* as a pesky weed that commonly infested the better pasturelands. A. W. Carter made mention of this perennial that could grow as high as 6 feet, with branches that terminate in panicles densely covered with lilac blossoms. An unrelated weed, Hāmākua *oʻī,* was commonly found in lower-elevation pasturelands. The Hāmākua variety was woodier and bore blue flowers.

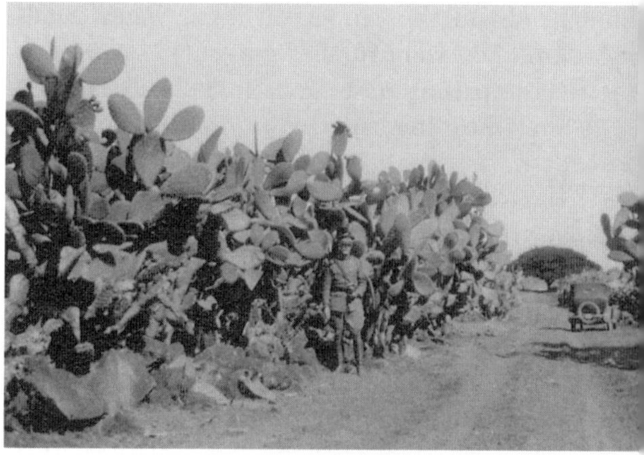
A lofty stand of cactus on the road through Keamuku. 1911. Hartwell Carter Collection.

Cactus and the Eradication Campaign

Cactus holds a special place in early ranch history. Called prickly pear *(Opuntia megacantha* and *O. cordobensis)* in mainland circles, *pānini* is the Hawaiian name. Francisco de Paula Marin *(Manini)* is credited with introducing it to Honolulu in 1809 from Acapulco, Mexico. *Pānini* quickly spread to the neighbor islands. On the Big Island, the ports of Honuʻapo in Kaʻū and Kawaihae in South Kohala appeared to be the points of dissemination, as adjacent lands of Naʻalehu and Punaluʻu in Kaʻū and Puakō and Māhukona in South Kohala were heavily infested by the turn of the twentieth century. On Parker Ranch, thick stands occurred on the coastal ranges between Puakō and Māhukona, inland from Kawaihae to lower Waimea, and from Waimea to Keamuku along the old road to Kona. In my early memory, *pānini* climbed up the lower Kohala Mountains *makai* of the road to the lookout point of Kawaihae Uka and beyond. Along the North Kohala shoreline, *pānini* could be found as far north as Kapaʻa and Puakea.

Cactus was valuable in early ranch years for providing moisture and forage for range cattle, especially in the pre-Carter era when water delivery was almost nonexistent. Ranch cowboys enjoyed slicing the *pānini* pear as a snack in the field. Feral swine also benefited by the feed, moisture, and cover it provided. The more innovative *paniolo* such as James Palaika made *pānini* swipe, a rudimentary form of *okolehao,* or white lightning.

As children on trips to Kona, we would stop to enjoy the *pānini,* only to have our fingertips painfully embedded with the tiny pear thorns *(heu)*. On our father's advice, we would run our fingers through his wavy, shiny *lākana* (auburn) hair as he drove on to Kona. Within minutes, we would be relieved of the infestation, while satiated by the *pānini* break. I never did ask my father what finally became of the *pānini* in his wavy hair, but I was always impressed with his lifetime sharing of such country remedies for a variety of afflictions, so common to little boys.

The *pānini* continued to flourish, and it had ravaged the lower plains of the ranch by around 1945. It covered the lowlands almost completely in many areas. *Pānini* boldly moved to higher elevations, affecting better pastures.

Cactus Eradication Campaign

As early as 1930, A. W. called for government action. With only rudimentary tractor equipment, mechanical destruction was too costly. Even with the D-6 bulldozers of the 1950s, mechanical control of *pānini* was not feasible. Between 1930 and 1949, A. W. was hard at work with the Australian government, learning of their success at biological control. He needed to convince both the Hawai'i Board of Agriculture and Forestry (now the Hawai'i Department of Agriculture) and other dryland ranchers of the need to conduct a program similar to the Australian method of eradication. Other ranchers seemed convinced that *pānini* was a necessary evil—even a benefit during times of drought.

While it took years for the ranchers to come around, it took only months for A. W. to enjoin the Hawai'i Board of Agriculture and Forestry in a quiet, well-documented laboratory experiment using fungus disease spores *(Fusarium oxysporum)*. Due caution was appropriate, since other cactus parasites on trial were found to be effective in destroying pineapple. While these were still in vitro (laboratory) experiments, the organisms were immediately destroyed. Even accidental release of these trial organisms could have had a catastrophic effect on the pineapple industry. The prudent research and firm laboratory controls of the Hawai'i Board of Agriculture and Forestry forged ahead. The *Fusarium* spores were proven effective against the red-fruited variety of cactus, but research found that both *Cactoblastis catorum* (Australian-Uruguay strains) and *Dactylopius opuntiae* (Australian-Mexican strain) were effective in destroying both cactus varieties. These cochineal insects were

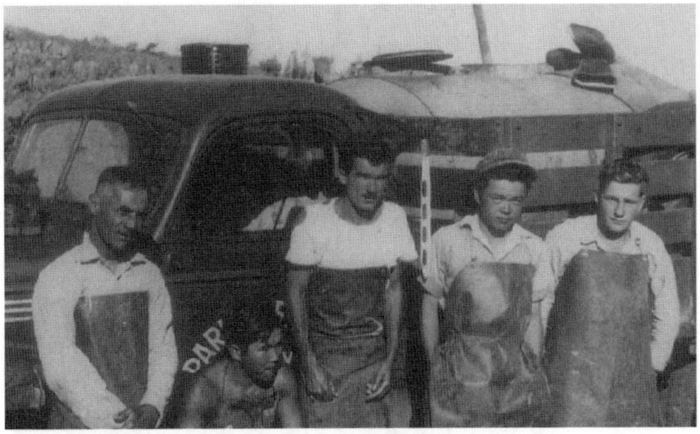

Kuakini's cactus eradication pesticide gang: Kuakini Lindsey, Sakae Hirata (sitting), Joe Cootey, Charles Kimura, and unknown. Hartwell Carter Collection.

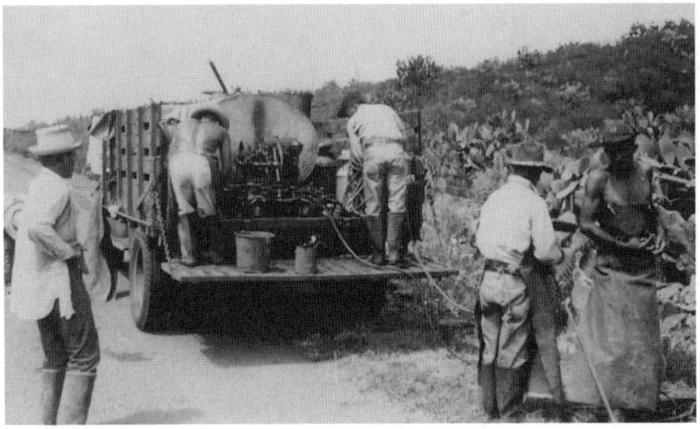

Pesticide crew with tanks and nozzles. Hartwell Carter Collection.

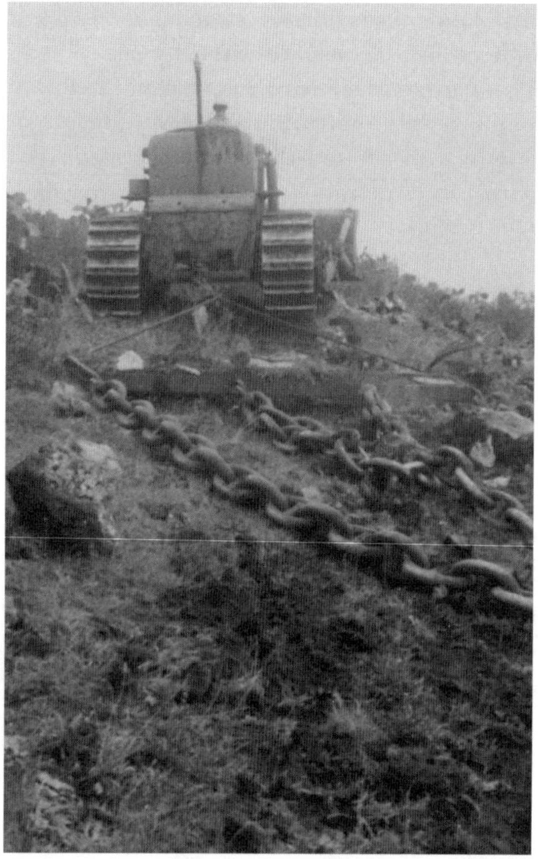

Bulldozer-mounted chain drags control both pānini *and* lākana *(cactus and lantana). Isami Nishie was the bulldozer operator. Hartwell Carter Collection.*

released in 1949, and by 1965 some 60,000 acres of Parker Ranch rangelands were freed of most cactus. Between 1954 and 1965, partial to complete destruction of cactus extended to the Keamuku Section between the 2,000- and 3,500-foot elevation. This incorporated the present site of the village of Waikoloa, *mauka* to the lower boundary of the Pōhakuloa training area. Similar progress was made between Kawaihae village and the town of Kamuela. Another cochineal insect, *Archlagocheirus funestus,* replaced the earlier species for effective cactus destruction along the Kohala Mountain Road. Similar progress in cactus destruction was observed in the Honu'apo (Na'alehu to Punalu'u) area of Ka'ū.

Concerns were legitimate about the negative effect of cactus eradication on forage availability, especially in naturally arid rangelands. Ni'ihau Ranch was denuded of cactus when the cochineal insects crossed the channel after being released on Kaua'i. The loss of the *pānini* forage was catastrophic for Ni'ihau cattle.

It was dramatic for me to have left Hawai'i for college in 1958, when cactus eradication efforts faced significant *pānini* entrenchment, then return home a decade later to find most of the prickly pear gone. Ranges I and II were cleared of *pānini,* along with Holoholokū and all of Keamuku. The Kohala Mountain Road, as it lines the cascade that sweeps *makai* to the shores from Puakō to Māhukona, offers a full panoramic view of expansive grasslands devoid of *pānini.*

Casual observation while driving through these areas of North and South Kohala reveals the minimal presence of cactus plants, which serve the biological purpose of hosting the cochineal insects that brought about eradication. One need only slice open a dried-up pad to view the presence of the squirming larva of these heroic insects. Kohala will always have some *pānini,* as these plants provide a host to colonies of these larvae, ever-present to quell any future outbreak of cactus. We can thank A. W. even for these worms.

12

A. W. CARTER'S LEADERSHIP PERSONNEL

1900

On assuming duties as trustee, A. W. developed a circle of dependable, loyal, and industrious field leaders. Some he brought in immediately, including Manuel Vierra, George Keola, and James Brighter, all products of his Moloka'i Ranch venture. Carter also drew from the ranks of the finer men of the five foundation families—notably the Lindsey and Purdy clans. Also, in the Waimea area, Carter looked to the Yamaguchi, Ishizu, Kawamoto, and Kimura families for their best and brightest prospects. From surrounding communities came the *'ohana* Akau from Kawaihae and the *'ohana* Kaniho from Hāmākua and Humu'ula, introducing names that became prominent in ranch history. Even from foreign lands came men of significance, such as Wilmot L. Vredenburg out of South Africa, who sired a family significant in Hawai'i's livestock industry.

In sequential order, Manuel Vierra was the first hand Carter recognized as having the characteristics he wanted in the men at his side. Vierra, originally an O'ahu resident, had worked on different Carter projects there while he maintained employment in the postal delivery service. As Carter developed the Moloka'i Ranch holdings, he took Manuel and his wife Frances. Manuel's mission was the building of windmills in this parched countryside. He served also as a shepherd of the imported sheep flocks. Vierra's loyal and industrious style was enhanced by his congenial nature that bode well in this Moloka'i venue. Vierra was a craftsman skilled in plumbing, carpentry, and building and fence construction assignments. His work was tidy and usually completed ahead of schedule, provided materials were available. In a pinch, Manuel was a cowboy. His first son, Manuel, was born on Moloka'i, and not long thereafter Carter called on the Vierra clan to join him in his new trusteeship—the Parker holdings. They moved in 1900, joined by another Moloka'i ranch hand, a man who later would become known as Palaika.

Manuel Vierra had thirteen children, five of them sons. Frank, Tony, Herbert, and George were born in Waimea. All of Manuel's sons worked for Parker Ranch at one time or another. Manuel Jr. worked in the Plaster Gang (utility), while Herbert worked briefly in the water department and motor pool but died young of a heart attack. George, the youngest, was nicknamed Pokole for his squat stature. He launched a truck-driving career that took him to Kapāpala Ranch, from which he retired in 1978.

Tony, like his older brother Frank, liked the cowboy life. Tony was a steady force as a roughrider in the Breaking Pen, but he was also a favorite groom of the Carters as he squired saddle mounts about the village for the convenience of family and friends. He was also a handy cowboy when John Samoa's cowboy crew was in need of help. Tony eventually moved to O'ahu.

Of the Vierra boys, Frank emerged as the lifetime cowboy figure, developing a reputation for quiet and steady progress at work but gregarious in nature *pau hana* (after work). Frank married Victoria Kaimuloa of Waimea and they had a son and daughter, neither of whom was connected with the ranch.

Frank's first job on the ranch was in 1918 doing yard work around the office and other facilities. His coworkers included Harry Kawai and Masao Yoshimatsu, both of whom developed important careers on the ranch. All three youngsters reported to Frank's father, and their experiences grew to include tree planting, pipe laying, and dam building high in the hills above Waimea. Both Harry and Frank gained invaluable experience with the myriad water lines reaching over the hills and plains. This exposure later served well in their respective ranch capacities. Both were sent to the Breaking Pen to become horse trainers under the guidance of James Palaika. Frank by this time had also gained mule-training experience with Joe Pacheco at the Pieper Stables.

By 1932, Frank was assigned to the prestigious Cowboy Gang under the foremanship of Willie Kaniho, where he remained a long while. When Kahuku Ranch was acquired, he was often part of the cowboy crew that would drive heifers from Humu'ula across to Kahuku Ranch on the old cattle trail. Market cattle would be returned on the same trail back to headquarters, often by the same cowboys.

In the twilight of his career, he became a *nānā'āina* man, reporting to Hartwell and eventually Dick Penhallow and Rally Greenwell. He retired on September 30, 1970, after more than fifty years of service.

Frank embraced the Hawaiian people beyond his marriage to Aunty Vicky. Even among the Native Hawaiian speakers, Frank was respected for his eloquent command of the *'ōlelo kanaka* (spoken Hawaiian). Frank knew all the verses of the old Hawaiian love songs, and he had the voice to express their charm. He knew the names of each hill on all sections of the ranch, including the remote Humu'ula, Kohala, and Kahuku holdings. He readily shared the background of the ancient names, as well as the recent history of hills like Buster Brown. I regret the lost opportunities I had to spend full days with him on horseback or in a vehicle, as Frank dearly wanted to pass this knowledge on before his quiet death in 1984.

Palaika

James Kaleima'ema'e Edmonds was born on August 24, 1871, one of three children of William Ellsworth Edmonds and Elizabeth Edmonds. While James left home mysteriously as a boy, his younger brother, William Ellsworth Jr., chose a religious life and moved to Hilo. Their sister Elizabeth "Beke" married Joseph Kaleimomimakahikienaenaokalani Cockett, and they had eight children. One of their sons, Edward Ambrose, moved to Hilo after marrying Alice Kalaninuipaopao Crabbe and worked for railroad, postal, and stevedoring enterprises. Edward and his uncle, the Reverend William Ellsworth Edmonds, became close, and this relationship led to the discovery of the long-lost James Kaleima'ema'e, a Parker Ranch cowboy who went by the name of Palaika.

Where did Palaika come from? How did the name "Palaika" come about? How did he come to Parker Ranch? It began with James running away from his Waikapū, Maui, home and reaching the shores of neighboring Moloka'i via canoe, intending never to look back. The family presumed him dead after several years of hoping to learn of his whereabouts. With a natural love for horses, he found work at Moloka'i Ranch under G. C. Munro, then manager, who took him in and nurtured his talents. James had cast away *(ho'oka'awale)* his family name despite Munro's efforts to learn it. During the twilight of Munro's management, A. W. developed an interest in buying Moloka'i Ranch, and he observed the young man's talent around horses and cattle. Under Carter's brief ownership of the Moloka'i operation, his respect for James Kaleima'ema'e was enhanced. George Keola, another ranch hand, also gained favor with Carter.

By 1900 Carter had sold his interest in Moloka'i Ranch as he assumed the trusteeship of the holdings of Thelma K. P. Smart. When this occurred, he brought Manuel Vierra and James Kaleima'ema'e to the Parker outfit. Keola followed shortly thereafter. There was a slight problem with James Kaleima'ema'e, however, who refused to reveal his last name. Carter was not to be outwitted and legally gave him the surname Brighter (so as not be confused with Waimea resident James K. Bright).

On arrival via schooner in Kawaihae, the men were presented with solid mounts for their ride *mauka* to their new careers. The Manuel Vierra family was provided a horse-drawn wagon. The reception by the *kama'āina* (native born) was understandably cool; after all, Parker Ranch was the cowboy capital of the Territory. But this apprehension was short lived, as the new ranch hands demonstrated their skills and strong work ethic. Brighter's new name was readily Hawaiianized to "Palaika," and his horsemanship skills brought him early respect. Carter moved him to the Horse Department within Palaika's first year.

Carter and Palaika shared a deep love for horses, especially Thoroughbreds. By 1907, Palaika was foreman of the entire horse operation, following the career of the highly respected William "Atau" Akau of Kawaihae. He had the efficient help of Pō'ai Nahale'a and George Keola and oversaw the early training of racing prospects, grooming them for the track, polo, and military markets.

Palaika and Keola married women of Waimea who were half sisters on their mother's side. Palaika wed Mary Ann (Maliana) and Keola married Kulili. Mary Ann brought a son into the marriage named Willie, whose father was Kiliona Alapa'i, the cowboy foreman at Pu'uwa'awa'a Ranch. Kiliona was the father of several fine *paniolo* of North Kona, including David "Bully" Simeon and George Alapa'i. Willie was another *paniolo* son to be proud of, and Palaika embraced the young boy as his own son, giving him the surname Brighter. He grew into a highly respected ranch cowboy and horse breaker. Mary Ann and Palaika had a dozen children of their own, none of

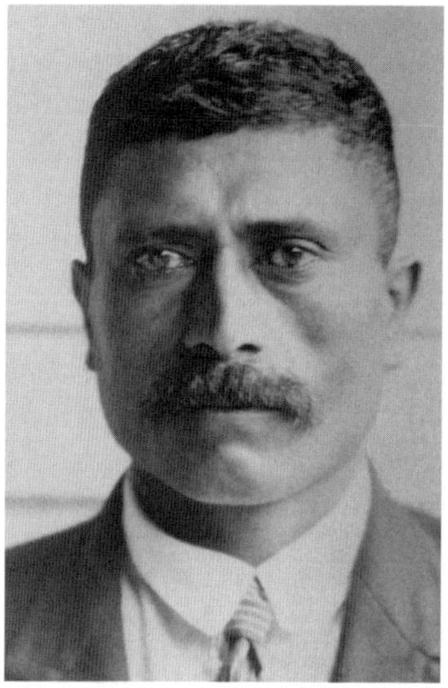

James K. Brighter (Palaika) at age forty-five. Bishop Museum Sullivan Collection, Negative CN 4538.

whom developed ties with Parker Ranch. Most of them moved to O'ahu in early adulthood.

It was likely a serious roping accident that bonded Carter and Palaika for life. On a long *nānā'āina* trek to Makahilinu'u, the two happened upon a rogue bull among some yearling heifers. In hot pursuit of the *pipi 'āhiu* (wild cattle), with lariat loops in full swing, they apprehended the bull. Palaika had roped the bull by the horns when A. W. came dangerously close to pitch his loop around the rogue's heels. The *pipi laho* (bull) spun around and charged Carter's horse. A. W. reflects on the episode in his memoirs:

> Palaika roped it and threw it, but the bull kept jumping up and plunging. So I roped its hind legs, but foolishly hadn't tightly cinched my saddle. I was thrown and jammed against the rocks, the horse took flight and I was rolled into the gulch. When I came to, I was lying on the ground. Palaika was straddling my body giving me lomilomi (massage) around the heart and tears were streaming down his cheeks. He said he thought I was dying. The fact was my leg was broken and we were five miles away from any road and it was mighty rough country. Palaika was a man about the same weight as I. He tied and left the horses and put me over his back. In that way he carried me all that distance to the road. He laid me down and went for a rig, which carried me back to the Ranch. My leg has always given me trouble.

While Carter's leg was a persistent discomfort, equally enduring was his faith in the strength, loyalty, and character of his *paniolo nui* (main cowboy), Palaika.

It was about this time—the early 1920s—that Palaika's brother, the Reverend Edmonds, together with his nephew Edward

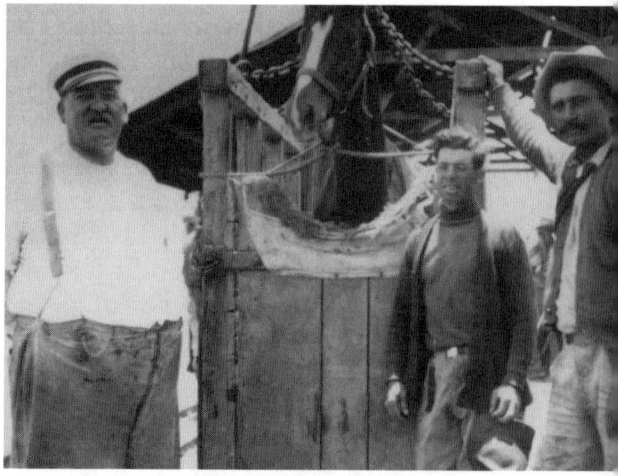

James Palaika (right) with Jonny Carroll and ship's captain on shipping day. A. W. Carter Collection.

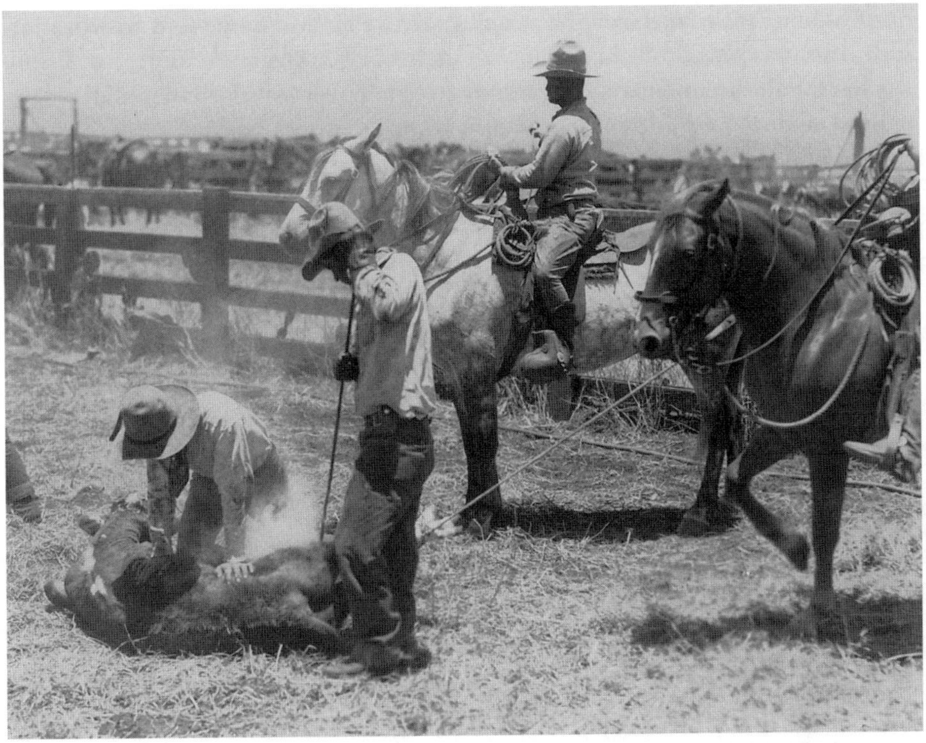

Willie Palaika on grey cow horse. Willie was nicknamed "Pokole" because of his short stature. A. W. Carter Collection.

Cockett, followed up on a lead that the lost Edmonds boy actually lived in Waimea under the given name Brighter. The Hilo men hired a driver to bring them to Waimea, where inquiry quickly led them to a large home in the foothills of Pu'u Hoku'ula. The Palaika homestead stands today as a 10-acre *kuleana* bridging Hoku'ula and Kapi'olani Roads in Waimea. Palaika was out in his round pen training a colt when Edmonds and his nephew stepped from the car. There was an outpouring of emotions between the brothers, extending into a two-day reunion. For Edward, now a parent of seven children, it was a meaningful bonding with his Maui *'ohana.*

In the ensuing years, the children of Edward and Alice Cockett spent full summers living with their newfound family. One of the Cockett girls, Maude, had an only child, Patricia Luellen Cockett. In 1962, Patricia and I were married. In our fortunate lives, Maude has lived with us and raised our extended family. She often recalls staying with *Tutu* (relative of a grandparent's generation) Palaika and his family. She also recalls Willie Palaika spending his evenings in the round pen, gently starting new colts. Maude described much of *Tutu* Palaika's nature.

As Palaika aged, his work became that of *nānā'āina*, riding the assigned paddocks and checking on water, grass, calving, and breeding. The men of the ranch, who felt that he had truly earned his dues, looked up to him. Palaika was a consummate Hawaiian, using little or no English in his daily routine. A. W. Carter, in fact, once advised his son Hartwell to have Palaika and Sam Puna Kauwe excused from jury duty. Carter stated, "Palaika and Puna speak so little English, they could no more understand the court proceedings than if it were delivered in French."

Palaika preceded Carter in death by two years. Upon hearing of Palaika's passing, A. W. wrote Hartwell: "I got an awful jolt the other day when I was informed that Old Palaika had died. He was one I could always rely upon, and who went to work for the Ranch 47 years ago. A more loyal and reliable man never lived than Palaika. He would have given his life for me or any of my children."

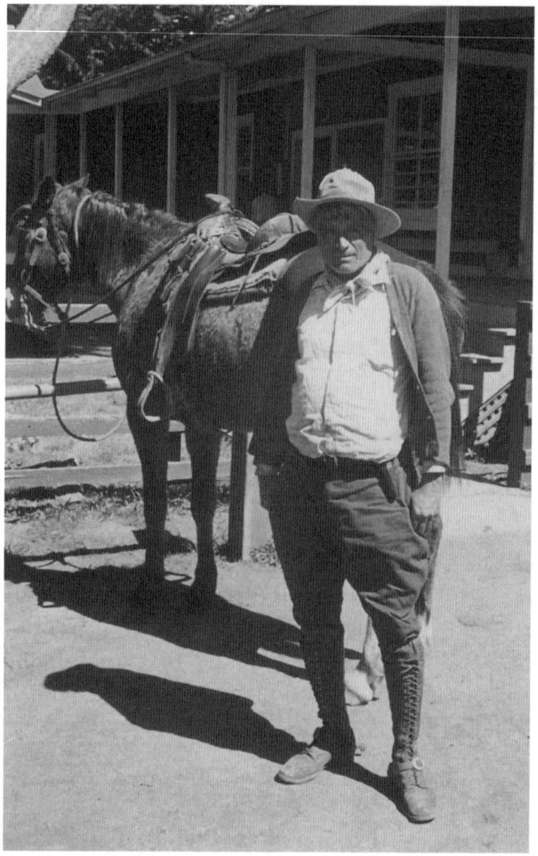

James Palaika near retirement at ranch headquarters. Hartwell Carter Collection.

For Palaika's funeral, Carter made charter arrangements to transport the children and grandchildren from Honolulu. To Hawaiian Airlines he wrote, "You are authorized to issue tickets and charge the Parker Ranch for passage from Honolulu to 'Upolu this afternoon." The manifest listed thirteen adults and eleven minors. These included eighteen Brighters and six Perrys. In closing, he wrote, "In the event more members of the family show up, I understand you can let them have room on the regular plane to 'Upolu at 1:15 pm."

It was Carter's way of delivering a final tribute to a genuine friend, loyal associate, and consummate Hawaiian.

William Paul Mahina'uli Akau

The appropriate place to begin to list the leadership contributions of the *'ohana* Akau is with its patriarch, William Paul Mahina'uli Akau of Kawaihae. He was born in Līhu'e, Waimea, in 1869. William Akau's prowess at sea was matched by his *paniolo* talent. He was a boat and canoe builder as well as weaver of the finest quality *'upena* (fishing nets). Part Chinese, his surname was Hawaiianized from Sen Kau Ching, who wed his mother Kamakahema Awa. William Paul Mahina'uli Akau's first marriage was to a Hawaiian woman named Kealohapauole Kalaluhi, a native of Kawaihae. Four children came from this marriage: two boys, Abraham and Theodore (Akeni), and two girls. William later married Lydia Awa'a of Kawaihae Uka.

Born in 1887 in Kahuā, Abraham married Alice Pualeialoha Ahina of North Kohala, and they also had two daughters and two sons. Solomon, the younger, became a fisherman and park caretaker, while William became a noted horseman and cowboy. William, known as Bill or "Atau" (from Francis ʻĪʻī Brown), was born in Kawaihae Uka. His father, Abraham, dropped the middle name Mahinaʻuli (dark moon), as it seemed to bode ill health. Bill Akau, as he became known, was the first ranch hand designated as the horse foreman, with responsibilities over the Breaking Pen and breeding bands, including the stallions. This occurred at the end of Paul Jarrett's management, just prior to A. W.'s assuming ranch trusteeship. Carter was fond of Bill Akau's dependable and knowledgeable nature and encouraged him to move to Kawaihae, where he was eventually appointed postmaster under federal service, as well as harbormaster under territorial service. In his marriage to Helen Kaohu of North Kohala, they had twelve children, none of whom became associated with the ranch.

Abraham's younger brother, Theodore Akeni, however, developed an important and enduring line of ranch leaders, while maintaining a lifetime role on the ranch, mostly as a stonewall builder of the finest order. Theodore had seven children, two of whom were prominent cowboys and eventual foremen. Theodore Akeni II, known as Sonny Akeni throughout his career on Parker Ranch, was the eldest. Abraham II was known as Manny.

Manny and Sonny were natural horsemen, though a little on the rough-and-ready side, but their consecutive assignments to

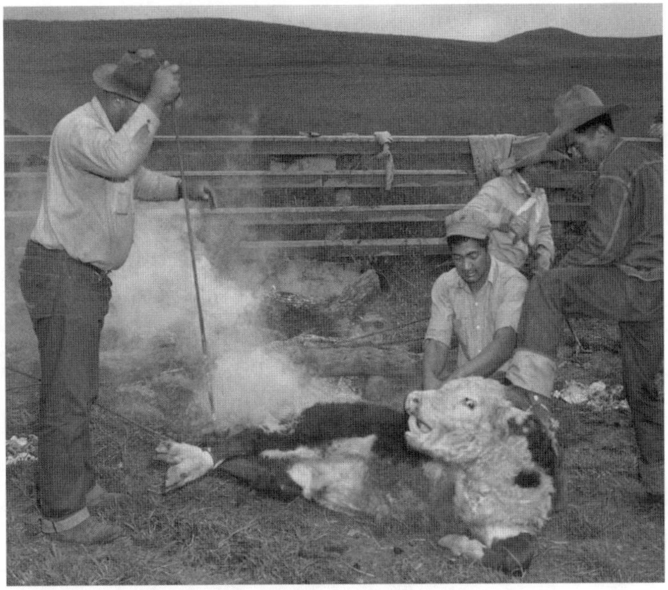

Sonny Akeni Akau holding down a calf at branding at Puʻu Mahaelua. David Kaula Sr. wields the branding iron. James Merseberg's foot is on calf's head and Bob Sakado is in the background. Hartwell Carter Collection.

the Breaking Pen for extensive periods brought out the best of their skills.

Manny left the ranch after a brief, tempestuous career and went to work for W. H. Shipman, Ltd., at Keauhou Ranch under foreman Tommy Lindsey. When that ranch was sold to W. W. Nobriga, Manny was transferred to Puʻuʻōʻō Ranch, working for Tom Bell. Tommy Lindsey later became the foreman at Puʻuʻōʻō Ranch.

Manny was an excellent horse trainer and had a splendid eye for good horseflesh. When I began to work summers at Puʻuʻōʻō Ranch in 1953, Manny had just left to become a cowboy at Kualoa Ranch at Kaʻaʻawa on Oʻahu. Some of his horses ended up on my string, and they were fine horses, gentle and well reined. Manny later became foreman at Kualoa Ranch and enjoyed a very colorful *paniolo* career. He is a highly revered *paniolo* and *kahuna pule* (Hawaiian priest). He is the pastor of a *Kalawina* (Calvinist/Congregational) Church in his home community near the Kualoa Ranch.

Theodore II (Sonny), Manny's older brother, remained at the Breaking Pen for several years and built a reputation for converting the progeny of the more broncky lines of horses—notably the Lindsley and Heine colts—into sound and productive workhorses. One of Sonny's close friends and coworkers was Joe Hui, also a Kawaihae product and respected horse trainer. Both men were elevated to the Cowboy Gang when openings occurred, taking with them some of the selected horses they trained.

Sonny was a natural astride horses. He was a genuine all-around cowboy, as good a roper as he was a horseman. He continued to train many young horses for others. Over the years, Sonny had some serious horse falls, and a chronic back injury denied him a continued cowboy career.

Sonny was transferred to the agronomy crew, a loosely arranged gang responsible for grass planting, weed and pest control, and reforestation. His boss was Charley Christensen, who had come to the ranch from the Kohala Agricultural Experiment Station of the University of Hawai'i. Eventually, when Hisa Kimura was appointed to head the formal Agronomy Department, he and Sonny worked very well together.

Sonny married Rita Puhi and had a family of five sons, all of whom worked for the ranch—two of them becoming foremen. Arthur, the eldest, was a truck driver who eventually became trucking foreman. His younger brother Arnold, an ex-Marine, became a welder on the ranch, eventually becoming foreman of the welding gang. Although he left the ranch in the January 14, 2002, Voluntary Separation Program, Arnold supervised all construction and maintenance activities in addition to supervising the welding crew. Arnold is highly respected in the leadership history of Parker Ranch. His hard-working, serious, and loyal character served the ranch well.

William Paul Mahina'uli Akau: Marriage II

Lydia Awa'a of Kawaihae Uka was the second wife of William Paul Mahina'uli Akau. She was the sister of Sam Awa'a Sr., who was with the horse department of the ranch. This marriage produced six children.

The eldest child, John, became a seaman aboard the SS *Humu'ula.* The second son carried a family name of Chinese origin—Apitai—and he developed a long career on the ranch doing fence work. With his first wife, he had Apitai Jr., who rose to become the Big Island fire chief. His sister, Jeanette Keala, married a ranch cowboy, Walter Puhi Sr. Their eldest son, Walter Puhi Jr., was also a ranch *paniolo* in the Keamuku Division and eventually became a foreman under the supervision of Walter Stevens. He later became the manager of Waiki'i Ranch. After his father passed away, his mother married Charles Stevens Jr., then the superintendent of the Mānā Division of Parker Ranch. Douglas, Alex, and Andrew were also sons of Apitai (Sr.), but did not develop lasting ranch ties. Most prominent of the sons of Apitai Sr. was Alexander Ahileka Akau. The remaining siblings developed no real ties to the ranch.

Alex Akau grew up to become a well-

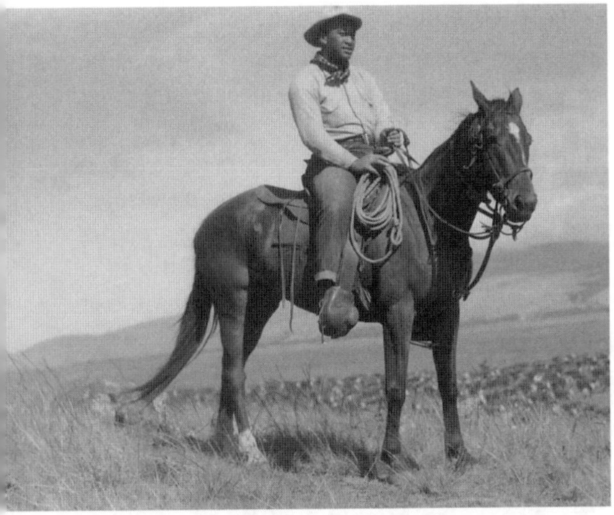

Alex Ahileka Akau mounted in Holoholokū. Hartwell Carter Collection.

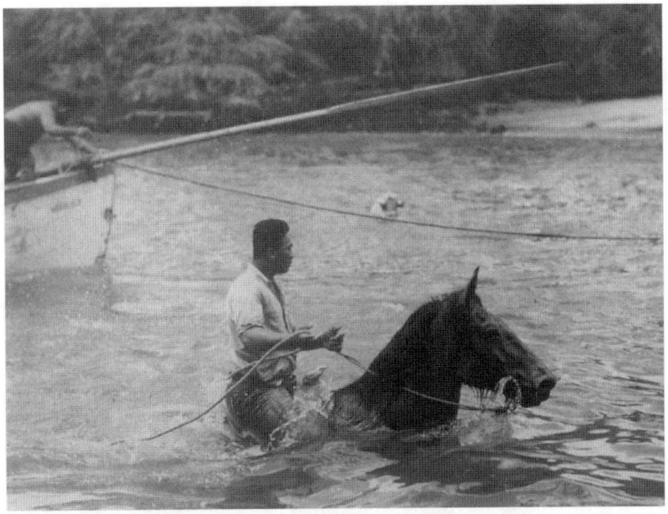

Alex Akau hōʻau pipi *(swimming cattle). Hartwell Carter Collection.*

known cowboy at both Parker and Kahuā Ranches. His early days on the ranch were spent at Humuʻula in the sheep operation, but Alex's talents brought him to the Cowboy Gang at the request of Willie Kaniho, for whom he had worked at the Sheep Station.

Alex was a big man whose brute strength was well controlled by his calm and sensible demeanor. He handled horses, cattle, and people with firm and congenial manners, but he was still considered a "bull" by his cowboy peers. Alex was a skilled roper, both in the branding corral and Kawaihae surf. He was one of Carter's chosen *hōʻau* ropers, leading the market cattle into the waves for shipment to Hawaiʻi Meat Company via the SS *Humuʻula.* Alex eventually transferred to Kahuā Ranch, where he spent the remainder of his distinguished cowboy career—a leader among his peers.

All told, William Paul Mahinaʻuli Akau has a lot to be proud of as the progenitor of outstanding leaders on Parker Ranch and other ranches.

The Yamaguchi Dynasty

Out of Hiroshima-ken came Hisamatsu and Satsu Yamaguchi to the shores of Kohala in the late 1880s. Not long thereafter, they made Waimea their home. By 1895 came their firstborn, a son named Matsuichi, destined for a short but spectacular cowboy career. Their tiny cottage sat along Waikoloa Stream in the heart of early Waimea, where Matsuichi (later called Matsu or Mack) developed a deep interest in horses.

By age fourteen, Matsu had attained American citizenship, while his parents remained subjects of the emperor of Japan. In effect, Matsu retained dual citizenship, but returning to his parents' homeland never entered his mind—the life of a cowboy had greater allure. His first job on the ranch was as a yardman at the sprawling

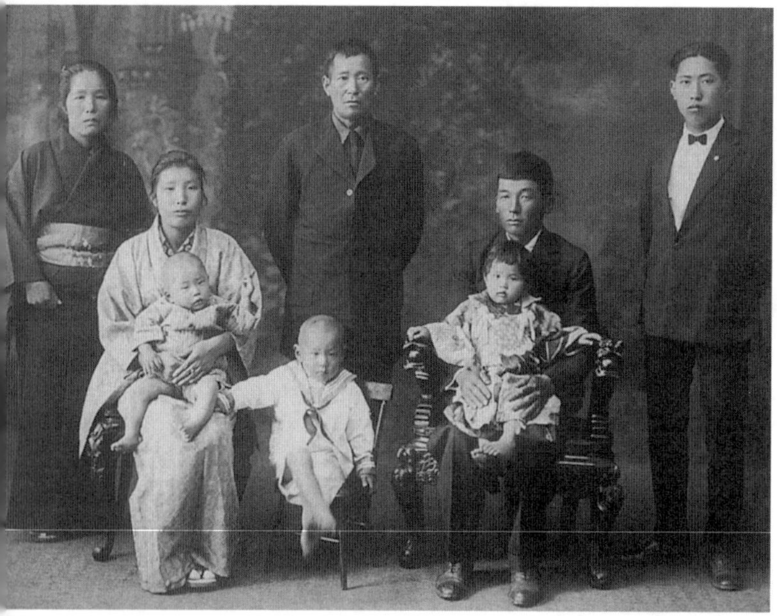

Above: *The Yamaguchi clan. Matsuichi and Harue Yamaguchi are seated; Harue's mother, father (Mr. and Mrs. Yoshikami), and Uncle Susumu stand behind. On Harue's lap sits son Jiro, with firstborn Ichiro next to him and daughter Amy on Matsuichi's lap. Ichiro Yamaguchi Collection.*
Below: *Matsu Yamaguchi as a stallion tender. A. W. Carter Collection.*

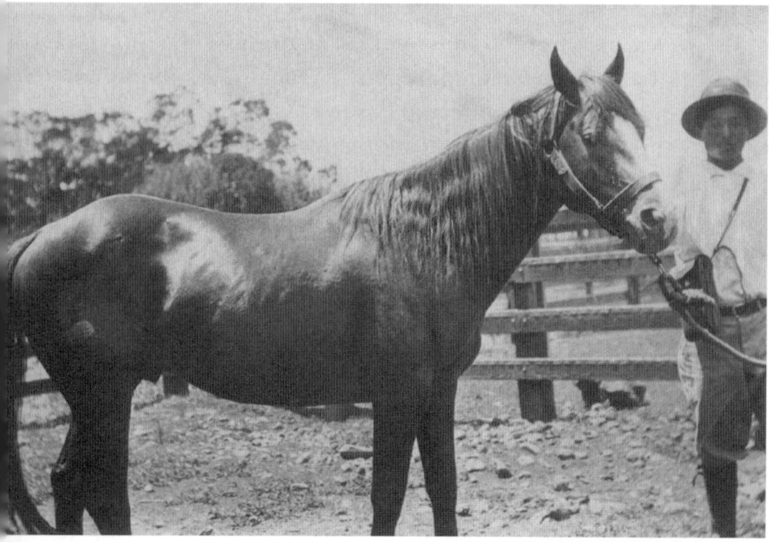

A. W. Carter home, and his living quarters were part of the cook's house. Well in tune with horseflesh, Matsu was already entering the children's races at the New Year's and Fourth of July Parker Ranch horse races. His horsemanship skills did not go unnoticed, and before long he was training racing prospects and handling breeding stallions.

Upon marrying Harue Yoshikami of Waimea, Makahālau became their first home. Matsu had been assigned to the purebred Hereford operation, a move fostered by A. W. in further developing the promising young man. The Yamaguchis had two sons who developed lasting ties to the Ranch. Ichiro was born (1922) at Makahālau, along

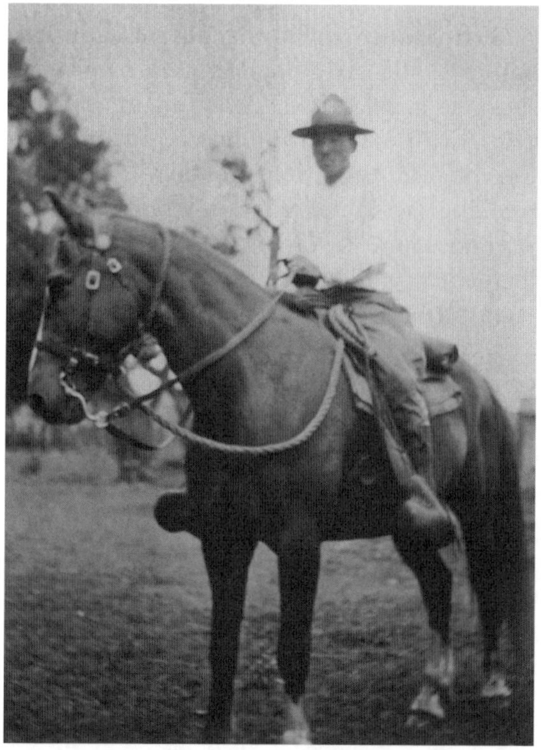

Matsu Yamaguchi as a cowboy (assistant foreman). Ichiro Yamaguchi Collection.

with a sister, Fumie. The family moved back to Waimea in the late 1920s when Matsu was appointed to the Cowboy Gang under Keoni Li'ili'i. In 1930 the family occupied the home of former teamster boss Henry Pieper in the area called Makahikilua, where North Hawai'i Community Hospital now sits. In Waimea, another boy and girl were born: Jiro and Matsuko.

Matsu made history as the first Parker Ranch cowboy of Japanese descent. Although many more nisei would follow in the ranch ranks, it is important to note that other ranches took pride in their crack cowboys being of Japanese descent—notably Okuda of Horner (Kūka'iau) Ranch and Muranaka of Pu'u'ō'ō Ranch (W. H. Shipman, Ltd.).

But Matsu set the tone for excellence in horsemanship, cattle handling, and leadership and A. W. had recognized the potential within him. Matsu was a quiet but congenial person, and these traits combined with his skills earned him the position of assistant cowboy foreman. This promotion was another first for the history of the ranch, having a non-Hawaiian lead the Cowboy Gang.

Matsu was fatally injured in a tragic horse accident in Kemole II. In pursuit of wild sheep, he suffered a severe head wound when his horse rolled forward and over with him at a dead run. In the rocky hills of Kemole II on the south slope of Mauna Kea, Matsu's life was at its ultimate ebb. His head wound alone was massive, but a complicated chest injury due to the crushing weight of the saddle horse likely would have killed the average man at the moment of impact.

Matsu struggled to upright his mount, which he managed to do, although both were in a dazed state. With incredible strength and determination, Matsu rode slowly in search of his fellow cowboys, who then quietly accompanied him back to Makahālau Camp. Matsu was driven to Waimea and, under the advice of Dr. Whitman, rushed on to Kohala Hospital. En route, Matsu remained conscious; hopes for his recovery were high but unrealistic. He died during the night from the massive head and chest wounds, but he demonstrated valiant strength until his final breath. This was Tuesday morning, January 28, 1936, two days after celebrating his *yakudoshi* (forty-first birthday).

A decade earlier, another noted Japanese cowboy leader, Edward Junichi Ishizu, foreman at Humu'ula, was killed in a similar accident. Ishizu was fording a raging gulch in the Laumai'a area when he was washed down. The *kahawai* is now known as Ishizu Gulch. Typical of A. W.'s caring ways, both families were kept on in ranch homes, maintaining the privileges of meat, milk, and poi as well as full medical care. Some of the children of both these nisei became significant employees of the ranch.

Harue had her hands full raising her little family. A. W. Carter accommodated them in a smaller but comfortable home on the Government Road in Waimea. Son Ichiro still lives there. None of his children had ties with the ranch.

Jiro grew up an image of his father, especially as a *paniolo.* An avid horseman and top roper, he was one of the most photographed *paniolo* in ranch history, given his photogenic and congenial nature. He spent many years on the Cowboy Gang and eventually became foreman of the Waiemi

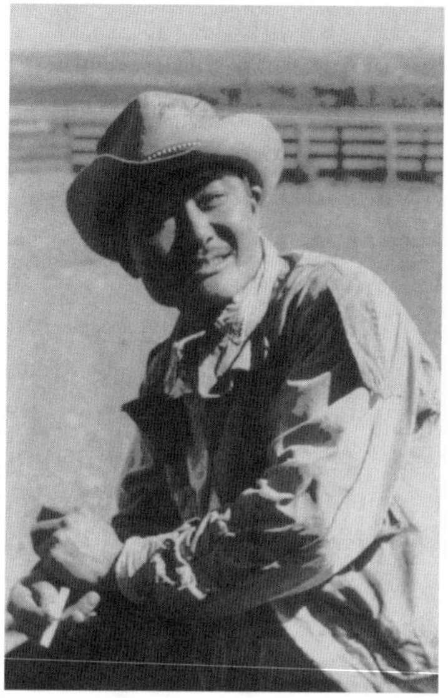

Jiro Yamaguchi, cowboy, at Pu'uhihale Corral, 1948. Jiro was one of the ranch's most photographed paniolo. *Jiro Yamaguchi Collection.*

Section. Jiro married Mollie Nishida and they had five children, two of them twin boys, Mark Haruo and Mack Matsuo. As with their father and grandfather, horse racing began in the children's pony races during the Fourth of July Parker Ranch festivities.

Mark went on to become a Parker Ranch cowboy, joining the ranks in 1974 and working in the Keamuku Division. By 1990 Mark was a ranch foreman, and history was again established. With three successive generations of Parker Ranch foremen, the Yamaguchi *'ohana* qualifies as a historic dynasty. Although Mark left the ranch with twenty-eight years of proud service ending with the 2002 Voluntary Separation Program, Matsu comfortably rests on the laurels of the contribution made in the name of Yamaguchi of Hiroshima-ken.

The Kimura Dynasty: Masajiro Kimura the Patriarch

Like the Yamaguchi family, the Kimura clan originated from Hiroshima-ken in Japan. In 1898, Masajiro emigrated to Hawai'i; his future wife, Hisamu, followed three years later. Masajiro was awarded a three-year contract to work for Ho'ea Sugar Plantation Company in North Kohala. He landed at Māhukona and immediately went to work in the cane fields as a laborer. He was disengaged from the contract in 1901, but stayed on as a yardman and buggy driver for John Hind, the plantation manager. He earned the trust and respect of Hind.

Hisamu arrived in Māhukona in October of 1901 for their prearranged marriage. Their first home was in Hāwī. Their first son, Masao, was born in November of 1903, about the time that Hind's brother Robert was looking for a dependable man. The small family moved to Pu'uwa'awa'a Ranch in North Kona, where a cottage was provided. Masajiro continued as a yard and buggy attendant, while Hisamu became a housemaid for Mrs. Hind. Working conditions were vastly improved, and the Kimura couple expanded their family to include Yutaka in 1905, Kaoru in 1906, and a girl, Jitsuno, in 1909.

In the summer of 1909, the Kimura family moved by horse-drawn wagon to Waimea, seeking better educational opportunities for their children. Waimea at that time had a population of about two hundred people, about 10 per cent of whom were immigrant Japanese. The Parker, Purdy, Spencer, and Lindsey families were the predominant residents, with few Portuguese and virtually no Filipinos.

Samuel Parker had contracted fifteen Japanese laborers, some of whom began families in Waimea. By the time the Kimura clan arrived, the Japanese Benevolent Society had been formed, and this helped to give

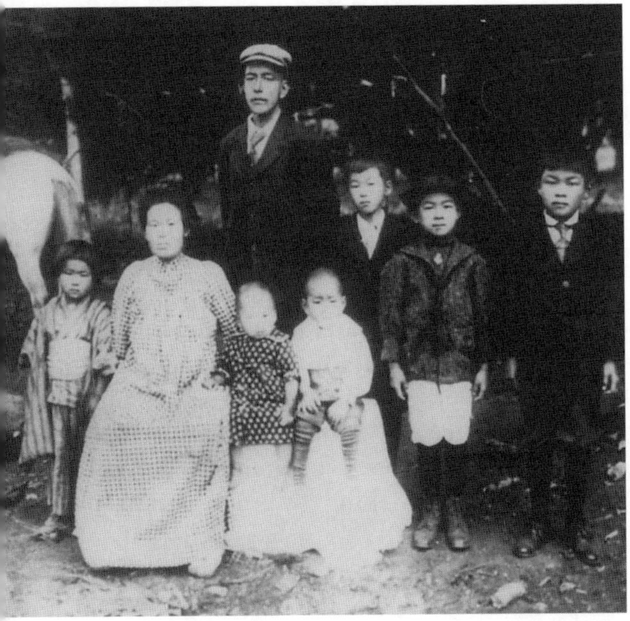

The Kimura clan. Front row: Jitsuno, Hisamu, Hisa, and Kazuo. Standing, back row: Masajiro, Masao, Wallace Kaoru, and Yutaka. Hisao Kimura Collection.

thought of by A. W. In a memo to Yoshio in 1943, he reflected: "My mind goes back to the time your father first came to Waimea. I always had the highest respect for him. He was a man of splendid character and reliability. Your mother was of the same character. The Kimura family I look upon as of the most reliable and loyal of those with whom I have been associated."

Yutaka Kimura and Sons

Yutaka entered the first grade at Waimea School, taught by Mrs. Margaret Kaimuloa. The strict and stern Robert Kihoi taught his second to fifth grade lessons. He was followed by Mrs. Nora Akina, wife of the territorial legislator Arthur Akina, who was far more pleasant. She was the last teacher he would have, as Yutaka left school after the sixth grade in 1917. He was naturally bright, like all the Kimura children, whose potential

them a welcome feeling in the community. But there was little or no work for Masajiro. He tried farming, fence work, and carpentry, eventually becoming a handyman on Parker Ranch. He served on one of the crews in building the original ranch racetrack, including grading, filling, and rail construction. By this time, the family began to expand, with Kazuo born in 1910, Hisao in 1912, sisters Tsugi in 1914, Sueko in 1916, and Yoshio in 1919.

Of the nine children born to the senior Kimura family, Yutaka, Hisao, and Yoshio all developed distinguished careers on the ranch. Yutaka went a generation farther in that his sons Samuel (Sonny), Charlie, and Tommy all had lifetime careers on the ranch. Masajiro and Hisamu were highly

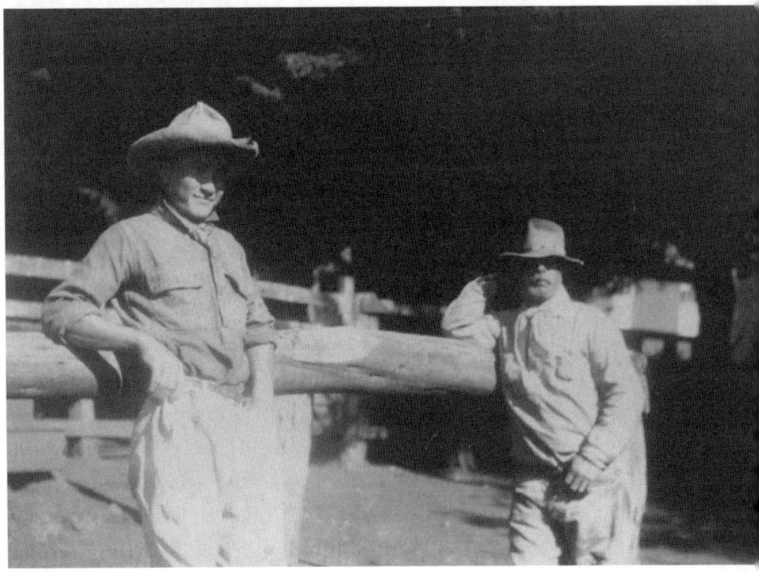

Yutaka Kimura with Francis Pakana Spencer Sr. Hartwell Carter Collection.

was limitless—but the family could not afford the luxury of a secondary education.

Yutaka's career started in 1918 at age thirteen, working at the sprawling Carter complex. Gardening and cow milking and delivery were his first obligations, the latter of which led him to a deep love for horses. His milk delivery routine comprised short rides to other management-level homes, including the Bloomfield, West, and Sorenson residences. O. L. Sorenson was Carter's assistant manager and also his brother-in-law (they married sisters of a family named Hartwell). Yutaka's mode of transporting the gallon milk cans was a pair of ponies, Moonlight and Skylark, with whom he developed strong bonds. Between the milk cows and the delivery ponies, Yutaka came to realize where his career path was headed—no doubt it would involve an emphasis on caregiving.

Yutaka then went to work in Waiki'i picking corn on a contract basis. A wagonload of twenty-five bushels paid $1.15. He enjoyed handling the draft horse teams and maintaining a relationship with livestock. Shifting to the Humu'ula Sheep Station on a contract basis, Yutaka took part in wool processing, including shearing, spreading, cleaning, and baling. Some fence work was also involved.

When the shearing season was done, Yutaka worked under the current foreman, Tom Lindsey, building a road across Kemole II. This led him to a better assignment as a result of his industrious nature. It was an upward career move to go to work for Martin (Makini) Campbell building redwood water tanks in Kemole I. Although he didn't savor more construction experience, these basic ranch chores were critical building blocks for a well-rounded ranchman in years to come.

In 1919, Yutaka went to work milking cows at the Old Dairy in Paliho'oūkapapa under foreman and later mentor Iwataro Ishii. This provided the foundation for a lifetime in animal husbandry, much of it with dairy cattle. An opportunity arose in the early 1920s to become a cowboy at the Makahālau Station under John Kaukokalani Purdy. His skills at riding, roping, and horse training were now in their basic stages, only to be perfected in years to come. Meanwhile, A. W. observed the gradual progression Yutaka was making toward becoming a meaningful ranch leader. It wasn't long before he became foreman—at age nineteen—of the new dairy at Pu'ukīkoni.

Yutaka married Haruyo Uyeda on February 7, 1925, and they made their home in a modest cottage near Pu'ukīkoni Dairy, where they lived until 1938. All six of their children were born there, three of whom were sons destined for lifetime careers with the ranch. The first son, Francis, died as an infant in 1925, but a year later Samuel Yutaka was born. Thomas Harume was born in 1928 and Charles in 1931. Sisters Mabel and Elaine were born in 1929 and 1933, respectively. In 1938, the Kimura family moved to Waimea into a home near Pu'u-'ōpelu that Hartwell Carter had especially built for them.

Samuel graduated from Honoka'a High School in 1943 and was immediately drafted by the U.S. Army. He joined the ranch ranks later and married Ella Lindsey, the daughter of Kuakini Lindsey. Sam's early career was in warehousing of ranch materials, working with Masao Yoshimatsu. Upon Yoshimatsu's retirement, Sam became warehouse foreman, with saddle-making personnel included. His leadership was exemplary. On Harry Kawai's retirement as ranch dispatcher/coordinator, Sam took the position and remained there until he retired after a distinguished forty-year career.

Thomas grew up with excellent carpentry skills and enjoyed a ranch career doing finish carpentry under the mentorship of his Uncle Hichiro "Lee" Uyeda (Haruyo's brother), a master cabinetmaker. Tommy retired from Parker Ranch with thirty-eight years of service.

Charles, who quickly became known as Charlie, developed an early and keen interest in his father's line of work, yet he progressed independently through the ranks as a stand-alone, able, and progressive leader. The quantum leap from his first job as a fence man to the livestock manager of Parker Ranch is the great American story with an Island flavor. After forty-seven years of service, Charlie retired in 1994.

Hisao, better known as Hisa, was inclined toward plant life, whether horticultural or forage in nature. In 1931, at age nineteen, Hisa embarked on a ranch career and, like many other fledgling young men, took any job offered as a means of getting a foot in the door. His early experiences were basic and broad, starting as a mechanics helper, then moving up to truck driving. Becoming a water pump operator broadened his exposure to the ranch, and eventually he went to work at Pu'ukīkoni Dairy with his brother Yutaka. Through these formative years, Hisa nurtured his interest in grasses and legumes. The dairy years gave him hands-on experience in the value of good feed relative to animal growth and milk production.

When the ranch hired Harold Baybrook as agronomist, it was an opportunity for Hisa to develop his natural instinct in this field. Eventually Charles Christensen, who remained only a few years, followed Baybrook. In 1972, Hisa was named ranch agronomist, and by this time he was well prepared for this leadership role. One of his mentors was a University of Hawai'i professor, Dr. Edward Hosaka. Hisa retired in 1978 with forty-seven years of service and enjoyed turning his attention to community services, nearly a full-time avocation.

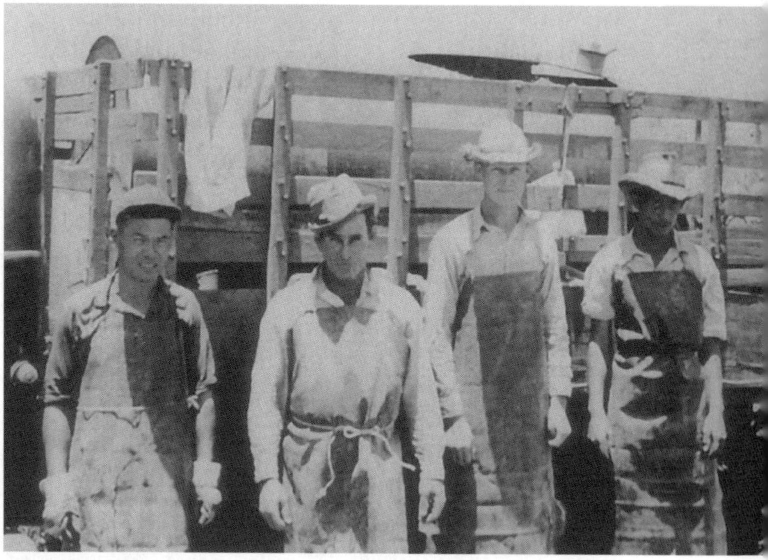

Hisao Kimura (at left) with cactus crew: Jacinth Deluz, unidentified, and Francis Kole Perreira. Hartwell Carter Collection.

Yoshio was the youngest of Masajiro's sons. Like Hisa, Yoshio was sent to Hilo for schooling and graduated from Hilo High School. In the summers and on Christmas vacations, he worked under his brother Yutaka at the Pu'ukīkoni Dairy.

A. W. Carter saw potential in Yoshio and offered him a starting role on the ranch as a carpenter's helper in 1938. Hartwell

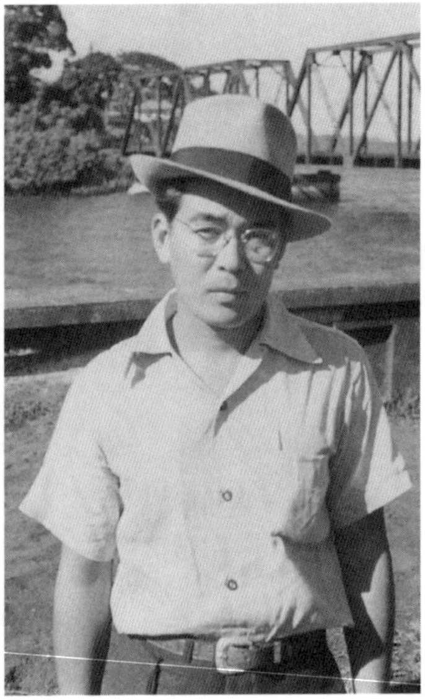

Yoshio Kimura at Hilo Meat Company near Hilo's Railroad Bridge over the Wailuku River. Hartwell Carter Collection.

Carter saw a different side of Yoshio—a talent for bookkeeping. In the interim, he often served as Hartwell's aide-de-camp and chauffeur when meetings took place throughout the Big Island during the post–World War II years. Yoshio became an account clerk in the ranch office and remained there until he was appointed manager of Pioneer Federal Savings and Loan in 1972. He left the ranch with thirty-three years of service.

Via Masajiro's three sons and three grandsons, all having valuable ranch careers, a total of 252 years of loyal service were given to the Parker Ranch, establishing a historic familial contribution to 'Āina Pakā—they were all loyal to the land.

The Vredenburg Clan

The Vredenburg family originated in South Africa and Jamaica. Orphaned in South Africa, Watson Vredenburg moved to his mother's homeland, Jamaica, where he prospered as a British merchant. In Buenos Aires, Argentina, he married Marie DuVinay, daughter of a prominent French couple. They had three sons: Edric Watson, Wilmot Laurence Andrade, and Ernest Watson.

Vredenburg, a British national, took up the law profession as a barrister and eventual consul general and judge in São Paulo de Loanda, a bustling city on the Bay of Bengo and the administrative center of the Portuguese colony of Angola. Of the three sons, Wilmot Laurence Andrade was destined for Polynesia, but his education came first. Graduating from a Jesuit college in southern France, he set sail for Hawai'i to deliver a shipload of Portuguese workers destined for the sugarcane fields. Fluent in seven languages, Portuguese was his forte.

On board ship, Wilmot met George Hind of North Kohala, whose family later developed the Puakō Plantation, a sugar venture. Interestingly, Wilmot found naturally growing sugarcane on this coastal plain and thus inspired the birth of the Puakō Plantation. He became a foreman of that operation, beginning an enduring relationship with the Hind family. He married a *hapa haole* (half Caucasian/half Hawaiian) native of Waimea named Norah Campbell, daughter of William Campbell and a pure Hawaiian woman surnamed Keawe'ehu. Norah's Hawaiian name reflected her mother's genealogy: Lilia'okahaku Kalunaho'oki'ekie Keawe'ehu. Some family members say Campbell had assumed the Scots surname for unknown reasons, as he was by some accounts a product of Schleswig-Holstein, then a province of Denmark. Family genealogists, however, claim he was a native of Scotland, having fled his motherland during the war years of Bonnie Prince Charles.

Wilmot and Norah had eight children—six boys and two girls: Marie, Wilmot II, Theodore, Edric, Ernest, Edmund, Norah II, and Paul. Wilmot II, Theodore, and Ernest worked for the ranch at various times and levels, while Edric, Edmund, and Paul developed off-ranch careers. Marie became a Catholic nun with the Sisters of Notre Dame du Nemours of Belmont, California, while Norah II worked on O'ahu for the Metropolitan Meat Company. A. W. was to play a big part in the upbringing of the eight children, as tragedy would later strike Wilmot.

After leaving Puakō Plantation in 1905, Wilmot had solicited A. W. for a position on the ranch, preferring the dairy or corn-farming ventures. In January of 1906, he was assigned to manage the dairy operation at Paliho'oūkapapa with orders from A. W. to double the production of butter. Wilmot was well endowed with technical skills and knowledge, and he rose to the occasion. In November of the same year, he asked A. W. to help him secure a new Packard to enhance butter deliveries to Hilo, the county seat. He later used the Packard to ferry mail between Pu'u'anahulu and Waimea.

In February of 1908, Carter reported that "There was much better work being done at the Dairy than there was previously and with much less expense."

Along the way, Wilmot became known as "Manuwai" (water bird), largely for his ability to find and preserve new water resources, some of which were developed in the hills above Waimea. As in the dairy venture, his technical expertise was applied to water resources.

At the end of 1908, A. W. directed Wilmot's efforts toward the corn-farming operation in Waiki'i. By this time, Waiki'i was a bustling little farming community with more than 1,500 acres in corn production and related activities. The original Hawaiian families were soon mixed with Japanese and Russian farm workers, introducing cultural blending with very little social strife. The Russians came with their families from Manchuria, fleeing the Bolshevik Revolution. By then, nearly a hundred people lived in Waiki'i village, and Wilmot's multiple language skills were put to the test.

While the Russians were predominantly used in the corn operation, the Japanese were responsible for fences, water systems, and the caring for the orchards, chickens, turkeys, and hogs. Wilmot's daily ledgers recorded activities in the following categories: orchard, plowing, hauling, corn crop, framing, pumps, weather, tanks, buildings, stock, pigs, paddocks, pasturage, and sundries.

Wilmot's technical interests had no bounds, and his concepts were decades ahead of their time. He requested a bone crusher to create bone meal as a mineral supplement for cattle, providing rich sources of calcium and phosphorus. Planting of trees in the form of orchards and windbreaks was a high priority, and vestiges of his work still stand in the hills of Waiki'i. On January 28, 1909, Wilmot gave A. W. his corn crop harvest figures: 8,206 bushel baskets averaging between 65 and 70 lbs. each. Alfalfa and wheat were also planted and harvested, although growing them was a struggle in the high, dry elevations of Waiki'i. Hay baling was also set up via a

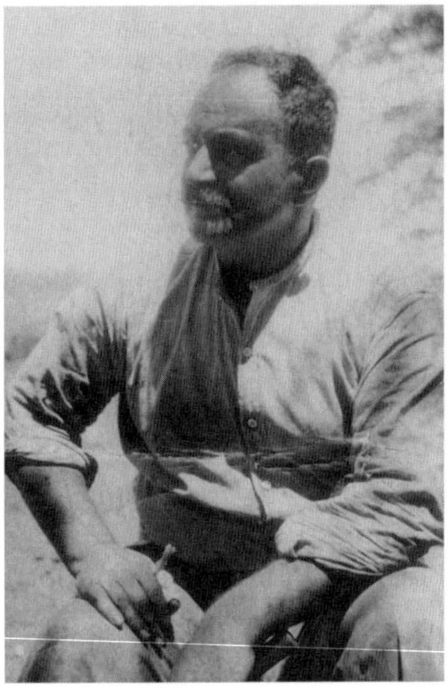

Wilmot Lawrence Andrade Vredenburg. Keawe Vredenburg Collection.

press installed by Wilmot and yielding a 140-pound bale every two minutes, producing 240 bales per day.

Wilmot had the foresight to request a trip to the Midwest to get a glimpse of stall (feedlot) feeding of stock. As he noted in a memo to A. W., "I firmly believe that the time is very near when we shall be turning out hundreds of corn fed steers. The ranches that can only turn out grass and range fed bullocks will take a back seat." In the next four decades, the American beef industry evolved into a feedlot/grain finishing production line and remains so today. A. W. appreciated his eccentricity but was unaware that Wilmot was accurately predicting where the beef industry was headed.

Tragedy struck on December 12, 1909. In a December 13 report to A. W., Donald Macalister, then Keamuku superintendent, wrote:

> I have just come in to Waimea and I hear no one has written you an account of Mr. Vredenburg's death. We, Mr. J. G. Searle, Searle's Boy Sam, Theo Vredenburg, Dolly [Norah] Vredenburg and myself were out looking for pigs mauka of Pu'u Lā'au corral; then we found a band and chased after them. Mr. Searle shot at one and wounded it. It ran into a gulch. Manuwai, Theo and myself went into the gulch. Sam Searle in the meantime had had a shot at another and wounded it. He came down and wanted a knife so I left Manuwai and Theo V. cutting up the pig, and went with Sam Searle. A little time elapsed and we heard a shot. We went down and found Vredenburg dead. We did not find another wounded pig, and when we heard the shot concluded that it was the others shooting the wounded pig. While cutting up the pig, Vredenburg's revolver slipped from his holster and going out shot him in the abdomen.

In a memo from Mr. Jamieson, Carter was told,

> I suppose ere this time you have received the news of Vredenburg's death. He was out hunting on Sunday mauka of Waiki'i, accompanied by his son, Searle and his son, and Macalister. Searle shot a pig, and Vredenburg went down into the ravine where it fell to cut it open, and in stooping to do so the revolver fell out of the holster, striking the hard ground in such a way, as the muzzle was pointing toward his chest, the bullet piercing him in the region of the heart. Searle was up on the bank at the time and at once rushed down to his assistance, but the only words he uttered were, "I am dead". His own son

was right by him when he fell. It's a tough business on widow and family.

In a December 13 memo to Sam Spencer, Carter wrote: "Your letter of the 12th was received, as also the wireless regarding Vredenburg's death. It is hard for me to believe that poor old Manuwai is down and out; a man full of life, ambition and hope; and it does seem cruel that he should be taken from his family."

In an October 30, 1948, memoir concerning Wilmot, Carter reflected as follows:

> Old man Vredenburg (Wilmot) was a genius. He came here with the first lot of Portuguese immigrants as interpreter. He spoke seven languages. He had wonderful conception, but was impractical and could never get ahead. He was very musical and would play a duet, using the left hand on the piano and the right hand playing a horn. He wanted to buy a Packard one-cylinder machine to go to Kona and get the mail. I got one for him, but the roads were terrible and it was always breaking down. He always carried a wire with him, and when he had a breakdown he would throw it over the telephone wire, get Central and tell her, "Tell Mr. Carter to send the mules." His wife, who is still living, was the mainstay of the family, a wonderful woman.

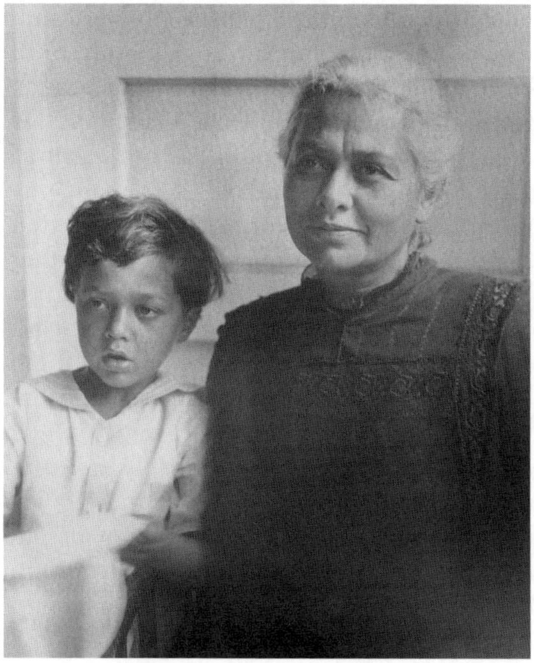

Norah Vredenburg and Wilmot Vredenburg III. Bishop Museum photo CN 4899.

In *John Hind of Hawi,* Robert Renton Hind further expounds on Wilmot's musical talents:

> Vredenburg was quite a musician. For many years, prior to the Puako days, he came by horseback to Kohala from his farm at Waimea to furnish music for the occasional dances held at Kohala Hall (successor to the Kaiopihi Hall). His forte was playing a cornet with his right hand and the accompanying chords on the piano, with his left. It was the nearest thing to an orchestra that Kohala could muster in those days. As he neared Hawi, down the long hill from the homesteaded lands back of the plantation, Manuwai would break out with his cornet which could be heard within a radius of a half-mile on a quiet evening and the laborers and their families would rush to the roadside to enjoy the impromptu concert. The cornetist reveled in this spontaneous appreciation.

A. W. immediately took the Vredenburg brood under his wing, assuring the health, education, welfare—and even employment—of the children. Especially grieved by the tragedy he witnessed was little Theodore, who was only twelve years old at the time. A. W. gave him much fatherly support. Young Wilmot moved home immediately after his father's death, going to work at Waiki'i under Donald Macalister, who was moved there to supervise that operation. Young Norah eventually went to work at the Metropolitan Meat Market. Youngest brother Paul worked summers at Humu'ula while attending the University of Santa Clara. He first worked at Salinas High School before getting a job as a civil engineer with the National Park Service. Paul joined the Army Corps of Engineers in San Francisco when World War II broke out. He served under General Patton in Europe, returning in 1946 and rejoining the Corps of Engineers, where he remained for the rest of his career. A. W. had a direct hand in the lives of all the children of Manuwai and Norah.

Eventually, young Wilmot moved to Pāhala, Kā'u, as a police officer. An avid fisherman, Wilmot had lost an eye in a dynamite-related accident in 1932. He enjoyed toying with his glass eye, plucking it out in the presence of unsuspecting company. In my boyhood years in Ka'ū, I enjoyed Wilmot's skilled *'upena kiloi* (casting of the fishing net) along the rugged shores from Kanenelu to Wai'ahukini. He would deftly pluck his glass eye out, placing it in the watch pocket of his *palaka* shorts before he dove into the surf after his bountiful net.

It was during these same childhood years that I befriended a classmate, Dexter Keawe'ehu Vredenburg—son of Edmund, younger brother of Wilmot—who worked at the Hilo tax office after working briefly as a bookkeeper at Parker Ranch. Brother Edric went on to become quite the inventor of critical components of ammunition production during the war years, but he never made Hawai'i his home.

Ernest had significant time with the ranch but eventually went to work for Kahuā Ranch in 1928. Later he moved to Lāna'i, where he managed the Lāna'i Ranch.

Of all of the sons, it was Theodore that A. W. groomed for leadership, and this mentoring began in Vredenburg's childhood years.

Theodore Vredenburg

During his early years, Theodore was exposed to horses and cattle partly as a result of A. W.'s continued stewardship and encouragement. A. W. also arranged for him to be educated at St. Louis College in Honolulu. Like many of the other ranch boys, working in the 'Ōpae Gang was his first paying job. Then Carter groomed Theodore for a leading role on the ranch by placing him on the Fence Gang as his first assignment. This was the boot camp that launched many careers on the Parker outfit.

Theodore had an opportunity to work as cowboy shortly after A. W. acquired Kahuku Ranch. Under Makini Martinson, he was immersed in the day-to-day ranch experiences that included the long cattle drives from Kahuku Ranch to Humu'ula via the Pu'u'ō'ō Trail. His cowboy peers included Wahi Keanu, Ben Pahinui, and the Akiu brothers, Charles and Fidelis.

When he was reassigned to the home ranch, Theodore became engaged to a charming *hapa haole* wahine named Irene

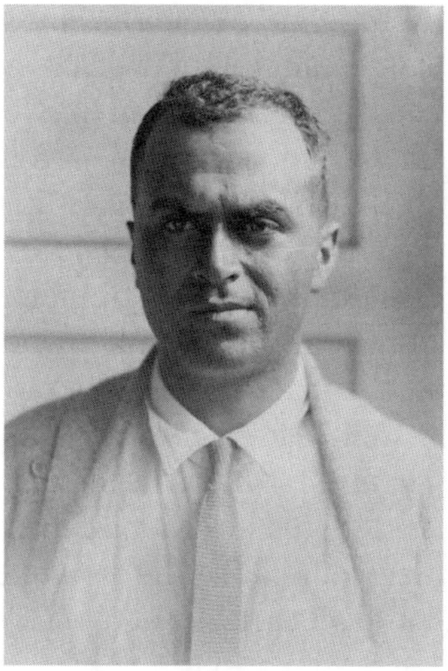

Theodore Vredenburg in 1920. Bishop Museum photo CN 4900.

Davison, daughter of a native Hawaiian woman of Lahaina and Dr. Davison, a physician. After their marriage, Theodore and Irene settled in Waimea. Their first child, Florence, was born a few years later.

When the military remount program was in its heyday, Theodore oversaw both horse and cattle production. For reasons unknown, he left Parker Ranch and moved to South Kona as a road superintendent, but it wasn't long before A. W. sought his return. In a cordial letter to Theodore in 1927, Carter invited him back to the ranch to run the Pu'ukīkoni Dairy. This assignment was described in a memo from A. W. to Theodore in which Carter stated that he "would be most delighted if he (Theodore) would agree to come back to the Ranch to run the dairy." Theodore returned to the ranch to manage Pu'ukīkoni Dairy, and his second daughter, Theone, was born there.

After passing on the dairy assignment to Yutaka Kimura, Theodore was given a larger mission: responsibility for the purebred Hereford herd at Makahālau. This was a significant promotion, as the purebred herd was not only the ranch's wellspring of breeding bulls—it was truly A. W.'s "baby" and had been developed to near perfection by Alexander "Sandy" Napier. This assignment honed Theodore's leadership skills further, and by 1929 he was brought back to headquarters to serve as assistant manager to Hartwell, who was gradually assuming more of A. W.'s responsibilities. As with the dairy, Yutaka Kimura succeeded Theodore at Makahālau. After this, Theodore effectively operated as general ranch superintendent, responsible for all operations—including water, marketing, fence maintenance, and livestock production.

From 1937 to 1941, the Kohala operations

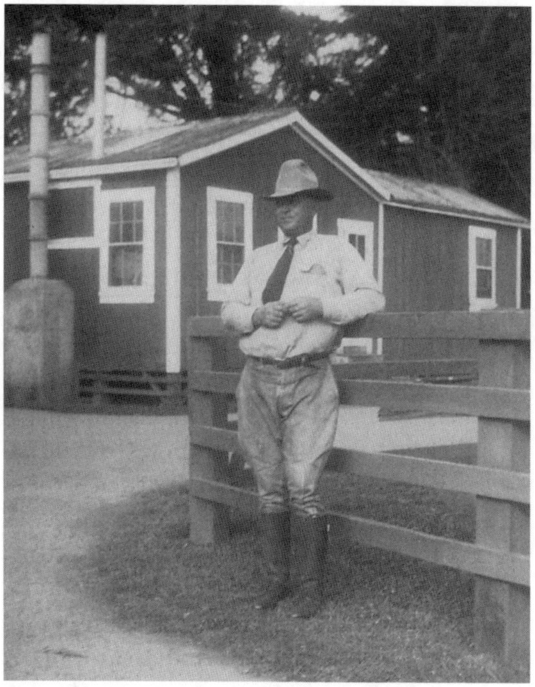

Theodore Vredenburg, ranch superintendent, taken behind the old ranch restaurant. A. W. Carter Collection.

of the ranch became Theodore's last Parker Ranch management assignment. He developed a good relationship with Johnny Kawamoto, who later assumed responsibility for both of the Kohala sections, Pu'uhue and, later, Puakea. Theodore left Parker Ranch for Hu'ehu'e Ranch, a trusteeship of A. W. and later Hartwell Carter. By 1945, the Vredenburg family was well settled at the North Kona operation, living in the former home of John Maguire, Hu'ehu'e Ranch's founder. During these years, daughter Florence (Coco) returned from mainland schooling. In her marriage to R. L. "Bobby" Hind II, they produced four children, the eldest of whom, R. L. "Robbie" Hind III, returned to his ancestral home, 'Āina Paka, as livestock manager in 1984.

Zenjiro Kawamoto

Zenjiro Kawamoto was another native of Hiroshima-ken. Born in 1874, he migrated with other countrymen to the shores of Kohala in 1894. He married Wasa Watabayashi, also an immigrant from Hiroshima, not long after he went to work for Puakea Plantation. Wasa was carrying a son from Kyujiro Karimoto when she wed Zenjiro. At birth, the baby was named Masato (Johnny) Kawamoto. Zenjiro raised Johnny as his very own.

Puakea Plantation began as a diversified farming operation, but livestock eventually became its mainstay, a field of work to which Zenjiro was well suited. The plantation became known as the James Wight Estate (Puakea Ranch), and Zenjiro was promoted to foreman in 1901, about the time his stepson Masato (Johnny) Kawamoto was born at the tiny Puakea village that was the Wight Estate headquarters. Dairying and butchering were adjunct affairs related to cattle ranching. The Wight outfit was a progressive one, importing blooded livestock including bulls and jackasses. In its early decades of cane farming and ranching, Wight's concentrated on water resource development, bringing channeled irrigation services to its lower elevations. Water storage capacity was also enhanced. Parker Ranch had become neighbors to the south *(mauka)* and had not yet progressed in water resource management to Wight's degree because the Pu'uhue Ranch was well endowed with rainfall.

Young Johnny grew up on the ranching side of the Wight enterprise and took on the cowboy role with enthusiasm. It was a small, diversified outfit, and Johnny's acquired skills included the mundane chores of milking cows and butchering fat cattle at the company abattoir (slaughterhouse). But he found time to build a solid foundation as a horseman. When Zenjiro retired in 1922 after thirty years of service with the Puakea company, Johnny was a hearty twenty-year-old ranch hand ready to assume many of the responsibilities; by this time the company was almost entirely in livestock production.

In 1932, Parker Ranch acquired Pu'uhue Ranch from the Woods family. About a dozen years later, Puakea Ranch, already under lease, was bought from the Wight Estate. On January 1, 1934, James "Kimo" Greenwell was sent by A. W. Carter to run the Kohala Ranch. By this time Johnny and his wife had two sons, Masatsu and Yoshio, who became promising young cowboys. Besides Johnny, other ranch hands included Peter Kainoa, Ernest Keli'ikuli, Keana Aniu, and Simeon Dela Cruz. Two more young cowboys, Johnny Kainoa and Albert Dela

Cruz, were sons of Peter and Simeon, respectively. In *One to Eighty-One: My Personal Story,* James Greenwell reflected positively on all the ranch hands, but he had special accolades for the four young lions—Masa, Yoshi, Johnny, and Albert—as showing great promise. History has borne this out. Three of the four men became territorial and state champion ropers, in addition to fine careers as cattlemen.

Masa and Yoshi were matched in their cowboy skills, and their sibling competition was intensive. As the years went by, both became foremen—Masa at Pu'uhue and Yoshi at Puakea. Johnny supervised the entire Kohala Ranch after Theodore Vredenburg moved to Hu'ehu'e Ranch in 1941.

The Kohala operation was run nearly independent of the main ranch. Johnny called Carter to report and receive orders, but he ran the operation as he saw fit. Frugal by nature, Johnny and his two crews made do with what raw materials they had. Johnny built a 40-by-60-foot storage shed using corrugated roofing, but the walls were made of metal panels that were fashioned from used oil drums flattened into 3-by-6-foot metal plates. The shed stands in Pu'uhue today, more than seventy years old, and it is in remarkably good shape.

Johnny utilized his farming skills by plowing and planting former pineapple fields in pigeon peas and mixed grasses and clovers. He used a 60-horsepower tractor with a cut harrow.

In the butcher plant, Johnny and Ernest Keli'ikuli slaughtered and processed four head of fat cattle each week out of the improved pastures. These carcasses were quartered and cut for a series of small meat markets scattered across Kohala's plantation communities from Niuli'i to Hāwī. These chores began at 1 A.M. and finished in time to begin the regular workday at 6 A.M. Masa and Yoshi became quite skilled at butchering during these years. These additional butcher chores were without added compensation for any of the men—just pure dedication. Masa and Yoshi went on to significant leadership roles on the ranch in the post-1950 era.

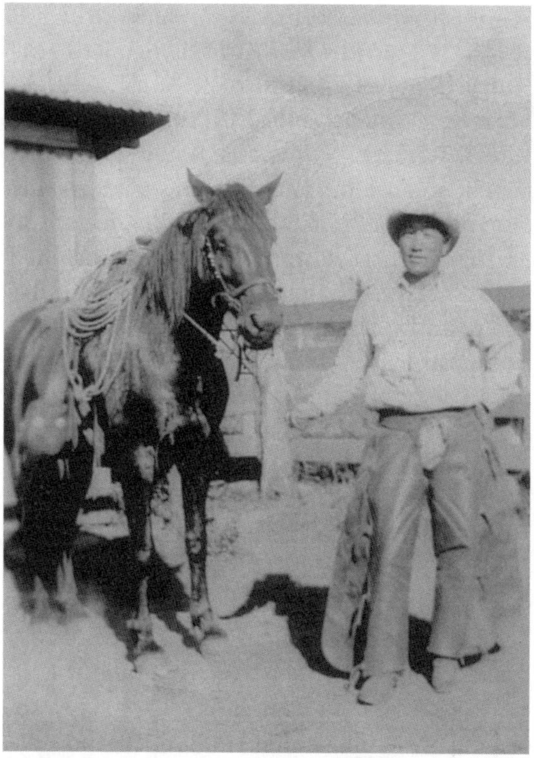

Johnny Kawamoto, foreman of the Kohala Section of the ranch, including Puakea and Pu'uhue. Edward Lewis Collection.

A. W. Carter had the foresight to develop a younger cadre of potential leaders for the long-range plans of Parker Ranch. Among these were Harry Kawai, William Kaniho, Joe Pacheco, and George Purdy.

A New Generation of Leaders

Harry Hi'ilei'ali'i Kawai

Born on November 30, 1908, the son of Richard "Dick" Paukeaho Kawai and Martha Zenobia Likolehua Spencer, Harry was destined for a lifetime career on the ranch, taking his first assignment in 1918 at age nine. He was assigned to the 'Ōpae Gang, with Frank Vierra and Masao Yoshimatsu as coworkers.

The 'Ōpae Gang was a crew of schoolchildren of ranch employees who worked vacations doing light utility and maintenance work. Due to their diminutive size, *'ōpae* (shrimp) was an apt description of the crew. Many of the ranch's finest had their start in the 'Ōpae Gang, notably Charlie Kimura, who rose to become the livestock manager.

Waimea Park as it is known today was originally hand graded and landscaped by the 'Ōpae Gang, and young Harry was supervised by Jimmie Spencer. The next assignment was a very thorough cleaning of the headquarters of recently acquired Humu'ula Sheep Station. This was followed by a Waiki'i corn-farming episode in which cutting corn suckers was the job at hand. Digging guava at Kawela paddocks and Pā'auhau also meant camping on-site. From the rain belt of those areas, Harry and his cohorts were moved to sultry Kawaihae Uka for the digging up of lantana. All through his school years, he worked in the early mornings and late evenings as a kitchen helper under Mr. Akana, who ran the Parker Ranch Restaurant.

Upon finishing school in 1925, Harry went directly to work for Parker Ranch under Martin Campbell, doing pipeline installation from the water heads in the forests above Waimea. On his own accord, Harry had developed significant horse experience beyond his innate horsemanship traits. He was now a member of the Plaster Gang, whose work was mostly construction oriented. In a variety of new assignments, Harry worked also under Foreman John Liana, who had been his boss in 'Ōpae Gang days, preparing 'Anaeho'omalu Beach for the building of the ranch cottage.

But Harry longed to be astride a horse, and that chance came when a position opened up on the elite Cowboy Gang with the death of Iakopa Kaha'ikupuna. Leading up to this assignment, Harry developed further horsemanship working at the Breaking Pen under James Palaika. His many and varied assignments over the whole spread served as meaningful experiences later in his career.

On February 17, 1930, Harry wed Lehua Brown, a Hilo native. Lehua had been a schoolteacher at Waiki'i village. Their daughter Puggy, who later married Rowell Yim, gave the Kawai family three splendid grandchildren: Harry, Edie, and Mary Alice.

Once on the Cowboy Gang, Harry's natural abilities surfaced readily in three distinct areas. Harry was first a consummate horseman, with a wonderful eye for picking a fine horse. His manner around horses was calm and gentle, and this was reflected in the horses he trained. Harry came along as a young cowboy when the ranch was peaking in horse production, and some of the progeny of certain stallions were noted for being mean buckers—for his purposes, exactly the challenge he sought. Turning those snorty broncs into sound, gentle workhorses, often loaned to the Carter family entourage, was exactly the kind of work at which Harry excelled, and a distinguished reputation as a fine horseman naturally followed. His fame as a bronc rider is legendary, especially given his disdain for using a *kaula hele*—the hobbling of the stirrups to each other, giving the rider significant advantage. This was not for Harry—he rode "freehand."

A second attribute came in the form of a fine eye for cattle—not just in quality but judging how they best could be handled. Handling vast herds of range cattle, he used the least amount of push or excitement to get the herd to go precisely where they needed to go. No yelling, no fanfare, absolutely no roughness. It was clear that A. W. was watching Harry carefully, after putting him on specific man-sized assignments to test his mettle. Carter was never disappointed, in fact, after assigning Harry the primary responsibility for the *hōʻau pipi* (cattle swimming) aspect of shipping the market cattle. It was on one of these Kawaihae drives that Harry worked the full day with a stomachache *(ʻeha ka ʻōpū).* After the long ride home to Waimea, his doctor rushed him to Kohala Hospital—his appendix had ruptured earlier in the day! Although Harry worked nearly an additional three decades, this setback in his health was just the beginning of a litany of abdominal crises.

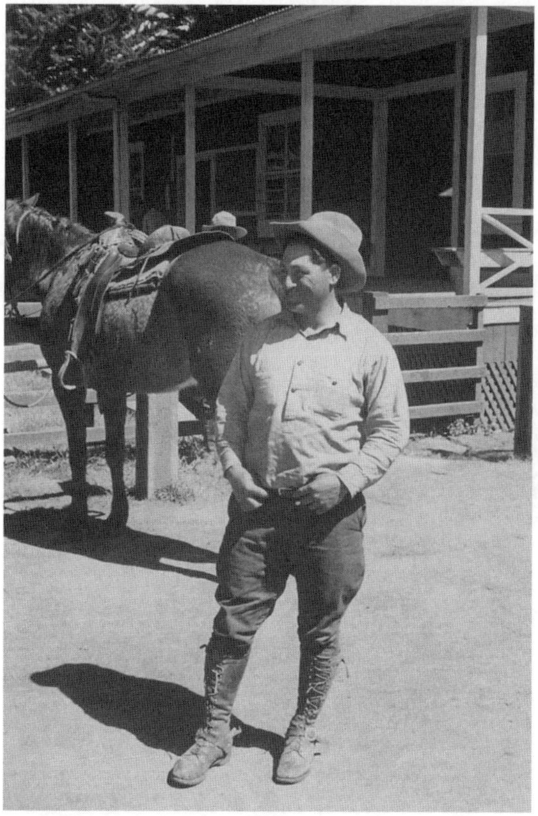

Harry Kawai, cowboy foreman, at the ranch office. Hartwell Carter Collection.

In his recuperative months, Harry worked as a special police officer for the U.S. Army, as by that time World War II was well underway. Shortly after returning to the Cowboy Gang, Harry was appointed to his first foremanship, that of the Breaking Pen, where he remained until the end of 1944. His leadership of the Roughrider Gang was then assumed ably by his brother-in-law and cherished friend, George Purdy. It was Harry's sister Annie who wed George, and they were a handsome couple.

From 1944 to 1947, Harry was cowboy foreman, and leadership was among the finest of his qualities. Richard Smart was quoted in retrospect in the December 1968 issue of the *Paka Paniolo* featuring Harry's retirement: "Harry had a special quality of leadership. First of all he was tops in his field, he was firm with the men, but a good leader. He was respected, always managing to get a job done with the least amount of commotion and fuss." These were Richard's words twenty-one years after Harry was asked by Carter to serve as general foreman of the entire ranch. During this period, he reported directly in succession to Hartwell Carter, Dick Penhallow, and Rally Greenwell, who all attested to Harry's being a cowboy's cowboy and a leader of leaders.

The position of general foreman was one that entailed sweeping authority over all ranch operations—including fencing, shipping, horse and cattle programs, trucking, water, and the motor pool. Much like his predecessor Theodore Vredenburg, Harry was well suited to the position as a result his formative ranch years. In this capacity, Harry clearly expressed his natural leadership talent in coordinating large operations with a myriad of strong personalities and intricate moving parts. He did this with distinction. After more than a half century of service to the ranch and the toll of decades on horseback—not to mention eleven subsequent surgical procedures—Harry chose to retire on November 30, 1968, his sixtieth birthday. On his final day of work, Harry's way of saying goodbye was a long ride checking the water lines and paddocks he so dutifully built and maintained over the last five decades. Harry was truly loyal to the land—Aloha 'Āina Paka. Sam Kimura succeeded him.

William "Willie" Kaniho

Born March 18, 1894, in Kalopā, Hawai'i, Willie's parents were William Kaniho Sr. and Kanaihola Papa. William Sr. worked for the German consortium of Hackfeld and Haneberg when they ran the Humu'ula Sheep operation in the early 1890s. Willie's mother's parents, the John Papa *'ohana,* also were employed by the German consortium. Willam Kaniho Sr. died as a young man and the John Papa family raised young Willie.

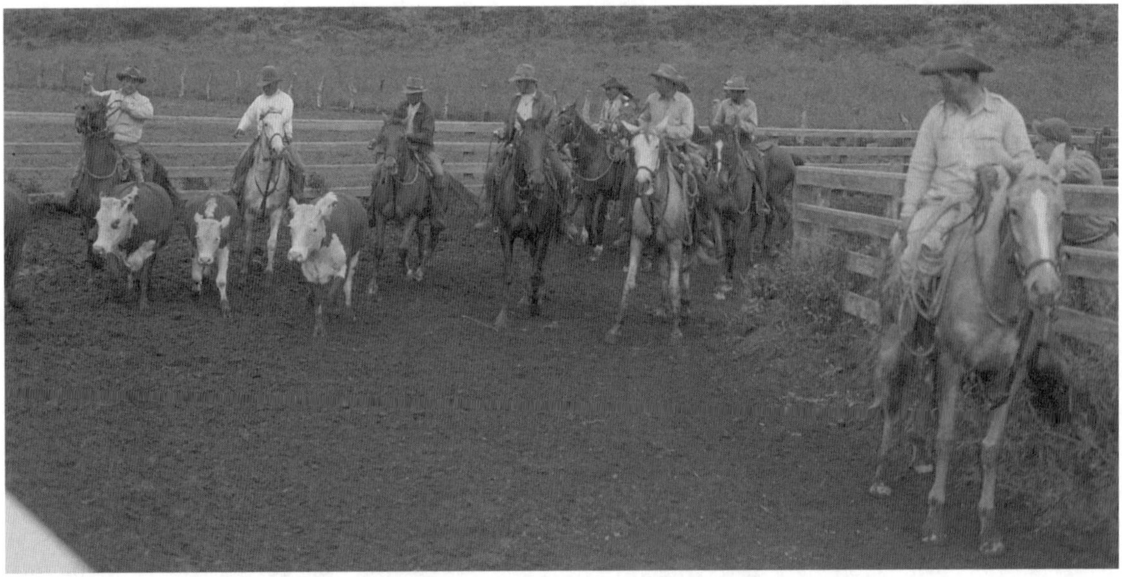

Harry Kawai, cowboy foreman, with crew. Hartwell Carter Collection.

He was educated in Kalopā, but his heart was in the mountains—Mauna Kea. At age fourteen, he left home and went to work at Humu'ula for Sam Parker Jr., who had assumed ownership. Ikua Purdy was the foreman of Humu'ula at the time.

Within a few years, Parker Ranch acquired the Humu'ula Sheep Station Co., and Purdy and Kaniho stayed on under A. W. Carter's management. Willie was sixteen when he became a Parker Ranch employee. Ikua remained foreman until he was offered an opportunity to take the same position at Princeville Ranch in Hanalei on Kaua'i. Willie joined Ka'inapau Kailikini and Ikua and moved to Kaua'i, going from a salary of $12 per month to $75 plus a bonus. Willie met and married his first wife, Dora Niau, while on Kaua'i.

The Princeville venture lasted only two years. Willie was offered $85 per month to go to Kahuā Ranch, then under the ownership of Frank Woods. Willie and Dora were blessed with their only child, Willie Jr. (Wili Keiki) shortly before Dora succumbed to influenza while in Honolulu.

Things at Kahuā did not work out for Willie, and A. W. Carter summoned him back to Parker Ranch. After some toe-to-toe dickering on wages and position, Willie was made foreman at Keamuku at a salary of $100 per month. Single again, Willie lived in a cottage on the Carter premises, where he became close to Hartwell, dining frequently with him in the evenings in formal attire. A young Rally Greenwell was also a tenant at the Carter complex.

Willie courted Daisy, daughter of Sam Awa'a Sr., a tall, brawny *paniolo* from Kawaihae Uka. Daisy was an able *paniola* herself, often doing the hog-tying when her spouse roped wild cattle. In a rapid series, several children were born who would become prominent. Kenneth had a full career with the Honolulu Police Department before relocating to Waimea. Dudu (his real name) also moved to O'ahu to enjoy a construction career. Sonny (Nicholas) cowboyed on the ranch before entering military service. Tommy was raised at Kahuku Ranch and cowboyed at Ka'alu'alu Ranch. A daughter, Dolly, moved to the mainland. Finally there was Daniel, who grew to match his father's reputation on Parker

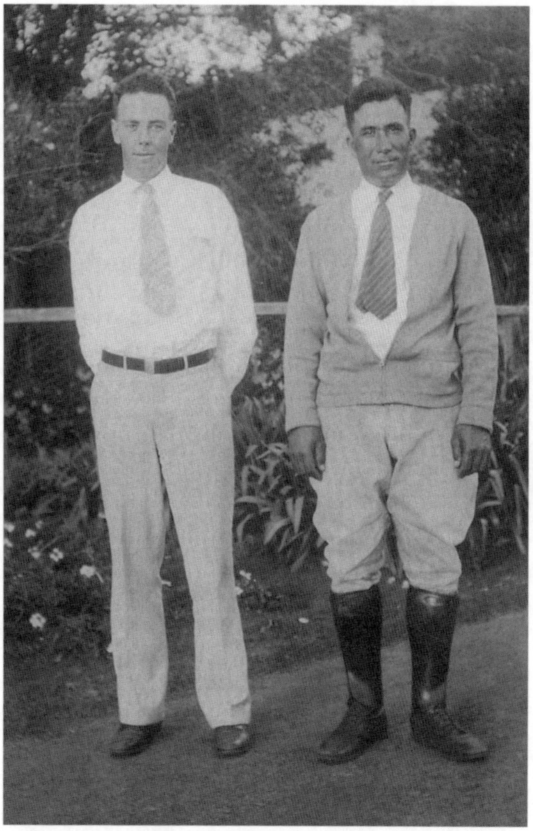

Rally Greenwell and Willie Kaniho in 1935, preparing to dine with A. W. and Hartwell Carter. A. W. Carter Collection.

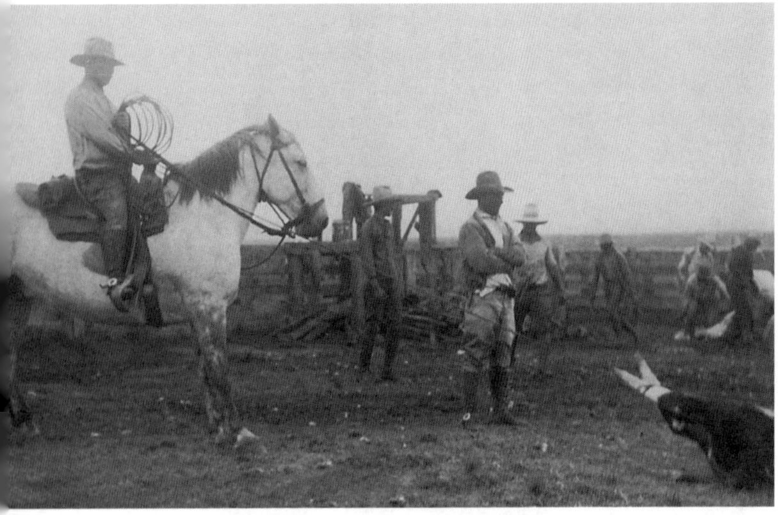

Willie Kaniho supervising a branding, with his arms typically folded. A. W. Carter Collection.

Ranch. Sadly, Daisy died in 1932, shortly after Daniel's birth. Aunt Betsy Aveiro, who raised him affectionately, named baby Daniel "Ipo"—Hawaiian for "sweetheart."

Willie completed a two-year foremanship at Keamuku and was transferred to Humu'ula following the death of foreman Edward Junichi Ishizu. Willie remained at the sheep station as foreman for eight years before he was called back to Waimea to run the Cowboy Gang. Keoni Li'ili'i had retired and A. W. chose Willie to be the new Cowboy Gang foreman. During the transition, Willie married his third wife, Mary, daughter of Martin (Makini) Martinson, foreman of Kahuku Ranch. They had three children: Jane, Mary, and a son, Martin, who relocated to Alaska as an adult.

In the transitional period between A. W. and Hartwell's management, Willie was reassigned to Humu'ula. At this point he launched an aggressive water resource development program, largely of his own design. Tomo Fujii, Isami Nishie, and David Kahikina did the bulldozing and grading of more than twenty earthen water reservoirs between Kalai'eha and Keanakolu.

Willie was a builder who increased ranch resources. Water and grass were his building blocks. The actual site selection, reservoir design, and landscaping were all done by the eye of the master, Willie, who knew every hill, valley, and ridge of the Humu'ula lands from Kipuka 'Āinahou to Pā Malani at the lowest elevations of the Waipunalei fee simple lands. He knew where fog hovered heavily, and he knew the paths of raging storm waters down the *kahawai* (gulches) that would only be lost to the ocean. Were it not for the well-designed bypass system, this precious resource would have been wasted; instead it was passed into an expansive series of waterholes via concrete and earthen dams.

Willie was brought back to the main ranch (Waimea) in 1960, where he was placed in charge of the Waiki'i and Keamuku Sections. He brought his fervor for water development with him and turned to the northeastern slopes of Mauna Kea, where an additional half dozen earthen reservoirs were built. In addition, Willie was a key player in the Pu'u'ōpelu to Pā'ali'i reservoir transport systems that fed water by gravity from high in the Kohala Mountains to the lower slopes of Mauna Kea. He further developed water storage in Kawaihae Uka II and III. He was also active in subdividing large paddocks to intensify grazing efficiency.

Of his sons, Daniel inherited the yen for building ranch resources, especially water. His Ka'ū son, Tommy, inherited his father's rough-and-ready cowboy skills.

When Willie retired after fifty-six years on the ranch, one of the statements he was most proud to make was that he "trained over a hundred young cowboys." From the standpoint of livestock production, Willie's greatest gift to the ranch was in water resource development, having built thirty large reservoirs with a total storage capacity of nearly 100 million gallons. This was quite a feat for a *kua'āina* (backcountry) cowboy. Willie was a magnanimous builder of Parker Ranch assets.

Joe Pacheco

Kona is the ancestral home of many of Parker Ranch's finest *paniolo,* and Joe Pacheco is no exception. Joe was born September 30, 1903, to Manuel Pacheco and Lavina Iokua Poepoe of Hōlualoa, Kona, Hawai'i. Manuel was a first-generation Hawai'i-born child of Portuguese immigrant G. Pacheco and his Hōlualoa native wife, Marian. Lahela and Miki Poepoe, Marian's parents, were originally from Lahaina, Maui.

Shortly after Joe's birth his parents moved to Puakō, where his father worked for the Hind Plantation. This led to a career with Parker Ranch, as Manuel and Lavina made their first home at Keamuku Station. Manuel was a skilled teamster of mule and draft horses, and his talent drew him to the corn-farming operation at Waiki'i. Little Joe was already attracted to horses at a very early age. As the family moved to Makahālau and then to Humu'ula to allow Manuel to continue his farming chores, Joe was exposed to the entire ranch. The move to Humu'ula Sheep Station probably had the most profound impact on young Joe, as he was exposed to the glorious years of Sam Parker's elegant lifestyle. While Joe never sought posh social circles as an adult, he was well aware of how important it was to be appropriately attired for any occasion.

Joe was in his early teens when his father received his first foremanship, the Hopuwai Section of the Humu'ula Sheep Station Co. By this time, Joe exhibited his love for horses by having his own little string of mounts, the worn-out cow horses of the older cowboys. Joe groomed these old horses as though they were show prospects. Their tails and manes were cleaned and curried. Their coats were kept free of fly sores with Joe's only medicament being *'āila pipi* (tallow) mixed with a little kerosene and axle grease. Their feet were trimmed neatly but not shod, as the deep soils of Hopuwai were soft and forgiving. Shoeing was reserved for the high mountain drives in pursuit of wild sheep. These opportunities gave Joe his first exposure to roping, a skill for which he was noted later in his ranch career.

At age fifteen, Joe was formally employed by the ranch under an emergency measure. The Great Fire that broke out at Horner's Ranch at Umikoa (Kūka'iau) had burned out of control, taking wide swaths of majestic stands of *koa* trees as it marched toward the Hanaipoe Section of the ranch. Elders claim that from beginning to end, the Great Fire burned for three years before it was finally extinguished, largely by hand, well into the Hanaipoe Section, reaching a hill called Keiki Pa'u'ūla.

Joe's diligence as a worker did not go unnoticed, and when his *kinai ahi* (firefighting) days were done he was assigned to the motor pool at ranch headquarters in Waimea. The adjacent teamster stables

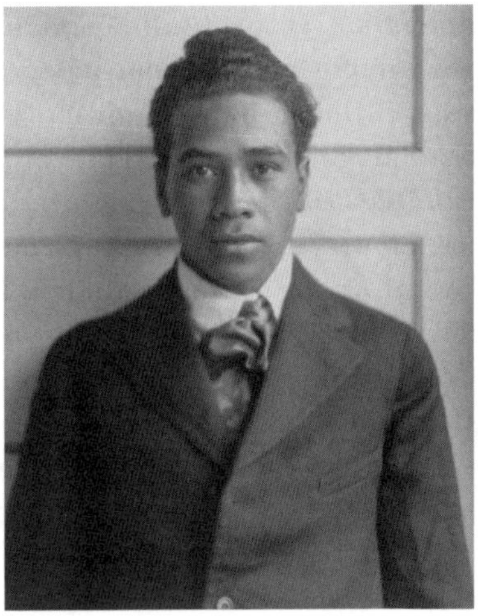

Joe Pacheco in his youth, 1920. Bishop Museum photo CN 4539.

beckoned to Joe, thanks to Foreman John Kahaikupuna's awareness of Joe's talent for working with horses. A. W. had long-range plans for Joe, and before long he was stationed at Pukalani Stables under Charley Duncan, caregiver of noted remount stallions including Eastertide, Frolic, Harvester, Gladiator, Gold Oak, Rodgers, and Roi Donovan. Joe's primary responsibility was to exercise these stallions on a daily basis when breeding season was over. During breeding season, Joe gained experience in breeding these studs to select remount mares.

Carter also exposed Joe to the more mundane but vital chores of ranching. Under Ernest Vredenburg, Joe built miles upon miles of new pipelines, strung as many lengths of fencing, and planted thousands of eucalyptus trees as windbreak and timber (fence post) resources. Carter was consistent in including water resource management as part of leadership development, emphasizing the basic role of water in animal husbandry.

Joe received an assignment to the Breaking Pen in 1922 with great enthusiasm. His horsemanship skills were refined by James Palaika, who spared nothing to

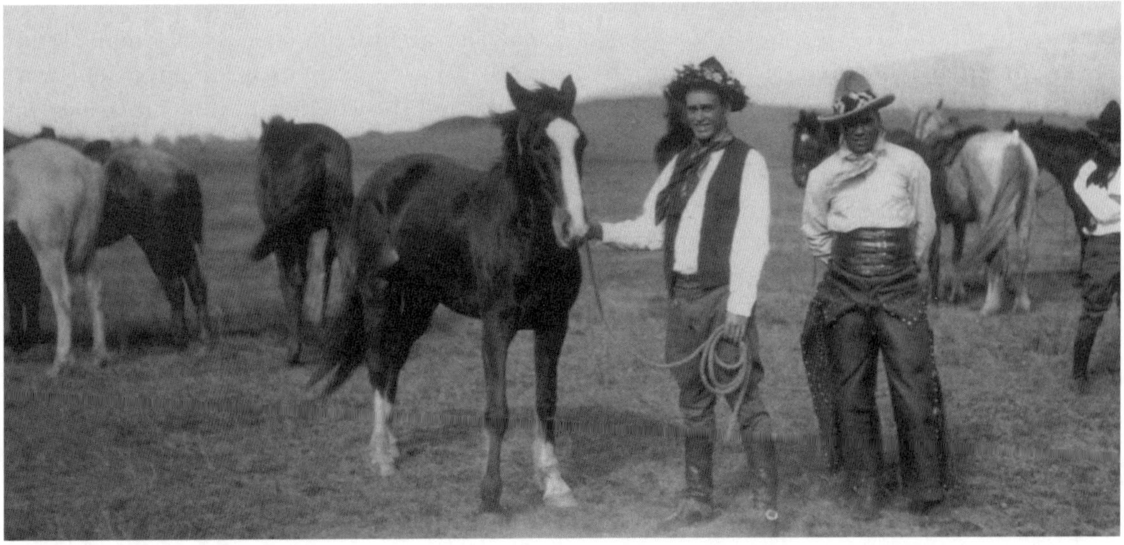

Tom Bell and Joe Pacheco in fancy cowboy attire at Parker Ranch Fourth of July horse races. A. W. Cater Collection.

reinforce Joe's exposure to the progeny of all the stallions—the good and the ornery. It was during this period that Joe fine-tuned his skills in putting a good mouth and headset to his mounts, and this reputation followed him throughout his ranch career.

In 1924, Joe was called to Humu'ula to work under Willie Kaniho, who recognized his potential. The sheep station was familiar country to Joe, and he brought with him a solid string of eighteen mounts to ride the expansive operation from Kipuka 'Āinahou to Waipunalei. Willie became dependent on Joe's loyalty and ability and brought him back to Waimea. Here Kaniho became cowboy foreman, and Joe was his assistant. After a two-year stint, Joe received his first foremanship at the Breaking Pen, where he remained for several years, passing on some of the talent and skills he had acquired to the younger generation of horsemen. Joe developed a lasting reputation as a firm boss who expected 110 per cent from his charges, but who was at the same time sincerely concerned with the health and welfare of these young men.

In a 1928 marriage to Betsy Leinani Ma'ina'aupō, daughter of Kaliko Ma'ina'aupō, three children were born. Mary, the eldest, married ranch cowboy Allen Hartwell Na'ea Lindsey, while Betty married another company employee, Yoneichi Nakamura. The only son, Joseph Jr., lived primarily on the mainland.

Just before the outbreak of World War II, Joe was reassigned as assistant Cowboy Gang foreman under Yutaka Kimura, a position he kept until 1945. At this time, Hartwell was impressed enough with Joe to make him foreman of the First Gate, Second Gate, and Pā'auhau Sections of Parker Ranch. Beyond running the cattle operations of these sections, Joe had a huge mission: planting the newly introduced pangola grass. This was all done by 1947, when more than 1,100 acres were in full bloom with this excellent forage imported from South America. Much of the planting work was expanded along the new belt highway in 1952 as it passed through Āhualoa and into Waimea in fields called the Baker and Vierra Paddocks. These paddocks became the showcase of the ranch's leadership in range management, and Hartwell and Richard Smart both credited Joe Pacheco with their success. In fact, Joe was promoted to Kalopā Section foreman and given 1,300 mother cows to look after while he was charged

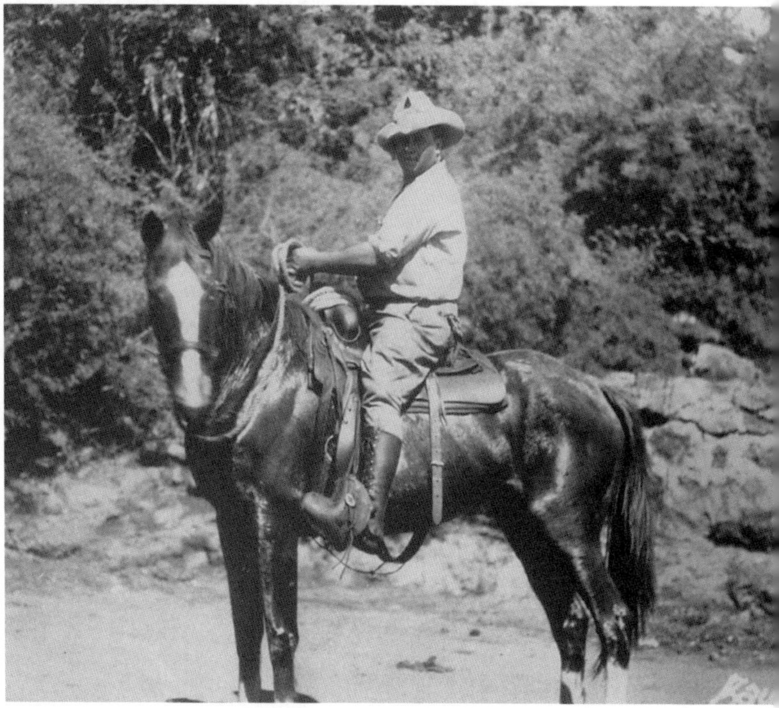

Joe Pacheco preparing to hō'au pipi at Kawaihae, 1930. A. W. Carter Collection.

with another mission of pasture improvements. The ranch built the family a spacious home there in Kalopā.

In 1958, Joe was reassigned to the First Gate, Second Gate, and Pā'auhau Sections, where he remained until his retirement in 1968.

Richard Smart, in dedicating the 1968 Fourth of July horse races to Joe, stated: "Joe Pacheco, the name of a man whose entire life has been wrapped up in the work and progress of Parker Ranch. Joe—with your ready smile, your strong handshake, your lusty singing and your happy spirit—these are part of Waimea. We will miss your ever loyal efforts and your hard work but we are glad that you will always be a part of our community, our good times and our Parker Ranch family." It was a fitting send-off for a deserving *paniolo kahiko* (cowboy of old).

13
BIRTH OF THE TERRITORIAL BEEF MARKET
1871–1950

As early as 1871, an entrepreneur named Gilbert Waller founded a beef marketing operation on O'ahu called the Metropolitan Market. In 1872, his nephew, Gilbert J. Waller, acquired the meat company and eventually took in as partners James Campbell, James Dowsett, and T. H. Foster, prominent ranch owners on the Island of O'ahu. This meat purveying operation grew rapidly over the next two decades, with Parker Ranch its main supplier of beef. In July 1900, the original partnership was transformed into a limited liability stock company and incorporated as the Metropolitan Meat Company, Ltd., with Gilbert J. Waller serving as manager/treasurer. As fate would have it, Samuel Parker was president of its board. Included among its membership were prominent O'ahu residents M. P. Robinson, Ed Ingham, A. J. Campbell, F. M. Swanzy, Cecil Brown, W. G. Irwin, and L. L. McCandless.

The Metropolitan Meat Company flourished at its King Street location, with forty-seven employees providing processing and delivery of slaughtered beef throughout Honolulu. The company eventually added a tannery as part of its broad diversification.

Bringing 40 to 50 percent return to its investors, it isn't any wonder that A. W.'s original overture to have Metropolitan Meat company allow greater ownership interest by the actual ranchers—many of whom were neighbor island graziers who shipped their market cattle to this consortium—was firmly rebuffed by the directors. After trying to work within the system as a Metropolitan Meat Company director, A. W. resigned and took a different tack.

A. W. and other neighbor island ranchers had legitimate concerns with Metropolitan Meat Company's autocratic and monopolistic way of doing business. Prices per pound of carcass weight were as low as 10 cents per pound when A. W. became the Parker Ranch trustee. Without notice, the price could drop as much as a penny per pound. Moreover, Metropolitan Meat Company could cancel cattle shipments to O'ahu at the last minute. All too often, Parker Ranch cattle would be held at Kawaihae for a day or two, only to be walked all the way back to Waimea because of the company's delayed orders. These draconian moves forced A. W. to take action. He served Metropolitan Meat Company notice that within six months, no market cattle would flow their way.

Hawai'i Meat Company Is Born

With the participation and support of other neighbor island ranchers such as the Greenwells of Kona, the Maguires of Hu'ehu'e, Horners of Kūka'iau, Kohala Ranch Company, and W. H. Shipman, A. W. announced plans to build a slaughterhouse on O'ahu. This announcement brought Metropolitan Meat Company directors to their senses. They offered their outfit for sale to A. W.'s consortium. Carter's group purchased Metropolitan Meat Company with a $100,000 cash deposit supported by a note signed by his partners, who were not required to put forth cash. The partners' only requirement was to fill the new marketing firm orders. In March of 1909, Hawai'i Meat Company was born, with the articles of incorporation signed by A. W., Robert Leighton Hind Sr., Albert Horner, J. A. Maguire, Maud W. H. Greenwell, J. D. Paris, J. F. Woods, A. Morrow, and R. A. Cooke. Gilbert J. Waller was named manager of the new company.

Under A. W.'s direction, total reorganization took place. Emphasis was placed on processing efficiency, underscored by the prudent treatment of all ranchers choosing to market their beef and sheep through Hawai'i Meat Company. He took pride in the company posture of not turning down any cattle offered and went as far as delaying Parker Ranch shipments to O'ahu to accommodate the shipping needs of other Big Island ranchers. Now the actual industry that produced the beef could participate in determining market pricing and, to a degree, actual shipping arrangements. A. W.'s views are summarized in Brundage's *Alfred W. Carter:*

In the year 1909 the graziers got together and organized a cooperative marketing company. Under the Articles of Association, no one could hold stock except those interested in beef cattle and further, dividends to the stockholders were limited to 7% and all profit made over and above that amount was divided pro rata according to the number of pounds of meat each grazier put into the market. This additional profit was divided equally with those who were not stockholders of the organization but who sent their cattle to the Hawaii Meat Co. to be slaughtered.

Before the time of the organization, we were getting ten cents for our beef and they cut us down to nine cents. We could only ship cattle when the Metropolitan called for cattle. On top of that, the Inter-Island would not have a steamer for us when we were ready to ship. Many times I have taken cattle down to Kawaihae and the steamer would not call for them.

I suggested to three or four of the partners of the Metropolitan that they should let the graziers in on it and give us some interest in the business. They were making forty or fifty percent on their money so did not wish to let anyone in. I informed the Metropolitan that I intended to start a slaughtering business and would supply beef to them for six months and, after that, we would slaughter our own cattle.

It was at this time that Metropolitan suggested that Mr. Carter purchase the stock of the company. Carter continues: "I said I would appoint someone, they should appoint someone, both of these men to

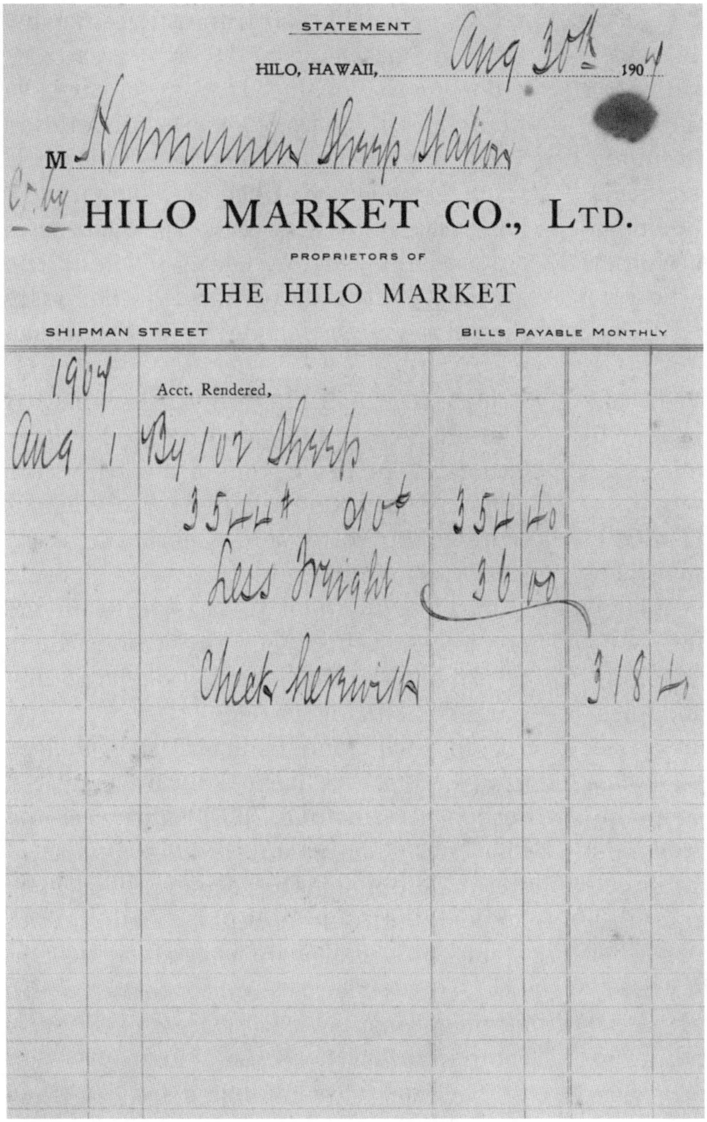

STATEMENT

HILO, HAWAII, Aug 30th 1907

M Humuula Sheep Station

Cr by

HILO MARKET CO., LTD.

PROPRIETORS OF

THE HILO MARKET

SHIPMAN STREET BILLS PAYABLE MONTHLY

1907	Acct. Rendered,			
Aug 1	By 102 Sheep			
	3544# @ 10¢	354 40		
	Less Freight	36 00		
	Check herewith			318 40

Hilo Market Company invoice.

appoint a third, to put a price on the business and we would pay it. That is how we got the Metropolitan Meat Co. I went to Hawaii and got the graziers together, outlined my plan and arranged with the bank to borrow $100,000. All the fellows who came in endorsed that note. No one put up a cent. Three-quarters of a cent per pound was taken out of every account sales to repay the loan and, thereafter, in raising the capital we never asked a stockholder for money."

In his efforts toward market development, A. W. also looked to the Big Island to develop its own market that could, in particular, serve the culled cow and bull resources. This came about in 1911 when A. W. participated in the reorganization of Hilo Market Company into the Hilo Meat Company, capitalized at three hundred shares of $100 per share value. The officers and directors included W. H. Shipman as president, A. W. as secretary, and A. Horner as treasurer, along with members who represented the Hawaiian Agricultural Company (Kapāpala Ranch).

Now A. W. had secured a steady Big Island resource in marketing Parker Ranch cattle, sheep, and hogs for the burgeoning Hawai'i Meat Company. Within its first year, with a capitalization of only $100,000, Hawai'i Meat Company profited more than 42 per cent.

No longer a packing house, Metropolitan Meat Company continued as a retailing concern on King and Bethel Street and finally closed its doors in 1950, a victim of the cash and carry trend.

Having successfully climbed the mountain of marketing, A. W. was eager to address another thorny issue: the transportation of cattle to market.

Developing a Livestock Transport System

A. W. dove into the livestock transportation issue with his usual vigor and emerged in his normal problem-solving mode. He knew the canoes of the Polynesian forebears of Parker Ranch hands had transported hogs, dogs, and chickens over thousands of miles of open ocean. He also knew that to a limited degree, similar canoes brought ashore the cattle, horses, sheep, dogs, and goats from the ships of Vancouver and Cleveland. Additionally, small interisland schooners were employed for such transport—notably the *Beretania,* built in 1793 by Vancouver's sailors for Kamehameha I. Some twenty-five years later there were reports of cattle on O'ahu, suggesting that offspring of the cattle given Kamehameha I had made the interisland voyage. Writing in 1828, Laura Judd reported that an interisland schooner trip included horses and cattle. By 1837, more than sixty freight schooners plied Hawaiian waters.

In July 1840, the brig *Clementine* left Kawaihae for Honolulu with a load of forty bullocks. The description of this trip is not atypical. They were snubbed up to a central spar by their horns, side by side. This was a safer way to travel than being loosely arranged with room to fall, owing to the rough seas on this trip. In bad weather, there is no comfortable ocean travel for any species, including humans.

In 1844, John Palmer Parker shipped nearly 10 tons of barreled beef aboard the schooner *Chinchilla.* That same year the king and government used an iron-sided steamship for transport among the Islands. In 1853, the *Akamai* was the first steamer to provide regular interisland service. In 1857, the schooner *Mary,* captained by Thomas King, provided regular cattle service to Kawaihae on a weekly basis.

In 1860, C. A. Williams and Company of Honolulu ordered a 137-foot brigantine with a 25-foot beam and 10-foot draft for the Hawaiian government, with an approved charter for interisland operations under the Hawaiian Steam Navigation Company. The actual builder of the vessel was Paul Curtis of Boston. The ship was a 400-ton wooden-screw steamer that sailed around Cape Horn to Hawai'i under Captain William G. Bush.

The vessel was christened the *Kilauea.* This steamer would have the sad distinction of bringing the physicians of a dying king, Lunalilo, back and forth between O'ahu and Hawai'i. On January 16, 1874, Lunalilo's doctor, Dr. George Trousseau, examined the monarch and concluded that he could live only a short time. The king was brought back to Honolulu, where round-the-clock nursing kept him alive for a little while longer. He died at 8:30 on the evening of February 3, 1874.

R. C. Janion, W. H. Green and Company, names familiar to the WGAC, were pivotal in early interisland transport, operating both the *Kilauea* and the *Annie Laurie* for a period. By 1866 both vessels were wrecked, the *Annie Laurie* off Kōloa, Kaua'i, and the *Kilauea* driven ashore at Kawaihae. The latter ship was later repaired. A historic and prophetic trip was taken in 1872 by the five-year-old A. W. Carter aboard the SS *Kilauea* as it called on ports that included Kawaihae, Māhukona, Laupāhoehoe, and Hilo, then around to Punalu'u and Honu'apo, and finally to Kona on the Big Island. Little Alfred spent time at the old Spencer home in Waimea, as he traveled in the company of the daughter of Francis Spencer, a Mrs. Beckerton.

Frank Vincent Jr., quoted in Brundage's *Alfred W. Carter,* made the following observations of bullock swimming in 1874 in his "Through and Through the Tropics":

> After having embarked on a schooner [at Kawaihae] preparatory to returning to Honolulu, I was much amused at the manner in which some bullocks were shipped. The schooner was anchored about 200 yards from the shore. The animals were first driven into a small enclosure [sic], situated on the beach and surrounded by a high stonewall. A rope, by means of which a large scow could be pulled to and fro, was stretched between the enclosure and the vessel. Two kanakas, on horseback, now set to work. One of them lassoed one of the cattle and twisted the lasso two or three times around the horn of his saddle while the other seized the animal's tail. Both horsemen then started for the scow [sic] and however obstreperous the bullock might have appeared at first, the logic of the lasso and the argument of the twisted tail soon proved to him the prudence of submission. It is said that the bone in a bullock's tail is easily fractured and that when thus caught, the animal will resist until fracture threatens after which he becomes quiet. Reaching the scow, the heads of cattle were secured and while two or three natives pulled the boat, the animals were compelled to swim to the schooner to which they were hoisted by means of a tackle and a broad girth around the belly. They were placed in double rows, their heads to the center and their horns firmly lashed, and in this manner were transported to the capital.

In 1878, Captain Thomas R. Foster developed a fleet of schooners providing coastal service throughout the Islands. He went on to build a comparable fleet of small steamships, the first of which was the *James Makee,* followed by the *C. R. Bishop* (formerly christened the *Kelpie*) and the *Iwalani.* These carried schooner-rigged sails

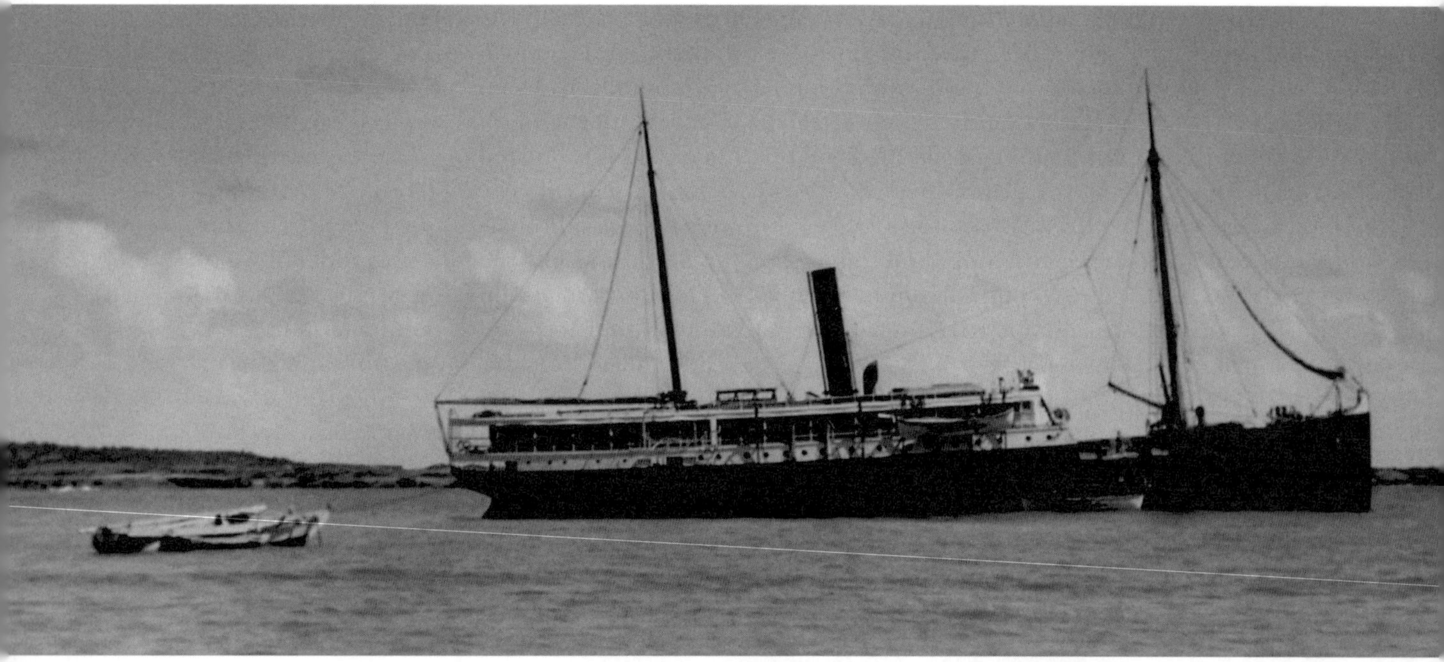

Top: *SS* Claudine *off O'ahu. She was built in Glasgow, Scotland, and named after the daughter of Claus Spreckels, sugar tycoon. The* Claudine *was the rescue ship for Captain Kini and his crew when the SS* Bee *struck a reef on Maui in 1924. They landed at Kawaihae. Bishop Museum Henry C. Ovendon Collection, Negative CD 20092.*

Bottom: *SS* Helene *off Kawaihae Bay. Also named after the daughter of a sugar magnate, W. G. Irwin, the* Helene *was a productive vessel for the sugar industry as well as a livestock transport for Parker cattle and others. A. W. Carter Collection.*

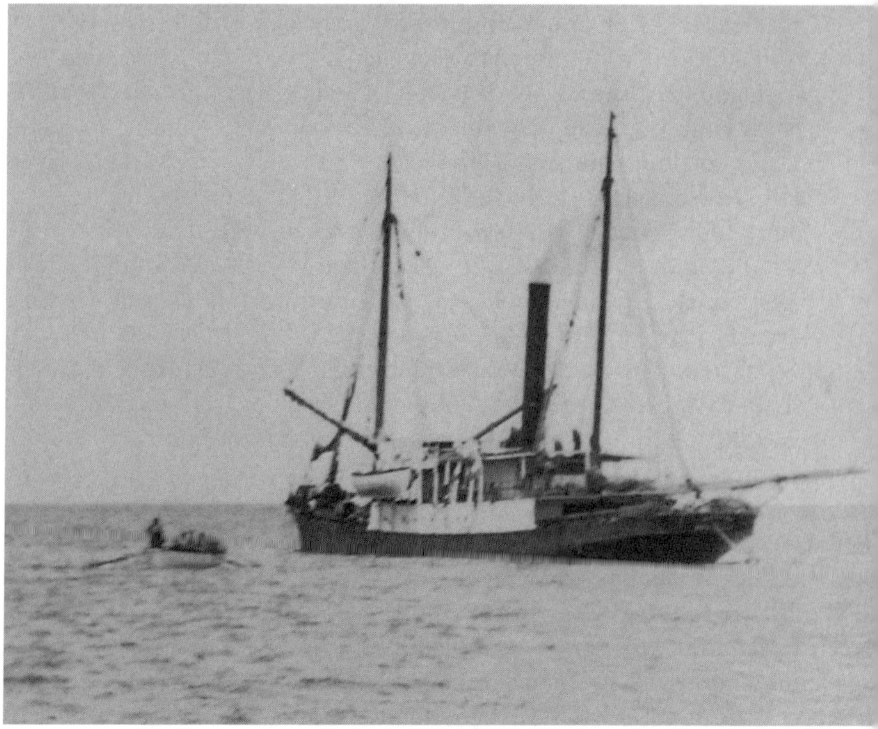

for auxiliary power and undoubtedly transported some livestock during their tenure in Hawaiian waters.

In 1879, British author C. F. Gordon-Cumming reported on life aboard the *Likelike:* "By the time they have shipped an indiscriminant mass of human beings, horses, cattle, luggage, timber, sugar, coffee, and sundries, there was not much elbow room to spare and certainly no possibility of luxury." In 1894 the *C. R. Bishop* went aground off Kaua'i, but the *James Makee* plied the channels of Hawai'i for more than forty years.

By 1898, the SS *Claudine,* SS *Maui,* and SS *Helene,* all of steel construction, were routinely carrying cattle from Kawaihae to Honolulu.

Captain Foster incorporated his consortium as the Inter-Island Steam Navigation Company as early as 1883. A year later, Samuel Wilder incorporated the Wilder Steamship Company on the same day that A. Frank Cooke chartered his group as the Pacific Navigation Company, the latter never becoming a main force in interisland shipment. The corporations of Wilder and Foster had a geographical understanding between them that avoided overlap and competition. Both companies were kept competitive by pressure from overseas shippers that made direct calls on plantation ports. A. W. knew that shipping cattle among the islands simply meant the development and evolution of appropriate vessels. He became familiar with the history of interisland shipping, eventually serving as the director of one such company, and he maintained an interest in the seaborne transportation of cattle.

By 1905 the Wilder Steamship Company merged into the Inter-Island Steam Navigation Company. This company became the target of A. W.'s concerns for its poor service to the livestock industry, especially out of Kawaihae. During the next decade, A. W. continued to work within the system for greater coordination of shipments and improved cattle handling practices, seeking to prevent the extensive bruising that periodically occurred. Direct complaints to management and directors of the Inter-Island Steam Navigation Company were usually rebuffed. While the Public Utilities Commission rendered successful attention to complaints, the complaints became moot when the territorial Supreme Court ruled that the commission acted outside its jurisdiction.

A. W., through Gilbert J. Waller of Hawai'i Meat Company, approached J. A. Kennedy, the head of Inter-Island Steam Navigation Company, with a suggestion that they switch to the SS *Helene,* which was better able to accommodate cattle shipments. Kennedy threatened to cut off all Parker Ranch shipments if he received any more visits from people advising him on how to run his shipping business. By this time, A. W. had had quite enough.

By 1915, the Hawai'i Meat Company Board of Directors was considering the acquisition of a transport vessel of its own. In 1920, the SS *Bee* was purchased in San Francisco, California, for more than $165,000. Retired Captain Robert V. "Skipper" Woods of Keauhou, Kona, sailed her home. Built in 1907, she displaced 622 tons and brought 750 tons of cargo from San Francisco on her original trip. Shipping arrangements and conditions were immediately improved. Gilbert J. Waller had the

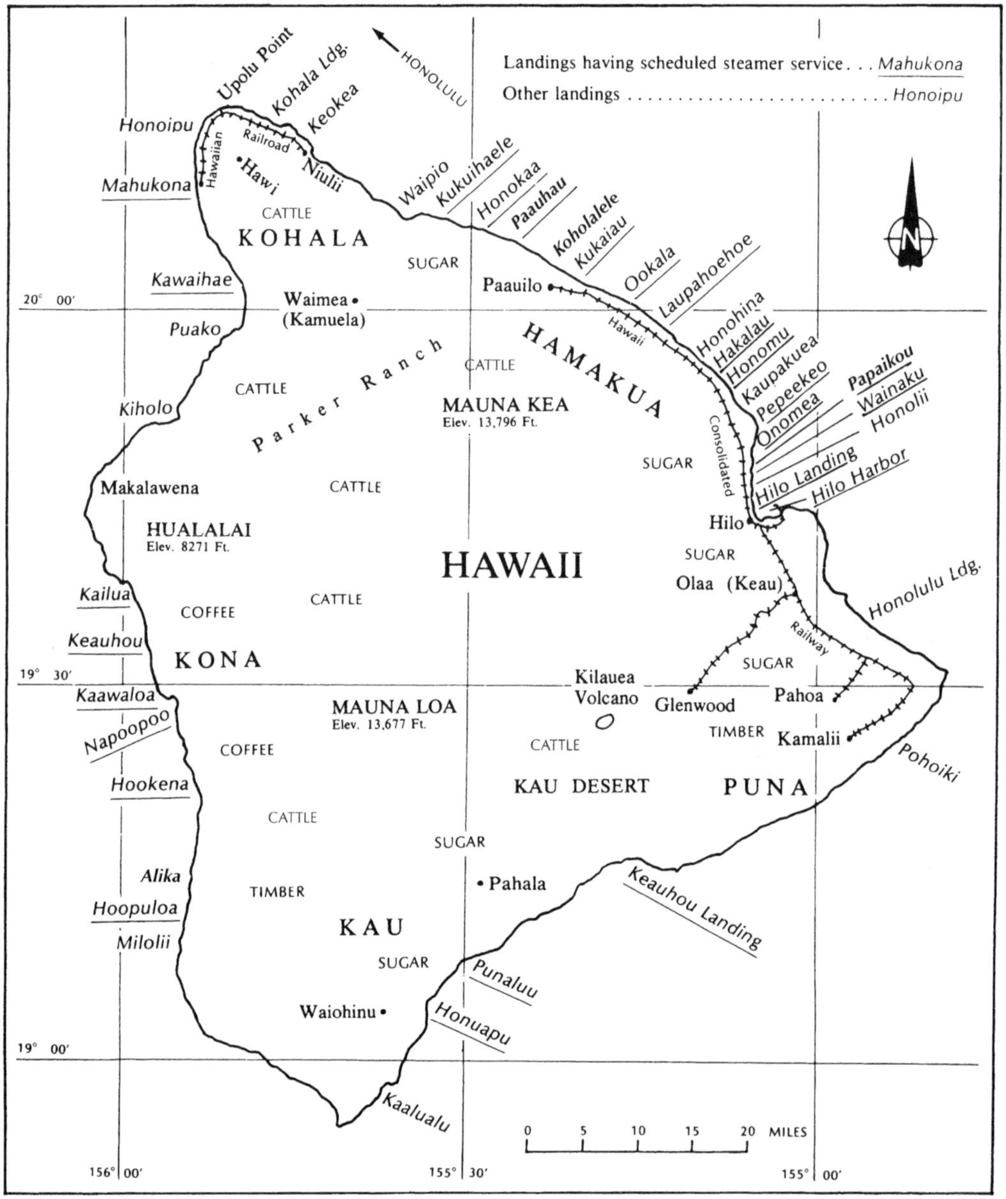

Map showing landings serving the cattle industry of the island of Hawai'i. Modified from Mifflin Thomas, Schooner from Windward: Two Centuries of Inter-Island Shipping. *Below the map is a list of vessels plying Hawaiian waters between 1828 and 1952.*

Akamai	*Dickenson*	*'Iwalani*	*Lehua*	*Mokoli'i*
Annie Laurie	*Haleakalā*	*James McKee*	*Likelike*	*Nettie Merrill*
Bee	*Hawai'i*	*Ka'imiloa*	*Mary*	*Onomea*
Chinchilla	*Helene*	*Ka'iulani*	*Maui*	*Wai'ale'ale*
Claudine	*Hornet*	*Kihalani*	*Mauna Kea*	*Wailele*
Clementine	*Hualālai*	*Kīlauea*	*Mauna Loa*	*W.G. Hall*
C.R. Bishop	*Humu'ula*	*Kīna'u*	*Mikahala*	

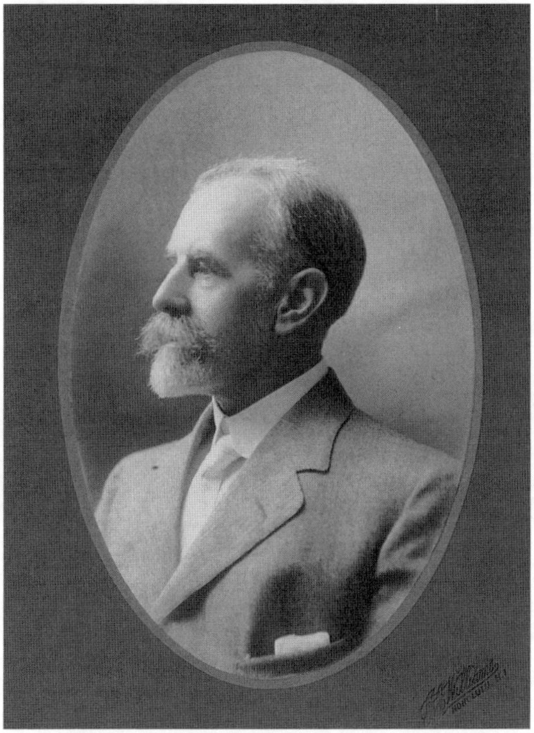

Skipper Wood, Captain R. V., was called out of retirement from his Keauhou Bay, Kona, home to bring vessels to Hawai'i from the West Coast. He and fourteen men were sent via the Matsonia *to bring the SS* Hornet *to Hawai'i in 1923. In November of 1924, he brought home the SS* Hawai'i *from San Francisco. Rally Greenwell Collection.*

175-foot wooden steam schooner thoroughly renovated, with accommodations for cattle shipments.

The SS *Bee* made her maiden interisland voyage to Māhukona and Kawaihae on November 8, 1920. By 1923, chill facilities were placed onboard to accommodate both beef and ice cream, which had become an interisland commodity. Cargo included other nonchilled goods for customers along the cattle coasts. The *Bee* continued to operate with reasonable profit, serving the coasts of the Big Island that included, in addition to Kawaihae and Māhukona, the shipping points of Puakō, Kīholo, Kailua, Keauhou, Nāpo'opo'o, Ka'alu'alu, and Hilo. Ni'ihau, Kaua'i, Lāna'i, and Moloka'i were single-service landings; however, Maui had three available landings.

In late 1923, the Hawai'i Meat Company

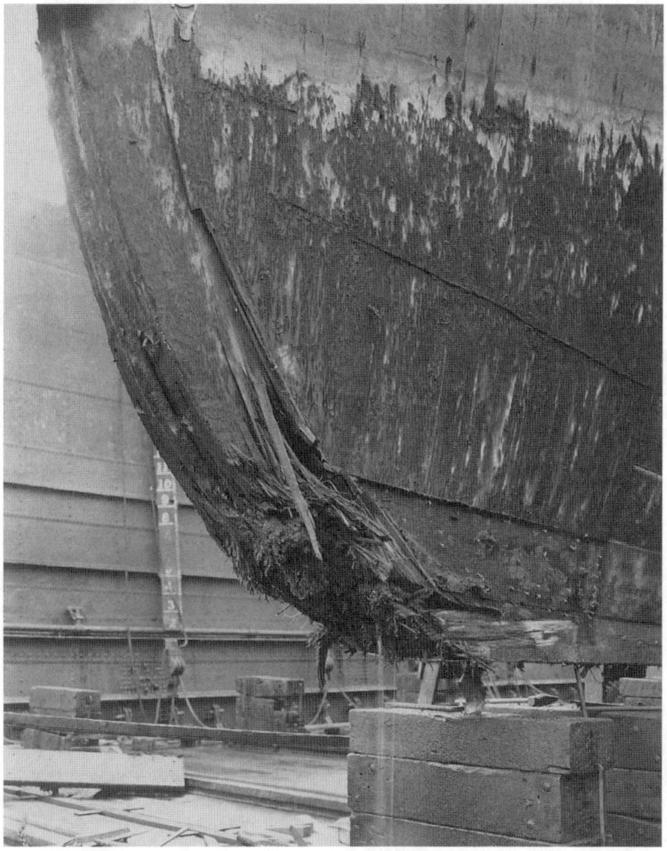

Hull damage to the SS Bee *after she ran aground on Maui. The* Bee *faithfully served the Hawai'i Meat Company through the ports of Māhukona, Kawaihae, Puakō, Kīholo, Kailua-Kona, Keauhou, Nāpo'opo'o, Ka'alu'alu, and Hilo. She also served Ni'ihau, Lāna'i, Moloka'i, and three landings. A. W. Carter Collection.*

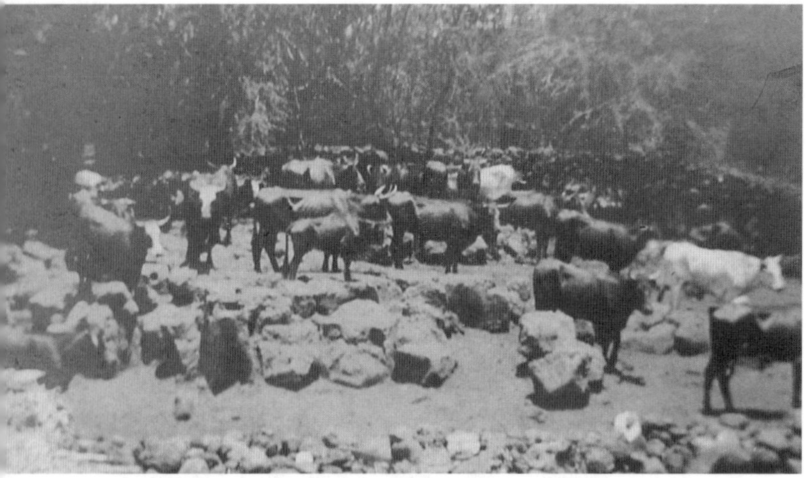

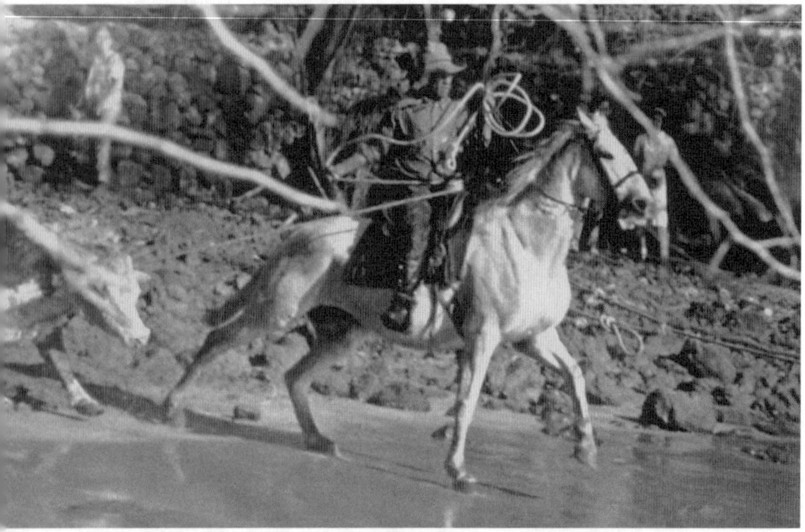

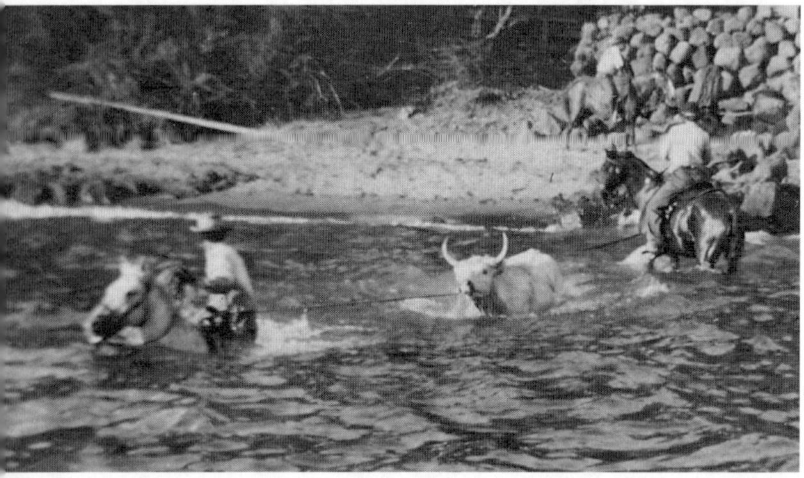

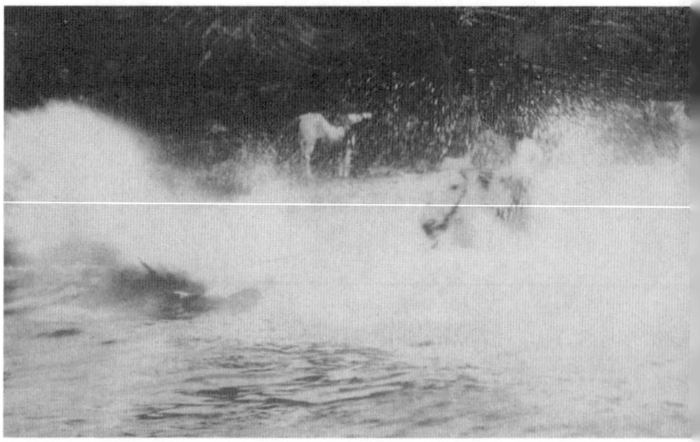

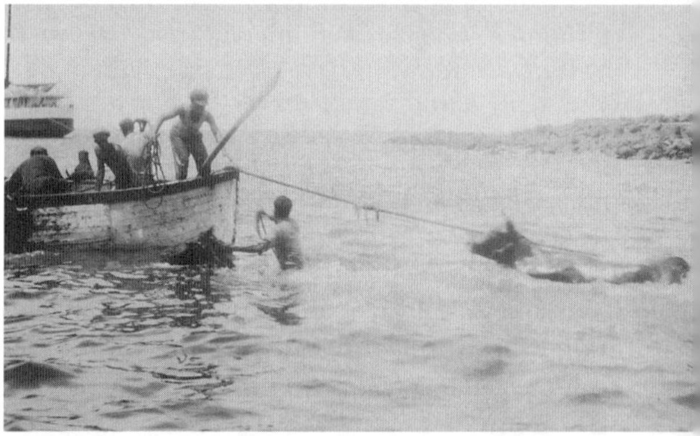

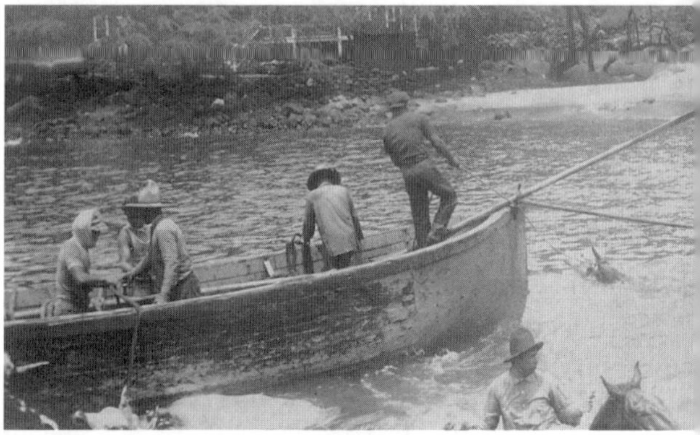

Top left: *Longhorn market cattle penned at Kawaihae. A. W. Carter Collection.* Middle left: *Peter Kainoa dallies up and leads a steer into the surf. A. W. Carter Collection.* Bottom left: *White longhorn steer tests the waters of Kawaihae. A. W. Carter Collection.* Top right: *Longhorn steers in tow, splashing through the waves. A. W. Carter Collection.* Middle right: *Cowboy pitches his slack to the tender. A. W. Carter Collection.* Bottom right: *The tender "reels in his catch." A. W. Carter Collection.*

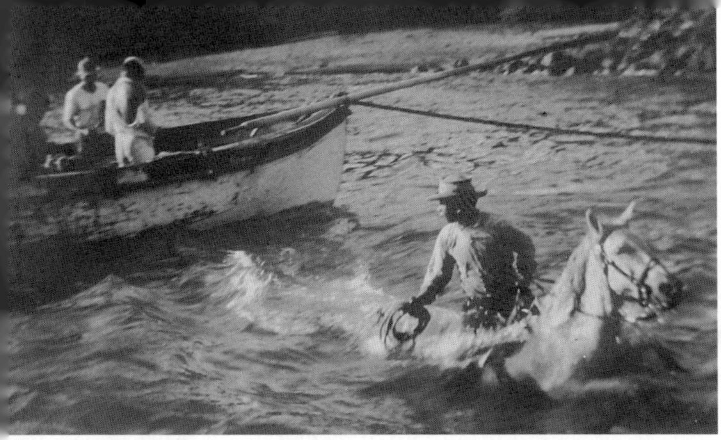

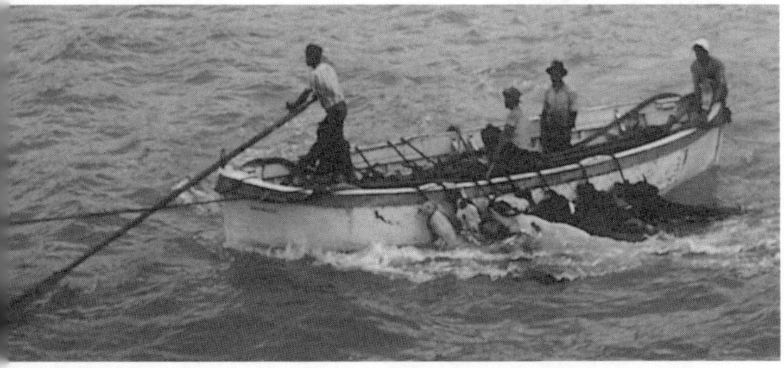

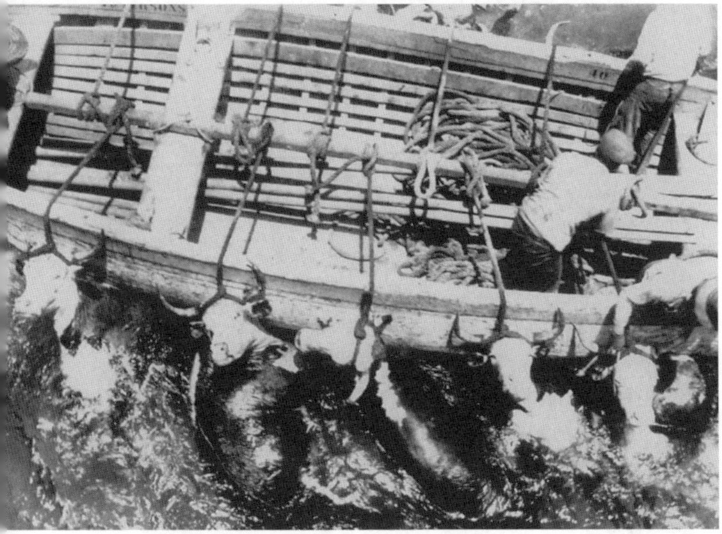

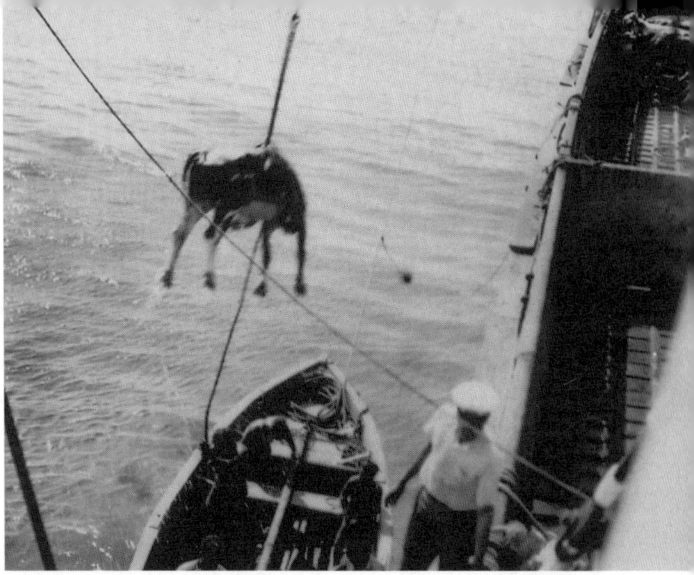

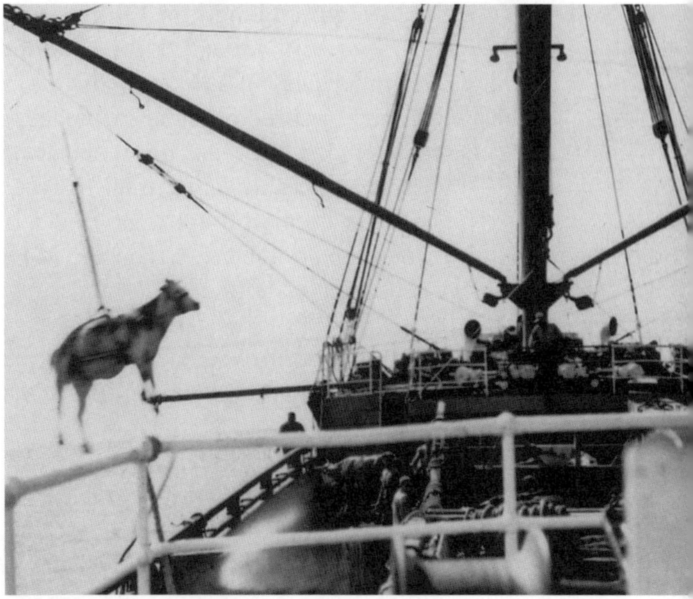

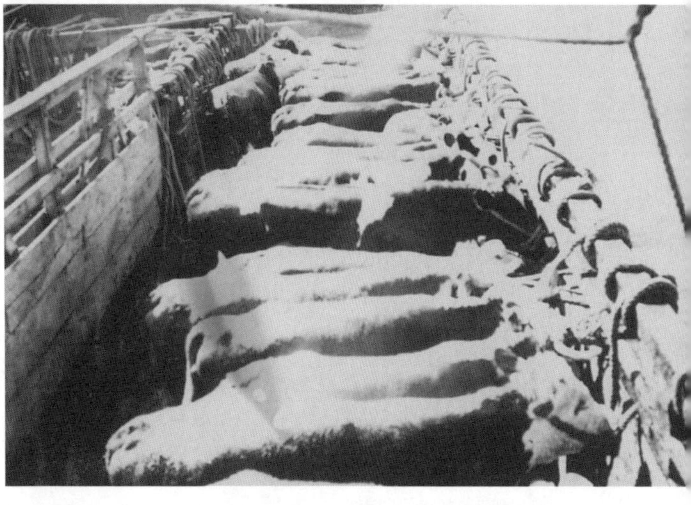

Top left: *The cowboy heads back to shore for another steer as the stock tender snubs up the steer. A. W. Carter Collection.* Middle left: *The loaded lighter, with twelve steers snubbed alongside, is towed back to the steamship. A. W. Carter Collection.* Bottom left: *Horned steer heads lashed to mid-spar. A. W. Carter Collection.* Top right: *Bellybanded steer is hoisted from the water. His tie rope dangles from his head halter. A. W. Carter Collection.* Middle right: *A steer is lifted on deck. A. W. Carter Collection.* Bottom right: *Steers snubbed to the gunwale spar for the trip to Honolulu. A. W. Carter Collection.*

directors pondered the need to build another vessel. The *Bee*, which occasionally suffered hull damage from reef collisions, was showing some fatigue but could continue as a backup vessel despite her years of service. The directors voted on the purchase price for a new vessel of $200,000, which was readily approved by Bishop Bank for financing. Cost overruns and additional amenities, however, brought the price closer to twice that amount. The newly ordered, all-steel-hulled vessel was slightly larger than the *Bee,* with a length of 190 feet and beam of 36 feet, displacing 851 tons and having a freight capacity of 444 tons. She was outfitted with a three-cylinder triple expansion steam engine and was christened the SS *Hawai'i*. As the ship's architect-in-fact, A. W. wanted a deep well-deck forward with machinery and passenger accommodations aft. In accommodating the needs of the plantations that would be served by the SS *Hawai'i,* accommodations for steerage passengers also were made.

The Hawai'i Meat Company vessel, the *Bee,* was increasingly depended upon for a variety of cargo needs at all of its ports of call. But misfortune called on April 8, 1924, when she ran aground and sank off Maui during a storm. Captain Kini and his crew were rescued by the *Claudine* and transported to Kawaihae—there were no lives lost and no injuries. The *Bee* was never seen again, but an insurance settlement of $600,000 was quickly concluded. The company set out to make arrangements for a replacement vessel, as they had done for the delivery of the SS *Hawai'i* from San Francisco.

In a much more compliant mood,

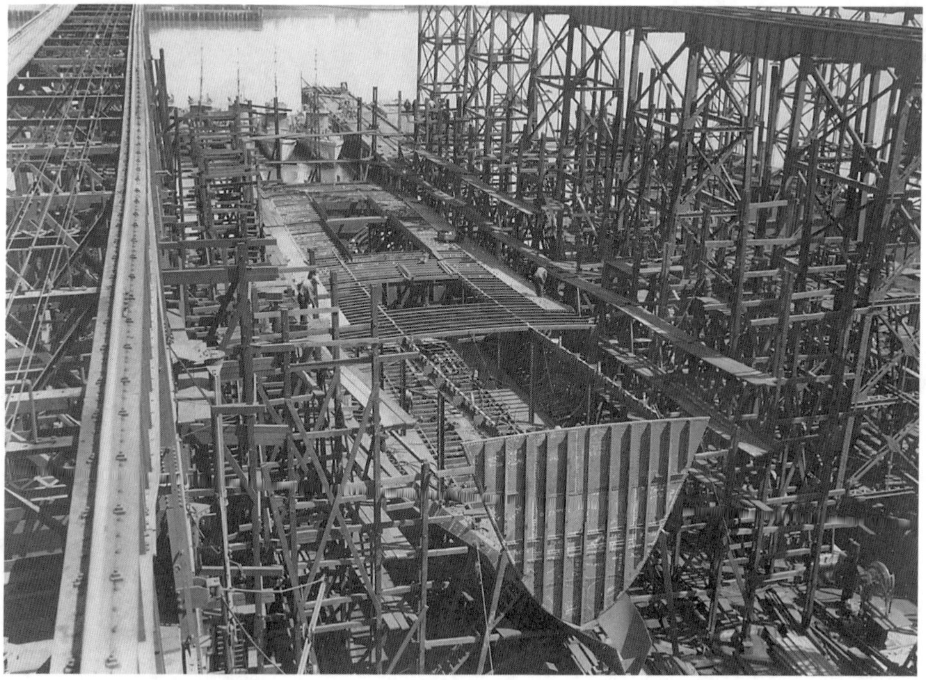

SS Hawai'i *construction at Bethlehem Shipbuilding Corporation in San Francisco. Midsection of the hull is complete. July 24, 1924. A. W. Carter Collection.*

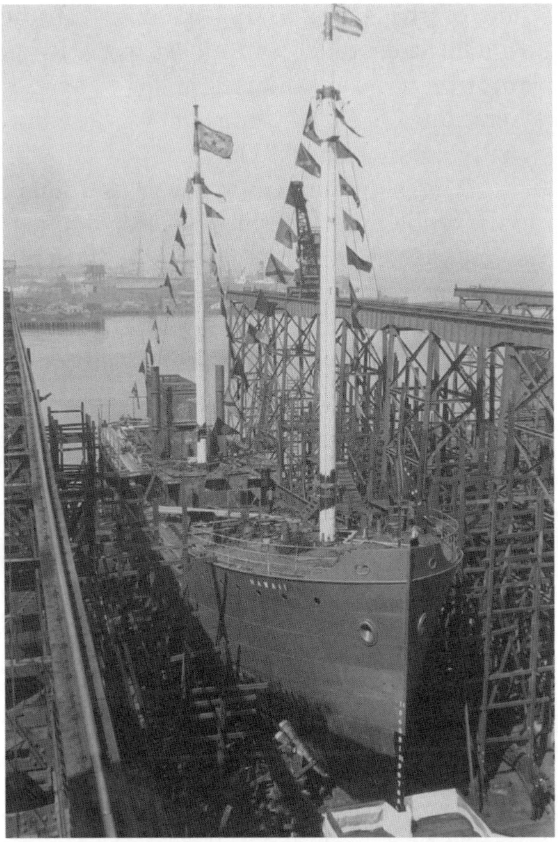

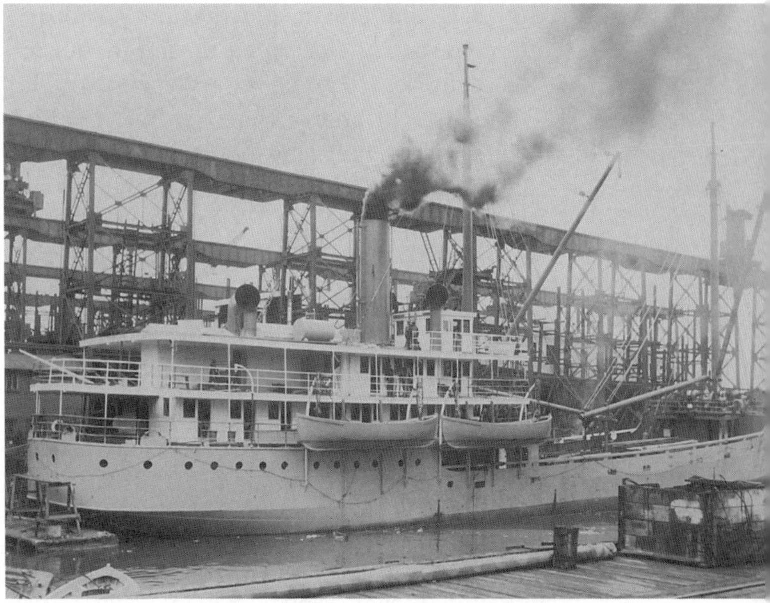

Top left: *SS* Hawai'i *nearing completion, September 17, 1924. The Plimsoll marks on the bow indicate load draft; at the base of the bow is the christening deck; the cargo booms are functional. A. W. Carter Collection.* Top right: *The engines of SS* Hawai'i *are fired up, October 16, 1924. A. W. Carter Collection.* Bottom: *SS* Hawai'i *leaving Honolulu Harbor for Kawaihae, December 15, 1924. A. W. Carter Collection.*

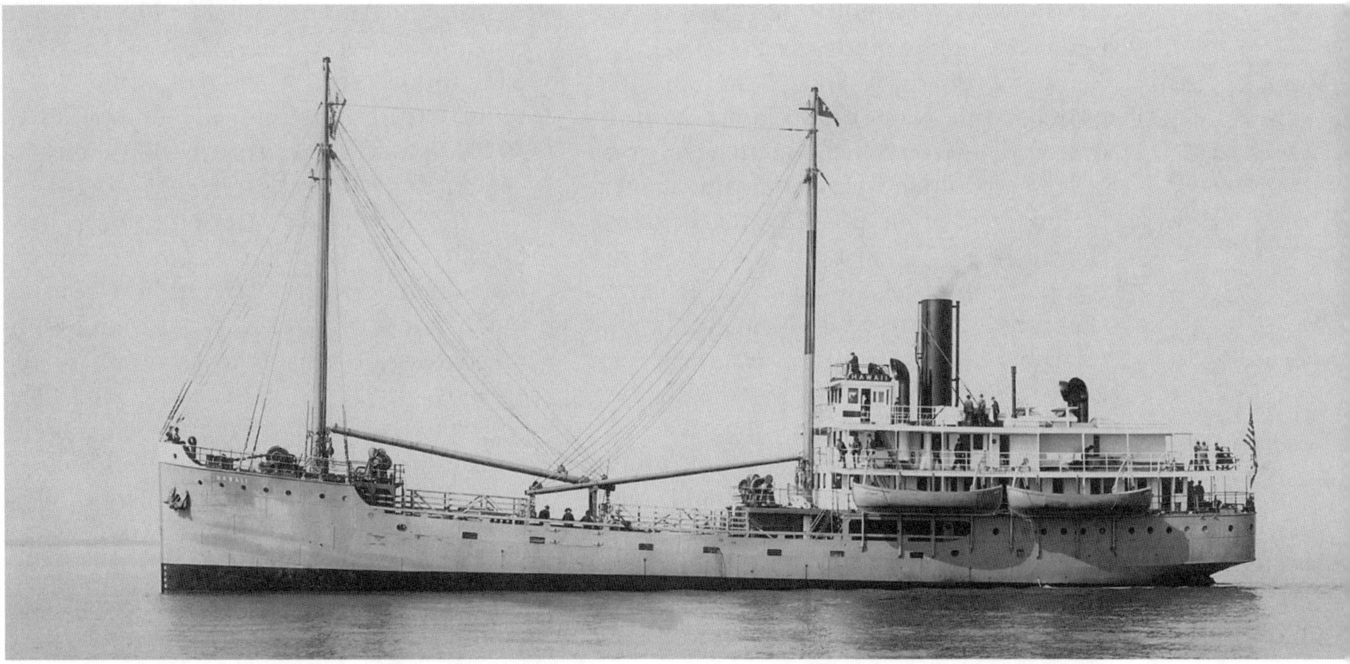

Inter-Island Steam Navigation Company offered its *Onomea* for cattle shipments and a yachtlike steamer, the *Dickenson,* for other freight and passengers. The *Dickenson,* under the Hawai'i Meat Company charter, made thirteen interisland trips between April 25 and June 26, 1924, immediately following the loss of the *Bee.*

Meanwhile, the Hawai'i Meat Company secured the sister vessel of the *Bee*—the *Hornet*—from San Francisco at a price of $73,000, which included renovations. The *Hornet* was no match for her sister's speed and efficiency and was relegated to the Maui ports of Lahaina, Kahului, and Hāna. She later called at Hilo Harbor, continuing to include freight of livestock and general cargo. Skipper Wood brought this vessel home to Hawai'i also.

The *Hawai'i* was an immediate success, making her maiden interisland voyage to Māhukona on November 20, 1924. The following June she carried more than 1,100 passengers in both classes, while achieving her primary mission of cattle transport. Later, *Hawai'i* would add Lahaina and Kaunakakai to her ports of call.

It was typical A. W. style to beat Inter-Island Steam Navigation Company at its own game—livestock shipment—just as he beat Metropolitan Meat Company at meat packing earlier. In the shipping business, Hawai'i Meat Company's success was a result of Inter-Island Steam Navigation Company's inept service and poor public relations. In both instances, however, A. W.'s early attempts to work with their respective systems were rebuffed. His strength and perseverance motivated him to do it his way. While the Hawai'i Meat Company thrived as an entity nearly to the end of the millennium—its doors at the Middle Street plant were closed in 1991—its role in interisland shipping ended much earlier. Hawai'i Meat Company was not formally defunct until 2002.

The shipping enterprise was fraught with problems concerning its seagoing personnel, freight spoilage due to fuel leakage, occasional groundings, a nagging problem with market animal bruising, and—finally—a collision with Inter-Island's *Helene.* All of these helped bring Hawai'i Meat Company to the realization that a proposed buyout of both the *Hawai'i* and *Hornet* should be entertained seriously. In June 1926, the shipping company directors accepted an Inter-Island Steam Company offer of $375,000 cash for the *Hawai'i* and *Hornet* that included five thousand shares of Inter-Island stock. A. W. was not only promised some scheduling control over the *Hawai'i* but was placed on the Board of Directors of Inter-Island Steam Navigation Company, where he served for more than a decade. The *Hornet* was immediately laid up and eventually scuttled off Lāna'i. The *Hawai'i* remained active in Inter-Island Steam Company service until it was sold to the De La Rama Steamship Company of Manila in 1946. It was wrecked in the Philippines a year later.

In the ensuing years of service, plans for the construction of the *Humu'ula* by the Inter-Island Steam Navigation Company incorporated many of A. W.'s design recommendations, especially with regard to the decking, pen space, chutes, and gates. The *Humu'ula* was well appointed, with thirty-eight cabins and room for fifty deck passengers, along with voluminous cargo holds. Built in San Francisco in 1929, she was a steel vessel, 961 feet long, with a single-gear steam turbine engine. She was named after the sheep station on Mauna Kea.

In his autobiography, *One to Eighty-One,* Jimmy "Kimo" Greenwell Sr. details the quarters for passengers: "Over the forward portion of the regular passenger section was the bridge and Captain's quarters. The galley and dining room for regular passengers were situated beneath the passenger section and [were] actually pretty respectable. At this level, passenger fare was all-inclusive and the food, to those not overcome with seasickness between the frequently rough channels of Molokai, Maui and Hawaii, was considered quite good."

Greenwell also describes food preparation and meals for the crew aboard the SS *Humu'ula:*

> The ship would troll for fish with two lines off the stern as she traveled—one line for the Captain and one for the crew. The Captain and crew usually had two or three hogs on deck in the cattle area, which absorbed the swill and leftover food from the galley. The kitchen merely had to dump its leftovers out the back door to the waiting pigs! In the same vicinity of that modest-size area, there would also be a fifty-gallon barrel of poi. The chef would prepare a huge stew; pour it into a large, deep pan at mealtime, placing it next to the poi barrel. The crew ate standing up and in fact, each man carried his own large spoon in his pocket—perhaps not the most sanitary but certainly, the simplest system. There were no dishes, no washing up, no dining room—unions did not yet prevail.

Hawai'i Meat Company used the SS *Humu'ula* for cattle shipments, while Kahuā Ranch and other ranches shipped cattle on the SS *Hawai'i.* Captain William Punahoa, originally on the *Claudine,* had been recruited by A. W. to captain the *Humu'ula.*

The *Humu'ula* was undoubtedly the most celebrated interisland steamer, particularly from the romanticized perspective of passengers and lei stand operators. But the livestock industry had no doubts as to the contribution this vessel and its crew made to the security and well-being of thousands of horses, cattle, sheep, goats, and hogs that graced her decks. Sailing the interisland seas between 1929 and 1952, the *Humu'ula* even found time to serve her country on military missions during World War II. Although wartime conditions disrupted interisland service, reasonable schedules were kept.

In 1936, reports were circulating that the Territorial Harbor Commission was planning a concrete pier that would provide wharfage extending considerably beyond the shallows of Kawaihae Bay, which by its natural configuration was not a deep-draft harbor. With an extended pier, ships could moor alongside and allow cattle to be loaded through an extendable chute directly onto the *Humu'ula* or any other ships that commonly served Kawaihae. This was a government (territorial and federal) project with a $50,000 price tag, but cost overruns put the final construction cost at about $80,000. During the planning stages, A. W. Carter was a key resource for advice on the technical points and was asked to be master of ceremonies for a *lū'au* blessing the projects. Parker Ranch, the project's grateful beneficiary,

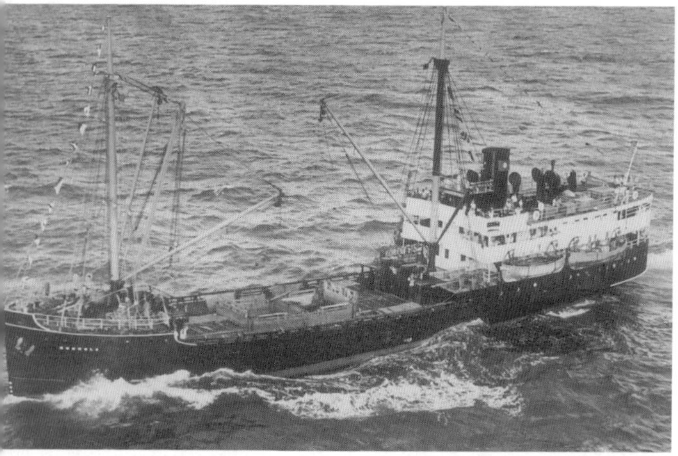

Top left: *SS* Humu'ula *steaming to Kawaihae with empty pens. Bishop Museum photo CP 59437.* Middle left: *Fat, modern Herefords are led out of a tire-padded pen. Hartwell Carter Collection.* Bottom left: *With gentle nudging, the fat steers enter the surf. Hartwell Carter Collection.* Middle right: *Fat cattle quietly leave shaded Kawaihae pens for the finger wharf. Hartwell Carter Collection.* Bottom right: Paniolo *quietly walk fat cattle to shipside. Hartwell Carter Collection.*

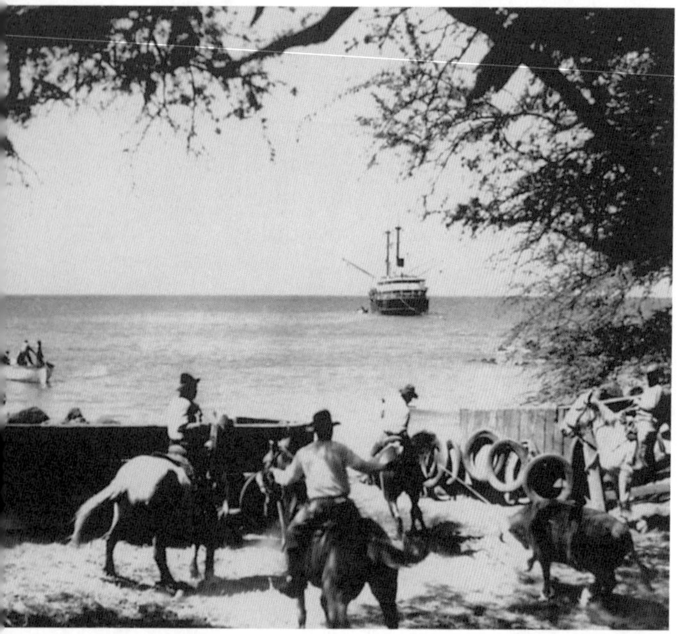

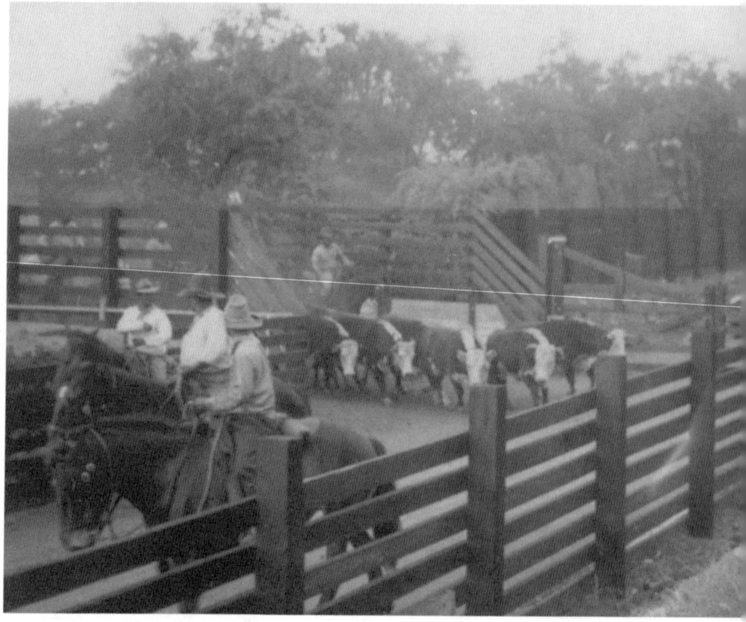

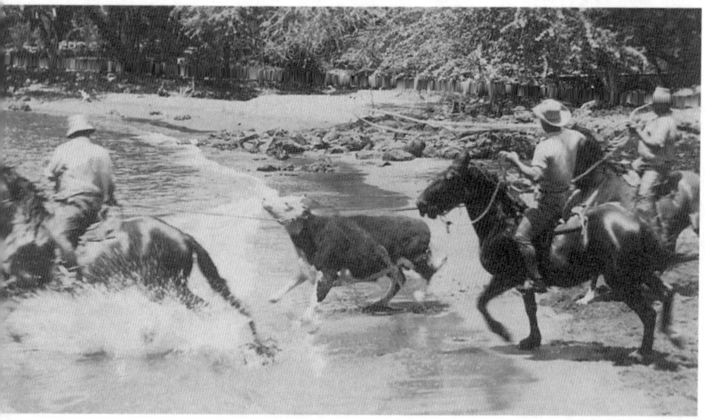

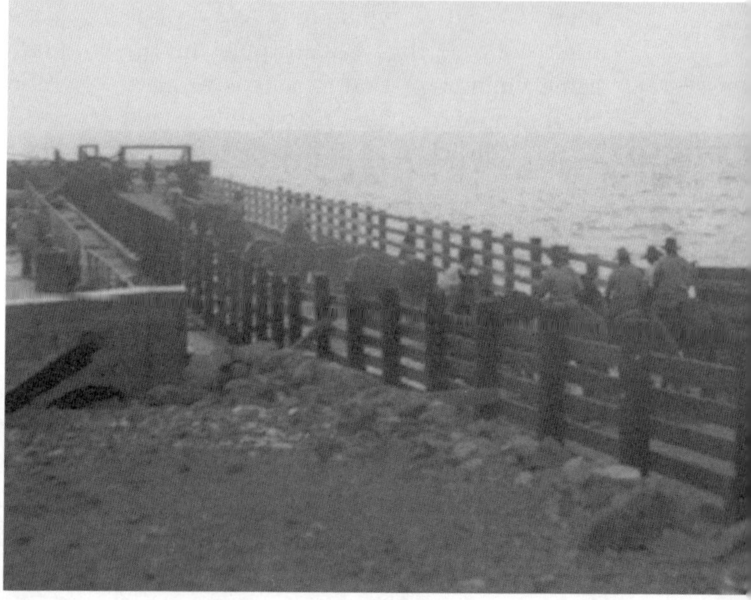

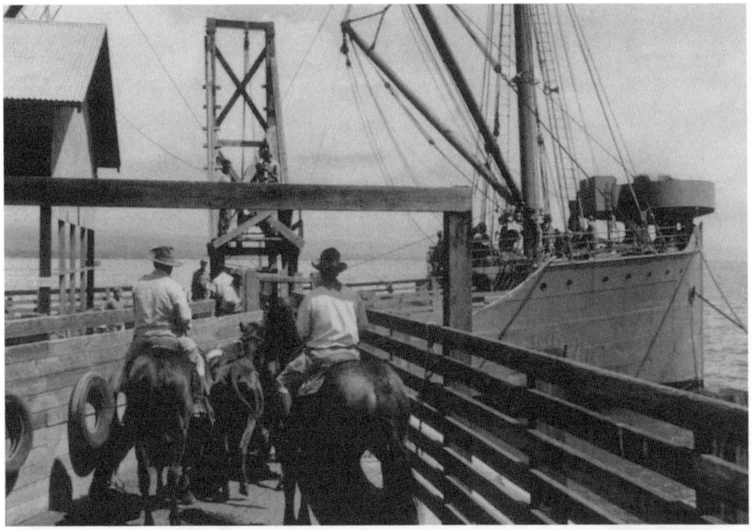

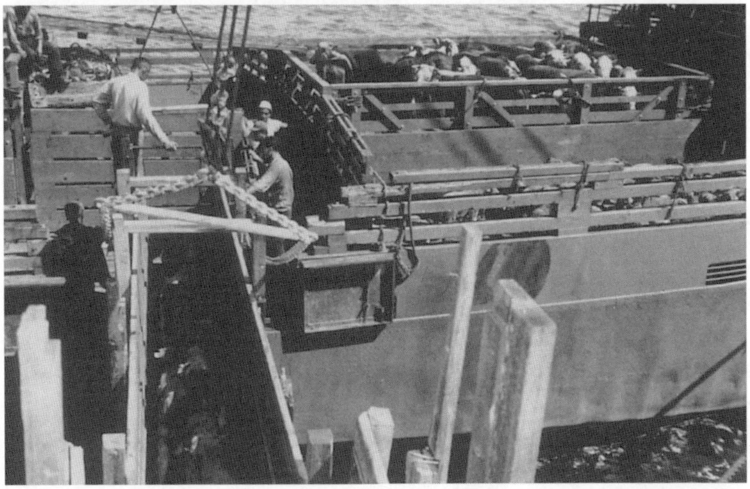

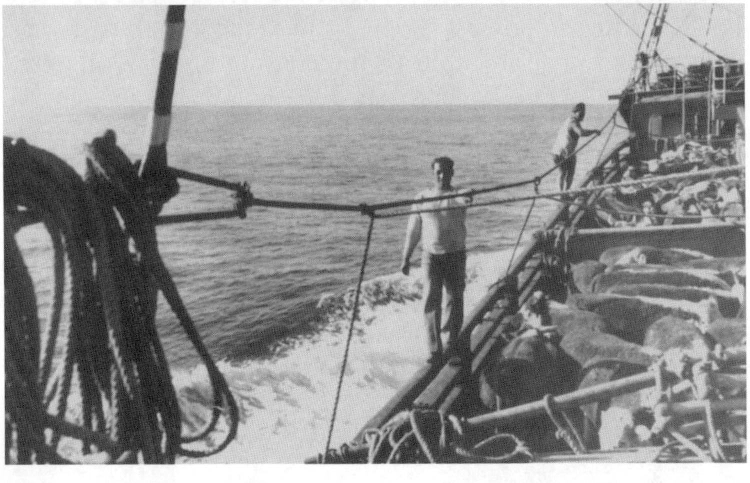

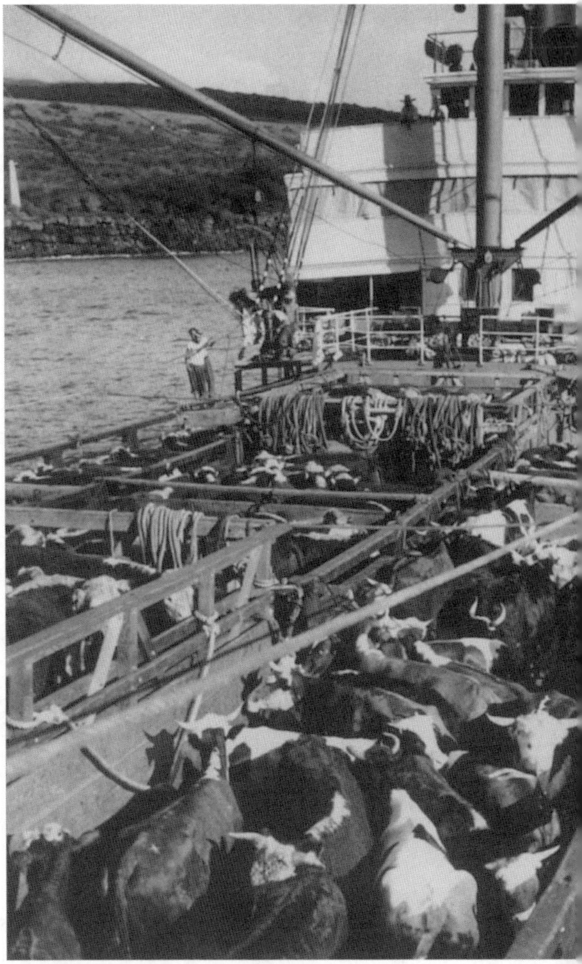

Top left: *Harry Kawai (left) and Frank Vierra gently load cows. Hartwell Carter Collection.* Middle left: *In the foreground, Humu'ula sheep are loaded. Hartwell Carter Collection.* Top right: *SS* Humu'ula *with Kawaihae lighthouse in background. Cattle are branded "K" (for Kahuā Ranch). Bishop Museum photo.* Bottom left: *SS* Humu'ula *loaded and underway from Kawaihae. Bishop Museum photo.*

hosted the *lū'au,* and as A. W. could not leave Honolulu for the Kawaihae blessing, he sent in his place Hawai'i Meat Company's Waller. The new pier ended the era of *hō'au pipi* (swimming of cattle).

By 1947, the Inter-Island Steam Navigation Company, struggling in a postwar economy, reduced its three vessels for interisland service to two: the *Humu'ula* and the *Hualālai*. The SS *Wai'ale'ale* was chartered to De Le Rama Steamship Company of Manila but laid up years later in Honolulu. The Inter-Island Steam Navigation Company began to look at tug and barge service, but Young Brothers Ltd. had quite a head start in the system. Despite continued Inter-Island Steam Company losses, in 1949 the *Humu'ula* still served Hilo and Kawaihae, providing Parker Ranch cattle with transport to market. It was during that year that A. W. passed away, probably not savoring the end of an era of a maritime livestock system he was so pivotal in creating nearly three decades earlier.

When the Inter-Island Steam Navigation Company decided to discontinue interisland shipping, it offered the *Humu'ula* for sale in November 1948. Hilo Navigation Company, incorporated in 1951, made the purchase and, in a limited fashion, provided interisland cattle transportation until the *Humu'ula* broke down off Kaho'olawe in March 1952 and was returned under tow to Honolulu Harbor. In 1957 she was sold and sailed with numerous mishaps to Miami, Florida, under the flag of the Dominican Republic. Her final service was via conversion to a barge.

The breakdown of the *Humu'ula* in 1952 marked the final voyage involving interisland transport of cattle via steamship. This ended one of the most colorful roles a Parker Ranch cowboy would play, as he splashed through the waves of Kawaihae with a fat Hereford steer in tow in the early years of cattle transportation.

Impact of War on Hawai'i Meat Company

World War I

Hawai'i Meat Company had been chartered and incorporated only a year when A. W., in 1911, sought military contracts for meat marketing. To fill the Marine Corps military contract for beef and mutton on O'ahu, he went beyond ranch boundaries. In the next two years, over and above ranch resources, Hawai'i Meat Company imported about 240 tons of beef from Australia and nearly 70,000 tons of mutton. Nearly 25 tons of California beef also was imported during that period.

Securing beef and mutton from Australia came with its problems. In 1913, due to an accident, the SS *Sonoma* failed to deliver 112 tons of beef to Hawai'i Meat Company. A year later the Australian government closed off all meat exports. Meat shortages in both cases meant that replacement orders to Hawai'i Meat Company from California came at higher prices. As a point of honor, in both private and military pricing, none of the additional cost was passed on to the consumer. A. W. successfully approached the quartermaster general in Washington, D.C., to lift the Australian embargo and, afterward, monthly shipments of 75 tons of beef were routine.

Hawai'i Meat Company, which campaigned for more cattle from neighbor island ranchers, also addressed the Marine Corps and navy's need for meat by offering even higher than market prices to the purveyors. Again, the company absorbed the difference by holding firm on the prices to contract as well as private local market consumers. Even the Australian imported meat came at 1.5 cents per pound loss. By 1919, the Hawai'i Meat Company needed a contracted resource of 1,100 head of cattle on a monthly basis. Private markets alone called for a monthly carcass need of 660 head, but none of this came from Australian resources because the embargo relief was based on military use only. Before 1920 arrived, Australia closed even this resource, and California became the sole source of imported meat—at considerable cost.

A. W.'s honorable treatment of the military came at a greater cost than just lost income to Hawai'i Meat Company. On the home ranch, the heightened need for carcasses came at a time when his superior imported bulls were producing dramatically improved heifers for breeding purposes. Under this pressure, the men of the ranch scoured the bushes for feral and substandard cattle. Eventually, these resources were exhausted, and replacement-breeding prospects were marketed.

Impact of World War II on Hawai'i Meat Company

As in the case of World War I, the U.S. government approached Hawai'i Meat Company well in advance of the war crisis to determine its viability as a dependable resource. In the case of World War II, it was a six-year prelude to the bombing of Pearl Harbor.

By 1936, A. W. was able to supply the following information to the U.S. Army: In routine reserves alone, Hawai'i Meat Company kept 160,000 pounds of beef; capacity for frozen beef was 400,000 pounds; chilled beef capacity was at 210,000 pounds and could be significantly expanded on short notice.

By this time, Hartwell Carter, A. W.'s son, was resident manager of the home ranch and A. W. routinely operated out of the Hawai'i Meat Company offices to oversee both operations. Later in the history of both companies, it was the consistent desire of a series of Hawai'i Meat Company presidents that they have similar pivotal control over both the ranch and packinghouse. This never occurred, as Parker Ranch's owner Richard Smart always emphasized the need for a system of checks and balances in his personal management style.

In times of a great war, the isolated situation of the Hawaiian Islands combined with the need to feed thousands of military personnel made pivotal control of both operations an absolute necessity.

Realizing the need for ready access to O'ahu beef carcasses when interisland flow of cattle might be halted due to war, storms, or strikes, Hawai'i Meat Company leased about 20,000 acres in the hills above Makakilo on O'ahu, where the Tongg Ranch presently lies. This became the Honouliuli Ranch. This property continued to be leased into the 1960s. A. W.'s vision was behind this venture. War would dramatically increase the need for beef for military consumption, and there would undoubtedly be commandeering of interisland ships. In August, before the December 7, 1941, bombing of Pearl Harbor, A. W. wrote: "I consider the situation extremely critical. Living conditions are as usual, however. Steamship companies will not guarantee the sailing of ships either before or after war is declared. Many of the ships now operating may be taken, and probably will be, for Government service. I do not anticipate trouble amongst the civilian residents and consider Hawaii a comparatively safe place."

While the actual bombing of Pearl Harbor four months later was not part of A. W.'s prophecy, the crisis in interisland shipping was as predicted. The Inter-Island Steam Navigation Company was early on appointed general agent of the War Shipping Administration. Under this command it had exclusive control of the *Hualālai, Hawai'i, Wai'ale'ale,* and *Humu'ula*—the workhorse cattle transports for the livestock industry. Fortunately, between the cattle kept at the Honouliuli Ranch and ready reserves of more than 160,000 pounds of beef, Hawai'i Meat Company forged its way through the first months of the war. In a short while, the SS *Hawai'i* and SS *Humu'ula* were made available to Hawai'i Meat Company for the Kawaihae run, and, to the extent possible, near-market and marketable cattle were moved to Honolulu in record numbers. Further, the small Big Island slaughterhouses—especially Hilo Meat Company's—were all on an accelerated schedule. They provided local markets as well as the Waimea military installations with beef. These were only marginally served by Parker Ranch's own plant, located where the present-day carriage and draft horse concession, Kohala Carriages, is headquartered.

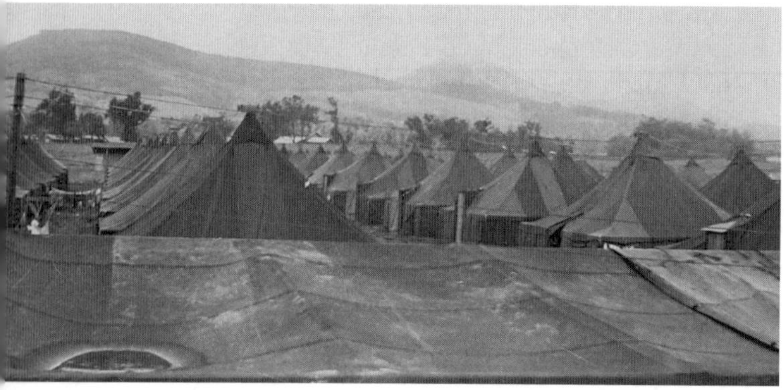

Impact of World War II on the Ranch

The ranch provided another accommodation to the military during these war years by housing 30,000–50,000 Marines in Camp

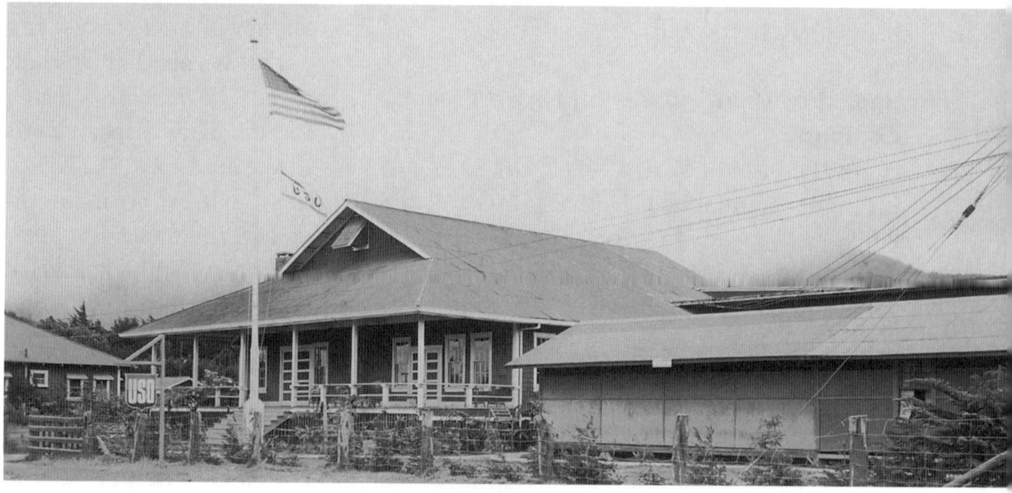

Top: *A tent-top view of Camp Tarawa's Tent City. The Second Marine Division, Fifth Amphibious Corps. (VAC) arrived early in December 1943 after the bloody battle at Tarawa Atoll, giving the Waimea camp its name. Hartwell Carter Collection.*
Right: *Barbara Hall, the social center of Parker Ranch, Waimea, which was converted to a USO facility. Hartwell Carter Collection.*

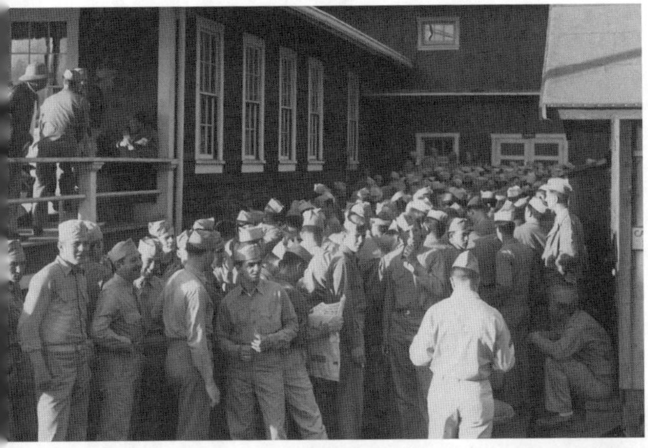

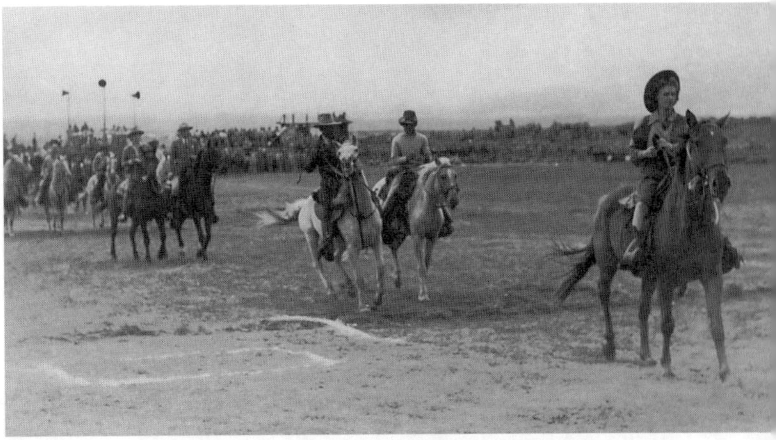

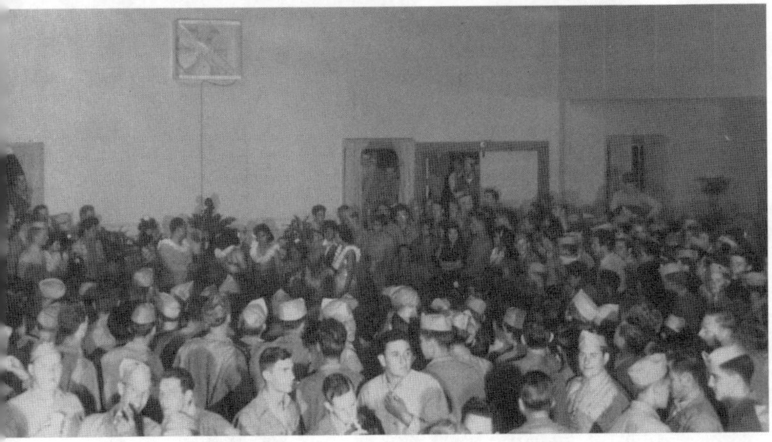

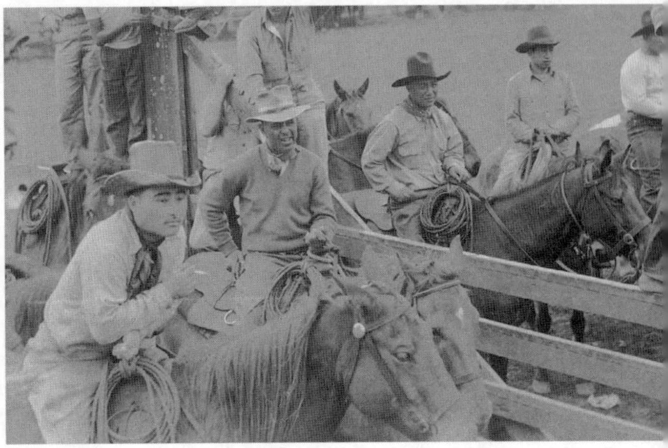

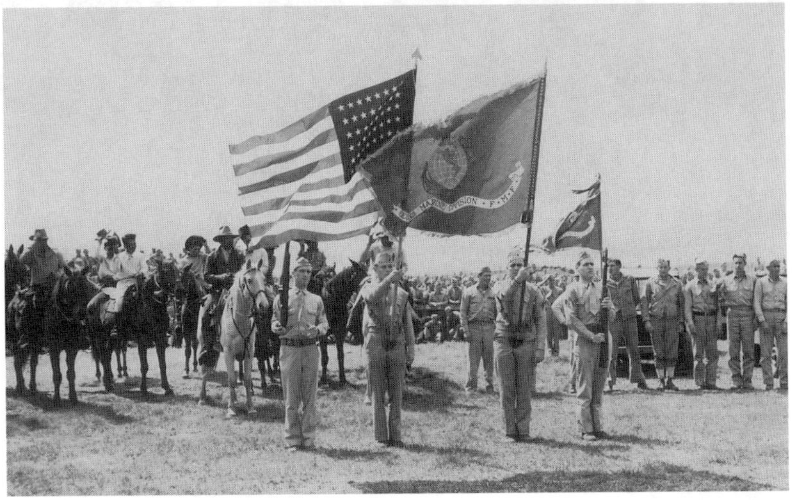

Top left: *Marines line up for admission to the movies. Hartwell Carter Collection.* Middle left: *Native Hawaiian entertainment for the Marines. Hartwell Carter Collection.* Bottom left: *Marine Color Guard at Marine Rodeo, Pu'uhihale Corral flats, 1944. Hartwell Carter Collection.* Top right: *Marine Rodeo Queen, Mrs. Marjorie Wilson, escorted by* paniolo *in the Grand Entry. Hartwell Carter Collection.* Above: *Ropers are ready. Mounted left to right: Gilbert Shinko Hokama, Albert Uiha Lindsey, Henry AhFong, Herbert Dead Lindsey, and Charlie Keiki Lindsey. Hartwell Carter Collection.*

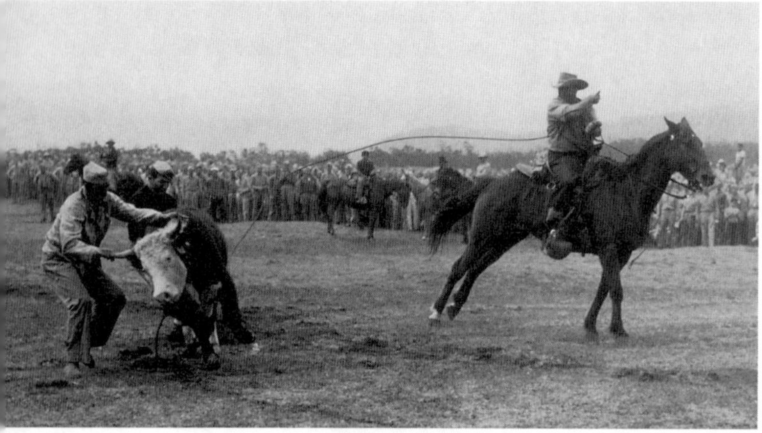

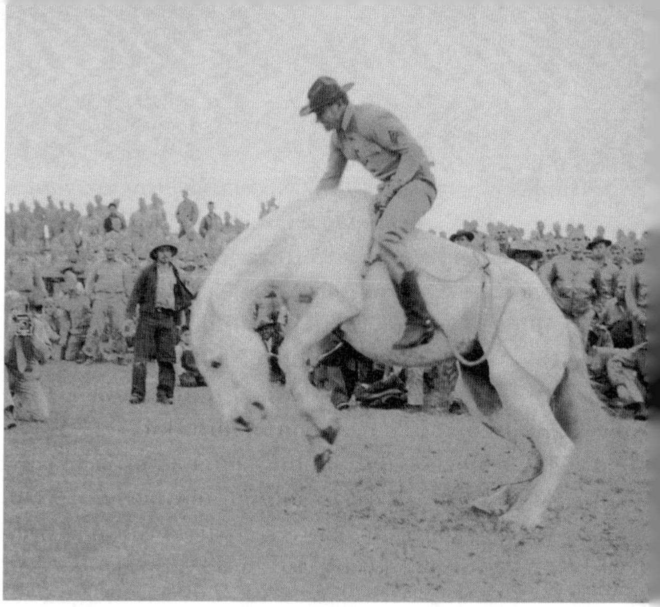

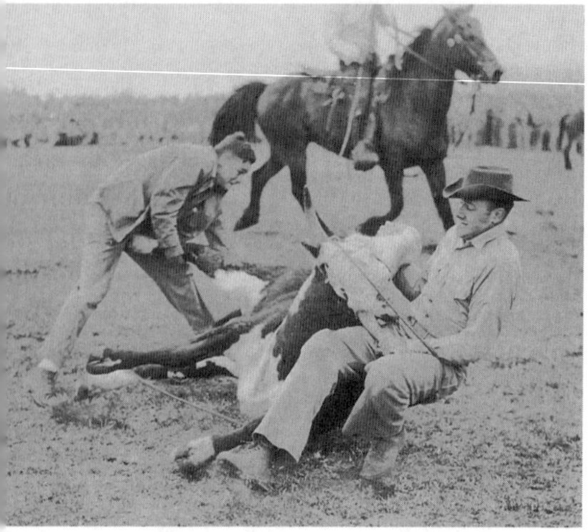

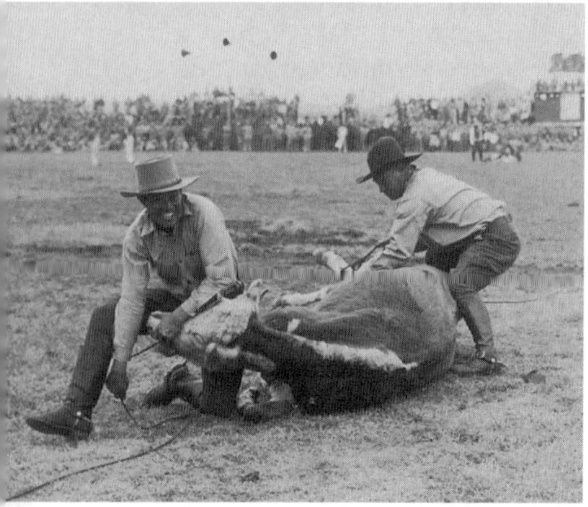

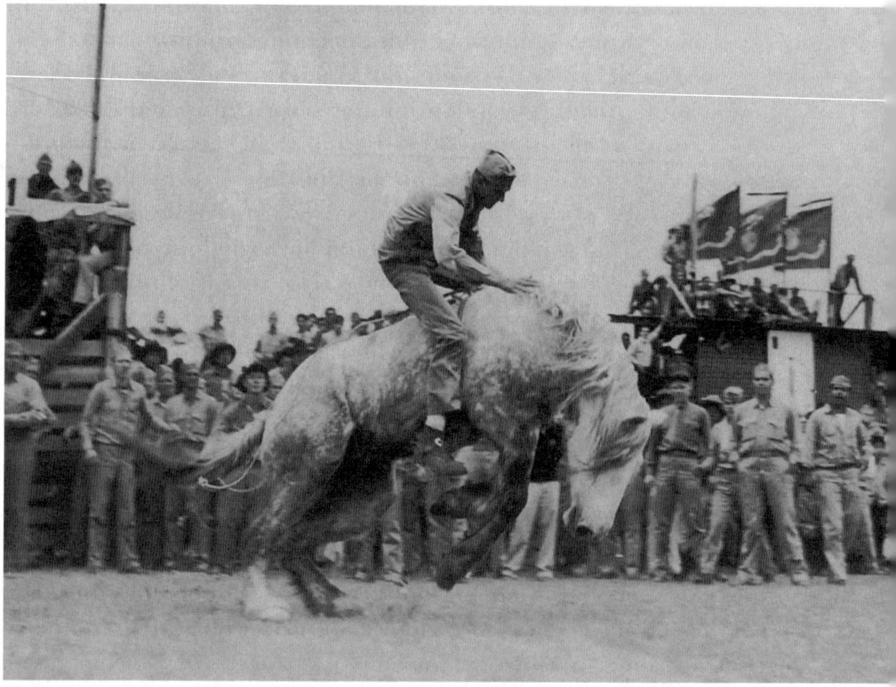

Top left: *Kaliko Ma'ina'aupō pitches his slack to jerk down the big Hereford in the steer-mugging event. Hartwell Carter Collection.* Middle left: *Marine cowboys struggle to tie down a Hereford steer Hartwell Carter Collection.* Bottom left: *John Samoa Lekelesa (left) and Henry AhFong untie the steers for Marine contestants. Hartwell Carter Collection.* Top right: *Marine sergeant bronc rider handling his half Percheron bronc. Hartwell Carter Collection.* Above: *This Marine seems to be losing his grip on the dappled grey Percheron bronc. Hartwell Carter Collection.*

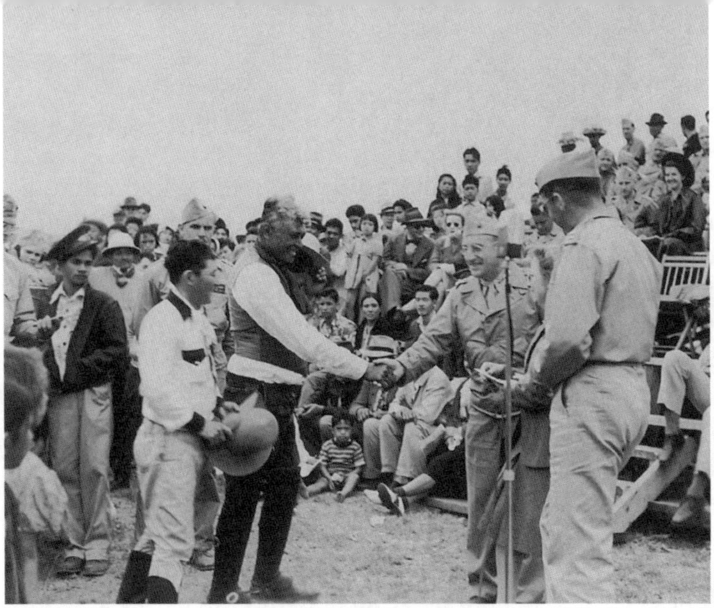

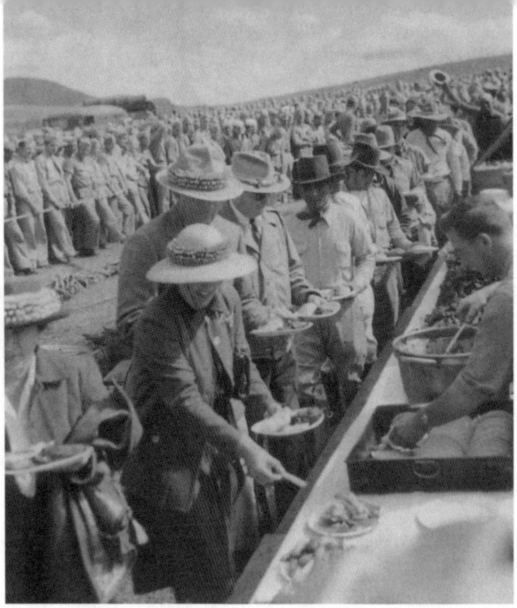

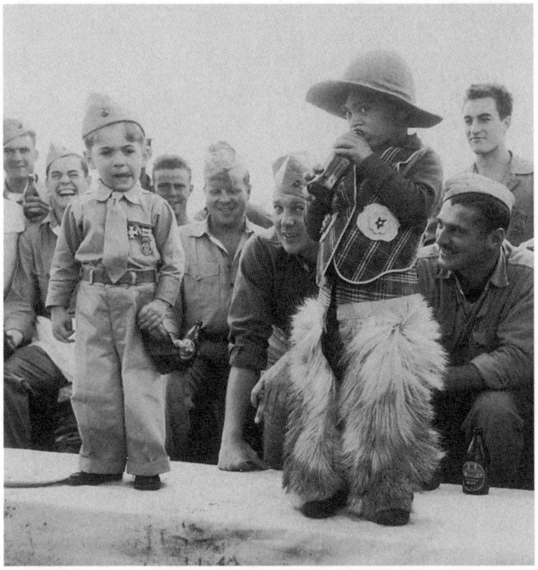

Top left: *Jiro Yamaguchi (on left with hat in hand) and James Palaika receiving awards from the brass. Hartwell Carter Collection.*
Top right: *Marine Rodeo food fest* lūʻau, *Marine style. Left to right with hat leis: Mrs. Hans L'Orange, Edith Carter, Hartwell Carter, Hans L'Orange, Sam Liana, Yutaka Kimura (face hidden by Sam's hat), Jacinth Paiva, next two faces hidden, Harold Baybrook, Hogan Kauwē, and waving the hat at the end of the line, James Palaika. Hartwell Carter Collection.*
Above: *A little Marine and* keiki paniolo *join the festivities. Hartwell Carter Collection.*

Tarawa and providing more than 50,000 acres for training and maneuvers. This took place in the large ranch paddocks called Range I, II, and Puʻu Pā, as well as the lands of Lālāmilo, ʻŌuli, and Kānehoa, which were referred to as the firing area. Both the campsite and training ranges, where vital pasture was once used, were made available to the U.S. government for the nominal fee of one dollar.

The massive influx of Marines into the small ranch town was overwhelming. Besides the expansive tent city at Camp Tarawa—which was located in the area now embracing Waiʻemi, Honolulu, and Līhuʻe paddocks—the military commandeered Puʻuʻōpelu for an officer's club. Barbara Hall, where Parker School is presently housed, became a recreation hall with a theater. Movies were also shown in the Wakayama family's historic Kamuela Liquor Store, with its corrugated metal roof. Dozens of Quonset huts went up, as well as long, narrow, lapstrake gable-roofed wooden buildings that served a variety of purposes, including quartermaster, medical, and administrative offices. Camp Tarawa's demographics brought electricity and an ice plant to Waimea town via a power plant. Several of the Quonset huts, wooden buildings, and other facilities were retained by the ranch and put to good use after the war.

In 1944, Parker Ranch staged a rodeo with events that included steer roping, riding, and mugging. Bronc riding was also featured. It was followed by an awards ceremony and a barbeque *lū'au* (feast). Twenty-five fat steers were butchered for the event, together with eighty-eight market hogs.

The firing range area required extensive combing after the war for live ammunition, duds, bombs, and grenades. The military assigned more than two hundred men to locate and detonate these munitions. Still, ordnance continues to surface, and—to the unknowing—severe and even fatal injuries have resulted. In 1947, Russell Iokepa and Theodore Lindsey were killed as they prepared to dig a posthole in Range I. Ke'ao Lindsey survived with substantial injuries. More recently, in 2002, five live grenades were discovered at Waimea Middle School as students were preparing the soil for a gardening project.

14

A NEW DAIRY IS BORN

1920–1949

Establishing a new purebred herd at Pu'ukīkoni would not only distance the herd from health problems at the Old Dairy but also expedite delivery of dairy products to town. This move represented A. W.'s typical state-of-the-art management style that paralleled the influx of some of the finest bloodlines of the Holstein breed and an animal health program designed by the world-renowned veterinarian, Dr. W. L. Williams. He arrived from San Francisco on June 28, 1920, aboard the SS *Manoa* and was received personally by A. W. Carter.

The arrival of Williams and the twenty-five imported Holsteins brought a major change in animal health on Parker Ranch. Credit must be given to A. W. for his foresight in preventive veterinary medicine. It was not that he hadn't tried earlier to secure routine territorial veterinary services for the ranch. In his early days on Parker Ranch (around 1900), he accessed the only veterinarian on the island, Dr. H. B. Elliot, MRCVS, from Hilo. Livestock work of the plantations and ranches of Puna, Hilo, and Hāmākua already consumed Dr. Elliot. By 1905, A. W. enjoyed the occasional services of Drs. A. R. Rowat and J. N. Onnissimenko of North Kohala. Although closer, they were also kept busy by intensive horse and mule demands of the north Hawai'i plantations from Niuli'i to Puakea. Both veterinarians moved to O'ahu shortly thereafter. In 1907, A. W. approached the Territorial Department of Agriculture and offered to share the cost of placing a territorial veterinarian in Waimea, with no results. Typically, A. W. then sent for the best when the need required it.

Dr. Williams plunged into the multifaceted animal health problems with enthusiasm. In addition to Dr. Mills, another

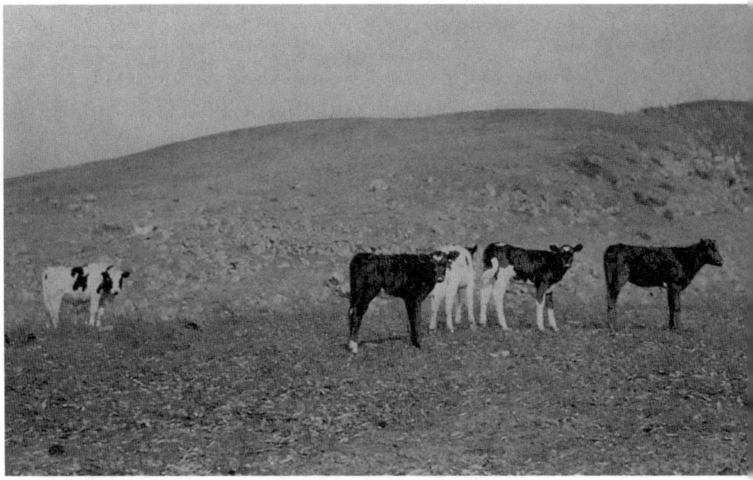

Thrifty Holstein baby calves grazing in Waimea. A. W. Carter Collection.

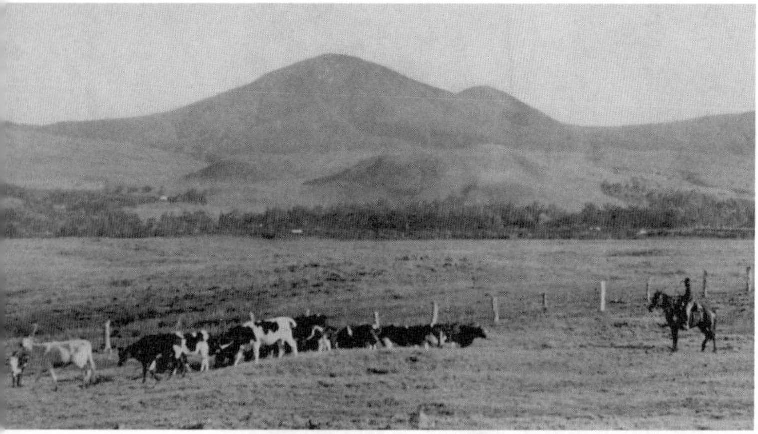

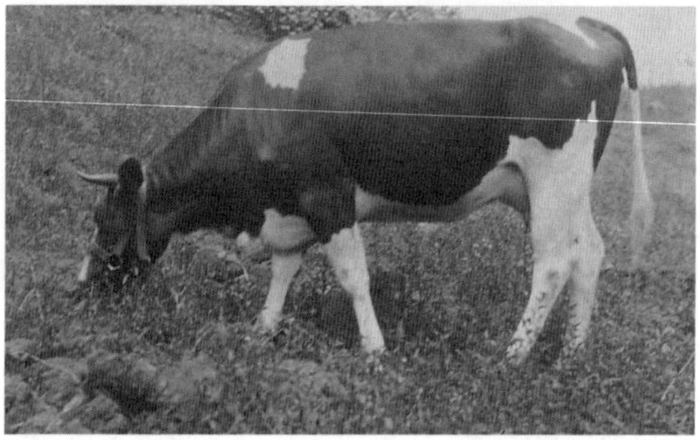

Top: *Yearling Holstein heifers. A. W. Carter Collection.*
Above: *An ideal type of Holstein springer (ready to calve) heifer. A. W. Carter Collection.*

veterinarian, Dr. Wallace, was brought onboard to strengthen the dairy product processing aspect.

The brucellosis infection of the dairy herd was brought under control, and reproductive performance was vastly improved by diligent individual examination and treatment of calving cows coupled with excellent record-keeping practices that measured conception rates, milk production, and calving percentages. At its peak, the dairy milked two hundred cows twice daily.

The new facility at Pu'ukīkoni, called the New Dairy, was an ideal venue for such progressive programs. The construction of this modern facility began in 1920 and was completed in 1924 at a cost of $35,000, under the direction of Martin Campbell. The main barn was built of concrete and steel in four sections. Each of the four sections boasted concrete feed troughs that are still evident today. The troughs ran the full length of twenty-five metal stanchions, each of which had galvanized metal shields to keep the cows from stealing grain from each other's manger. The shield folded out of the way for easy trough cleaning. Large enough to accommodate a hundred cows at a single milking, all floors were of well-graded concrete, providing excellent drainage.

The creamery was housed separately, where the pasteurizer, cold storage plant, bottling facilities, and equipment for butter, cottage cheese, and cheddar cheese production were also accommodated. A wood-burning boiler produced steam for pasteurization and cleansing the dairy equipment. Refrigeration was powered by a Fairbanks-Morse diesel engine. Electricity to run the pasteurizer and to light the barn and creamery was provided by a Koehler generator.

The processing plant was cutting edge in design, fixtures, and appliances. Adjoining holding pens and pastures were well graded and grassed. Pasteurization vastly improved milk, cream, and cheese processing and quality. By 1944 the dairy was shipping 1,100 quarts of milk to Hilo four times a week, in addition to supplying the ranch employees and the Waimea community.

For the team of Williams, Mills, and Wallace, training dairy personnel was a major responsibility. While Yutaka Kimura,

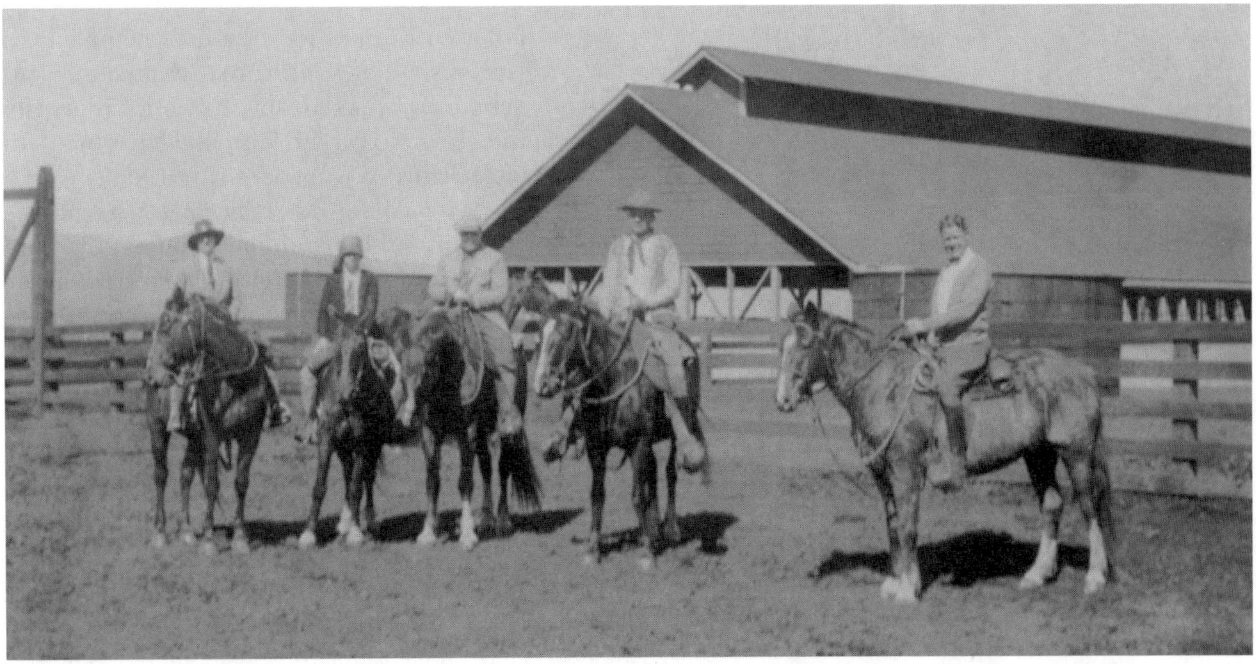

Hartwell Carter proudly touring the new dairy with some mainland guests on a trail ride. A. W. Carter Collection.

the dairy foreman, spent every moment under the diligent tutelage of Dr. Williams, Mills and Wallace introduced the dairy workers to the advanced methods of milk processing for quality production of butter, cheese, and even cottage cheese. Sanitation of the entire plant was earnestly indoctrinated, along with animal hygiene, milking techniques, and routine medical procedures. Production of milk soared, as did its quality. Conception and calving rates returned to normal levels, and cases of mastitis (infection of the mammary glands) were greatly reduced.

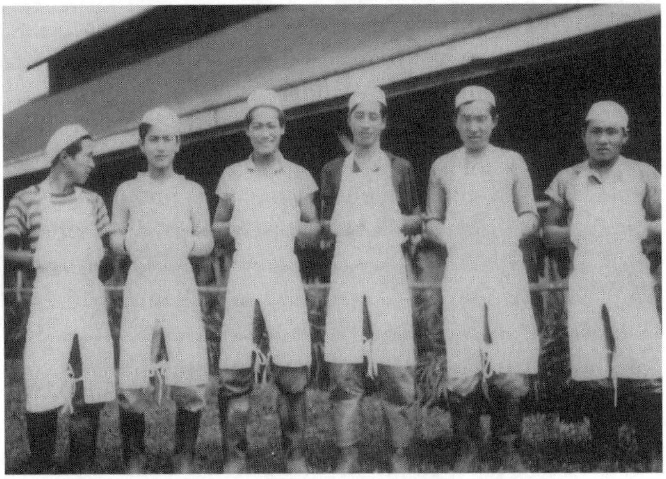

Dairy boys, working under highly sanitary conditions, proudly show their working apparel, with hands folded neatly behind the aprons. Left to right: Shigeo Yagi, Yoshiharu Goya, Shigeo Kuba, Niroku Dochin, Morito Ota, and Ichiro Yamaguchi. A. W. Carter Collection.

Early dairy administrators included John Benny, a Swiss dairyman who had operated the Kamehameha Schools farm on O'ahu. Theodore Vredenburg and then Yutaka Kimura followed him. Paul Gantt followed Kimura for a brief period before a University of California graduate, Paul Schroeder, was brought onboard. Eventually the reins of Parker Ranch's dairy program were returned to Yutaka Kimura.

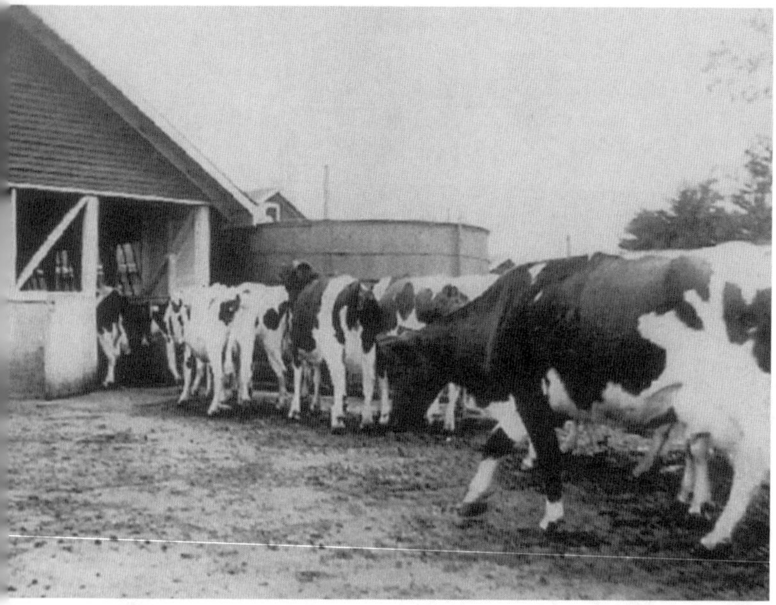

Dairy cows entering the milking parlor of Pu'ukikoni Dairy facility. A. W. Carter Collection.

Pu'ukīkoni Dairy launched the careers of many of its workers, most of whom remained on the ranch. Most of the dairy staff were dependable nisei, and they were referred to affectionately as the Dairy Boys. More than twenty-four men are listed, and more than half these dairymen stayed on the ranch in later years, serving in different capacities. In their heyday, the Dairy Boys were noted for team sport victories that included baseball (the Parker Ranch Creams) and relay horse racing on New Years and Fourth of July ranch events. After being Dairy Boys, their roles became diversified.

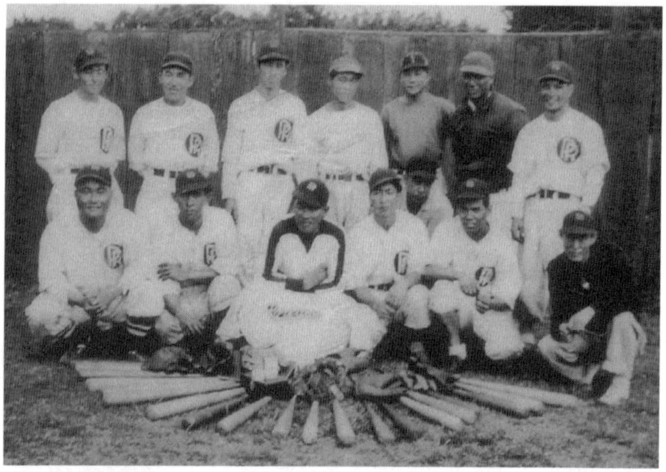

The Parker Ranch Creams baseball team. Standing left to right: Shige Uyeda, Yoshi Haru, Fumiyuki Kiyota, Alex Hokama, Jiro Yamaguchi, Henry AhFong (coach), and Hatsuichi "Tommy" Kaneshiro. Front row: Gilbert "Shinko" Hokama, George Kaya, Yutaka Kimura, Chiso Nakanishi, John Rickard, Hisa Kimura, and Hichiro "Lee" Uyeda. The Parker Ranch logo is on the jerseys. Hartwell Carter Collection.

Shiro Fujitani was a dependable cowboy. Nicknamed Pāpale, he became a skilled leather worker, especially in making *tapaderos*. He eventually went to work for Hartwell Carter at Homestead Farm until his retirement.

Tetsuji "Dempsey" Harada moved on to become a strong force on the ranch carpentry crew, where he retired as foreman with forty-one years of service. Tetsuo Hiromasa became a career ranch plumber. Gilbert "Shinko" Hokama moved up to become an assistant foreman in the main Cowboy Gang until the ranch reorganized in the Rubel/Lent era. Shinko became foreman of the Welding Gang, from which he retired with forty-nine years of service. Hisao Kimura followed his interest in horticulture and soil science, first in victory garden production during the war years and eventually becoming superintendent of the ranch agronomy department. Fumiyuke

"Long John" Kiyota, retiring with forty-nine years of service, spent most of his later years as a "pump man" in the water department. Chiso Nakanishi, like Dempsey, chose to develop a ranch career as a carpenter and retired with fifty-one years of service—a career paralleled by Hichiro "Lee" Uyeda, who served more than fifty-three years. Both men were noted finish carpenters, doing cabinetry on ranch homes, the ranch office, and Pu'u'ōpelu, home of Richard Smart.

Shigeo Yagi, born at Waiki'i, also joined the ranks of the Pu'ukīkoni crew. He was part of the sheep-shearing team at Humu-'ula and eventually settled as one of the ranch truck drivers. He died in a tragic accident at Kauna'oa Beach.

Both Ichiro and Jiro Yamaguchi were products of the Pu'ukīkoni team. Ichiro worked his way through the Utility Gang to the Welding Gang, from which he retired after forty-four years. Jiro went on to become a cowboy in the Cowboy Gang and eventually foreman of the Wai'emi Section of Mauna Kea Division under Henry Ah Sam during the Lent/Hanson eras. He retired with fifty-three years of service and remains one of the most photographed Parker Ranch cowboys, always striking a distinct pose upon a horse. Jiro died on November 16, 2002, leaving as his legacy a profoundly colorful career on the ranch.

The Pu'ukīkoni crew enjoyed their leisure time also, forming that successful baseball team, the Creams, with Yutaka as coach. The Pu'ukīkoni crew also fielded relay teams for the Fourth of July and New Years horse races. Several of the men were recruited later into the horse department as jockeys and trainers of the Thoroughbreds. Kenichi Inafuku, nicknamed "Japan" (pronounced "Jahpahng") was prominent among these horsemen.

The Nakata Men

The brothers Yoshio, Tadao, Tsuruo, and Yasuo Nakata are important contributors to Parker Ranch history, especially as caregivers. Their birthplace was the Old Dairy at Paliho'oūkapapa, where they also were raised.

Yoshio, the eldest, under the supervision of Yutaka, oversaw the milkers, who included at times his brother Tadao, Tom Kadooka, Isami Ishihara, and Masayuke Chinen. Yoshio left the ranch and moved to the mainland early on. Tsuruo (Tsuru) and Yasuo (Yanko) started as utility workers at Pu'ukīkoni.

Tadao emerged as Yutaka's protégé, having spent endless hours at his mentor's side, delivering calves and treating sick cows and calves. When Yutaka retired from service, Tadao continued to use his obstetrical experience in the field working beside me. He prepared himself to become a herdsman for the purebred herds under Dick Penhallow at the Wai'emi Station adjacent to Pu'u'ōpelu and under the management of Charlie Kimura back at the Makahālau Station. Tadao was also a skilled rawhide craftsman, braiding skin ropes and bullwhips for the cowboys. He loved young people, especially children, and was a natural teacher of animal care and craftsmanship to the interested person. In his twilight years on the ranch, Tadao oversaw the orphan calf pen at Pukalani Stables until his retirement after fifty-two years of animal welfare on the ranch. Part of his duty was caring for horse patients at the Surgery Barn, located behind the stables.

Tsuru also became a horseman after the dairy days, riding Thoroughbred colts for Teddy Bell, who then headed up the horse program out of the racetrack stables. While Tsuru also capped off his Parker Ranch career caring for the orphan calves, the bulk of his career was in the agronomy department, where he enjoyed his interest in horticulture. In retirement, he continues as a part-time orphan calf attendant.

Yanko, the youngest, evolved into a cowboy for the Makahālau Section, spending most of his career under direct or indirect supervision of Charlie Kimura, who eventually rose to become livestock manager from 1978 to 1984.

Although Tadao and Yanko are both deceased, their craftsmanship has left an indelible mark on ranch cowboy gear. Tadao, as already noted, was a master braider of rawhide ropes and whips; Yanko enjoyed bit and spur making.

Pu'ukīkoni Dairy men enjoyed pride and camaraderie that instilled a high level of dairy hygiene, indoctrinated by the Cornell specialists.

Dr. Williams and Yutaka developed a very close relationship, first under intensive technical training during the first six months, then on an on-call consultative basis over the next decade. Yutaka gained invaluable experience that can be attributed to his inquisitive nature and thirst for knowledge, and there was much to be gained from a noted practitioner like Williams. Yutaka carried on with the high standards of dairy husbandry instilled in him by Drs. Williams, Mills, and Wallace and, although none of the New York contingent remained in Hawai'i, they should be

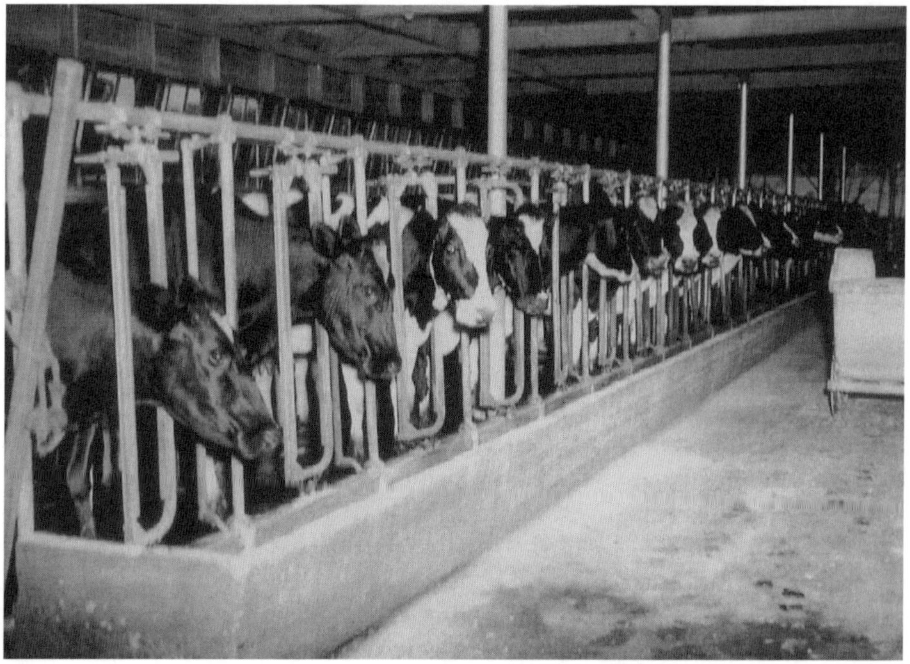

The new dairy facility at Pu'ukapu, 1949. Dairy cows are comfortably stanchioned in the new milking parlor, waiting for feed. Dick Penhallow Collection.

credited for raising the bar. The Parker Ranch dairy program became the model of the Territory, an ever-increasing force in merchandizing quality dairy products and superb dairy genetics through the sale of bulls and heifers. The basis of the Parker Ranch Standard indeed can be ascribed to the dairy. The dairy was a vital component of the community, especially during the World War II years.

By 1946 the facility at Puʻukīkoni approached obsolescence and was scheduled for closing. A new dairy was built in Puʻukapu on a 20-acre parcel fronting the highway near the entrance to Mānā Road. It became operational in 1949. The transition from the Puʻukīkoni Dairy program to that of the Puʻukapu Dairy was remarkable for its detailed coordination and vast leap in technology—from hand milking to an automated milking system.

The morning at Puʻukīkoni started at 3 A.M. with the last hand milking in ranch history. As twenty-four select top producers were finished, trucks stood by for delivery of the cows to the Puʻukapu facility, where very gentle indoctrination and training of the animals and staff were practiced. The cows were placed in their new stanchions to receive the cylindrical suction cups—one under each of four teats—that transmitted milk to a large 5-gallon stainless steel kettle suspended below via a rubber-belted surcingle that encircled the entire belly. Pulsating negative pressure was applied only momentarily as the exercise was one of training, not actual milking. By mid-afternoon the cows were acclimated to the new barn and automatic milking parlor, which was ready for the evening shift to begin.

With the new system in place, emphasis shifted to supplying milk and other dairy products to the public school system in the region as a gesture of strong support for the community. As in the past, milk was supplied free to all employees, as was beef and poi. At nominal cost, milk, cream, and cottage cheese were made available to the general citizenry of Waimea.

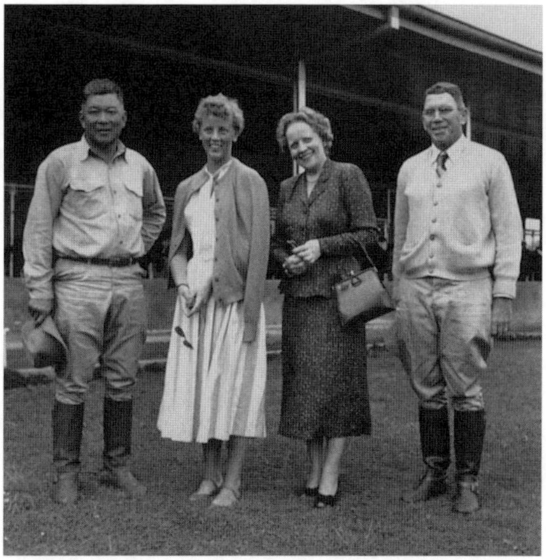

Yutaka Kimura (left) and Dick Penhallow proudly display the new Puʻukapu Dairy facility to British guests. Dick Penhallow Collection.

Although architects professionally designed the new facility at Puʻukapu, the practical design aspects were clearly created by Yutaka, who took even deeper pride in the dairy barn and its accessory buildings. Parker Ranch Dairy bottles bore a unique logo that reflected the nature of the community. Milk quality was of the highest order and the quantity produced continued to meet needs under Yutaka's management, but despite these facts, the ranch made a business decision to concentrate only on beef cattle. The dairy was sold in 1962.

The Parker Ranch Hog Operation

Parker Ranch phased out of commercial hog production after World War II, but the pig retained a place in its history for no other reason than the sheer enjoyment and pride it gave a ranch cowboy to rope a fat *laho'ole* (castrated male or barrow) to take home for the family oven or smokehouse.

The small hog that came with the Polynesians (weighing around 50 pounds), when mated with introduced European breeds and, later, highly improved British breeds, evolved into the present-day feral pig that can commonly reach 100–200 pounds.

Early foreign settlers in Hawai'i joined in the Hawaiian tradition of respect for pigs, quickly observing their many uses. By the time John Palmer Parker built his home at Mānā in 1853, the ranch was focused on breeding, feeding, and processing swine for ranch families and for limited sales to purveyors. During Parker's tenure and that of his son, John Palmer Parker II, there was a satisfaction with the status quo and little effort was made in upgrading and expanding the hog operations. But under managers Paul Jarrett and Fred Carter, a larger market for hogs developed, and this aspect of livestock diversification received more attention. Shortly after 1901, Fred Carter imported four registered Berkshire boars—a breed known for its short, almost pug snout, but one noted for its efficiency in converting feed to flesh. The feed itself was garbage, milk, and corn from the limited corn production on the ranch.

World War I brought about increased pork production to meet the needs of the military; the ranch expanded accordingly, but kept some of its best hogs for breeding instead of the market. The ranch in 1916 planted more than 2,000 acres of corn, more than doubling it the following year, leading to even higher production in 1918 to meet wartime demands for corn meal as well as feed for the livestock.

In the heyday of Waiki'i corn production, this section naturally became the headquarters for swine production. By 1929, the ranch had more than 1,000 purebred Berkshire hogs and a substantial number of feral crosses also on feed. The latter were used primarily for ranch consumption, while the purebreds went to markets in Hilo and Honolulu. With the outbreak of World War II, hog production took place near Camp Tarawa in Waimea.

In the war years, A. W. requested germination tests on existing Waiki'i corn and considered alternative strains to maximize production. His worst moments came during droughts; still, in 1943 there were 1,700 acres in corn—a bumper crop year. Stalks averaged two excellent ears and stood 7 feet tall.

One of the swine facility tenders was Ralph King, who shared with me how the hog operation went full circle. The daily military garbage provided the dietary needs of large number of pigs, whose ultimate fate was to provide for the stomachs of hungry Marines. Remnants of the cast-iron swill cookers can still be found near the Wai'emi Stable, where the Keamuku cowboys used to saddle up. When World War II ended, so did Parker Ranch's commercial hog production.

15

EVOLUTION OF ALLEN HARTWELL CARTER IN RANCH MANAGEMENT

1937–1950

The gradual passing of ranch management from A. W. to his son Hartwell began in 1937. Alfred Hartwell Carter, born December 1, 1896, in Honolulu, the first of three children and only son, completed primary school at the Valley School in Nu'uanu and graduated from Punahou School in 1915.

Raised on O'ahu by his mother Edith, Hartwell's upbringing had strong Honolulu influences. Mrs. Carter never really called Waimea her home and saw to it that Hartwell and his younger sisters Barbara and Edith (Edie) were exposed to the educational and social circles that only Honolulu could provide. Nevertheless, periodic extended stays on the Big Island gave the three Carter children a real love and feel for the ranch. By age twelve, Hartwell could frequently be found on Parker Ranch alongside his father or ranch hands during school vacations. In 1915, Hartwell left home for Harvard, where he received a Bachelor's Degree in 1919.

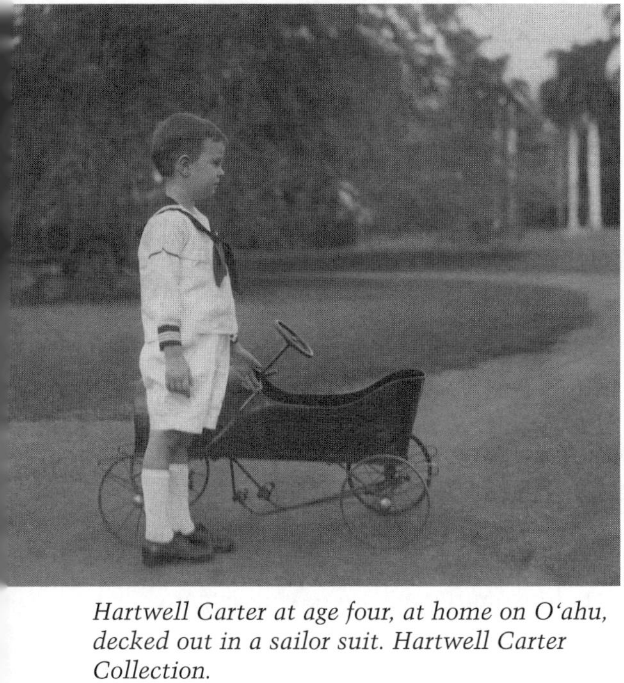

Hartwell Carter at age four, at home on O'ahu, decked out in a sailor suit. Hartwell Carter Collection.

Exposed to key ranch leaders since childhood, it followed that young Hartwell was the logical choice to succeed A. W. To his advantage, Hartwell surrounded himself with a young group of his peers who, with maturity, emerged to form his supervisory team. Notable among this group were Rally Greenwell, Willie Kaniho, Theodore Vredenburg, Albert Uiha Lindsey, Harry Kawai, Henry Ah Fong Ah Sam, Alex Bell Sr., Joe Pacheco, George Purdy, Johnny Kawamoto, and Yutaka Kimura.

Other contemporaries within Hartwell's family had been under consideration to succeed A. W. One of them was Hartwell's

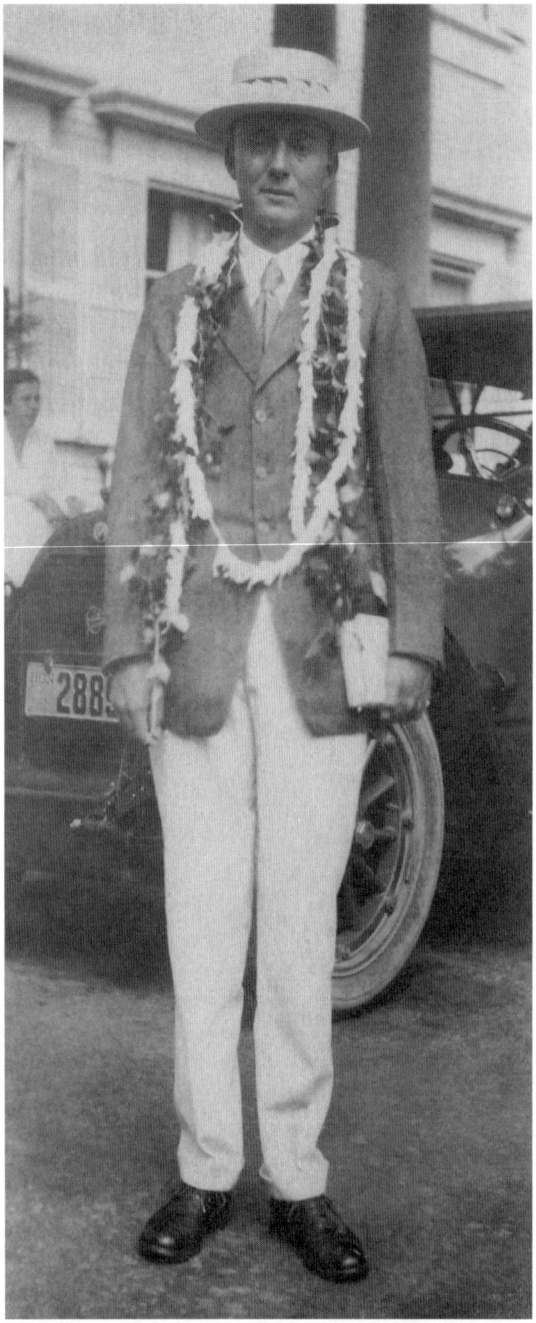

Hartwell Carter after graduation ceremonies at Punahou School, 1915. Hartwell Carter Collection.

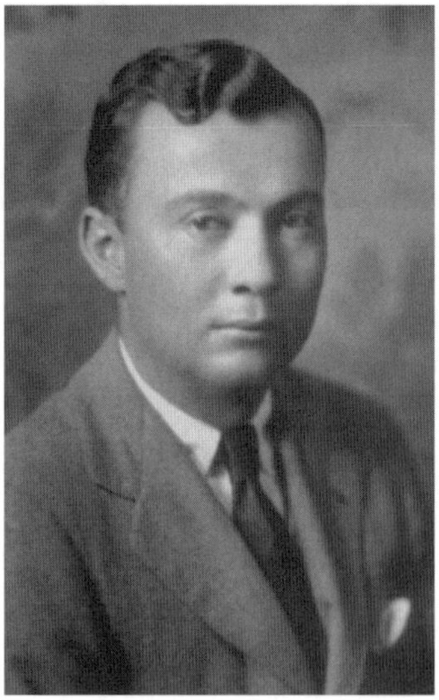

Hartwell Carter in the early days of his Parker Ranch career, 1922. Hartwell Carter Collection.

cousin, Alfred Carter "Pete" Young, who was not only A. W.'s namesake but someone in whom A. W. saw a great deal of potential. Pete had a hands-on style that fit a ranch managerial position. This was not to be however, as Hartwell's mother exercised her maternal influence to ensure the grooming of her son as his father's successor. In the interim, Hartwell continued to mature while at Harvard, before returning home to spend the next three years working for an auditing firm and trust company. His formal career at Parker Ranch, poised for launching, began in 1922.

There also were individuals onboard who could have placed Hartwell's succession at risk. Rally Greenwell and his younger brother, James—both of whom lived at the Parker Ranch manager's sprawling estate known today as Hale Kea—were members of a *kamaʻāina* (native-born) Kona ranching family and were well experienced. James ("Kimo") moved to cover the Puʻuhue and Puakea Sections in Kohala, assuming management of the Kohala component of the Ranch. Rally, who continued to live at the Carter home for a decade, became foreman of the Ke'āmuku Section. Kimo in time became manager of Hawai'i Meat Company, where A. W. had concentrated his efforts. Rally moved on to become livestock manager of Parker Ranch in 1962, but he had served under Hartwell in a loyal capacity until Carter's retirement in 1960.

A. W. eventually decided to move the base of operations to O'ahu. The decision had both practical and social considerations. Hawai'i Meat Company had territory-wide marketing concerns; slaughtering, processing, and shipping coordination were Honolulu functions. Furthermore, Mrs. Carter was firmly committed to continued residence on O'ahu. As a result, Hartwell became comfortably ensconced on the ranch as an eligible bachelor. In 1931, he married Mary Emilie Donahue, a woman prominent in California social and livestock circles. This union quietly terminated within a short time. Mary Emilie came with a love for the outdoors—seemingly appropriate for her role as wife of the emerging protégé of ranch management—but other pressures eroded the foundation of their marital relationship. Mary Emilie returned to California, taking with her a small retinue of her own horses.

A. W.'s day-to-day influence on his son's management was strong and constant. They communicated frequently and intensively, routinely closing memos with respectful words of mutual love. Cattle shipments occurred weekly and, eventually, biweekly. Water development, personnel matters, and animal health issues paralleled market cattle discussions in depth and intensity. Hartwell, even in emergency situations, took no risks in making decisions without his father's advice and consent. For practical purposes, this relationship gave the ranch stability and continuity. Hartwell seized the great opportunity to be mentored thoroughly, learning all the historic intricacies that made the ranch work. The transition could have been chaotic if Hartwell had a stubborn streak that nudged him to do things his way. Because Hartwell also surrounded himself with trusted and knowledgeable men of the ranch, the transition went smoothly.

In the years leading up to the time Hartwell assumed day-to-day ranch management, conditions on the ranch ensured its continued success. Hartwell had his own style of doing business. He was soft-spoken and worked diligently to avoid confrontation. He also depended on his key leaders to keep him informed. Ranch personnel, accustomed to A. W.'s brusque hands-on approach, had to adjust to the change in leadership style.

Hartwell, frequently seen on the ranch on horseback and later in his Ford sedan, followed up personally on some matters, despite daily and often twice-daily reports from his key foremen keeping him advised and informed. He was not as trusting as his father in delegating responsibility to others. This was a shift in management style in which a few capable men, under close scrutiny, ran the ranch operations while Hartwell managed the broader picture, which included coordinating shipping and marketing concerns. While A. W. showed care, thoughtfulness, and confidence in all levels of ranch employees down to the ranch hand, Hartwell limited his contacts to his circle of foremen and a few informants. To his credit, he demonstrated specific kindnesses, such as the time he ordered handmade Monel spurs for the entire Cowboy and Roughrider Gangs. To horsemen of Hawai'i during those years, spurs of the quality made by master loriner Manuel Sebastian Santos of Maui were comparable to fine ones made by G. S. Garcia of Capriola and Sons of Elko, Nevada.

Although 1937 marked the transfer of management responsibility from A. W. to Hartwell, the transition began a dozen years earlier and continued through the decade that followed. Between 1946 and 1947, A. W. underwent two surgical procedures to stem the spread of cancer. By the end of 1947, he continued his involvement practically from his bedside until his death in 1949. During those years he continued to give his son thorough training in livestock management, and Hartwell spent the first half of his tenure executing the mandates of his father, becoming a stand-alone manager only after his father's death. His solo flight, however, lasted just a decade.

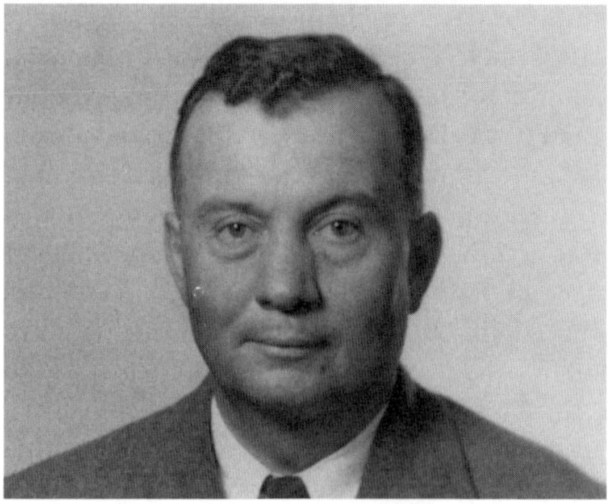

Hartwell portrait, 1949. Hartwell Carter Collection.

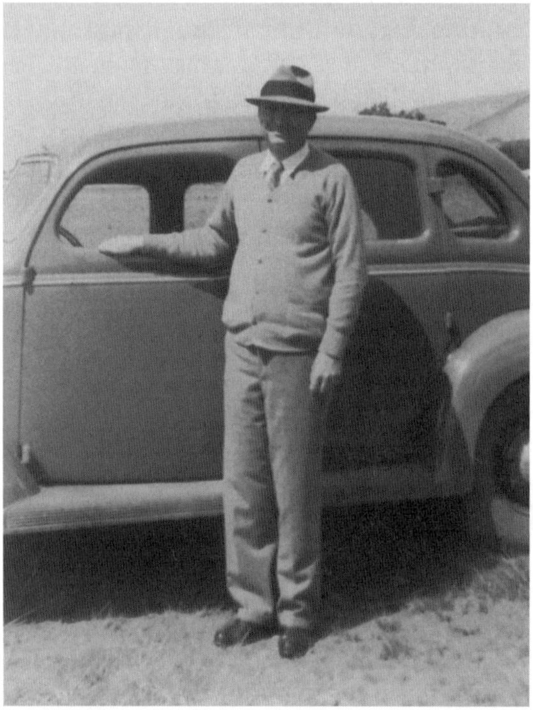

Hartwell alongside his 1939 Ford sedan. Hartwell Carter Collection.

As 1950 loomed before him, Hartwell became acutely aware of the impending influence of ranch owner Richard Smart on the day-to-day operations. In 1948, Richard returned home for an extensive engagement in two Honolulu theatrical productions featuring the Straw Hat Players of San Francisco, followed by the Fiftieth State Revue. During this trip, Richard intimated his longing to permanently return to the home ranch and immerse himself in the land of his roots, 'Āina Paka. In a most spiritual sense, Richard invoked his standard of life—*aloha 'āina.* The year 1959 marked his return home.

In the decade between 1950 and 1960, Hartwell faced daunting challenges that merit attention. At a high priority level, Hartwell mandated that the cowherd be rounded out to twelve thousand mother cows of uniform polled Hereford breeding, conformation, and constitution. He paralleled these efforts with an equine counterpart, redesigning the Horse Program from an obsolete amalgam of Thoroughbred/Standardbred/Morgan mount into an early prototype of the American quarter horse—the steed of the future.

The most serious challenge he encountered came from the first major reduction in ranch acreage since building the 327,000-acre empire of his father's time. The Hawaiian Homes Commission, anxious to award 300-acre ranch lots to fifty-two qualifiers, called for the return of the commission's 20,000 acres of prime cow country leased by the ranch over several previous decades. Equally challenging was Hartwell's mandate to relocate military activities from nearby Lālāmilo to the remote plains of Pohakuloa. He faced these challenges directly and admirably.

A second volume will open with the stand-alone decade of Hartwell's tenure and accomplishments. The profound adjustment from his father's paternal oversight to Richard's constant ranch presence will be detailed.

A. W. knew the land, livestock, and people of the ranch intimately. As he groomed Hartwell as his successor, he based his decisions and advice on pure facts and figures underscored by decades of experience. At the same time, compassion came naturally to him and enhanced his overall character.

By contrast, Richard employed deep-seated sensitivity based on aesthetic, artistic, and humanitarian values that did not necessarily convert to sound business decisions. There appeared to be an opportunity for Hartwell to rise to the occasion and become a firm, guiding mentor of the ranch owner destined to return home from Broadway with a plethora of glorious plans for 'Āina Paka. But Richard did not avail himself of Hartwell's overtures of mentorship, wanting to be in direct control of the fate of his estate. Dutifully, Hartwell carried on as manager, ever loyal to the land and its owner.

Thus the prodigal son returned, anxious for change and to take the reins of his entitlement. Hartwell came to realize that a greater force—the ranch owner—had replaced the paternal hand of his father. Little did he realize the magnitude of the changes that would take place. In its lone descendent, Richard Smart, the strong Parker influence returned to the land and its people, grown from a king's mandate into an organization that set the standard of excellence for the cattle industry of Hawai'i and the American West. Like his pioneering forebears, Richard Smart had enthusiasm, stamina, and vision for the future, and he held Hawai'i's ranching future in his hands. The challenges, tragedies, and triumphs lay before him. He forged ahead to build an empire grounded in an ethic that the land is a precious commodity and its people highly valued.

The story of Parker Ranch is an unending saga. The values upon which it is based, so eloquently stated in Richard Smart's mantra—aloha 'Āina Paka, endearment to the lands of Parker—are applicable today. Ultimately, stewardship of the ranch land is the only path that will allow the Parker legacy to endure for future generations.

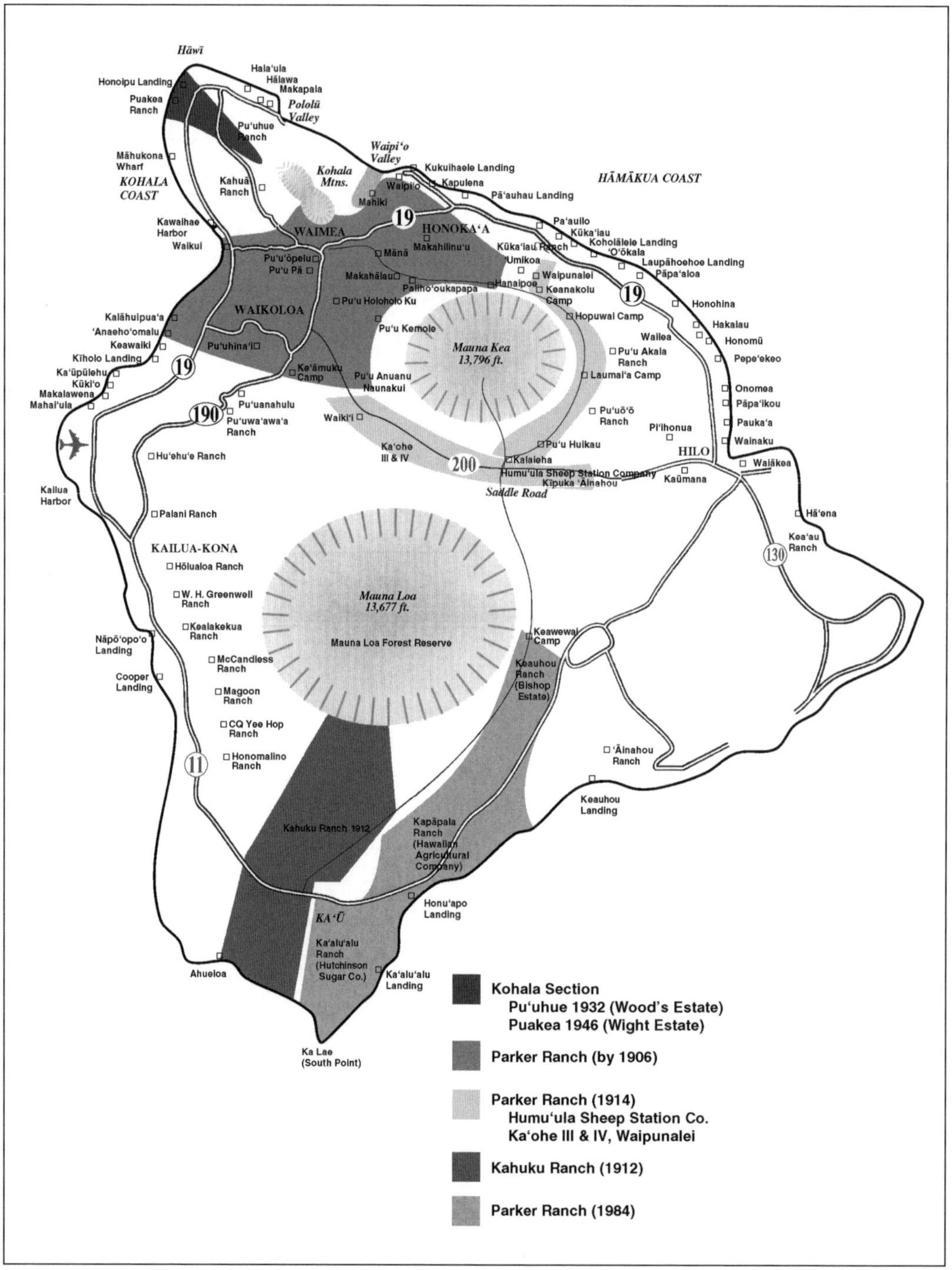

Big Island properties held by Parker Ranch over its history.

GLOSSARY

While Hawaiian has been the spoken language of the *paniolo,* their everyday language was unique to them and to their work environment. They frequently substituted or shortened words to get a point across quickly. Purists would argue that the terms they used are incorrect. While that may be true, because this book is about the *paniolo,* this glossary uses the colloquial Hawaiian language of the Hawaiian cowboy.

Pukui and Elbert's *Hawaiian Dictionary* (1971) served as a resource for the Hawaiian diacritical markings—*kahakō* (macron) and *ʻokina* (glottal stop)—used in this book. When information was not available as to correct placement of these macrons and glottal stops, they were not included to avoid changing the intended meaning.

Cassell's *Spanish English Dictionary* (1969) and *Farm and Ranch Spanish* by George W. and Rex R. Kelly (1960) served as resources for Spanish words. When corresponding Spanish terms were not known, they were not included in the glossary.

Hawaiian words are presented in alphabetical order, followed by corresponding Spanish terms with similar meanings.

ʻāhaʻi – escapan arancar contigo: to run away, bolt, flee, as a horse out of control

ʻahawa – nacimiento (agua): waterhead, intake at beginning of a watercourse

ʻahele – lazo: to snare; a method of roping livestock with an overhead lariat loop

ʻāhiu – salvaje: wild, untamed, or feral livestock

ahuʻili – a small land section within a larger parcel such as an *ahupuaʻa*

ahupuaʻa: a land section extending from the uplands to the sea; boundary markers *(ahu)* were stone heaps surmounted by the image of a pig *(puaʻa),* or where pigs were offered in payment of tax to chiefs

ʻaila pipi – sebo: beef tallow, suet

ʻaila puaʻa – manteca: lard

ʻainakini: navy blue cotton cloth

ʻĀina Paka: Parker lands

ʻai paʻa: cooked taro pounded into a hard mass not yet mixed with water, usually contained in ti leaf bundles called *holoʻai*

akahele – prudente: prudent, cautious, careful, safety oriented in performing daily tasks *alaka'i – delantera:* to lead, guide, direct; horsemen *alaka'i* livestock with a lead rope or lariat

'ala 'ōhi'a nei: the special fragrance of Waimea produced by the *'ōhi'a lehua* tree and flower, as described by Reverend Lorenzo Lyons

ali'i – rey: royalty, chief, chiefess, monarch, king, queen, prince, princess

aloha 'āina: love or reverence for the land or one's country

'āmala – herrero: armorer, blacksmith; something made by a blacksmith

'āmana: a Y-shaped post used to snub up a cow

anana – brazada: a fathom; distance between fingertips of outstretched arms

'āpana: a section of land, several of which can be contained within a larger section of land known as a *kuleana*

'auhuhu: a slender shrub used as a fish poison

'auwae pu'u – boludo: lumpy jaw in cattle due to a bacterial abscess often requiring lancing

awāwa: groove, gulch; crack in the lava

'awe'awe: braided rawhide rigging of the "Hawaiian tree" cowboy saddle

'eha ka 'ōpū: hurt stomach, stomachache, colic

'ehu kai: sea spray

hainakā 'ā'ī – panuelo: neckerchief; sometimes *hai nakā lei*

hale – casa: house, building, lodge, station Hale Mānā: Mānā house, Parker home in Mānā

hale mau'u: grass house

hale uahi – fumador: smokehouse; structure used for smoking beef, pork, mutton, or fish

hālua – hueco: hollow, depression, dip, furrow

hāmale – martillo: hammer

hāmana – campaña: fork of a saddle, beneath the pommel and attached to the bars

hānai: foster child, adopted child

hao kuni: branding iron

hapa – mestizo: portion, fragment, part, fraction

hapa haole: half Caucasian ancestry; usually refers to Hawaiian-Caucasian ethnicity

hapa lala – anquerito: literally half of a *lala,* the apronlike leather covering the saddle seat used as rear cinch rigging

hāwele or *hāwili – dar la vuelta:* to wrap one's rope or lariat around a saddle horn, post, or tree to restrain or snub up livestock

heiau: pre-Christian place of worship or shrine

heu – borroso: fine hair, fuzz, down, cactus bristles, cane tassels, spliced rope ends

he'upueo: a reedlike bent grass *(Agrostis avenacea)* indigenous to Hawai'i, found at about 4,000 foot elevations; also another name for *pili uka* grass

hi'aloa: Hawaiian shrub of medicinal value, with low palatability to livestock

hili – trenza: to braid, plait, or to whip or strike as a hammer

hili 'eono: braiding with six strands

hili hā: braiding with four strands

hili kolu: braiding with three strands

hīmeni – himno: hymn

hinahina – gris: grey color; also forage belonging to the Filaree group

hina pū – caída maro mío: to fall forward, often as a horse and rider falling forward

hōʻoalu: to slacken or loosen a lariat

hōʻau – flotar, nadada: swim

hōʻau pipi: process in which cattle are led by a cowboy on horseback from shore into the ocean and swum to a waiting boat called a lighter

honohono: wandering Jew *(Paspalum dilatatum),* a grass common to wetter areas, of marginal nutritional value

hoʻo kaʻawale – desechan: to exorcise, cast off, to separate oneself from something

hoʻokū: to release a tethered animal from a tree or post

hoʻōla – currar, salve: to cause to heal or live, save

hoʻomoʻa – asado: to bake, roast

hoʻopalaki – broza, monte: brush, as a horse's coat before saddling

hoʻopōmaikaʻi – bendición: to bless

hoʻoponopono – resolución: to set right, to correct

huelo hili – trenza: braided pigtail

hui – sociedad: group, club, association, partnership

humuhumu – coser: to sew, suture

iʻa kōpī – carne solobre, carne salado, salpresar: to salt meat, meat sprinkled with salt

iama – tablas: side bars of the saddle

ihu – nariz: nose, snout

ihu keko – chato: tapadero in a pug-nosed or snub-nosed design resembling a monkey's face; also called *ihuʻū*

ihu puaʻa: stirrup cover *(tapadero)* in the shape of a hog's snout

ʻili – cuero, piel: skin, hide, bark

ʻili eʻe: wild plumbago, a medicinal shrub used for tattoo ink (sap)

ʻili hope – contrareata: a braided strap joining the *ʻaweʻawe* rigging rings from behind the cantle that controls rigging placement

ʻili huki – latigo: cinch pull strap

ʻili keʻehi – aciones or *arzones:* leather stirrup straps

ʻili maʻawe – tientos: rawhide or leather saddle strings

ʻīlio – perro: dog

ʻīlio poli: a plump young dog raised on human breast milk, reserved for consumption by royalty

ʻili pipi – beef hide, cattle hide, or cowhide

ʻili waena: a keeper strap between the front and rear saddle cinches to prevent sliding back into the flank area of the horse

ʻili wahī – cubierta: sheath, knife case of leather or rawhide; *funda:* holster

imu: underground oven

ʻiole – ratón: rat

ipu – copa: pot, crown of a hat, gourd

i waho: outside, beyond

kāʻai – faja: sash, belt, girdle; a vestige of the Mexican *vaquero* apparel, the red mid-sash is still worn by traditional Hawaiian entertainers

ka'e – al de sombrero: brim of a hat, brim of a hole or crater

kahakai – playa: coastal plains or seacoast

kahawai – quebrada, rambla: wet or dry stream *(arroyo),* river *(rio),* ravine, or gulch

kahiko – antiguo: ancient, old, antique; representing the olden days; can also mean dressing oneself in fine fashion

kahuna pule – sacerdote: priest, minister

kā konakona: a native variety of panicum grass *(Panicum torridum)* known for its hardiness

kalamālō: a thatch grass used as building material for roof and walls, coarse and unpalatable to livestock

Kalawina – Calvinista: Calvin, Calvinist religion, Congregational Church

kama'a puki – bota: boots, shoe boots; *bota alta:* jackboots

kama'āina – indigena: native born

Kamapua'a: the pig demigod

kanaka makua: one that behaves and speaks as an adult; usually refers to a young person

kanaka maoli: full-blooded Hawaiian person

kani ka ihu: to snort, blow one's nose

kapa'āpana: Hawaiian quilt with appliquéd designs

kapu: taboo, prohibited, forbidden

kapuahi – hogar, estufa: stove, fireplace

kapuahi 'aila māhu: furnace

kapua'i mahele 'ia – pie hendido: cloven hoofed, as cattle, hogs, goats, sheep, deer, etc.

kauhale – pueblecito: village, group of houses, such as the cowboy camps of Waiki'i, Pu'uhue, Puakea, Makahalau

kauka: doctor

kauka lio: horse doctor, veterinarian specialized in the horse, equine practitioner

kāula: a prophet or seer

kaula 'ā'ī: neck rope

kaula 'auwae – barbiquejo: chin strap of a sombrero

kaula 'auwae – varvada: chin (curb) strap of a bridle

kaula hele: a rope used to hobble the lower saddle stirrups to each other below the girth, giving the rider an advantage

kaula ho'ohei – lazo, la reata: lariat, roping rope, lasso

kaula 'ili – soguilla: skin rope, rawhide lariat

kaula keko: rear cinch

kaula lae: brow band of a bridle

kaula 'ōpū – cincha: front saddle cinch

kaula papālina: cheek strap of a bridle

kaula po'o: poll strap of a bridle

kaula pu'u: throat strap of a bridle

kaula waha – freno, cabezada: a bridle, also called a headstall

kaula waha hao – bocado del freno: the bit or horse's mouthpiece, part of the bridle

kaula waha hao kēkake: refers to a snaffle bit, "jackass bit," or "broken bit"

kaupu: seating a child in front of the rider on a saddled horse

kelamoku: sailor, ship sailor

keleawe – cobre: brass, bronze, copper

kēpā – espuela: spur

kēpā pele – campana or pajados: jingle bobs: two plummet-shaped bobs on spurs that create bell-like chime sounds in motion

kiawe: the algarroba or carob *(Prosopsis pallida),* a large leguminous tree that makes excellent fence posts, firewood, and lumber, brought to Hawai'i in 1828 from Peru; known as mesquite in the southwestern United States

kīhāpai – jardin: garden, cultivated patch, orchard

kīhei – chal: shawl, serape

kīhei pili: a clinging bed covering or blanket often used for wrapping babies for travel

kīhele – lazo, trampa: to hook or ensnare, as a roper placing a loop over the head of a calf and snaring its hind legs as the animal passes through the loop

kī hō 'alu: slack key method of guitar tuning and playing

kīkānia – coquina: the cocklebur weed *(Xanthium);* a type of spur rowel (wheel) with fine needlelike points that resemble a cocklebur

kila: steel, sharpening steel used in cutlery; can also refer to a knife blade or chisel

kiloi: throw, cast away

kiloi kaula – lazo: to throw a lariat loop

kinai ahi – bombero: firefighter

kīpuka – arroyo: canyon or land definition; clear oasislike opening among lava fields or forests; *kīpuka* is also the *honda* or loop of the lariat opened in preparation of roping

koa: largest of native Hawaiian trees *(Acacia koa),* belonging to the legume family; excellent wood for canoes and furniture, but too heavy for a saddletree

koai'a: native Hawaiian tree *(Acacia koai'a)* used for saddles, staffs, furniture, etc.

kōlea – avefria: Pacific golden plover, a migratory bird that arrives in Hawai'i about the end of August and leaves in May for Siberia and Alaska

koloa – pato: Hawaiian duck

kolohe: rascal

kolū – cola: glue; also a thorny shrub *(Acacia fornesiana),* sometimes spelled *klu*

konohiki: head man of an *ahupua'a* land division under the chief

ko'oko'olau: all varieties of a shrub called beggar-ticks, used medicinally by Hawaiians and commonly as a tea

kōpī – salpicar: to sprinkle, as salt, sugar, sand; to salt meat or fish

kua'āina – palurdo: country, backcountry, outback countryside; said of someone who is from the country

kua hao – yunque: anvil

kua'i lole: to dress out, to eviscerate and skin an animal for food

kua lio: the back of a horse

ku'i hao – herrador: blacksmith

kui lā'au: dowel pin

kuka 'aila – gabardina: raincoat

kūkae pua'a: crabgrass (*Digitoria pruriens*)

kula – llano: plain open field, pasture; open country

kulaīwi hanau – nacimiento: native land, land of birth

kuleana – finca solariega: right, privilege, responsibility; estate, homestead, small piece of property

kuni – hierro: to brand, stamp, etch

kuni pipi – hierro vaca: to hot brand cattle

kūpeʻe – maniota: to hogtie; also hobbling of a horse

laho: testicle, male of any species

laho ʻāhiu: wild bull, wild stallion

laho ʻole: a neutered male of any livestock species

laina – reinda: line, reins (*laina kaula waha*)

lākana – alazán: chestnut (dark sorrel) color of a horse; auburn haired; also used when referring to lantana

lakeke – chaqueta: jacket

lala – mochilla: an apronlike saddle covering of leather; called a "Mother Hubbard" in the old West

lala ʻihope – anquera: rear saddle covering, rump covering; vestige of the *vaquero anquera* or *conquistator anquitera*

lau kuni – ruedo: stamping, carving, or etching, usually of the leather edges of a saddle

laulele: milkweed *(Asclepias curassavica)*, a palatable herb; wild dandelion *(Taraxacum officinale)*

lawe hānai – adoptar: to adopt or raise a child as one's own

likini wāwae – polainas: leggings, usually of leather, worn by the working cowboy

lima: hand

lima ʻākau: right-handed

lima hema: left-handed

lina – argolla: ring, as in a cinch ring

lina ʻaweʻawe: rigging rings

lio – caballo: horse

lio kauʻō – caballo de tiro: draft horse

lio laho: stallion

lole: clothes, clothing, dressing

lole wāwae: pants or trousers

lole wāwae pōkole – sota or calzoneras: breeches

lole wāwae pūhaʻuhaʻu – pantolones: bloomers

lūʻau – fiesta: Hawaiian feast

luna nui – major domo or *patrón:* head, boss, lead cowboy, head cowboy, foreman, superintendent, manager

maʻawe – tientos: a strand, fiber, thread; strips of leather or cowhide

mahaʻoi – entremetido: bold, impertinent, impudent, offensive, rude

makaʻāinana – plebeyo: commoner, people in general

makai – almar: seaward, toward the sea

makinikela – pretal: martingale; device used to steady a horse's head from upward movement

makuahine – madre: mother

makuakāne – padre: father

malihini – forastero: newcomer, stranger, guest

malumalu – sombrío: shady, gloomy

māmane: native leguminous tree *(Sophora chrysophylla)* that grows high on Mauna Kea and Mauna Loa, excellent for fence posts, firewood, and as a building material

mana: spiritual power

Mānā: Hale Mānā, Parker ancestral home

mānienie: Bermuda grass *(Cynodon dactylon),* found in lawns and pastures at lower elevations

ma'ū – fresca, sereno: cool, as in weather; damp, moist

mauka – tierra ascendente: toward the mountains; upslope, inland

mehameha – solitario: alone, lonely, solitary

moa – gallina: chicken

moa nui: large flightless bird of old Hawai'i and New Zealand

mumuku: strong wind of the Kawaihae, Kawaihae Uka, and Waimea plains

naio: bastard sandalwood, a native wood with sandalwood-like scent

nā lina keleawe – las argolla de cobre, argolla latón: brass rings

nānā'āina – campiando: to watch over the land

nā pohā: sinkholes

nēnē: Hawaiian goose

nīele – entremetido: persistent curiosity; to quiz, pump for information; question to the point of annoyance

niho palaoa – marfil: whale's ivory, whale tooth; usually a symbol of royalty

noho – silla: sit; seat

noho lā'au – fuste: the wooden saddletree

noho lio: saddle for a horse

'oala – repara: buck as a horse; also spelled *'owala*

'ohana – familia: family, kin

'ōhi'a: native tree (*Metrosideros collina*) dominant in wet upland areas

'ōhi'ā-'ai: mountain apple (*Eugenia malaccensis)*

'ōhi'a lehua: flower of the *'ōhi'a* tree

'ōhiki – pico: hook, probe, pry

'ōhiki lio – pico de casco: hoof pick, tail-cleaning pick

oī, ōwī, or *ha'uōwī:* joe (or joe-pye) weed, a weedy pest of rangeland that is also used for cuts and bruises

'oki hao – aserrar cuerno: to dehorn livestock

'oki huelo: to cut the tail switch of livestock for identification purposes

'ōkolehao: distilled liquor made from the ti root

'ōkumu – cabeza: saddle horn, pommel

'ōkupe – traspié: to stumble, trip as in walking

'ōlelo – habla: speak or say, language

'ōlelo kanaka – habla Hawaiiano: to speak the Hawaiian language

'ō'ō – barra: digging stick, spade bar

'o'oma: bonnetlike cover

'o'oma noho – tapadera: tapadero, leather covering of stirrups

'ōpae – camarón: shrimp; often referred to youth workers ('Ōpae Gangs)

pā'eke – corral: corral, livestock enclosure, pen

pā hale – yarda: houselot, yard

pahi – cuchillo: knife

pahi olo – aserrar: saw, saw knife

pāhu'a – ilanos: flat or clear areas in the fields

pa'i 'ai: hard; pounded but undiluted poi

Paipala Hemolele – biblia: Holy Bible

palaka: block; also refers to block-print fabric

palaka laina: block lines; rope reins threaded through snaffle bit rings and tied back to the rigging rings on both sides

pale: blanket *(sudadero)*; guard *(fendera)*; skirt *(faldones)*

pale ke'ehi: stirrup fender *(fendera)*

pale lio: horse or saddle blanket *(sudadero)*

pā loa: wing fence to guide cattle to the corral, holding pen for cattle

pālulu – tapadera, tapadero: protective bonnetlike leather covering of stirrups

pani maka – tapaojo: blinders, used to cover eyes of fractious horses when saddling, mounting, dismounting, etc.

pānini – nopal: cactus fruit, prickly pear *(Opuntia megacantha)*

paniolo – vaquero: cowboy

paniolo nui: head cowboy, lead cowboy, foreman, cowboy boss

pāpale – sombrero: hat

pāpale 'ili – vaqueteada: leather hat

pāpale lauhala: woven pandanus hat; straw hat

pā pōhaku: stone wall

pau: finished, completed, over

pau hana: end of work, usually the end of the work day

pepeiao – oreja: ear

pepeiao hoki – oreja de mulo: mule ear; refers to elongated *tapaderos*

pepeiao lāpaki – oreja de conejo: rabbit ear; refers to elongated *tapaderos*

piha'ā – tronco: log, fallen tree trunks in forests and plains

pi'i – ascendente: to go upland, inland; to climb, to mount; also a triangular piece of cloth in the legs of bell bottomed trousers to add width

piko – ombligo, navel: umbilicus, point, end point, poll; end of a rope, point

piko holo: slip knot

pilipili: a plant of the beggar tick family that is quite palatable for livestock but troublesome sheep forage as the awns bind the wool

pili uka: stiff tufted native grass *(Trisetum glomeratum)* that provides excellent livestock forage in high altitude ranges; also called *he'upueo*

pini – cabestro: partner, pin ox, companion

pipi – vaca: beef cattle

pipi 'āhiu – salvaje, buiso: wild cattle

pipi kaula'i – charqui: dried, jerked beef; commonly called *pipi kaula*

pipi kauō – buey: draft oxen

pipi keiki – becerro: calf

pipi laho 'āhiu – orejano: wild bull

pipi miko – salpresar: salted, seasoned beef

pō'ai – argolla: circle, ring (cinch), hoop; also refers to a cinch sore on a horse

poe – teja: rounded, usually contoured seat of the saddle, the cantle

pōlena: bowline knot

po'o waiū: tethering of livestock by horns, head, or neck to a forked post *(āmana)*, tree, or root

pua kala: native prickly plant of medicinal value

pualele: the sow thistle *(Sonchus oleraceus)*, a weedy herb of the daisy family having livestock palatability *puka:* opening, hole

puka pā – puerta: gate, fence opening

pūkiawe: black-eyed Susan *(Abrus pecatorius),* a slender, climbing legume with limited forage value

pulu: wet, kindling; humus that provides moisture and nutritional value when grown around tree bases

pūnuku – cabestro: horse halter, hackamore *(jaquima),* noseband of a bridle *(bosal)*

'ua'u or *'uwa'u:* dark rumpled petrel, a seabird; considered by some an *'aumakua* or family god

uaua: hardened or calloused, as in the mouth of an abused horse

ukana: baggage, luggage, gear, supplies

ukana lio – montura: horse gear duffel

uku – regalo: gift, offering, payment

'uku – pulga: insect; flea, louse *(piojo),* tick, mite *(ácaro),* external parasite

'ukulele: small guitar of Portuguese origin popularized in Hawai'i

uluhe: false staghorn fern

'umi: ten, tenth

'umi 'umi: beard, mustache, whiskers

'upena: fishing net, web, trap

'upena kiloi: to throw a net, as in fishing

'u'ulei: Hawaiian shrub of medicinal value

waihona – armario: closet, cabinet; a safe-keeping depository

wao one – yermo: desert

REFERENCES

Ahlborn, Richard E. *Man Made Mobile: Early Saddles of Western North America.* Washington, D.C.: Smithsonian Institution Press, 1980.

Alexander, W. D. *A Brief History of the Hawaiian People.* New York: American Book Co., 1891.

Belich, James. *Making Peoples: A History of the New Zealanders: From Polynesian Settlement to the End of the Nineteenth Century.* Honolulu: University of Hawai'i Press, 1997.

Bergin, Brady James. "The Birth of Hawai'i's Cattle Industry." History term paper, Hawai'i Preparatory Academy, Kamuela, HI, 1995.

Bergin, Pat. "The Making of the Hawaiian Tree Saddle." Unpublished paper for course credit, University of Hawai'i at Hilo, 1976.

Brennan, Joseph L. *The Parker Ranch of Hawaii.* New York: John Day Co., 1974.

———. *Paniolo.* Honolulu: Kupa'a Publishing, 1978.

Brundage, Lucille. *Alfred W. Carter: Hawaii's Dean of Cattlemen and Notes on Hawaiian Livestock.* Hong Kong: Cathay Press, 1971.

Buck, Peter H. *Vikings of the Sunrise.* New York: F. A. Stockes Co., 1938.

Callaghan, F. R. *Science in New Zealand.* Wellington: A. H. and A. W. Reed, 1957.

Carter, A. H. Taped interview by Clemson Lam. Kamuela, HI, 1983.

Carter, A. W. Personal correspondence, 1900–1949.

Cordy, Ross. *Exalted Sits the Chief.* Honolulu: Mutual Publishing Co., 2000.

Dary, David. *Cowboy Culture: A Saga of Five Centuries.* Lawrence: University Press of Kansas, 1981.

Doyle, Emma L. *A. W. Carter by Emma Lyons Doyle: The Story of Alfred Wellington Carter, 1867–1949.* Compiled from original sources by Emma Lyons Doyle. Privately printed. Honolulu: Honolulu Star-Bulletin, 1951.

Dorrance, William H., and Francis S. Morgan. *Sugar Islands: The 165-Year Story of Sugar in Hawaii.* Honolulu: Mutual Publishing Co., 2000.

Dutton, Brian, L. P. Harvey, and Roger M. Walker. *Cassell's Spanish and English Dictionary.* New York: Macmillan Publishing Co., 1969.

Eyre, David L. *By Wind, By Wave: An Introduction to Hawai'i's Natural History.* Honolulu: Bess Press, 2000.

Gonzalez, Fidel J., Jr. *The World's Largest Island Ranch.* Waimea, HI: Self-published, 1945.

Greenwell, James. M. *One to Eighty-One: My Personal Story.* Hilo: Petroglyph Press, 2000.

Greenwell, L. Radcliffe. Personal interviews by Dr. Billy Bergin, 2002.

Handy, E. S. Craighill. *Ancient Hawaiian Civilization.* Rutland, VT: Charles E. Tuttle Co., 1965 (revised edition).

Henke, L. A. "A Survey of Livestock in Hawaii." Honolulu: University of Hawai'i Bulletin No. 5, 1929.

Hind, R. Renton. *John Hind of Hawi: His Memoirs (1858–1933).* Manaoag, Pangasinan, Philippine Islands, 1951.

'Ī'ī, John Papa. *Fragments of Hawaiian History.* Honolulu: Bishop Museum Press, 1963.

Jordan, Terry G. *North American Cattle Ranching Frontiers: Origins, Diffusions and Differentiation.* Albuquerque: University of New Mexico Press, 1993.

Kamakau, Samuel M. *Ka Po'e Kahiko: The People of Old.* Honolulu: Bishop Museum Special Publication 51, 1964.

———. *Ruling Chiefs of Hawaii.* Honolulu: Kamehameha Schools Press, 1992 (revised edition).

Kay, E. Allison. *A Natural History of the Hawaiian Islands: Selected Readings II.* Honolulu: University of Hawai'i Press, 1992.

Kelly, George W., and Rex R. Kelly. *Farm and Ranch Spanish.* Self-published, 1960.

Kennedy, Malcolm J. *Hauling the Loads: A History of Australia's Working Horses and Bullocks.* Melbourne: Melbourne University Press, 1992.

Kent, Harold Winfield. *Treasury of Hawaiian Words in One Hundred and One Categories.* Honolulu: Masonic Public Library of Hawaii, 1986.

Kuykendall, Ralph S. *The Hawaiian Kingdom [Vol. 1] 1778–1854: Foundation and Transformation.* Honolulu: University of Hawai'i, 1938 (reprinted 1947 by University of Hawai'i Press).

———. *The Hawaiian Kingdom. Vol. 3. 1874–1893: The Kalakaua Dynasty.* Honolulu: University of Hawai'i Press, 1967 (first published in 1938).

Lawrence, Mary S. *Old Time Hawaiians.* Honolulu: Patten Co., 1939.

Lindmeir, Tom, and Steve Mount. *I See by Your Outfit.* Glendo, WY: High Plains Press, 1996.

Lindsey *'Ohana* Association. 1999. "Na Mamo Aloha 'a Thomas John Weston Lindsey."

Linnekin, Jocelyn. *Children of the Land: Exchange and Status in a Hawaiian Community.* New Brunswick, NJ: Rutgers University Press, 1950.

Low, Sam, and E. P. Low. "Puuwaawaa: Eben Parker Low's Memoirs." South Natick, MA: Personal manuscript, 1999.

Malo, David. *Hawaiian Antiquities (Moolelo Hawaii).* Honolulu: Hawaiian Gazette Co., 1903.

Maly, Kepa. *Mauna Kea—Kuahiwi Ku Ha'o ī ka Mālie: A Report on Archival and Historical Documentary Research, Ahupua'a Humu'ula, Ka'ohe, Districts of Hilo and Hāmākua, Island of Hawai'i.* Hilo: Kumu Pono Associates, 1997.

Martin, Lynn J. *Nā Paniolo o Hawai'i.* Honolulu: Honolulu Academy of Arts, 1987.

Martin, Ned, and Jody Martin. *Bit and Spur Makers in the Vaquero Tradition: A Historical Perspective.* Nicasio, CA: Hawkhill Press, 1997.

Meyer, Phillip A. *Niihau.* Ni'ihau: Hoomana Ia Iesu Church, 1998.

Mora, Jo. *Californios: The Saga of Hard Riding Vaqueros, America's First Cowboys.* Ketchum, ID: Dober Hill, Stoecklin Publishing, 1994.

Morris, Ernest. *El Vaquero.* Wilton, CA: Vaquero Enterprises, 1989.

Nakano, Jiro. *Parker Ranch Paniolo: Yutaka Kimura.* Honolulu: United Japanese Society of Hawaii (distributed by University of Hawai'i Press), 1992.

Norgaard, Victor A., D.V.S. Reports of the Division of Animal Industry, Bureau of Agriculture and Forestry, Territory of Hawai'i. Honolulu: Honolulu Gazette Bulletin Publishing Co., 1905–1918.

Oliver, Douglas L. *Native Cultures of the Pacific Islands.* Honolulu: University of Hawai'i Press, 1989.

Olmstead, Frances Allyn. *Incidents of a Whaling Voyage.* New York: D. Appleton Co., 1841 (reprinted 1969).

Pukui, Mary Kawena. *'Ōlelo No'eau: Hawaiian Proverbs and Poetical Sayings.* Bishop Museum Special Publication 71. Honolulu: Bishop Museum Press, 1983.

Pukui, Mary Kawena, and Samuel Elbert. *Hawaiian Dictionary.* Honolulu: University of Hawai'i Press, 1971.

Pukui, Mary Kawena, Samuel Elbert, and Esther T. Mookini. *Place Names of Hawaii.* Honolulu: University of Hawai'i Press, 1974.

Rojas, Arnold R. *California Vaquero.* Fresno, CA: Academy Library Guild, 1953.

———. *These Were the Vaqueros.* Bakersfield, CA: Hall Letter Shop, 1974.

Schmitt, Robert S. *Firsts and Almost Firsts in Hawai'i.* Honolulu: University of Hawai'i Press, 1995.

Shinseki, Kyle. "El Pueblo Mexicano de Hawai'i." Master's thesis, University of California at Los Angeles, 1997.

Simpson, MacKinnon. *The Lymans of Hawaii Island.* Hilo: Orlando Lyman Trust, 1993.

Sinclair, Keith. *The Oxford Illustrated History of New Zealand.* Oxford: Oxford University Press, 1990.

Slatta, Richard W. *Cowboys of the Americas.* New Haven: Yale University Press, 1990.

Tabor, James. *Alfred Wellington Carter.* Self-published, 1978.

Thomas, Mifflin. *Schooner from Windward: Two Centuries of Hawaiian Interisland Shipping.* Honolulu: University of Hawai'i Press, 1983.

Wallis, Michael. *The Real Wild West.* New York: St. Martin's Press, 1999.

Ward, Fay E. *The Cowboy at Work: All About His Job and How He Does It.* Norman: University of Oklahoma Press, 1987 (originally published by Hastings Howe Publishers, 1958).

Wellman, Bernard Brian. "The Parker Ranch: A History." Ph.D. dissertation, Texas Christian University, 1969. Ann Arbor, MI: General University Microfilms.

INDEX

About the Author

Dr. Billy Bergin was born and raised on the Big Island of Hawai'i. He attended Kansas State University, receiving his Doctor of Veterinary Medicine degree in 1967. Dr. Bergin established the first private large animal veterinary practice on the Big Island in 1968 and served as the chief veterinarian of the Parker Ranch from 1970 to 1995. From 1971 to the present, he has been appointed the Veterinary Medical Officer with the Livestock and Disease Control Division, State of Hawai'i Department of Agriculture. He also maintains an active private veterinary practice in West Hawai'i. Dr. Bergin belongs to numerous professional and community organizations. He served a four year appointment as a member of the University of Hawai'i Board of Regents. In 1999 he organized the Paniolo Preservation Society, a nonprofit organization dedicated to the heritage of the Hawaiian cowboy. Dr. Bergin and his wife, Patricia Cockett Bergin, live in Kamuela on the Big Island in the shadow of the Parker Ranch.

Production Notes for Bergin/LOYAL TO THE LAND

Cover and interior designed by Santos Barbasa Jr.
in Trump Mediaeval-Hawn, with display type in
Davison Americana.

Composition by Santos Barbasa Jr.

Printing and binding by Thomson-Shore, Inc.

Printed on 60 lb. Joy White Offset, 440 ppi.